WATER BOUNDARIES

WATER BOUNDARIES:
Demystifying Land Boundaries Adjacent to Tidal or Navigable Waters

Bruce S. Flushman

Member of the Firm of Stoel Rives LLP

JOHN WILEY & SONS, INC.

This book is printed on acid-free paper. ⊗

Copyright © 2002 by John Wiley & Sons, Inc., New York. All rights reserved.

Published simultaneously in Canada.

No part of this publication may be reproduced, stored in a retrieval system or transmitted in any form or by any means, electronic, mechanical, photocopying, recording, scanning or otherwise, except as permitted under Sections 107 or 108 of the 1976 United States Copyright Act, without either the prior written permission of the Publisher, or authorization through payment of the appropriate per-copy fee to the Copyright Clearance Center, 222 Rosewood Drive, Danvers, MA 01923, (978) 750-8400, fax (978) 750-4744. Requests to the Publisher for permission should be addressed to the Permissions Department, John Wiley & Sons, Inc., 605 Third Avenue, New York, NY 10158-0012, (212) 850-6011, fax (212) 850-6008, E-mail: PERMREQ@WILEY.COM.

This publication is designed to provide accurate and authoritative information in regard to the subject matter covered. It is sold with the understanding that the publisher is not engaged in rendering professional services. If professional advice or other expert assistance is required, the services of a competent professional person should be sought.

Library of Congress Cataloging-in-Publication Data:

Flushman, Bruce S.
    Water boundaries : demystifying land boundaries adjacent to tidal or navigable waters / by Bruce S. Flushman.
        p. cm.
    Includes bibliographical references and index.
    ISBN 0-471-40391-1 (cloth : alk. paper)
    1. Water boundaries—United States. 2. Land titles—United States. 3. Real property—United States. 4. Surveying—Law and legislation—United States. I. Title.

KF639 .F59 2001
346.7304'3—dc21

2001026514

Printed in the United States of America.

10 9 8 7 6 5 4 3 2 1

To my wife, Bette, and my sons, Michael, Zak, Josh, and Eli.
Their continual and continuing encouragement, support, and love were and are my bedrock.

And in memory of my longtime colleague and friend,
the late Louis Claiborne.
Louis was the very best of his profession. Not only was he both an honorable adversary and a brillant advocate, he was also a completely genuine person, generous with his insight and time,
full of humor and love of life.
I wish he were here to enjoy this book and to continue, over a glass of cabernet, our continuing debate that began in the Supreme Court.

CONTENTS

Preface   xvii

Acknowledgments   xxvii

Table of Authorities   xxix

1   The Basics of Land Title   1

    1.1   Introduction / 1
    1.2   United States' Acquisition of the Public Domain / 3
    1.3   Recognition of Title Granted by Prior Sovereigns / 4
        1.3.1   Legal Basis for Recognition of Prior Sovereign Grants / 6
        1.3.2   Confirmation of Former Sovereigns' Land Grants—Land Claims Act / 6
    1.4   The Public Domain and the United States' Survey of the Public Lands / 9
        1.4.1   Background of Public Land Surveying System / 10
        1.4.2   Rectangular System of Public Land Surveying / 11
        1.4.3   Effect of Public Land Surveys / 13
        1.4.4   Advantages and Disadvantages of the Public Land Survey System / 14
    1.5   Authority to Dispose of Public Lands / 15
    1.6   Public Land Dispositions—In General / 16
    1.7   Swamp and Overflowed Land Grant to the States / 17
        1.7.1   In General / 18
        1.7.2   Definition of Swamp and Overflowed Lands / 19
        1.7.3   Identification of Swamp and Overflowed Lands / 19
        1.7.4   Title Problems in California and Attempts at Their Resolution / 21
    1.8   State Sovereign Lands / 23
        1.8.1   Equal Footing Doctrine / 23
        1.8.2   Public Trust / 23

1.9 Grants of State Lands / 25
1.10 Limitations on State Disposal of Lands Adjacent to Tidal and Navigable Waters / 26
    1.10.1 Constitutional and Statutory Prohibitions Against Alienation of Tidelands—In General / 27
    1.10.2 Constitutional Prohibition—Effect on Boundary Line Agreements / 28
    1.10.3 Common Law Public Trust—Restriction on Alienation / 29
    1.10.4 Common Law Public Trust Compared with Constitutional Prohibition Against Sale of Tidelands / 31
    1.10.5 Effect of the Existence of the Public Trust Easement / 34
1.11 Adverse Possession / 35
    1.11.1 Color of Title / 36
    1.11.2 Character of Possession / 37
    1.11.3 Payment of Taxes / 37
1.12 Agreed Boundaries / 38
1.13 Estoppel / 39
1.14 Adverse Possession, Estoppel and Like Doctrines Inapplicability to Sovereign Lands or Land Held in Trust / 39
1.15 Exception: When Estoppel May Be Applicable to the Sovereign / 42
1.16 Chains of Title / 44

2 What is the Choice-of-Law?   46

2.1 Introduction / 46
2.2 Interstate Political Boundary / 49
2.3 Interstate Boundaries—Private Titles and Boundaries / 51
2.4 Land Title and Boundary Disputes—In General / 53
2.5 Land Title and Boundary Disputes When the United States Is Not a Disputing Landowner / 54
2.6 Land Title and Boundary Disputes When the United States Is a Disputing Landowner / 56
    2.6.1 The Wilson Case / 56
    2.6.2 The Humboldt Spit Case / 59
2.7 Later Cases / 62
2.8 Effects of These Choice-of-Law Cases in Other Situations / 64

CONTENTS   **ix**

3   Basic Legal Principles Defining Property Boundary Movement   68

3.1   Introduction / 68
3.2   Variable Nature of Title and Boundary Disputes / 70
3.3   Importance of the Location of the Property Boundary Along Tidal or Navigable Waterways / 71
3.4   Ordinary High-Water Mark Property Boundary—In General / 73
3.5   Purpose of the Ordinary High-Water Mark—To Separate Arable from Nonarable Lands / 75
3.6   Definition of the Ordinary High-Water Mark / 78
3.7   Illustration of the Components of the Dynamic Ordinary High-Water Mark / 78
3.8   Surveying Techniques to Locate the Ordinary High-Water Mark—Meander Lines / 81
  3.8.1   Public Land Surveys / 82
  3.8.2   Effect of Meander Lines on Property Boundaries / 85
3.9   Exceptions to Rule—When a Meander Line Can Be Treated as a Property Boundary / 87
3.10 Basic Legal Terms Describing the Process of Property Boundary Movement / 90
  3.10.1 Study of a Property Boundary Movement Case / 91
3.11 Legal Terms Describing the Process of Change in Physical Location / 92
3.12 Property Boundary Consequences of a Change in Geographic Location of the Boundary Watercourse / 94
  3.12.1 Background of Rules / 95
  3.12.2 Property Boundary Effect of Accretion and Erosion / 95
  3.12.3 Property Boundary Effect of Reliction and Submergence / 97
  3.12.4 Property Boundary Effect of an Avulsion / 97
  3.12.5 Accretion or Avulsion? That Is the Question / 98
3.13 Presumptions and Burden of Proof in Title and Boundary Litigation / 99

4   Property Boundary Determination Along the Open Ocean Coast   101

4.1   Introduction / 101

4.2 Early Case Discussion of the Physical Indicia of the Location of the Ordinary High-Water Mark / 104
4.3 Introduction of Tidal Measurements to Physically Locate the Ordinary High-Water Mark / 105
4.4 Use of Indicia Other Than Tidal Measurements—Vegetation or Erosion Lines / 107
4.5 Refinement of the Use of Tidal Measurements as the Physical Indicia of the Ordinary High-Water Mark / 108
4.6 Explanation of the Tides, Technical Terms and Expressions / 110
    4.6.1 Tide-Producing Forces / 110
    4.6.2 Variability of Tide-Producing Forces / 111
    4.6.3 Moon's Phase—Spring Tides and Neap Tides / 111
    4.6.4 Distance from Earth—Perigean and Apogean Tides / 112
    4.6.5 Declination—Tropic Versus Equatorial Tides / 112
    4.6.6 Effect of These Variables—18.6-Year Cycle / 113
    4.6.7 Time and Type of Tides / 113
4.7 The "Neap Tides" Confusion / 114
4.8 Foundation of Tidal Datums / 115
    4.8.1 Background / 115
    4.8.2 Methodology of Tidal Observations / 116
    4.8.3 Need for Long-Term Tidal Observations / 116
4.9 Tidal Datums / 117
    4.9.1 Use of Tidal Datums in Coastal Mapping / 118
4.10 Mean High-Water Line Adopted as the Physical Location of the Ordinary High-Water Mark—The *Borax* Cases / 119
4.11 The California Aberration Dispelled—The *Kent Estate* Case / 122
4.12 The Location of the Ordinary Low-Water Mark / 122
    4.12.1 Use of Submerged Lands Act Definition / 123
    4.12.2 Mean Low-Water Line Adopted as Physical Location of the Ordinary Low-Water Mark / 125
4.13 The Open Coast Mean High-Water Line—Fluctuation of the Landform / 126
4.14 Property Boundary Effect of the Geographic Movement of the Open Coast Shoreline / 129
    4.14.1 In General / 129
    4.14.2 Effect of Artificial or Human-Induced Changes on Coastal Property Boundaries / 129
    4.14.3 Effect on the Property Boundary of Filling by the Littoral Owner / 132

	4.14.4	Effect on the Property Boundary of Artificial Erosion / 132
	4.14.5	Effect on the Property Boundary of Reemergence and Reliction / 133
	4.14.6	Effect of Great Storms or Hurricanes on the Property Boundary / 133
4.15	Effect of Presumptions on Coastal Property Boundaries / 134	
4.16	Examples of Types of Proof in Open Coast Property Title and Boundary Litigation / 135	
4.17	Effect of the Burden of Proof on the Outcome of Coastal Title and Property Boundary Litigation / 138	

5 Property Boundary Determination In Estuarine Areas   141

5.1 Introduction / 141
5.2 Difficulty and Confusion in Title and Boundary Determination in Tidal Marshes / 147
5.3 Title Derivation of Tidal Marshlands / 148
  5.3.1 Ordinary High-Water Mark—Boundary of Tidelands and Swamp and Overflowed Lands / 148
  5.3.2 Segregation of Swamp and Overflowed Lands from Tidelands by Visual Observation of Their Physical Character / 149
  5.3.3 Judicial Guidance for Segregation of Swamp and Overflowed Lands / 149
  5.3.4 Legal Effect of the United States' Segregation and Patent of Tidal Marshlands as Swamp and Overflowed Lands / 151
5.4 Means to Remedy Uncertainties of Title to Tidal Marshlands / 158
  5.4.1 Doctrines of Estoppel and Adverse Possession / 158
  5.4.2 Curative Acts: End to the Uncertainty of Title to Tidal Marshlands? / 160
  5.4.3 Navigability Versus Tidality—Must the Waters Overlying Tidelands Be Navigable, Not Merely Tidal Waters? / 161
  5.4.4 The Impact of Spanish/Mexican Land Grants on the Location of the Ordinary High-Water Mark / 163
5.5 Mean High-Water Line—the Physical Location of the Ordinary High-Water Mark Property Boundary in Tidal Marshes / 168
5.6 Legal Character of Tidal Marshlands Determined at the Time of the Swamp Lands Act Grant; Customary Property Boundary

xii CONTENTS

    Principles Determine Effect of Change in Physical Location of the Ordinary High-Water Mark / 170
5.7 The Tidal Marsh Regime and Changes to That Regime / 170
5.8 Property Boundary Effect of Changes in the Tide Marsh Regime / 172
  5.8.1 Effect on the Property Boundary If the Activities Took Place on the Tidal Marsh / 172
  5.8.2 Effect on the Physical Location of the Property Boundary of Activities Not Directly on the Tidal Marsh / 174
5.9 Proof of the Legal Character of Tidal Marshes and the Physical Location of the Historic Ordinary High-Water Mark / 177
  5.9.1 Means to Establish the Physical Character of Tidal Marsh in Historic Times / 178
  5.9.2 USCS and USC&GS Mapping / 178
  5.9.3 Method of Preparation of USCS and USC&GS Maps / 179
  5.9.4 Interpretation of a USCS Map / 181
  5.9.5 Cautions on the Use of and the Value of USCS and USC&GS Maps for Property Boundary Purposes / 182
  5.9.6 Other Types of Mapping / 185
  5.9.7 Other Sources of Factual Information / 186
  5.9.8 Suggestions for Exhibits / 186
  5.9.9 Types of Physical Investigations / 187

6 Property Boundary Determination Along and In Tidal River Regimes 189
 6.1 Introduction / 189
 6.2 The Delta Regime and Delta Meadows / 195
 6.3 The Impact of Human Activities on the Delta / 198
 6.4 The Impact of Human Activities on Delta Meadows / 200
 6.5 Title to Delta Marshlands / 201
 6.6 Tidality of the Adjacent Watercourse—Consequences and Proof / 204
 6.7 The Impact of Property Boundary Movement Principles / 208
 6.8 Proof of the Physical Location of the Historic Ordinary High-Water Mark—Use of USGS Maps / 209
 6.9 Proof of the Physical Location of the Historic Ordinary High-Water Mark—Use of Physical Measurements / 210
  6.9.1 Establishing Initial Water Levels / 211
  6.9.2 Re-creation of Historic Water Levels / 213
  6.9.3 Establishing the Location of the Perimeter Watercourse / 215

6.9.4 Establishing the Historical Physical Conditions of Delta Meadows / 217
6.9.5 Establishing the Topography and Prominent Physical Features and Attributes of Delta Meadows / 218
6.9.6 Establishing Historic Vegetation and Vegetation Patterns / 220
6.9.7 Establishing the Soils Types Present / 222
6.9.8 Re-creation of the Historic Topography of Delta Meadows / 224
6.9.9 Re-creation of the Location of the Historic Mean High-Water Line / 228
6.9.10 Physical Justification of Location of the Historic Mean High-Water Line / 229
6.10 Concluding Thoughts and Suggestions / 230

7 Property Boundary Determination Along and In Navigable, Nontidal Rivers and Streams Regimes   232

7.1 Introduction / 232
7.2 The Matter of Navigability / 235
    7.2.1 Significance of Navigability in Title and Boundary Disputes Along Nontidal Water Bodies / 235
    7.2.2 Questions About the Concept of Navigability / 236
    7.2.3 Federal Courts Establish Test for Navigability / 237
    7.2.4 Relevance of State Tests of Navigability / 238
    7.2.5 Basic Parameters of Navigability / 239
        7.2.5.1 Navigability—A Question of Fact / 240
        7.2.5.2 Date to Establish / 241
        7.2.5.3 Natural and Ordinary Condition / 241
        7.2.5.4 Effect of Obstructions or Seasonal Flows / 241
    7.2.6 Actual Use Is Not Required, Only Susceptibility for Navigation Is Essential / 242
    7.2.7 Use for Commerce or Transportation Will Establish Susceptibility / 243
    7.2.8 Evidence of Navigability / 244
7.3 Property Boundary Consequences of River Movement—In General / 246
7.4 Rules to Avoid Uncertainty in Deciding the Property Boundary Consequences of River Movement / 246
    7.4.1 Understand the Physical Process; Retain Expert Consultants / 246
    7.4.2 Understand the Property Boundary Movement Issue / 248

7.5 Property Boundary Consequences of River Movement—Accretion as Compared with Avulsion / 252
    7.5.1 Purposes of Accretion and Avulsion Rules / 253
    7.5.2 Three Conflicting Cases / 254
        7.5.2.1 The *Jefferis* Case—Application of the Accretion Rule / 255
        7.5.2.2 The *Rutz* Case—Application of the Avulsion Rule / 257
        7.5.2.3 *Nebraska v. Iowa*—Accretion Found Once Again / 261
    7.5.3 The Flexible Approach to the Accretion Versus Avulsion Issue / 262
7.6 Impact of the Burden of Proof on the Property Boundary Consequences of River Movement / 263
7.7 The "True" Meaning of "Gradual and Imperceptible" / 264
7.8 Significance of the Cause of the River's Change in Geographic Location on the Property Boundary / 268
7.9 Reemergence / 271
7.10 Physical Indicia of the Ordinary High-Water Mark in a Riverine Environment / 272
    7.10.1 Physical Location of Ordinary High-Water Mark Property Boundary of Federal Lands Probably Determined by Federal Law / 273
    7.10.2 The *Howard v. Ingersoll* Test—Geomorphic Features / 273
    7.10.3 Variants of *Howard v. Ingersoll* Test Including Use of Stage Data as Indicia of Ordinary High-Water Mark / 275
7.11 Physical Indicia of the Ordinary Low-Water Mark in a Riverine Environment / 277
7.12 Proof in a Riverine Environment—Case Studies / 278

8 Property Boundary Determination Along Navigable Lakes   282

8.1 Introduction / 282
8.2 Character of Title to the Beds of Navigable Lakes / 285
8.3 Resolution of the Quality and Character of State Title to the Beds of Navigable Lakes / 287
8.4 Navigability of Lakes / 291
8.5 What Is a Lake? / 293
8.6 Location of the Ordinary High-Water Mark Property Boundary of Littoral Lands / 295
    8.6.1 In General / 295

- 8.6.2 Value of Meander Lines / 296
- 8.6.3 Physical Indicia of the Ordinary High-Water Mark as Described by the Courts—In General / 299

8.7 Vegetation/Erosion Line Test / 301
- 8.7.1 Problems with Vegetation/Erosion Test in the Case of Lakes / 303
- 8.7.2 Other Court-Approved Indicia of the Ordinary High-Water Mark in the Case of Lakes / 306

8.8 Physical Indicia of Ordinary Low-Water Mark Property Boundary of Littoral Lands / 307

8.9 Property Boundary Consequences of Littoral Shoreline Changes—In General / 311

8.10 The Mono Lake Recession Case—Application of the So-Called Federal Common Law Rule / 311
- 8.10.1 The So-Called Federal Law of Reliction / 312
- 8.10.2 The Reliction Rule Possibly Inapplicable to Intentional Filling or Drainage / 314
- 8.10.3 The Imperceptibility Necessary for the Application of the Reliction Doctrine / 315
- 8.10.4 The Great Salt Lake Case—The Requirement That Reliction Be Permanent / 317
- 8.10.5 So-Called Federal Law Misapplied in the Mono Lake Case / 319
- 8.10.6 Property Boundary Consequences of Littoral Shoreline Changes—Submergence / 320

8.11 Proof in Lake Boundary and Title Cases / 321

Glossary / 326
Appendix A / 338
Appendix B / 350
Index / 373

PREFACE

"In the beginning God created the heaven and the earth."[1] With this immortal sentence, the Bible memorialized the first and longest-running boundary dispute. This book will not delve into the whys and wherefores of the ever continuing dispute between church and state, not only for fear of attempting to explain why there was a pope in Rome at the same time there was a pope in Avignon. Even so, some readers will accuse this author of being much like a medieval alchemist attempting to find a "philosophers' stone" that will somehow magically convert obscure arguments, cases, texts, articles and sources to immutable and clear principles governing boundary disputes.

For example, who could possibly be interested in whether land exposed by river meandering is exposed suddenly or gradually and imperceptibly over a period of many years? After all, new land has been uncovered; the end result is still the same regardless of how the physical change occurred. And is there a problem caused by the fact that 100 years before, the site of your clients' proposed outlet mall was a fecund tidal marsh that had been conveyed to their predecessors by the state under authority of legislation authorizing the sale of salt marsh, tidelands and swamp and overflowed land? Finally, what possible difference would it make if the mean high-water datum is determined by averaging all of the high tides for a 9.3-year as opposed to an 18.6-year period? Indeed, just what is a mean high water datum and what does it have to do, if anything, with the property boundary of land adjacent to tidal or navigable water bodies?

Answers to such seemingly abstruse and, perhaps, obtuse queries can determine the ownership or boundaries of many valuable waterfront acres. These questions, plus the myriad of other still-to-be-asked questions, and the methods and procedures used to obtain answers have incessantly confounded and befuddled countless generations of surveyors, lawyers, judges, and shoreline property owners. This confusion and frustration has perhaps been best expressed by Supreme Court Justice Hugo Black who called the task of property boundary determination concerning lands underlying or adjacent to tidal or navigable waters "an unjudicial job."[2]

---

1. Genesis 1.1.
2. ". . . Settling and identifying boundaries on land is a surveyor's job; he must go to the land with his instruments and mark it off. Identifying [a water] boundary . . . is a much more complex

Indeed, as if confirming Justice Black's worst fears, witness a case concerning the Red River border between Texas and Oklahoma.³ In that lengthy boundary dispute, Oklahoma landowners successfully claimed that the land then on the Texas side of the river had really been part of Oklahoma and therefore rightfully belonged to the Oklahoma landowners. It was estimated that this dispute concerned a 440-mile stretch of the river; more than 100,000 acres of land might have changed hands from Texas to Oklahoma landowners.⁴

As a further example, consider those persons who lived in the modest homes that had once been surrounded by New Jersey's Hackensack meadowlands. It was reported that one of those home owners committed suicide rather than take what she considered to be a grossly unfair purchase offer for her land. That offer was based on the State of New Jersey's claim that it owned two-thirds of her land because it was once land below the "ordinary high water-mark."⁵

Finally, what could be more emblematic of the intractability of boundary disputes than the controversy over ownership of the original and the now-filled portions of Ellis Island. This storied place, where this country welcomed my grandparents, four constituents of the "huddled masses yearning to breath free," was the source of a decades-long controversy between New York and New Jersey.⁶

Such examples should dispel any claim that study of boundary and title disputes concerning lands along tidal or navigable water bodies, while of

---

job; it takes much time by surveyors, cartographers, photographers, and oceanographers, a knowledge of angles, tides, rolling waters, higher mathematics, etc. Shorelines are constantly changing, and thus . . . even this painstaking work cannot provide a means of marking the boundary for all time. I cannot accept the argument that Congress ever intended to impose on this Court such an unjudicial job." *Louisiana Boundary Case* (1968) 394 U.S. 11, 85 (Black, J., dissenting). Justice Black's frustration aside, the United States Supreme Court has accepted and, arguably, even broadened the role of federal courts in this area. *California ex rel. State Lands Comm'n v. U.S.* (1982) 457 U.S. 273 (boundary and title disputes between the state and federal government not involving international boundaries are decided under "federal common law").

3. "Landowner Suits Stirring Red River 'Border War'" (February 25, 1985) Los Angeles Times; *James v. Langford* (10th Cir. 1983) 701 F.2d 123, *cert. den.* (1984) 104 S. Ct. 702.

4. As noted by the Court:

> In our view the trial court correctly applied the rules and concluded that the changes since 1923 in the area in question came about through at least several floods.
>
> . . .
>
> . . . The trial court thus found that the portion of the stream bed . . . "was not added to land of Defendants . . . by accretion or reliction as said Defendants contend." The trial court instead found that the movement of the channel of the stream and the change in the stream bed were the result of the avulsion during the 1908 flood and the several subsequent large floods. James, supra, 701 F.2d at 125–126.

5. "A Tidal Wave of Claims" (July 12, 1982) National Law Journal, Vol. 4, No. 44, p. 1.
6. *New Jersey v. New York* (1998), 523 U.S. 567, 140 L.Ed.2d 993.

academic interest, has little practical impact.[7] In fact, these disputes are prosecuted and defended at a level of intensity[8] and with a tenacity rivaled in nature by the to-the-death conflict between a mongoose and a cobra, and in politics by the disputes between the president and Congress over their respective prerogatives.[9] This is a reflection of the elemental, scarce and valuable nature of the subject matter: land adjacent to water.

This almost talismanic attachment to land does not alone explain the intensity and longevity of these disputes, however. An important element promoting such passion is a lack of understanding of the technical legal principles. Many property owners react in anger and disbelief that principles drawn from dusty lawbooks by a smooth-talking lawyer, and applied by a judge used more to accepting pleas in relatively mundane criminal matters, could deprive a person of years of work and a family's patrimony. The consequent stubborn, relentless search for the "right" answer has resulted in many reported cases, including a large number that eventually wound up in the Supreme Court. In those property boundary cases brought before the Supreme Court, vital and important legal principles by which the nation governs itself were either decided or reaffirmed.[10]

This book is written for a particular, but (it is hoped) broad audience. Persons who comprise that audience are both professionals—the surveyor, the

---

7. As a result of the great earthquake in the San Francisco Bay Area in October 1989, the location of historic shoreline boundaries may have more than a property boundary significance. Many buildings or roadways damaged in that seismic event were located on former tidal marshlands, or in portions of the bay or its distributaries that had been filled and reclaimed. Kreiger, L., "Location Key Factor in Withstanding Quake" (S.F. Examiner 10/22/89), p. 14.

8. *New York v. New Jersey, supra,* 140 L.Ed.24 1006("Having waste no words, the noble grantor all but guaranteed the succession of legal fees and expenses arising from interstate boundary disputes, now extending into the fourth century since the conveyance of New Jersey received its seal"); *Phillips Petroleum Co. v. Mississippi* (1988) 484 U.S. 469, 481-482, reh. den. (1988) 486 U.S. 1018 (claims allegedly upset expectations and interests matured since 1817); *S.D. Wildlife Federation v. Water Mgt. Bd.* (S.D. 1986) 382 N.W.2d 26, 35 (Wuest, J., dissenting) (emotional, heated opposition).

9. Witness the episodic hearings concerning congressional oversight (control) of presidential prerogatives such as foreign policy. Incarnations of this phenomenon span the decades. Witness the Iran-Contra hearings in 1987 and the hearings conducted in 1999 and 2000 regarding lack of attention to alleged Chinese spying in our nuclear weapons laboratories and Chinese most-favored nation status.

10. *Pollard's Lessee v. Hagan* (1845) 44 U.S.(3 How.) 212; *New Orleans v. United States* (1836) 35 U.S. (10 Pet.) 622 (argued by Daniel Webster); *Borax, Ltd. v. Los Angeles* (1935) 296 U.S. 10; *United States v. California* (1947) 332 U.S. 19; *State Land Board v. Corvallis Sand & Gravel Co.* (1977) 429 U.S. 363; *California, ex rel. State Lands Comm'n v. United States, supra,* 457 U.S. 273. In the famous *Pollard* case, it was stated:

> . . . [T]his is deemed the most important controversy ever brought before this court, either as it respects the amount of property involved, or the principles on which the present judgment proceeds. . . . *Pollard's Lessee, supra,* 44 U.S. at 235 (Catron, J., dissenting).

lawyer, and the expert consultant—and nonprofessionals—concerned property owners, interested local citizens, public governing bodies and regulatory agencies and their staffs. Their common ground will be a desire to acquaint themselves with the basic legal principles of property boundary and title determination concerning lands adjacent to tidal or navigable water bodies. Any person attempting to understand or resolve property boundary and title disputes concerning such lands should be in firm possession of the legal principles created over time by the courts and legislatures in an effort to resolve such disputes. These persons should also possess a basic understanding of the regime, the physical attributes, characteristics, and processes of both the water body and the landform that form and define the property boundary or title problem.

The author will provide descriptions of certain of the physical processes and features pertaining to lands along or underlying tidal or navigable waters. But this book does not pretend to be a primer on such physical features or processes. Explanation of the physical world is better left to those with specialized training and experience. Among the relevant physical sciences are coastal oceanography, geology, geomorphology, soils morphology, river or stream hydraulics, potomology, and limnology.

On the other hand, the author expects some commonsense knowledge of these processes. For example, the author assumes the reader understands that the geographical position of a river on the face of the earth will, over the course of years, tend to meander, or to migrate in geographic or physical location, where physical conditions allow. Nonetheless, as will become apparent as one reads on, common sense does not found all of the rules in this special area of property law.

Before letting the reader step directly into the "Twilight Zone" of property titles and boundaries of lands along tidal or navigable water bodies, a preliminary word of advice and a brief explanation are both fair and necessary. The author relies on reported case[11] authority to stimulate and support discussions and explanations. This fact has two related consequences.

First, given the nature of humanity, there may be as many interpretations of such cases as there are case interpreters.[12] Not every person who reads the cases will glean the same information or draw the same lessons from them.

---

11. For United States Supreme Court cases citation is to the official United States Reports, where available, or to the Lawyer's Edition for current, not yet officially reported, United States Supreme Court opinions. For California cases, citation is made to the official state reports. Citations to (non-California) cases are made to the national reporter. Parallel citations are not supplied.

12. The author was told a story about the late, great trial lawyer Edward Bennett Williams. Williams began his opening statement in what seemed, after the completion of an exceptionally well presented prosecution case, to be a hopeless defense with the following: "Ladies and gentlemen of the jury, an old prospector once told me a fact I'll always remember: 'No matter how thin you make a pancake, it always has two sides.'"

Nevertheless, the interpretation of cases in this book is intended to be neutral, with the one following admonition. The author viewed his task in writing this book much the same as that confronting the cartographer who prepares coastal navigation charts. Like that cartographer, who is enjoined to craft nautical charts to protect the mariner, the author has also intended to emphasize potential obstacles, difficulties, or problems to provide some caution to landowners and their advisers and consultants in this largely uncharted area.

Second, while many cases are cited, this work is not intended to be a complete legal treatise with each and every last case catalogued. Rather, the author hopes this work will stimulate thought and point its readers in the right direction for any more in-depth research they may desire or require.

That advice aside, a brief explanation of the author's conception of the land-water boundary will set the stage for the introduction to this work. Whether in case reports or in learned treatises, the discussion and conception of property boundaries of lands along water bodies is often one dimensional. That is, the principal focus of such discussion is largely on the water surface elevation. To avoid this myopia, the reader of this work is asked to take a less crabbed and narrow-minded view of the physical process and features that compose and create property boundaries of lands along water bodies.

The land-water boundary is not one dimensional as some have thought. In fact, that boundary has several components. Those components must be understood before one may reasonably attempt to resolve property boundary and title disputes about lands along or beneath tidal or navigable water bodies. Two relevant components are the water surface elevation of the boundary water body and the relative elevation and the topography (or bathymetry) of the land adjacent to the water body or submerged thereby. Thus, knowing a precise and exact water level of the boundary water body is important, but may not be sufficient. The topography and relative elevation of the land to that bounding water body must also be defined.

But even knowing these two elements is not enough. What makes this boundary determination infinitely more complex is a third factor: the dynamic nature of both the water body and, to a certain, but important, extent, the topography as well. The non-static nature of these components perhaps can best be understood by focusing on just one of those elements.

When a water body meets the land, if the water level of the water body fluctuates vertically, rises or falls in elevation relative to the land, that vertical movement also has a horizontal component.

As shown in Figure A, as the water level rises from x′ to x′ + w′, the water body encroaches over a greater area of land than was submerged prior to its rise. Because the slope is relatively flat, a relatively minor rise in the water surface has a substantial horizontal effect. If the angle of the slope is more acute, however, the horizontal component is not as great. As shown in Figure A, the rise in the water level from x′ + w′ to x′ + y′ does not submerge a correspondingly great area of land as the rise from x′ to x′ + w′. Thus, both the slope of the land and the water level, both of which may

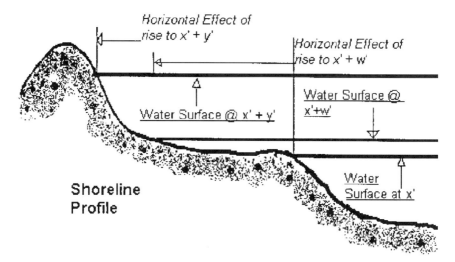

Figure A Shoreline profile

fluctuate, determine the extent of encroachment or recession. The inability to conceptualize this dynamic physical process or the disregard of one of the physical components of the boundary in the reasoning leading to the decision has been the downfall of many a court, lawyer, and surveyor.

The dynamic interrelationship of these elements accentuates the difficulty, confusion, and likelihood of contradictory outcomes that have occurred and may result by trying to force these processes and components to fit into neat and rigid legal rules. Indeed, many of the legal principles discussed in this work were developed on a case-by-case basis to fit a particular physical situation or occurrence. When subsequent generations of lawyers or surveyors attempt to apply such "rules" to a somewhat different situation, a confusing or contradictory decision may apparently result.

To avoid at least some of this confusion, perhaps the best advice that can be given to the reader is to keep a piece of scratch paper nearby. Then the reader may sketch out the physical locale and processes being described. In so doing, the reader's understanding of the applicability (or inapplicability) of the legal principle to the physical situation described or the correctness (or incorrectness) of the conclusions reached in the cases will be immeasurably assisted.

Without an understanding of the principles of land title a surveyor, lawyer, or concerned property owner will lack the very basis and foundation for any water boundary determination. In fact, it makes no difference whether the property is along a watercourse or in the middle of a desert. The property's boundaries cannot be located until one knows the intended extent of ownership that can be conveyed by the title instrument. Property ownership is dependent on two basic elements: from what source the land ownership is derived, and how and to what extent that ownership has been transferred after the initial conveyance. Consequently, Chapter 1 of this book explains the derivation of land titles and the value and use of chains of title.

Yet knowing the source of ownership and how title has been transferred over the years is only the beginning of the inquiry. These crucial facts also have an important role in deciding the kind of law, state or federal, that controls property boundary location. While this issue may, at first blush, appear to be of significance only to lawyers, surveyors and others should not stop reading here. It is important that surveyors be able to spot a potential legal problem, point it out to their clients, and suggest that they obtain competent advice concerning that issue. It is also important for landowners to recognize that the wrinkles in property boundary determination concerning lands adjacent to navigable waters require the application of specialized legal training and skill. The "choice-of-law" discussion in Chapter 2 is designed to provide basic guidance to the surveyor and a, by no means comprehensive, chronicle of cases and other sources where a lawyer might begin to research the field.

With these cardinal principles set forth, Chapter 3 focuses on the elemental nature of water boundaries. There are many in the surveying and legal community who would, if they could, emulate King Canute of English and Norse legend, who is reported to have commanded the tide to cease advancing.[13] It is the recognition of the impossibility of obedience to that command, no matter whether along the ocean or along other fluctuating water bodies such as rivers or lakes, that is the very essence of understanding the ambulatory character of the property boundary of lands adjacent to and along water bodies.[14] Simply as a matter of pure physical fact, the shoreline along tidal waters and rivers changes instant by instant.[15] Shoreline property owners, their bankers, title insurers, and lawyers all abhor this uncertainty. The legal principles of property boundary determination should be considered as a scheme forged by lawyers and judges to, in the eyes of blind justice, command the "tides" to stand still.

The crux of that scheme is the elusive ordinary high-water mark. This legal phrase, intended to describe the property boundary of riparian and littoral properties, will be explained in Chapter 3. That chapter describes some accepted methods of locating that legal property boundary as a physical and geographical position on the face of the earth. The author also discusses surveying principles applicable to government surveys of water boundaries and explains how those principles may or may not assist in the location of

---

13. 4 Collier's Encyclopedia (1961 ed.) 356
14. The phrase "the boundary of property adjacent to a water body" accurately and precisely describes what some more colloquially or generically call a "water boundary." The author has adopted the colloquial term for the title, but in the body uses the more precise phrase as his convention.
15. Lakes also vary in elevation, mainly in accordance with the seasons of the year.

boundaries along or adjacent to water bodies. The key legal terms that describe property boundary movement, *accretion, avulsion, reliction*, and *submergence*, are defined and their legal effect discussed. Where instructive, case examples of the application of these principles are provided.

Thereafter, Chapters 4 through 8 focus on water boundary determinations as they concern different characters and regimes of watercourses and topography. Not all water boundaries are treated the same in the eyes of the law. Different principles may be applied, depending on whether the property boundary is along the open coast (Chapter 4), or adjacent to "inland waters" that are tidal (Chapter 6), or on nontidal navigable waters (Chapter 7 (rivers) Chapter 8 (lakes)). To a certain extent the differences are based in the physical differences in the regimes of the boundary waters and the associated terrains. The principles to determine the location of the property boundary along a beach on the open coast, constantly battered by waves and tides, naturally would appear to be different than the principles used to determine the location of the boundary of land in Iowa along the Mississippi River, where tides have no effect. In addition, the boundary principles are also affected by the legal character of the boundary being located. As we shall see, the legal principles used to establish the offshore boundary between state and federal ownership are different from those used to establish the property boundary between competing private owners or private owners and the state.

Accomplishment of property boundary determinations in these dissimilar and complicated circumstances requires a variety of physical investigations. Some of those investigations will be suggested in this work. In addition, the author will identify the types of consultants or experts necessary to carry out such investigations, interpret and then convincingly explain to the client or the trier of fact the resulting conclusions.[16]

---

16. Expert witnesses hope that your reaction to their testimony may be the same as that described by a well-known mystery writer:

> ... [W]hen he stood in the box and recited, with slightly world-weary formality and in his actor's voice, the details of his formidable qualifications and experience, they gazed at him with the respectful admiration of men who know a distinguished consultant when they see one and have no intention of being so disobliging as to disbelieve what he might choose to tell them. James, P.D., *Death of an Expert Witness*.

On the other hand, some have a more cynical view. One court called expert opinion ". . . an ordinary guess in evening clothes." *Earl M. Kerstetter, Inc. v. Commonwealth, Department of Highways* (Pa. 1961) 171 A.2d 163, 165. An early California case was even more graphic:

> ... [I]t must be painfully evident to every practitioner that these witnesses are generally but adroit advocates of the theory upon which the party calling them relies, rather than impartial experts, upon whose superior judgment and learning the jury can safely rely. Even men of the highest character and integrity are apt to be prejudiced in favor of the party by whom they are employed. And, as a matter of course, no expert is called until the party calling him is assured that his opinion will be favorable. Such evidence will be

One bit of warning is appropriate, however: Only the most basic investigations are suggested. Boundary and title determination and litigation are highly sophisticated matters. Only a foolish author would ever presume to hand either the surveyor or the lawyer a "cookbook" in such a complex subject. And that is not the author's purpose here. Rather, the author intends to give substance to technical rules of law by placing them in their physical context. The author submits this approach will provide a broader and fuller understanding of the nature and complexity of boundary and title disputes along navigable water bodies.

One final word is required. This book is not meant to be the last word you ever read on this subject, though for some it may be more than enough. The book will provide guidance, much the same as a compass gives direction in a forest. It intends to place readers on the right path, but not lead them by the hand. This book has three modest goals: to provide a basic understanding of the land title and boundary system, to provide a framework for analyzing disputes about the boundary of property adjacent to a water body, and to equip its readers to recognize when it is necessary to consult with or engage those more experienced and familiar with the standards, means and methodology of resolving such disputes and to make such engagements useful and productive.

---

received with great caution by the jury, and never allowed, except upon subjects which require unusual scientific attainments or peculiar skill. *Grigsby v. Clear Lake Water Co.* (1870) 40 Cal. 396, 405-406.

ACKNOWLEDGMENTS

During the more than ten years that this book was under construction, my four sons grew up and (mostly) completed college. Besides managing the lives of these active young men, my wife has had her own career as an infant development specialist at Children's Hospital in Oakland, California. She is a much respected member of that caring community. Without my family's support (which allowed me time and concentrated effort), good humor, and tolerance (when there were difficult moments), I could not have undertaken, much less completed, this effort.

This book was a labor of love. I came to love this speciality through the good auspices of two remarkable people, N. Gregory Taylor, former Assistant Attorney General of the State of California, and John Briscoe, former Deputy Attorney General and my current law partner. I owe both of them much: John, for his insights and skillful and gentle prodding of my writing while we were in the Attorney General's Office and thereafter, and many feats of levitation; Greg, for his steadfast support and patience during some very difficult cases, some of which are referenced in this book.

My former colleagues in the California Attorney General's Office, Lands Section, were and are now a group of marvelous and talented lawyers as well as wonderful people. Many provided matchless sounding boards and were reliable and creative collaborators. I cannot mention all, but Dennis Eagan, Rick Frank, Joe Barbieri, and Chris Greene (paralegal extraordinary) were always there when needed. Dennis, although a painfully slow base runner, is one of the best and most fluid writers with whom I have been privileged to work. Joe is both dedicated and tenacious and has an uncanny ability to find legal support for his novel theories.

Along with my lawyer colleagues, I was fortunate to work with and have some of the very best instructors and finest collaborators in the land management, surveying and physical consultant community. Present and former California State Lands Commission boundary determination officers (surveyors) Bud Uzes, Roy Minnick, Mike McKown and Rich Hanson; State Lands' land managers Lance Kiley and Les Grimes and private surveyor Bill Wright lead me through some significant thickets through their persistence and nonpareil insights. Bud Uzes took some of his valuable time and read, commented on, and provided many crucial and helpful suggestions on this work. Mike

McKown was of critical help in recalling the existence of various maps some of which are included herein.

During the white-hot times of trial preparation as well as during trial, the late Jerry Elliott, Dr. Warren Thompson and the late Dr. Ray Krone were indefatigable and ever-creative in their efforts. All provided critical postgraduate education in the physical sciences to this history major and, as well, were willing and exceedingly clever advisers regarding proof at trial. Much of their legacy is contained in this book.

While I was in the California Attorney General's Office, my client was the California State Lands Commission. Clair Dedrick and Jim Trout in the executive office at State Lands and Bob Hight (currently head of the California Department of Fish and Game) and Jack Rump (currently Chief Counsel for State Lands) all provided thoughtful suggestions and policy perspective.

Ned Washburn, Sean McCarthy, and David Ivester, my former adversaries and current partners, are completely professional advocates. Their insistent and intelligent advocacy when we were in opposition was a daunting professional challenge. In fact, some of the contentions made by the litigants in reported cases found a good portion of this book. In the more than ten years since I left the Attorney General's Office, Ned, Sean, David, and I have become close colleagues and, more important, friends. We may not always agree, but I highly value their opinions and insights.

I could not have undertaken and completed this effort without the assistance, guidance and good offices of many other people, most especially including my wonderful secretaries, Maggie Melton in the Attorney General's Office and Pamela Ceccanti-Harris and Carol Bergeron at my firm. For others I have not mentioned, I thank and humbly acknowledge my debt to you all.

My good friend, Sandy Prusiner, read and helped me edit this book at a particularly difficult time in her life. I am grateful beyond words for her perseverance, insights, humor, and courage in undertaking that task.

Finally, although many provided comments or assistance during the preparation of this book, whatever errors there may be are solely my own.

San Francisco, California July 1, 2001.

TABLE OF AUTHORITIES

Federal Cases   **Page(s)**

Alabama v. Georgia (1859) 64 U.S. (23 How.) 505 275
Alabama v. Texas (1954) 347 U.S. 272 16
Alaska v. United States (9th Cir. 1985) 754 F.2d 851, *cert. den.* (1986) 474 U.S. 968 .... 237, 245
Alexander Hamilton Life Ins. Co. v. Gov't of V.I. (3rd Cir. 1985) 757 F.2d 534 .... 69, 131–132, 139–140, 175, 208, 231, 281
Anderson-Tully Company v. Walls (N.D. Miss. 1967) 266 F.Supp. 804 258
An-Son Corporation v. Holland-America Insurance Company (10th Cir. 1985) 767 F.2d 700 205
Arizona v. California (1963) 373 U.S. 546 51
Arkansas v. Tennessee (1918) 246 U.S. 158 .... 49, 52, 93, 96–97, 173, 254, 260, 266, 270–271, 280, 334
Arkansas v. Tennessee (1925) 269 U.S. 152 280–281
Arkansas v. Tennessee (1940) 310 U.S. 563 38
Baer v. Moran Bros. Co. (1894) 153 U.S. 287 162
Banks v. Ogden (1864) 69 U.S. (2 Wall.) 57 .... 90, 95–96, 283, 313
Barakis v. American Cynamid Co. (N.D. Tex. 1958) 161 F.Supp. 25 ... 316
Barker v. Harvey (1901) 181 U.S. 481 .... 4, 6, 333
Barney v. Keokuk (1876) 94 U.S. 324 .... 48, 53, 64, 73, 77, 236, 257, 269, 272
Bauman v. Choctaw-Chickasaw Nations (10th Cir. 1964) 333 F.2d 785, *cert. den.* (1965) 379 U.S. 965 .... 94, 327
Bear v. United States (8th Cir. 1987) 810 F.2d 153 96
Bear v. United States (D. Neb. 1985) 611 F.Supp. 589, *aff'd* (8th Cir. 1987) 810 F.2d 153 .... 93, 313, 334
Beard v. Federy (1865) 70 U.S. (3 Wall.) 478 .... 6, 8, 167
Beaver v. United States (9th Cir. 1965) 350 F.2d 4, *cert. den.* (1966) 383 U.S. 937 .... 70, 265, 269
Block v. N.D. ex rel Bd. of Univ and Sch. Lands (1983) 461 U.S. 273 .... 42, 66, 159, 312
Bonelli Cattle Co. v. Arizona (1973) 414 U.S. 313 .... 53–55, 62, 97, 236, 253, 256, 315, 319
Borax, Ltd. v. Los Angles (1935) 296 U.S. 10 ... xix, 9, 64, 71, 73–74, 77, 103, 108, 117, 119–121, 149, 156–157, 161, 168, 204, 273, 300–301, 330
Borough of Ford City v. United States (3rd Cir. 1965) 345 F.2d 645, *cert. den.* (1965) 382 U.S. 902 .... 74, 272, 274, 296, 301–303

Federal Cases   Page(s)

Botiller v. Dominguez (1889) 130 U.S. 238 6–8
Buxton v. Traver (1889) 130 U.S. 232 14
California Oregon Power Co. v. Beaver Portland Cement Co. (1935) 295 U.S. 142 44
Cal. ex rel. State Lands Com'n v. United States (9th Cir. 1986) 805 F.2d 857, *cert. den.* (1987) 484 U.S. 816. .... 47, 49, 62–63, 93, 97, 256, 269, 273, 283, 290, 292–293, 295, 304, 311–313, 318–319, 334
California, ex rel. State Lands Comm'n v. United States (1982) 457 U.S. 273 .... xviii, xix, 49, 54, 59–61, 71, 96, 130–131, 136, 161, 234, 256, 268–269, 273, 312–313
California v. Arizona (1979) 440 U.S. 59 .... 51, 261
California v. Deep Sea Research (1998) ___ U.S. ___, ___ S.Ct. ___, 140 L.Ed.2d 626 283
California v. United States (1978) 438 U.S. 645 .... 65, 67, 98
Chitimacha Tribe of Louisiana v. Harry L. Laws Co. (5th Cir. 1982) 690 F.2d 1157, *cert.den.* (1983) 464 U.S. 814 .... 4, 6
Chouteau v. Molony (1853) 57 U.S. (16 How.) 203 5
City of Centralia, Wash. v. F.E.R.C. (9th Cir. 1988) 851 F.2d 278 245
City of Los Angeles v. Borax Consolidated Limited (9th Cir. 1935) 74 F.2d 901, *aff'd* (1935) 296 U.S.10 ..... 76, 107, 115, 117, 119–121, 168, 186
City of Los Angeles v. Borax Consol. Ltd. (9th Cir. 1939) 102 F.2d 52, *cert. den.* (1939) 307 U.S. 644 120
City of Los Angeles v. Borax Consol. Ltd. (S.D. Cal. 1933) 5 F.Supp. 281 119
City of Los Angeles v. Borax Consol. Ltd. (S.D. Cal. 1937) 20 F. Supp. 69 119
Coburn v. San Mateo County (9th Cir. 1896) 75 Fed. 520 .... 77, 87
Commissioner of Oklahoma Land Office v. United States (8th Cir. 1920) 270 F.2d 110 97
Cooper v. Roberts (1855) 59 U.S. (18 How.) 173 17
County of St. Clair v. Lovingston (1874) 90 U.S. (23 Wall.) 46 ...... 5, 68, 87, 92–96, 175, 253–254, 265, 268–269, 312–313, 316–317, 326
Cox v. Hart (1922) 260 U.S. 427 13–14
Coyle v. Smith (1911) 221 U.S. 559 23
Cragin v. Powell (1888) 128 U.S. 691 14
Daubert v. Merrill Dow Pharmaceuticals (1993) 509 U.S. 579 137
Davis Warehouse Co. v. Bowles (1944) 321 U.S. 144 47
De Boer v. United States (9th Cir. 1981) 653 F.2d 1313 96
DeGuyer v. Banning (1897) 167 U.S. 723 .... 77, 119, 126, 165
Downes v. Bidwell (1901) 182 U.S. 244 4
Durfee v. Duke (1963) 375 U.S. 106 52–53
Economy Light & Power Co. v. United States (1921) 256 U.S. 113 242
Erie R. Co. v. Tompkins (1938) 304 U.S. 64 .... 57, 66
Executive Jet Aviation, Inc. v. City of Cleveland (1972) 409 U.S. 249 205
Federal Power Com. v. Oregon (1955) 349 U.S. 435 .... 9, 44

Federal Cases Page(s)

Fremont v. United States (1854) 58 U.S. (17 How.) 541 .... 8, 164
French v. Fyan (1876) 93 U.S. 169 .... 15, 155
Georgia v. South Carolina (1990) 497 U.S. 376 .... 173, 175, 270, 316
Goodtitle v. Kibbe (1850) 50 U.S. (9 How.) 471 .... 103, 330
Great Salt Lake Special Master's Report I, adopted in *Utah v. United States*
 (1971) 420 U.S. 304 317–319
Hardin v. Jordan (1891) 140 U.S. 371 ... 14, 73, 77, 82, 85, 272, 283, 288, 298, 315
Hardin v. Shedd (1903) 190 U.S. 508 269
Harrison v. Fite (8th Cir. 1906) 148 F. 781 .... 241, 274, 301, 304
Hassinger v. Tideland Electric Membership Corporation (4th Circ. 1986)
 781 F.2d 1022, cert. den., sub. nom. *Coast Catamaran Corp. v. Hassinger* (1986) 478 U.S. 1004 205
Hay v. Bruno (D. Ore. 1972) 344 F.Supp. 286 102
Herron v. Choctaw & Chickasaw Nations (10th Cir. 1956) 228 F.2d
 830 .... 93, 96, 334
Hobart v. Hall (D. Minn. 1909) 174 F. 433 53
Horne v. Smith (1894) 159 U.S. 40 89
Howard v. Ingersoll (1851) 54 U.S. (13 How.) 381 .... 4–5, 76, 81, 233, 272–276, 301
Hughes v. Washington (1967) 389 U.S. 290 ...... 54–55, 60, 130, 269, 319
Hynes v. Grimes Packing Co. (1949) 337 U.S. 86 126
Ickes v. Fox (1937) 300 U.S. 82 44
Idaho v. United States (2001) _____ U.S. _____, _____ S. Ct. _____, _____
 L. Ed. 29 _____, 2001 Lexis 4665 9
Illinois Central Railroad v. Illinois (1892) 146 U.S. 387 .... 24, 30–31, 72, 102–103, 105, 148, 161, 282, 285–286, 295
Indiana v. Kentucky (1890) 136 U.S. 479 .... 38, 49, 279
Iowa v. Illinois (1892) 147 U.S. 1 49
Jacobellis v. Ohio (1964) 378 U.S. 184 74
James v. Langford (10th Cir. 1983) 701 F.2d 123, cert. den. (1984) 1045 S.
 Ct. 102 xviii
Jefferis v. East Omaha Land Co. (1890) 134 U.S. 178 ...... 82, 92, 94–98, 100, 233, 253, 255–256, 264, 272, 317, 326
Johnson v. McIntosh (1823) 21 U.S. (8 Wheat.) 543 3–4
Johnston v. Jones (1862) 66 U.S. 209 313
Joint Trib. Coun. of Passamaquoddy Tribe v. Morton (1st Cir. 1975) 528
 F.2d 370 57
Jones et al. v. Johnston (1855) 59 U.S. (18 How.) 150 .... 283, 313
Jones v. Soulard (1860) 65 U.S. (24 How.) 41 .... 36, 77, 95, 233
Joy v. St. Louis (1905) 201 U.S. 332 .... 53, 64, 269
Kansas v. Merriwether (8th Cir. 1910) 182 F. 457 269
Kansas v. Missouri (1944) 322 U.S. 213 .... 50, 99, 233, 258, 263, 267
Knight v. U. S. Land Association (1891) 142 U.S. 161 .... 74, 115, 161, 169, 210
Kumho Tire Company, Ltd. v. Carmichael (1999) 526 U.S. 137, 143
 L.Ed.2d 238. 137

Federal Cases **Page(s)**

Lake Mich. Fed. v. U.S. Army Corps of Engineers (N.D. Ill. 1990) 742
  F.Supp. 441 .... 25, 31, 286–287
Leslie Salt v. Froehlke (9th Cir. 1978) 578 F.2d 742 71
Littlefield v. Nelson (10th Cir. 1957) 246 F.2d 956 177
Live Stock v. Springer (1902) 185 U.S. 47 .... 88–89, 283, 298–299
Los Angeles Milling Co. v. Los Angeles (1910) 217 U.S. 217 .... 8, 65
Louisiana Boundary Case (1968) 394 U.S. 11 xviii
Marine Ry. Co. v. United States (1921) 257 U.S. 47 268
Martin v. Waddell (1842) 41 U.S. (16 Pet.) 367 .... 5, 77, 144, 149, 205,
                                                                                                238
Massachusetts v. New York (1926) 271 U.S. 65 .... 3, 5, 49, 51
Matthews v. McGee (8th Cir. 1966) 358 F.2d 516 281
Mikel v. Kerr (10th Cir. 1974) 499 F.2d 1178 271
Milwaukee v. Illinois (1981) 451 U.S. 304 66
Mississippi v. Arkansas (1974) 415 U.S. 289 .... 100, 263–264, 266,
                                                                                                278–279
Missouri v. Kansas (1909) 213 U.S. 78 233
Missouri v. Kentucky (1870) 78 U.S. (11 Wall.) 395 ..... 48, 233, 279–280
Missouri v. Nebraska (1904) 196 U.S. 23 .... 233, 260, 265–266
Mitchell v. Smale (1891) 140 U.S. 406 .... 86–87, 89, 263
Mobile Transportation Company v. Mobile (1903) 187 U.S. 479 205
Montana v. United States (1981) 450 U.S. 544 ..... 53, 66, 162, 236, 244,
                                                                                                248, 312
More v. Steinbach (1888) 127 U.S. 70 8
Mormon Church v. United States (1890) 136 U.S. 1 4
Moss v. Ramey (1916) 239 U.S. 538 258
Native Village of Eyak v. Daley (1998) 153 F.3d 1090; *cert. den.* (1999)
  527 U.S. 1003. 123
Nebraska v. Iowa (1892) 143 U.S. 359 .... 49–50, 72, 94, 97–98, 233,
                                                    253, 256, 261–262, 265–267, 270, 327
Nebraska v. Iowa (1972) 406 U.S. 117 .... 51–52, 261
Nebraska v. Iowa (1972) 409 U.S. 285 261
New Hampshire v. Maine (2001) ____ U.S. ____ S.Ct. ____, ____
  L.Ed.2d ____, 2001 U.S. Lexis 3981 .... 41, 44, 291
New Jersey v. New York (1998) 523 U.S. 767, 140 L.Ed.2d
  993 .... xviii, xix, 38, 100, 139, 267, 279
New Orleans v. United States (1836) 35 U.S. (10 Pet.) 662 ...... 6, 87, 96,
                                                                                                233, 236, 246, 253
Newhall v. Sanger (1875) 92 U.S. 761 .... 4, 333
Niles v. Cedar Point Club (1899) 175 U.S. 300 .... 89, 298–299
Nollan v. California Coastal Com'n (1987) 483 U.S. 825 .... 101–102, 107
North Dakota v. Andrus (D. N. Dak. 1981) 506 F. Supp. 619, *aff'd* (8th Cir.
  1982) 671 F.2d 271, overruled *sub nom. Block v. N.D. ex rel. Bd. of
  Univ. and Sch. Lands* (1983) 461 U.S. 273 41
Ohio v. Kentucky (1980) 444 U.S. 335 317
Oklahoma v. Texas (1922) 258 U.S. 574 .... 241, 301

Federal Cases   **Page(s)**

Oklahoma v. Texas (1923) 260 U.S. 606 ...... 50, 263, 265, 267, 274–275, 278–279, 301, 304
Oklahoma v. Texas (1925) 268 U.S. 252 .... 265, 275
Omaha Indian Tribe, Treaty of 1854, etc. v. Wilson (8th Cir. 1978) 575 F.2d 620, 622–627, *vacated and remanded* (1979) 442 U.S. 653 ..... 91, 247, 251, 254, 256, 264–265, 279, 335
Omaha Indian Tribe v. Wilson (8th Cir 1980) 614 F.2d 1153, *rehg den., cert. den.* (1980) 449 U.S. 825 .... 60, 251, 265
Oregon v. Riverfront Protective Ass'n (9th Cir. 1982) 672 F.2d 792 ... 237, 245
Packer v. Bird (1891) 137 U.S. 661 .... 73, 164, 257–258, 272
Paine Lumber Co. v. United States (W.D. Wis. 1893) 55 F. 854 .... 277, 308
People of the State of California v. United States (Ct.Cl. 1955) 132 F.Supp. 208 27
Peterson v. Morton (C.D. Nev. 1979) 465 F.Supp. 986, *vacated, sub nom Peterson v. Watt* (9th Cir.) 1982) 666 F.2d 361 266
Peterson v. United States (9th Cir. 1964) 327 F.2d 219 165
Philadelphia Co. v. Stimson (1911) 223 U.S. 605 ...... 53, 56, 93, 97, 265, 267, 317, 329
Phillips Petroleum Co. v. Mississippi (1988) 484 U.S. 469, 484, *rehg. den.* (1988) .... xix, 66, 72, 95, 146, 162–163, 169, 237, 240, 287, 289
Pollard's Lessee v. Hagan (1845) 44 U.S. (3 How.) 212 ... xix, 23, 49, 77, 144, 149
Puget Sound Power and Light Co. v. Federal Energy Regulatory Commission (9th Cir. 1981) 644 F.2d 785, *cert. den.* (1981) 454 U.S. 1053 .... 238, 245
Puyallup Indian Tribe v. Port of Tacoma (9th Cir. 1983) 717 F.2d 1251, *cert. den., sub nom. Trans-Canada Enterprises Ltd. v. Muckleshoot Indian Tribe* (1984) 465 U.S. 1049, *rehg. den.* (1984) 466 U.S. 954 .... 63, 270, 316
Railroad Company v. Fremont County (1869) 76 U.S. (9 Wall.) 89 21
Railroad Company v. Schurmeir (1868) 74 U.S. (7 Wall.) 272 ...... 11, 13, 48, 82, 84–86, 233
Rank v. Krug (S.D. Cal. N.D. 1950) 90 F.Supp. 773 .... 43, 67, 233
Rank v. (Krug) United States (S.D. Cal. 1956) 142 F.Supp. 1, 105, *aff'd in part* and *rev'd in part, sub nom. California v. Rank* (9th Cir. 1961) 293 F.2d 340 .... 43, 48, 199
Ritter v. Morton (9th Cir. 1975) 513 F.2d 942 89
Rodrigues v. United States (1863) 68 U.S. (1 Wall.) 582 6
San Francisco Savings Union v. Irwin (Cir. Ct. D.Cal. 1886) 28 F. 708, *aff'd* (1890) 136 U.S. 578 .... 19, 123, 150, 333
San Francisco v. Leroy (1891) 138 U.S. 656 .... 7, 73, 115, 150
Saulet v. Shepherd (1866) 71 U.S. (4 Wall.) 502 255
Sawyer v. Grey (1913) 205 F. 160 13–14
Sawyer v. Osterhaus (9th Cir. 1914) 212 F. 765 205

Federal Cases **Page(s)**

Scott v. Lattig (1913) 227 U.S. 229 258
Shapleigh v. United Farms Co. (5th Cir. 1938) 100 F.2d 287 ..... 100, 263
Shulthis v. McDougal (1912) 225 U.S. 561 53
Shively v. Bowlby (1894) 152 U.S. 1 ... 3, 23, 72, 149, 162, 205, 248, 257, 288
Smelting Co. v. Kemp (1881) 104 U.S. 636 .... 15, 156
Smith v. United States (10th Cir. 1979) 593 F.2d 982 89
Solid Waste Agency of Northern Cook County v. U.S. Army Corps of Engineers (2001) ____ U.S. ____, 121 S.Ct. 675, 148 L.Ed.2d 576 .... 144, 146
Soulard v. United States (1830) 30 U.S. (4 Pet.) 511 167
St. Louis v. Rutz (1891) 138 U.S. 226 .... 86, 233, 257–260, 268–269, 316
State Land Board v. Corvallis Sand & Gravel Co. (1977) 429 U.S. 363 .... xix, 23, 46–47, 50, 53–58, 60–64, 149, 236, 267
State of Alaska v. Ahtna, Inc. (9th Cir. 1989) 891 F.2d 1401 243–245
State of Alaska v. Babbitt (9th Cir. 2000) 201 F.3d 1154 .... 159, 241, 243
State of Alaska v. U.S. (D. Alaska 1987) 662 F.Supp. 455 .. 235, 241, 243, 245
State of Cal., etc. v. United States (N.D. Cal. 1981) 512 F. Supp. 36 ... 41, 145, 158–159
State of North Dakota v. Andrus (8th Cir. 1982) 671 F.2d 271, *rev'd on other grounds, sub. nom. Block v. North Dakota ex rel. Bd. of Univ. and State Lands* (1983) 461 U.S. 840 242–245
State of Utah (1963) 70 Int. Dec. 27 .... 305, 315
Steel v. Smelting Co. (1892) 106 U.S. 447 155–156
Stewart v. United States (1942) 316 U.S. 354 .... 8, 76, 165–167
Summa Corp. v. California ex rel. Lands Comm'n (1984) 466 U.S. 198, *rehrg. den.* (1984) 467 U.S. 1231 .... 7–9, 24–25, 35, 73, 146, 168, 333
Swift v. Tyson (1842) 41 U.S. (16 Pet.) 1 53
Tarshis v. Lahaina Investment Corporation (9th Cir. 1973) 480 F.2d 1019 107
Texas Boundary Case (1969) 394 U.S. 1 133
The Daniel Ball (1870) 77 U.S. (10 Wall.) 557 .... 95, 237, 240–243
The Montello (1874) 87 U.S. (20 Wall.) 430 .... 238, 242, 244
The Planter (1833) 32 U.S. (7 Pet.) 324 .... 206, 330
The Propeller Genesee Chief et al. v. Fitzhugh et al. (1851) 53 U.S. (12 How.) 443 .... 15, 237, 287
The Steamboat Thomas Jefferson (1825) 23 U.S. (10 Wheat.) 428 237
Thomas B. Bishop Co. v. Santa Barbara County (9th Cir. 1938) 96 F.2d 198, *cert. den.* 305 U.S. 623 85–86
Thompson v. United States (1959) 8 Ind. Cls. Comm. 1 5
Tubbs v. Wilhoit (1890) 138 U.S. 134 .... 15, 20–22, 152, 156
Twombly v. City of Long Beach (9th Cir. 1964) 333 F.2d 685, *cert. den.* (1964) 379 U.S. 904 23
Union Pacific RR v. Harris (1910) 215 U.S. 386 9
United States Gypsum v. Uhlhorn (E.D. Ark. 1964) 232 F.Supp. 994, *aff'd.* (8th Cir. 1966) 366 F.2d 211, *cert. den.* (1967) 385 U.S. 1026 ...... 73, 257, 288

Federal Cases Page(s)

United States Gypsum v. Uhlhorn (8th Cir. 1966) 366 F.2d 211, *cert. den.*
(1967) 385 U.S. 1026 271
United States v. 1.58 Acres of Land (D. Mass. 1981) 523 F.Supp.
120 27
United States v. 11.037 Acres (N.D. Cal. 1988) No. C-83-4605 JPV 27
United States v. 222.0 Acres of Land (D. Md. 1969) 306 F.Supp. 138 .. 268
United States v. Appalachian Electric Power Co. (1940) 311 U.S.
377 .... 238, 244
United States v. Aranson (9th Cir. 1983) 696 F.2d 654, *cert. den. sub nom.*
Colorado Indian Tribes v. Aranson (1983) 464 U.S. 982 ..... 63, 70, 96,
265, 268
United States v. Arredondo (1832) 31 U.S. (6 Pet.) 691 4–6
United States v. Ashton (9th Cir. 1909) 170 F. 509, *app. dsm'd., sub.nom.*
Bird et al. v. Ashton (1911) 220 U.S. 604 11
United States v. Benson (9th Cir. 1895) 70 F. 591 15
United States v. California (1947) 332 U.S. 19 .... 40–42, 54, 77, 123,
158–159, 335
United States v. California (1965) 381 U.S. 139 .... 61, 124–125
United States v. California (1977) 432 U.S. 40 62
United States v. California (1978) 436 U.S. 32 .... 54, 62, 124, 335
United States v. California (1980) 447 U.S. 1 .... 130, 330
United States v. California (9th Cir. 1980) 655 F.2d 914 67
United States v. Cameron (M.D. Fla. 1978) 466 F.Supp. 1090 ..... 78, 272,
275–277, 279, 301
United States v. Carmack (1946) 329 U.S. 230 27
United States v. Champlin Refining Co. (10th Cir. 1946) 156 F.2d
769 237
United States v. Chandler-Dunbar (1908) 209 U.S. 447 258
United States v. Chicago B&Q R. Co. (7th Cir.) 90 F.2d 161, *cert. den.*
(1937) 302 U.S. 714 304
United States v. Ciampitti (D. N.J. 1984) 583 F.Supp. 483 207
United States v. Claridge (9th Cir. 1969) 416 F.2d 933, *cert. den.* (1970)
397 U.S. 961 .... 269, 313
United States v. Claridge (D. Ariz. 1966) 279 F.Supp. 87, *aff'd* (9th Cir.
1967) 416 F.2d 933, *cert. den.* (1970) 397 U.S. 961 303–304
United States v. Coronado Beach Co. (1921) 255 U.S. 472 ...... 8–9, 166,
168
United States v. Curtiss-Wright Corp. (1936) 299 U.S. 304 4
United States v. Estudillo (1862) No. 234 "San Leandro" 150
United States v. Fallbrook Public Utility Dist. (S.D.Cal. 1952) 109 F.Supp.
28 .... 233, 276
United States v. Fossatt (1858) 62 U.S. (21 How.) 446 8
United States v. Georgia Pac. (9th Cir. 1970) 421 F.2d 92 41
United States v. Gerlach Live Stock Co. (1950) 339 U.S. 725 .... 16, 71,
199, 233, 276, 289
United States v. Hall (1889) 131 U.S. 50 .... 15, 96
United States v. Halleck (1863) 68 U.S. (1 Wall.) 439 .... 8, 164

Federal Cases   Page(s)

United States v. Harvey (9th Cir. 1981) 661 F.2d 767, *cert. den.* (1982) 459
   U.S. 883 .... 63, 96–97, 173
United States v. Holt State Bank (1926) 270 U.S. 49 ... 161, 237, 241, 244
United States v. Kimbell Foods, Inc. (1979) 440 U.S. 715 58–59
United States v. Louisiana (1950) 339 U.S. 699 123
United States v. Miller No. 66–75 (Ct. Cl. 1978), Slip Opinion filed June
   14, 1978 69
United States v. Morrison (1916) 240 U.S. 192 14
United States v. New Mexico (1978) 438 U.S. 696 .... 10, 67
United States v. O'Donnell (1938) 303 U.S. 501 .... 6, 8–9, 118, 148–149,
                                                                      156, 165
United States v. Oregon (1935) 295 U.S. 1 237
United States v. Otley (9th Cir. 1942) 127 F.2d 988 .... 14–15, 53–54, 82,
              88–89, 221, 283, 292, 298–300, 304, 306–307, 309, 336
United States v. Pacheco (1864) 69 U.S. (2 Wall.) 578 .... 73–74, 77, 105,
                                                   123, 151, 164–166, 205, 332
United States v. Pappas (9th Cir. 1987) 814 F.2d 1342 .... 66, 269
United States v. Perrin (1889) 131 U.S. 55 15
United States v. Ray (5th Cir. 1970) 423 F.2d 16 .... 79, 334
United States v. Ruby Co. (9th Cir. 1978) 588 F.2d 697, *cert. den.* (1975)
   423 U.S. 947 .... 89, 313, 317
United States v. San Francisco (1940) 310 U.S. 16 16
United States v. Stewart (9th Cir. 1941) 121 F.2d 705, *aff'd, sub. nom.*,
   Stewart v. United States (1942) 316 U.S. 354 .... 9, 166
United States v. Texas (1950) 339 U.S. 707 .... 4, 123
United States v. Utah (1931) 283 U.S. 64 .... 161, 237, 240–244
United States v. Utah (1975) 420 U.S. 304 .... 241, 275
United States v. Washington (9th Cir. 1961) 294 F.2d 830, 834, *cert.den.*
   (1962) 369 U.S. 817 272
United States v. Wilson (8th Cir. 1980) 614 F.2d 1153, *cert. den.* (1980)
   449 U.S. 825 59
United States v. Wilson (8th Cir. 1983) 707 F.2d 304, *rehg den., cert. den.*
   (1984) 465 U.S. 1101 251
United States v. Wilson (N.D. Iowa 1977) 433 F.Supp. 57 251
United States v. Wilson (N.D. Iowa 1977) 433 F.Supp. 67 251
United States v. Wilson (N.D. Iowa 1981) 523 F.Supp. 874, *rev'd* (8th Cir.
   1982) 707 F.2d 304 .... 59, 251
United States v. Wilson (N.D. Iowa 1984) 578 F.Supp 1191, *cert. den.*
   (1984) 465 U.S. 1025, 1101 .... 59, 251
Ussery v. Anderson-Tully Co. (E.D. Ark. 1954) 122 F.Supp. 115 266
Utah Div. of State Lands v. United States (1987) 482 U.S. 193, 107 S.Ct.
   2318 .... 20, 283, 285, 288–289, 291–292, 334
Utah v. United States, Special Master's Report reproduced in 1976 Utah L.
   Rev. 1 .... 71, 241, 275, 297–298, 304, 306
Utah v. United States (1971) 403 U.S. 9 .... 237, 241, 243–244
Utah v. United States (1971) 420 U.S. 304 317

Federal Cases **Page(s)**

Vermont v. New Hampshire (1933) 289 U.S. 593 308
Vermont v. New Hampshire (1934) 290 U.S. 579 .... 277, 308
Waring et al. v. Clarke (1847) 46 U.S. (5 How.) 440 206–207
Washington v. Oregon (1908) 214 U.S. 205 .... 51, 265
Weber v. Harbor Commissioners (1873) 85 U.S. (18 Wall.) 57 ...... 3, 23, 41, 87, 144, 149, 161
Whitaker v. McBride (1905) 197 U.S. 510 258
Willis v. United States (S.D. W.Va. 1943) 50 F.Supp. 101 275–276
Wilson v. Omaha Indian Tribe (1979) 442 U.S. 653 ... 7, 9, 54, 56–59, 61, 63, 91, 233, 248, 251
Wittmayer v. United States (9th Cir. 1941) 118 F.2d 808 96
Woodruff v. North Bloomfield Gravel Min. Co. (9th Cir. 1884) 18 F. 753 .... 171, 199–200, 330
Work v. Louisiana (1925) 269 U.S. 250 .... 18, 149
Work v. United States (9th Cir. 1927) 23 F.2d 136 .... 148, 165
Wright v. Mattison (1855) 59 U.S. (18 How.) 50 .... 36, 328
Wright v. Roseberry (1887) 121 U.S. 488 .... 18, 20, 152

State Cases **Page(s)**

Abbot Kinney Co. v. City of Los Angeles (1959) 53 Cal.2d 52 .... 121, 127
Abbott v. City of Los Angeles (1958) 50 Cal.2d 438 .... 34, 289, 334
Adams v. Frothingham (Mass. 1807) 3 Mass. 352 266
Allen v. McMillion (1978) 82 Cal. App. 3d 211 38
American Water Co. v. Amsden (1856) 6 Cal. 443 287
Anderson v. Trotter (1931) 213 Cal. 414, *cert. den.* (1932) 284 U.S. 686 89
Appeal of York Haven Water & Power Co. (Pa. 1905) 62 A. 97 .... 277, 308
Appel v. Berrman (1984) 159 Cal.App.3d 1209 38
Application of Ashford (Haw. 1968) 440 P.2d 76 107–108
Aptos Seascape Corp. v. County of Santa Cruz (1982) 138 Cal.App.3d 484, *app. dism'd* (1983) 464 U.S. 805 .... 122, 167
Arkansas Land & Cattle Co. v. Anderson-Tully Co. (Ark. 1970) 452 S.W.2d 632 .... 218, 247, 263–264, 279
Arnold v. Mundy (1821) 6 N.J.L. 1 .... 5, 144
Arraington v. Liscom (1868) 34 Cal. 365 .... 1, 336
Atwood v. Hammond (1935) 4 Cal.2d 31 27
Beach Colony II v. California Coastal Com. (1986) 151 Cal.App.3d 1107 173
Benne v. Miller (1899) 50 S.W. 824 258
Bess v. County of Humboldt (1992) 3 Cal.App.4th 1544 ..... 102, 242, 276
Best Renting Co. v. City of New York (1928) 162 N.E. 497 183
Board of Public Works v. Larmar Corporation (Md. 1971) 277 A.2d 427 145

State Cases **Page(s)**

Board of Trustees v. Sand Key Associates (Fla. 1986) 489 So.2d 34, *cert. den.* (1986) 475 U.S. 1094 .... 49, 64–65
Board of Trustees, etc. v. Mediera Beach Nom., Inc. (Fla. 1973) 272 So.2d 209 .... 76, 87, 90, 131
Bohn v. Albertson (1951) 107 Cal.App.2d 738 ..... 97, 161, 190, 244, 254, 271
Bone v. May (Iowa 1929) 225 N.W. 367 .... 263, 266, 271
Boone v. Kingsbury (1928) 206 C. 148, *cert. den.* and *app. dism. sub. nom. Workman v. Boone* (1929) 280 U.S. 517 .... 7, 118, 123, 335
Boorman v. Sunnuchs (1877) 42 Wisc. 233 316
Boston Waterfront Dev. Corp. v. Com. (Mass. 1979) 393 N.E.2d 356 ... 24, 76–77, 102–103
Bott v. Com'n of Natural Resources, etc. (Mich 1982) 327 N.W.2d 838 .... 239, 292
Bradford v. Nature Conservancy (Va. 1982) 294 S.E.2d 866 ...... 126, 142
Brainard v. State of Texas (1999) 12 S.W.3d 6 .... 175, 227, 268–270, 275
Breese v. Wagner (Wisc. 1925) 203 N.W. 764 285
Brundage v. Knox (Ill. 1917) 117 N.E. 123 .... 268–269, 314
Bryant v. Blevins (1994) 9 Cal.4th 47 38–39
Bryant v. Peppe (Fla. 1969) 226 So.2d 357 159
Bryant v. Peppe (Fla. 1970) 238 So. 2d. 836 159
Burket v. Krimlofski (Neb. 1958) 91 N.W.2d 57 269
Callahan v. Price (Idaho 1915) 146 P. 732 .... 258, 272, 295
Camping Com'n of Meth. Ch. v. Ocean View Land, Inc. (Wash. 1966) 421 P.2d 1021 126
Carpenter v. Board of Commissioners (Minn. 1894) 58 N.W. 295 ..... 275, 283, 300, 302–304
Carpenter v. City of Santa Monica (1944) 63 Cal. App.2d 772 .... 65, 71, 96, 107, 127, 130, 174–175, 268, 312
Carpenter v. Lewis (1897) 119 Cal. 18 37
Carpenter v. Ohio River Sand & Gravel Corp. (W.Va. 1950) 60 S.E.2d 212 .... 247, 275, 277, 280, 308
Carr v. Moore (Iowa 1903) 93 N.W. 52 .... 41, 295–296, 316
Chowchilla Farms, Inc. v. Martin (1933) 219 Cal. 1 .... 176, 321
Churchill v. Kingsbury (1918) 178 Cal. 554 .... 71, 151, 300, 312
Cimpher v. City of Oakland (1912) 162 Cal. 87 27
Cinque Bambini Partnership v. State (Miss. 1986) 491 So.2d 508, *aff'd sub nom. Phillips Petroleum Co. v. Mississippi* (1988) 484 U.S. 469 ..... 26, 107, 158–159, 162, 173, 286
City of Berkeley v. Superior Court (1980) 26 Cal.3d 515, *cert. den. sub.nom. Santa Fe Land Improv. Co. v. Berkeley* (1980) 449 U.S. 840 ... 24, 26, 34, 72, 76, 102–103, 122, 145, 148, 161, 168, 203, 282, 286, 336
City of Coachella v. Riverside County Airport Land Use Com. (1989) 210 Cal.App.3d 1277 39
City of Corpus Christi v. Davis (Tex. 1981) 622 S.W.2d 640 .... 127, 134–135, 138

State Cases   **Page(s)**

City of Daytona Beach v. Tona Rama, Inc. (Fla. 1973) 294 So.2d 73 ... 102
City of Long Beach v. Mansell (1970) 3 Cal.3d 462 .... 27, 29, 32, 39,
                                                              42–43, 122, 125, 158, 173
City of Los Angeles v. Aitken (1935) 10 Cal.App.2d 460 .... 283, 293, 311,
                                                                                    315
City of Los Angeles v. Anderson (1929) 206 Cal. 662 ... 92, 119, 129, 268,
                                                                                    326
City of Los Angeles v. Duncan (1933) 130 Cal.App. 11 121
City of Los Angeles v. Morgan (1951) 105 Cal.App.2d 726 36
City of Los Angeles v. Venice Peninsula Properties (1982) 31 Cal.3d 288,
    ovrld on other grnds, *sub. nom, Summa Corp. v. California Ex Rel. State
    Lands Com'n* (1984) 466 U.S. 198 164
City of Missoula v. Bakke (Mont. 1948) 198 P.2d 769 .... 255, 269
City of Newark v. Natural Resources Council (N.J. 1980) 414 A.2d 1304,
    cert. den. (1980) 449 U.S. 983 .... 78, 162, 187
City of Newport Beach v. Fager (1940) 39 Cal.App.2d 23 27
City of Oakland v. Buteau (1919) 180 Cal. 83 .... 94, 172
City of Oakland v. Wheeler (1917) 34 Cal.App. 442, *error dism'd* (1920)
    254 U.S. 659 118
City of Peoria v. Central Nat. Bank (Ill. 1906) 79 N.E. 296 297
Coastal Industrial Water Author. v. York (Texas 1976) 532 S.W.2d
    949 .... 93, 97, 271, 329, 335
Colberg, Inc. v. State of California, ex rel. Dept. of Pub. Wks (1967) 67
    Cal.2d 408, *cert. den.* (1968) 390 U.S. 949 .... 25, 87
Com. v. Morgan (Va. 1983) 303 S.E.2d 899 .... 44, 103, 108
Conant v. Jordan (Me. 1910) 77 A. 938 .... 23, 102, 289
Conoley v. Naetzker (Fla.App. 1962) 137 So.2d 6 314
Conran v. Givrin (Mo. 1960) 341 S.W.2d 75 .... 77, 99, 241, 247,
                                           257–258, 276–278, 281, 288, 306, 309
County of Lake v. Smith (Cal. 1991) 238 Cal.App.3d 214 ..... 40, 44, 289,
                                                              308–310, 322–323
County of Los Angeles v. Berk (1980) 26 Cal.3d. 201, *cert. den.* (1980) 449
    U.S. 836 102
County of Orange v. Heim (1973) 30 Cal.App.3d 694 .... 28, 32
Curry v. Port Lavaca Channel & Dock Co. (Tex. 1930) 25 S.W.2d
    987 268
Dana v. Jackson Street Wharf Co. (1866) 31 Cal. 118 269
Dartmouth College v. Rose (Iowa. 1965) 133 N.W.2d 687 .... 100, 256,
                                                                      258, 263–266
Deering v. Martin (Fla. 1928) 116 So. 54 108
DeGuyer v. Banning (1891) 91 Cal. 400 119
Den v. Spalding (1940) 39 Cal.App.2d 623 .... 85–86, 331
Department of Nat. Res. v. Mayor & C. of Ocean City (Md. 1975) 332
    A.2d 630 .... 121, 129, 132
Dept. of Natural Resources v. Pankratz (Alaska 1975) 538 P.2d 984 97
Determining Natural Ordinary High-water Level (Minn. 1986) 384 N.W.2d
    510 .... 295, 302, 304, 306–307, 325

State Cases Page(s)

Diana Shooting Club v. Husting (Wisc. 1914) 145 N.W. 816 301–302
Dickson v. Sandefur (La. 1971) 250 So.2d 708 267
Dillon v. San Diego Unified Port Dist. (1972) 27 Cal.App.3d 296 27
Dow v. Electric Co. (N.H. 1899) 45 A. 350 275
Dreyfuss v. Badger (1895) 108 Cal. 58 156
Driesbach v. Lynch (1951) 234 P.2d 446 107
Durfee v. Kieffer (Neb. 1959) 95 N.W.2d 618 .... 52, 316
Earl M. Kerstetter, Inc. v. Commonwealth Department of Highways (Pa. 1961) 171 A.2d 163 xxiv
Eichelberger v. Mills Land, etc. Co. (1908) 9 Cal.App. 628 105–106
Eltman v. Harvey (N.Y. 1978) 403 N.Y.S.2d 428 .... 3, 327
Ernie v. Trinity Lutheran Church (1959) 51 Cal.2d 702 99
Esso Standard Oil Co. v. Jones (La. 1957) 98 So.2d 236, *aff'd on rehg* (1957) 98 So.2d 244 .... 266, 269
F.A. Hihn Co. v. City of Santa Cruz (1915) 170 Cal. 436 107
Flisrand v. Madson (S.D. 1915) 152 N.W. 796 .... 93, 235, 239, 254, 258, 277, 291–292, 295, 306, 308, 315, 318, 334
Fogerty v. State of California (1986) 187 Cal.App.3d 224, *cert. den.* (1987) 484 U.S. 821. .... 38, 234, 254, 320, 322–324
Forestier v. Johnson (1912) 164 Cal. 24 .... 149, 157–158, 169
Forgeus v. County of Santa Cruz (1914) 24 Cal.App. 193 .... 107–110, 118–119, 173, 268, 327
Foss v. Johnstone (1910) 158 Cal. 119 82
Frank v. Smith (Neb. 1940) 293 N.W. 329 269
Freeman v. Bellegarde (1895) 108 Cal.179 .... 84, 297, 335
Fullerton v. State Water Resources Control Bd. (1979) 90 Cal.App.3d 590 206
Furlong Ent. v. Sun Exploration & Prod. (N.D. 1988) 423 N.W.2d 130 .... 23, 236, 254, 264, 267–268, 270–271, 285, 287
Gaines v. Dillard (Tex. 1976) 545 S.W.2d 845 264
Garrett v. State (N.J. 1972) 289 A.2d 542 .... 173, 268, 314, 316
Gaskill v. Cook (Mo. 1958) 315 S.W.2d 747 264
Gion v. City of Santa Cruz (1970) 2 Cal.3d 29 102
Golden Feather Community Association v. Thermalito Irrigation District (1989) 209 Cal.App.3d 1276 289
Gratt v. Palangi (Me. 1958) 147 A.2d 455 289
Gray v. Reclamation District No. 1500 (1917) 174 Cal. 622 ..... 197–200, 218, 227–228
Grigsby v. Clear Lake Water Company (1870) 40 Cal. 396 xxv
Gulf Oil Corp. v. State Mineral Bd. (La. 1975) 317 So.2d 576 76
Hager v. Specht (1878) 52 Cal. 579 36
Hall v. Brannan Sand and Gravel Company (Colo. 1965) 405 P.2d 749 100
Hancock v. Moore (Tex. 1941) 146 S.W.2d 369 265
Harkins v. Del Pozzi (Wash. 1957) 310 P.2d 532 107
Hawkins v. Walters (Miss. 1981) 402 So.2d 336 100

TABLE OF AUTHORITIES **xli**

State Cases **Page(s)**

Hazen v. Perkins (Vt. 1918) 105 A. 249 285
Heckman v. Swett (Cal. 1893) 99 Cal. 303 .... 267, 272
Hellman v. City of Los Angeles (1899) 125 Cal. 383 .... 168, 185
Helvey v. Sax (1951) 38 Cal.2d 21 99
Herminghaus v. Southern California Edison Co. (Cal. 1926) 200 Cal. 81, cert. dismd. (1927) 275 U.S. 486 276
Herschman v. State Department of Natural Resources (Minn. 1975) 225 N.W.2d 841 .... 93, 318, 334
Hillebrand v. Knapp (S.D. 1937) 274 N.W. 821 .... 239, 277, 292, 318
Hirsch v. Maryland Department of Natural Resources (Md. 1979) 416 A.2d 10 107
Hitchings v. Del Rio Woods (Calif. 1976) 55 Cal.App.3d 560 .... 238–239, 292
Honsinger v. State (Alaska 1982) 642 P.2d 1352 .... 97, 129, 174
Horry County v. Woodward (S.C.App. 1984) 318 S.E.2d 584 ...... 93, 133, 334
Houghton v. RR Co. (Iowa 1877) 47 Iowa 370 274
Hous. Auth. of City of Atlantic City v. State (N.J. 1984) 472 A.2d 612 137–138
Hudson House, Inc. v. Rozman (Wash. 1973) 509 P.2d 992 97
Hughes v. State (1966) 410 P.2d 20, 26, rev'd. *Hughes v. Washington* (1967) 389 U.S. 290 107
Hunt v. Barker (1915) 27 Cal.App. 776 .... 87, 100
Idaho For. Indus. v. Hayden Lk. Watershed Imp. (Idaho 1987) 733 P.2d 733 .... 285, 292, 305
In re Judicial Ditch Proceeding No. 15 (Minn. 1918) 167 N.W. 1042 318
In re Water of Hallett Creek Stream System (1988) 44 Cal.3d 448, cert. den. (1988) 488 U.S. 824 .... 9, 40–42
In the Matter of the Determination of the Ordinary High-water Mark and Outlet Elevation for Beaver Lake (So. Dak. Sixth Judicial Circuit, 1990) Civ. No. 89-94 294
Internal Imp. Tr. Fund v. Sand Key Assoc. (Fla. 1987) 512 So.2d 934 .... 71, 131
Irwin v. Phillips (1855) 5 Cal. 140 16
Island Harbor Beach Club Ltd. v. Dept. of Nat. Res. (Fla. 1986) 495 So.2d 209, rev. den. (Fla. 1987) 503 So.2d 327 126
Jackson v. Burlington Northern, Inc. (Mont. 1983) 667 P.2d 406 ... 92–93, 329
James v. State (Ga.Ct.App. 1911) 72 S.E. 600 316
Johns v. Scobie (1939) 12 Cal.2d 618 36
Joslin v. Marin Mun. Water Dist. (1967) 67 Cal.2d 132 .... 87, 276
Joyce-Watkins Co. v. Industrial Commission (Ill. 1927) 156 N.E. 346 .... 233, 308
Just v. Marinette County (Wisc. 1972) 201 N.W.2d 761 24
Kelley, ex rel. MacMulan v. Halledin (Mich. 1974) 214 N.W. 2d 856 292

State Cases   Page(s)

Kellogg v. Huffman (1934) 137 Cal. App. 278 37
Kentucky Lumber Co. v. King (Ky. 1901) 65 S.W. 156 308
Kernan v. Griffiths (1864) 27 Cal. 87 .... 15, 21, 156
Kimball v. Macpherson (1873) 46 Cal. 103 105
Kitteridge v. Ritter (Iowa 1915) 151 N.W. 1097 .... 48, 88, 252, 256, 264, 330
Klauber v. Higgins (1897) 117 Cal. 451 156
Klein v. Caswell (1948) 88 Cal. App. 2d 774 37
Klevin v. Gunderson (Minn. 1905) 104 N.W. 4 82
Kootenai Environ. Alliance v. Panhandle Yacht (Idaho 1983) 671 P.2d 1085 286
Kruse v. Grocap, Inc. (Fla. 1977) 349 So.2d 788 107
Lakefront Trust, Inc. v. City of Port Arthur (Tex. 1974) 505 S.W.2d 606 268
Lakeside Boating and Bathing Inc. v. State (Iowa 1984) 344 N.W.2d 217 .... 285, 311, 314
Lamprey v. Metcalf (Minn. 1893) 53 N.W. 1139 .... 235, 239, 244, 254, 291–292, 313–314
Leabo v. Leninski (Conn. 1981) 438 A.2d 1153 107
Lechuza Villas West v. California Coastal Com. (1997) 60 Cal.App.4th 218 .... 121, 126–128, 138
Leese v. Clark (1862) 18 Cal. 535 167
Lindberg v. Department of Natural Resources (Minn. 1986) 381 N.W.2d 494 324–325
Littleton v. State (Haw. 1982) 656 P.2d 1336 107
Littoral Development Co. v. San Francisco Bay Conservation etc. Com. (1994) 24 Cal.App.4th 1050 72
Lorino v. Crawford Packing Co. (Tex. 1943) 175 S.W.2d 410 ..... 130, 268
Los Angeles v. San Pedro etc. R.R. Co (1920) 182 Cal. 652, *cert. den.* (1920) 254 U.S. 636 .... 86, 119, 162, 165
Lummis v. Lilly (Mass. 1982) 429 N.E.2d 1146 .... 130, 330
Lusardi v. Curtis Point Prop. Owners Ass'n (N.J. 1981) 430 A.2d 881 101–102
Luscher v. Reynolds (Ore. 1936) 56 P.2d 1158 237
Lux v. Haggin (1884) 69 Cal.255 .... 79, 87, 167
Lyon v. Western Title Ins. Co. (1986) 178 Cal.App.3d 119 300
Marks v. Whitney (1971) 6 Cal.3d 251 .... 24, 87, 100, 122, 125
Mammoth Gold Dredging Co. v. Forbes (Cal. 1940) 39 Cal.App.2d 739 .... 276, 306
Martin v. Busch (Fla. 1927) 112 So. 274 ..... 71, 148, 151, 285, 287, 295, 298–299, 303, 314
Matcha v. Mattox on behalf of the People (Tex. 1986) 711 S.W.2d 95, 98, *cert. den.* (1987) 481 U.S. 1024 .... 102, 107, 121, 129
Mather v. State (Iowa 1972) 200 N.W.2d 498 269
Matter of Ownership of Bed of Devils Lake (N.D. 1988) 423 N.W.2d 141 .... 93, 285, 288, 296–299, 302, 311–313, 315, 317, 324, 328

State Cases Page(s)

Matthews v. Bay Head Imp. Ass'n (N.J. 1984) 471 A.2d 355, *cert. den.*
   (1984) 469 U.S. 821 .... 24, 102
McBain v. Johnson (Mo. 1900) 55 S.W. 1031 265
McCafferty v. Young (Mont. 1964) 397 P.2d 96 .... 263, 279, 316
Mehdizadeh v. Minier (1996) 46 Cal.App.4th 1296 38
Mesnick v. Caton (1986) 183 Cal.App.3d 1248 38
Miami Corporation v. State (La. 1937) 173 So. 315, *cert. den.* (1937) 302
   U.S. 700 295
Miramar Co. v. City of Santa Barbara (1943) 23 Cal.2d 170 ...... 93, 127,
                                                        129, 132, 134, 174
Montana Coalition for Stream Access, Inc. v. Curran (Mont. 1984) 682
   P.2d 163 292
Mood v. Banchero (Wash. 1966) 410 P.2d 776 .... 86, 241, 275, 299
More v. Massini (1860) 37 Cal. 432 .... 103, 105, 330
Morrison v. First Nat. Bank of Skowhegan (Me. 1895) 33 A. 782, ... 272,
                                                                       297
Muchenberger v. City of Santa Monica (1929) 206 Cal. 635 .. 28, 130, 330
Narrows Realty Company v. State (1958) 329 P.2d 836 107
National Audubon Society v. Superior Court (1983) 33 Cal.3d 419, *cert.
   den. sub. nom. Los Angeles Dept. of Water & Power v. National
   Audubon Soc.* (1983) 464 U.S. 977 .... 24, 169, 236, 238, 290, 311
Natural Soda Prod. Co. v. City of L.A. (Cal. 1943) 23 Cal.2d 193, *cert.den.*
   (1944) 321 U.S. 793, *rehg den.* (1944) 322 U.S. 768 ..... 269, 271, 312,
                                                                       321
Nesbitt v. Wolfkiel (Idaho 1979) 598 P.2d 1046 .... 263, 316
New Jersey Sports & Exposition Authority v. McCrane (N.J. 1971) 292
   A.2d 580, *app. dismd* (1972) 409 U.S. 943 144
New York v. Wilson & Co. (N.Y.App. 1938) 15 N.E.2d 408, *reh. den.*
   (1938) 16 N.E.2d 850 314
New York, N.H.H.R. Co. v. Armstrong (Conn. 1922) 102 A. 791 286
Newcomb v. Newport Beach (1943) 7 Cal.2d 393 .... 155, 240
Newman v. Cornelius (1970) 3 Cal. App. 3d 279 37
Nolte v. Sturgeon (Oklahoma 1962) 376 P.2d 616 .... 94, 98, 327
Noyes v. Collins (Iowa 1894) 61 N.W. 250 .... 297, 314
Oakland v. Oakland Water Front Co. (1897) 118 Cal. 160 .... 33, 147–148
O'Neill v. State Highway Department (N.J. 1967) 235 A.2d 1 ..... 99, 119,
                                                                121, 145, 331
Opinion of the Justices to the House of Representatives (Mass. 1974) 313
   N.E.2d 561 126
Ord Land Co. v. Alamitos Land Co. (1926) 199 Cal. 380 154
Otay Water District v. Beckwith (Cal. 1991) 1 Cal.App.4th 1041 38
Padgett v. Central and Southern Florida Control Dist. (Fla.App. 1965) 178
   So.2d 900 314
Pacific Gas & Electric Co. v. Superior Court (1983) 145 Cal.App.3d
   253 239

State Cases
Page(s)

Pannell v. Earls (Ark. 1972) 483 S.W.2d 440 263
Patton v. City of Los Angeles (1915) 169 Cal. 521 40–41
Payette Lakes Protective Ass'n v. Lake Reservoir Co. (Idaho 1948) 189 P.2d 1009 .... 276, 283, 306
Payne v. Hall (Iowa 1921) 185 N.W. 912 271
Peabody v. City of Vallejo (1935) 2 Cal.2d 351 173
Peat, Marwick, Mitchell and Co. v. Superior Court (1988) 200 Cal.App.3d 272 218
People ex rel. Baker v. Mack (1971) 19 Cal.App.3d 1040 ... 161, 238–240, 292
People ex rel. Blakslee v. Commissioners of the Land Office (N.Y. 1892) 32 N.E. 139 268
People ex rel. Dept. Pub. Wks. v. Shasta Pipe Co. (1968) 264 Cal.App.2d 520 .... 90, 275
People ex rel. State Water Resources Control Bd. v. Forni (1976) 54 Cal.App.3d 743 87
People v. California Fish Co. (1913) 166 Cal. 576 .... 26, 32–35, 72, 103, 105, 123, 147–149, 154–158, 160–161, 169, 204, 286, 333
People v. Chambers (1951) 37 Cal.2d 552 40
People v. Department of Housing and Community Development (1975) 45 Cal.App.3d 185 158–159
People v. Emmert (Colo. 1979) 597 P.2d 1025 238
People v. Gold Run D. & M. Co. (1884) 66 Cal. 138 .... 171, 199–200, 236
People v. Hecker (1960) 179 Cal.App.2d 823 ..... 107, 127, 129–130, 138
People v. Leahy (1994) 8 Cal. 4th 587 137
People v. Los Angeles (Cal. 1950) 34 Cal.2d 695 312
People v. Massey (Mich. 1984) 358 N.W.2d 615 .... 283, 285
People v. Morrill (1864) 26 Cal. 336 .... 76, 104–105, 148–149
People v. Shirokow (1980) 26 Cal.3d 301 40
People v. Ward Redwood Co. (1964) 225 Cal.App.2d 385 ..... 97, 99, 258–259, 276
People v. Wm. Kent Estate Co. (1966) 242 Cal.App.2d 156 .... 71, 119, 121–122, 126–127, 129, 316, 328, 331
Perry v. State of California (1956) 139 Cal.App.2d. 379 .... 78, 176
Peterson v. Harpst (Mo. 1952) 247 S.W.2d 663 258
Pierce v. Central National Bank (Ill. 1906) 79 N.E. 296 274
Plummer v. Marshall (Tex. 1910) 126 S.W. 1162 263
Port Acres Sportsman Club v. Mann (Texas 1976) 541 S.W.2d 847 .... 93, 335
Posey v. Bay Point Realty Co. (1932) 214 Cal. 708 37
Provo City v. Jacobsen (Utah 1947) 176 P.2d 130 ..... 234, 241, 275, 277, 279, 292, 295, 299, 302, 304, 307
Ray v. State (Tex. Civ. App. 1941) 153 S.W.2d 660 314
Rench v. McMullen (1947) 82 Cal.App.2d 872 99
Richards v. Page Inv. Co. (Ore. 1924) 228 P. 937 272

State Cases   Page(s)

Riverland Co., Inc. v. McAlexander (Ark. 1983) 661 S.W.2d 451 270
Roberts v. Taylor (N.D. 1921) 181 N.W. 622 294
Robinson v. Forrest (1865) 29 Cal. 317 .... 13, 19
Rondell v. Fay (1867) 32 Cal. 354 147
Rutten v. State (N.D. 1958) 93 N.W.2d 796 .... 206, 302, 306, 318
S.D. Wildlife Federation v. Water Mgmt. Bd. (S.D. 1986) 382 N.W.2d
   26 .... xix, 288–289, 302, 304, 325
Safwenberg v. Marquez (1975) 50 Cal.App.3d 301 37
Sage v. Mayor of City of New York (N.Y. 1897) 47 N.E. 1096 314
Sapp v. Frazier (La. 1899) 26 So. 378 .... 93, 318, 334
Schafer v. Schnabel (Ala. 1972) 494 P.2d 802 .... 129, 131
Seaman v. Smith (Ill. 1860) 24 Ill. 521 274
Sieck v. Godsey (Iowa 1962) 118 N.W.2d 555 .... 265, 269
Siesta Properties, Inc. v. Hart (Fla. 1960) 122 So.2d 218 134
Slauson v. Goodrich Transp. Co. (Wisc. 1897) 69 N.W. 990 ..... 255, 277,
                                                                     281, 308
Smith v. State (Ga. 1981) 282 S.E.2d 76 ..... 102, 107, 109–110, 121, 337
Smith Tug & Barge v. Columbia-Pacific Tow. Corp. (Wash. 1971) 482 P.2d
   769, *cert. den.* (1971) 404 U.S. 829 234
Snow v. Mt. Desert Island Real Estate Co. (Me. 1891) 24 A. 429 103
Solomon v. City of Sioux City (Iowa 1952) 51 N.W.2d 472 269
Southern Idaho Fish & Game Ass'n v. Picabo Livestock Co. (Idaho 1974)
   528 P.2d 1295 238–239
St. Joseph Land, etc. v. Florida State Bd. (Fla. 1979) 365 So.2d
   1084 121
State Bd. of Trustees, etc. v. Laney (Fla. App. 1981) 399 So.2d
   408 .... 86, 90
State Engineer of Nevada v. Cowles (Nev. 1970) 478 P.2d 159 269
State ex rel. Brunquist v. Bollenbach (Minn. 1954) 63 N.W.2d 278 237,
                                                                 284, 292, 301, 324
State ex rel. Buckson v. Pennsylvania Railroad Co. (Del. 1969) 267 A.2d
   455 .... 77, 103
State ex rel. Buckson v. Pennsylvania Railroad Co. (Del. 1971) 273 A.2d
   268 187
State ex rel. Thornton v. Hay (Ore. 1969) 462 P.2d 671 ..... 102, 105, 121
State of Cal. ex rel. State Lands Com. V. Superior Court (*Lovelace*) (1995)
   11 Cal.4th 50 ...... 47, 49, 64, 66, 96–97, 130–133, 171–176, 208–209,
                                                                 227, 268–270
*State of California, ex rel. Public Works Board v. Southern Pacific
   Transportation Company* (1983) Sacramento County Superior Court No.
   277312, Statement of Decision 157, 161, 177, 183, 188, 190,
                                                                 203–204, 332
State of California v. Superior Court (Lyon) (1981) 29 Cal.3d 210, *cert.
   den., sub. nom. Lyon v. California* (1981) 454 U.S. 865 .... 34, 73, 122,
                                                                 272, 285–290

State Cases
Page(s)

State of Kansas ex rel. Meek v. Hays (Kan. 1990) 785 P.2d 1356 238
State of Wisconsin v. Trudeau (Wisc. 1987) 408 N.W.2d 337, *cert. den.*
  (1988) 98 L.Ed.2d 652 .... 283, 285, 287, 291, 295, 302–303, 324–325, 330
State v. Balli (Tex. 1944) 190 S.W.2d 71, *cert. den.* (1946) 328 U.S.
  852 134
State v. Bonelli Cattle Co. (Ariz. 1971) 489 P.2d 699, *rev'd.* (1973) 414
  U.S. 313, *ovrld.* (1977) 429 U.S. 363 316
State v. Bunkowski (Nev. 1972) 503 P.2d 1231 240
State v. Cain (Vt. 1967) 236 A.2d 501 ... 77, 233, 277, 290–291, 309, 330
State v. Contemporary Land Sales (Fla.Ap. 1981) 400 So.2d 488 314
State v. District Court of Kandyiohi County (Minn. 1912) 137 N.W.
  298 314
State v. Edwards (Wash. 1936) 62 P.2d 1094 274
State v. Faudre (W.Va. 1903) 46 S.E. 269 297
State v. George C. Stafford & Sons, Inc. (N.H. 1954) 105 A.2d 569 268
State v. Gunther & Shirley Company (Ariz. 1967) 423 P.2d 352 ... 94, 327
State v. Holston Land Co. (S.C. 1978) 248 S.E.2d 922 126
State v. Korrer (Minn. 1914) 148 N.W. 617 .... 87, 285, 289, 309
State v. Longfellow (Mo. 1902) 69 S.W. 374 274
State v. Longyear Holding Co. (Minn. 1947) 29 N.W.2d 657 ...... 71, 293, 314, 318
State v. McFarren (1974) 215 N.W.2d 459 .... 295, 302, 308
State v. McIlroy (Ark. 1980) 595 S.W.2d 659, *cert. den.* (1980) 449 U.S.
  843 .... 241, 292
State v. Parker (Ark. 1918) 200 S.W. 1014, *cert. den.* (1918) 247
  U.S. 512 .... 295, 321
State v. Pennsylvania Railroad Company (Del. 1967) 228 A.2d 587 ...... 5, 117, 126
State v. Pennsylvania Railroad Company (Del. 1968) 244 A.2d 80 126
State v. Placid Oil Company (La. 1973) 300 So.2d 154, *cert. den.*, (1975)
  419 U.S. 1110, *rehg den.* (1975) 420 U.S. 956 .... 247, 268, 277, 294, 314
State v. Sause (Oregon 1959) 342 P.2d 803 ... 96, 98, 131, 253, 266, 268–269, 315
State v. Slotness (Minn. 1971) 185 N.W.2d 530 309
State v. Sorenson (Iowa 1937) 271 N.W. 234 321
State v. Southern Sand & Material Co. (Ark. 1914) 167 S.W. 854 286
State v. Superior Court (Fogerty) (1981) 29 Cal.3d 240, *cert. den.*, *sub.
  nom.*, *Lyon v. California* (1981) 454 U.S. 865 ...... 158, 174–176, 208, 254, 271, 283–285, 289, 320–321
State v. Thomas (Iowa 1916) 155 N.W. 859 .... 285, 301
State v. Thompson (Iowa 1907) 111 N.W. 328 318
Stover v. Jack (Pa. 1869) 60 Pa. 339 277

State Cases   Page(s)

Strand Improvement Co. v. Long Beach (1916) 173 Cal. 765 ..... 119, 129
Sun Dial Ranch v. May Land Co. (1912) 119 P. 758 274
Surfside Colony, Ltd. v. California Coastal Commission (1991) 226
   Cal.App.3d 1260 .... 102, 129
Swarzwald v. Cooley (1940) 39 Cal.App.2d 306 .... 68, 119, 121, 126, 331
Tatum v. City of St. Louis (Mo. 1894) 28 S.W. 1002 269
Teschemacher v. Thompson (1861) 18 Cal 11 .... 115, 164–165, 167
Thomas v. Sanders (Ohio 1979) 413 N.E.2d 1224 286
Trustees of Freeholders, etc. v. Helner (N.Y. 1975) 375 N.Y.S.2d 761 .. 162
Trustees of Internal Improvement Fund v. Wetstone (Fla. 1969) 222
   S.2d 10 275
Tucci v. Salzhauer (N.Y. 1972) 329 N.Y.S.2d 825 102
Union Sand & Gravel Co. v. Northcott (W.Va. 1926) 135 S.E. 589 277,
   308
United Plainsmen Assoc. v. North Dakota State Water Conservation Comm.
   (N.D. 1976) 247 N.W.2d 457 291
Utah State Road Commissioners v. Hardy Salt Company (Utah 1971) 486
   P.2d 391 .... 285, 297–298, 300, 304–305, 313, 315–316
United States Gypsum Co. v. Reynolds (Miss. 1944) 18 So.2d 448 263
United States v. State Water Resources Control Bd. (1986) 182 Cal.App.3d
   82 .... 199–200, 213
Valder v. Wallace (Nebraska 1976) 242 N.W.2d 112 98
Van Calbergh v. Easton (1917) 32 Cal. App. 796 38
Velsicol Chemical v. Environmental Prot. Dept. (N.J. 1982) 442 A.2d
   1051 .... 99, 145, 188
Wade v. Kramer (Ill. 1984) 459 N.E.2d 1025 .... 282, 286
Ward Sand & Material Co. v. Palmer (N.J. 1968) 237 A.2d 619 .. 129, 133
Ward v. Mulford (1867) 32 Cal. 365 147
Warren S. & G. Co., Inc. v. Commonwealth, Dept. of E.R. (Pa. 1975) 341
   A.2d 556 145
Waterman v. Smith (1859) 13 Cal. 373 .... 8, 164
Wemmer v. Young (Neb. 1958) 93 N.W.2d 837 271
Wernberg v. State (Alaska 1973) 516 P.2d 1191 87
White v. Hughes (Fla. 1939) 190 So. 446 .... 24, 69, 101–102
White v. State of California (1971) 21 Cal.App.3d 738 .... 147, 162
Whyte v. City of St. Louis (Mo. 1899) 54 S.W. 478 269
Wilbour v. Gallagher (Wash. 1969) 462 P.2d 232, *cert. den.* (1970) 400
   U.S. 878 311
Wilcox v. Pinney (Iowa 1959) 98 N.W.2d 720 .... 271, 274
Williams v. City of San Pedro, etc. Co. (1908) 153 Cal. 44 156
Winberger v. Passaic (N.J. 1913) 86 A. 59 314
Witter v. County of St. Charles (Mo. 1975) 528 S.W.2d 160 ...... 270, 316
Wright v. Seymour (1886) 69 Cal. 122 .... 94, 162
Wycoff v. Mayfield (Ore. 1929) 280 P.340 100
Yuba River Sand Co. v. Marysville (1947) 78 Cal.App.2d 421 36
Yutterman v. Grier (Ark. 1914) 166 S.W. 749 266

Constitutions and Statutes   Page(s)

Federal

U.S. Const., Art. III, § 2	40
Abandoned Shipwreck Act of 1987 (1988) P.L. 100-298, 102 Stat. 432	283
Act of May 18, 1796, 1 Stat. 464.	11, 16
Act of August 7, 1797, 1 Stat. 50	23
Act of April 24, 1820, 3 Stat. 566, now found in 43 U.S.C. 672	16
Act of March 3, 1831, 4 Stat. 492	8
Act of March 2, 1849, 9 Stat. 352	18
Act of September 9, 1850, 9 Stat. 452	18
Act of September 28, 1850, 9 Stat. 519, found in 43 U.S.C. § 982, et seq	18, 154
9 Stat. 452 (1850)	236
Act of March 3, 1851, 9 Stats. 631	6–8
Act of March 3, 1853, 10 Stat. 244	17
Act of March 22, 1855, 10 Stat. 634	21
Act of March 3, 1857, 11 Stat. 251	21
Act of July 2, 1862, 12 Stat. 503	17
Act of May 10, 1872, 17 Stat. 94	14
Act to Quiet Land Titles in California, Act of July 23, 1866, 14 Stat. 218, now found in 43 U.S.C. § 987	22
Homestead Act, Act of May 20, 1862, 12 Stat. 392	14
Interstate Compact Defining the Boundary Between the State of Arizona and the State of California (1966) 80 Stat. 340	51, 261
Northwest Ordinance of 1787, 1 Stat 50 (1789)	236
Preemption Act, Act of September 4, 1841, 5 Stat. 453	14, 17
Rivers and Harbors Act of 1899, 33 U.S.C. § § 401 et seq.	71
Submerged Lands Act, Act of May 22, 1953, 67 Stat. 29, set forth in 43 U.S.C. § § 1301, et seq.	23, 54, 77, 124, 335
28 U.S.C. § 1251	40
28 U.S.C. § 1332	52
28 U.S.C. § 1345	40
28 U.S.C. § 1346(f)	47
28 U.S.C. § 2409a	159
28 U.S.C. § 2409a(a)	159
33 U.S.C. § 403	205
33 U.S.C. § 1344	144, 205
43 U.S.C. § 31(a) and (b)	210, 337
43 U.S.C. § 700	11
43 U.S.C. § 751	11, 13–14
43 U.S.C. § 857	17
43 U.S.C. § 982	18
43 U.S.C. § 987	22
43 U.S.C. § 1301	272
43 U.S.C. § 1301(a)	124, 285
43 U.S.C. § 1301(c)	125
43 U.S.C. § 1311	272

Constitutions and Statutes   Page(s)

43 U.S.C. § 1311(a)(1) and (2) 23, 124, 236, 285
43 U.S.C. § 1313(a) 61
Treaty Between the United States and the French Republic, April 30, 1803, 8 Stat. 200 4
Treaty of Amity, Settlements and Limits, Between the United States of America and His Catholic Majesty, February 22, 1819, 8 Stat. 252 4
Treaty of Peace, Friendship, Limits and Settlement Between the United States of America and the Mexican Republic, February 2, 1848, 9 Stat. 922 4
Treaty with Great Britain, June 15, 1846, 9 Stat. 869 4
Treaty with Mexico, December 30, 1853, 10 Stat. 103 4
Treaty with Russia, March 30, 1867, 15 Stat. 539 4
Convention on the Territorial Sea and the Contiguous Zone, March 24, 1961, 15 U.S.T. 1606, T.I.A.S. No. 5639 .... 61, 125
33 C.F.R. § 323.2(b) 294
33 C.F.R. § 328.3(7)(b) 144
33 C.F.R. § 328.3(e) 302
33 C.F.R. § 328.3(f) 205–206
33 C.F.R. § 329.1 205
33 C.F.R. § 329.4 205
40 C.F.R. § 230.3(s)(1) 205
Fed. Rules of Evid. § 201 (b) 206, 330

State

Cal. Const. Art. X, 4 101
Cal. Const. Art. X, § 3 .... 27, 122
Act to Provide for the Reclamation and Segregation of Swamp and Overflowed, Salt Marsh and Tide Lands Donated to the State of California by Act of Congress, Act of May 13, 1861, Cal. Stats. 1861, Ch. 352 .... 22, 25, 155, 157, 204
California Coastal Act of 1976, Cal. Pub. Res. Code §§ 30000, et seq. 101
Cal. Civ. Code § 670 308
Cal. Civ. Code § 830 .... 73, 77, 205, 236, 272, 288–289, 291, 308
Cal. Civ. Code § 1007 .... 38, 40
Cal. Civ. Code § 1014 .... 92, 172, 174–175, 268–269, 311
Cal. Civ. Code § 1015 .... 94, 173, 271
Cal. Civ. Code § 1091 39
Cal. Civ. Code § 1624 39
Cal. Civ. Code § 3538 90
Cal. Code of Civ. Proc. § 321 38
Cal. Code of Civ. Proc. § 324 37
Cal. Code of Civ. Proc. § 998 247
Cal. Code of Civ. Proc. §§ 1250.310(c) 194

TABLE OF AUTHORITIES

Constitutions and Statutes   Page(s)

Cal. Code of Civ. Proc. § 1260.220 194
Cal. Corp. Code § 1002 36
Cal. Evid. Code § 115 .... 99, 327
Cal. Evid. Code § 450–460 .... 206, 330
Cal. Evid. Code § 600(a) 99, 333
Cal. Evid. Code § 623 39
Cal. Evid. Code § 662 99
Cal. Evid. Code section 1605. 164
Cal. Gov. Code 53035 101
Cal. Gov. Code 66478.11 101
Cal. Gov. Code 66478.3 101
Cal. Gov. Code § 66610 72
Cal. Harbors & Nav. Code §§ 101—106 240
Cal. Pub. Res. Code 7552 160
Cal. Pub. Res. Code § 6102 .... 7, 335
Cal. Pub. Res. Code § 7601 .... 312, 314
Cal. Pub. Res. Code § 7991 .... 28, 122
Cal. Water Code § 12220 190
Fla. Stat. (1981) § 161.051 131
Minn.Stat. § 105.37, subd. 16 (1984) 302
So. Dak. Civ. Code § 289 291
SDCL 43-17-20(2) 302

Legislative Materials   Page(s)

Annual Report of the State Surveyor General for the Year 1855 156
Report of the Joint Committees on Swamp and Overflowed Lands and Land
    Monopoly (1874) 203
Report of the Superintendent of the U.S. Coast Survey, 39th Cong., 1st
    Sess., House Ex. Doc. No. 75, Appendix 22 (1865) 179, 181
Report on the Subject of Land Titles in California (1850) 31st Cong., 1st
    Sess., Ser. 589, Doc. 18 5
S. Rep. No. 133, 83rd Cong., 1st Sess. (minority views) 10, 15–18,
    reprinted in 1953 U.S. Code Cong. & Admin. News 15343,
    1549–1551 62

Miscellaneous Texts and Authorities   Page(s)

1 Bowman, *Ogden's Revised California Real Property Law*
    (1974) 35–37, 39, 328
1 Ogden's California Real Property Law (1956) 3, 7, 11, 13, 327
1 Shalowitz, *Shore and Sea Boundaries* (Dept. Comm. Pub. 10-1
    1962) 47, 63, 78, 103, 110–111, 115, 117–118, 328, 330, 335
1 S. Wiel, "Water Rights in the Western States" (1911) 87
14 Encyc. Brit. (14 Ed. 1929), ... 2

Miscellaneous Texts and Authorities   Page(s)

1890 Manual of Instructions 185
19 Collier's Encyclopedia (1961) 2
2 Ogden's Revised California Real Property Law (Cont. Ed. Bar 1975) 167
2 Shalowitz, *Shore and Sea Boundaries* (Dept. Comm. Pub. 10–1 1962) 26, 74–75, 84, 109, 116–119, 122, 138, 179, 181–184, 210–211, 215, 225, 327–329, 331–333, 336–337
3 Miller and Starr, California Real Estate Law 3d, § 8.30 36
3 Shalowitz, Reed, M., *Shore and Sea Boundaries* (Dept. of Comm. Pub. NOAA/CSC/20007-PUB 2000) 47, 63, 103, 123–125
4 Collier's Encyclopedia (1961 ed.) 356 xxiii
4 Tiffany, *The Law of Real Property* (3d Ed. 1939) 93–94
43 Ops. Cal. Atty. Gen. 291 (1964) 277, 308, 311
5 West's Federal Practice Guide (1970) § 5235 .... 210, 337
6 Miller and Starr, California Real Estate Law 3d, § 16.23 37
6 Miller and Starr, California Real Estate Law 3d, § 16.3 .... 36, 328
6 Powell, The Law of Real Property (3d. Ed. 1939) 93
63A Am.Jur.2d Property, § 30 .... 1, 336
63A Am.Jur.2d Public Lands, § 3 4
65 Am.Jur.2d Quieting Title and Determination of Adverse Claims, § 78 139
65 Am.Jur.2d Quieting Title and Determination of Adverse Claims, § 79 139
73A C.J.S. Public Lands, § 2 .... 4, 333
78 Am. Jur. 2d Waters § 406 .... 92–94, 326–327, 329
78 Am. Jur. 2d Waters § 411 96–98
78 Am. Jur. 2d Waters § 412 95
78 Am. Jur. 2d Waters, § 421 .... 93, 97, 335
78 Am. Jur. 2d Waters § 427 100
78 Am. Jur. 2d Waters § 410 175
9 C.J. 271 263
Althaus, Helen F., "Public Trust Rights" (U.S. Fish and Wildlife Service 1978) 75–77, 167, 235
Amicus Curiae Brief of the California Land Title Association at 3–5, *Summa Corp. v. California ex rel. State Lands Com'n* (1984) 466 U.S. 198 25
Angell, *A Treatise on the Right of Property in Tide Waters* (2d Ed., 1847), ... 94
Annot. (1975) 63 A.L.R.3d 249 175
"A Tidal Wave of Claims" (July 6 12, 1982) National Law Journal Vol. 4, No. 44 xviii
Atwater, B.F, et al, "History, Landforms, and Vegetation of the Estuary's Tidal Marshes," *San Francisco Bay: The Urbanized Estuary* (AAAS 1979) 142, 170–171, 187
Atwater, B.F., "Ancient Processes at the Site of Southern San Francisco Bay: Movement of the Crust and Changes in Sea Level," San Francisco Bay: The Urbanized Estuary (AAAS 1979) 196

Miscellaneous Texts and Authorities Page(s)

Atwater, B.F., "Geologic Maps of the Sacramento—San Joaquin Delta, California" (U.S.G.P.O. 1982) 196
Ballentine's Law Dictionary (3d Ed. 1969) 93, 130, 337
Bascom, *Waves and Beaches* (Anchor Books 1964), ... 79–80, 127
Black's Law Dictionary (4th Ed. 1957) 000
Black's Law Dictionary (5th ed. 1979) 47, 69, 93, 327–328, 331, 334
Branch, Taylor, *Parting the Waters, American in the King Years, 1954-1963* (Touchstone 1988) 95
Briscoe, J., *Surveying the Courtroom* (Second Edition, John Wiley & Sons, New York, 1999) 99–100, 206
Briscoe, J., "The Use of Tidal Datums and the Law" (1983) 43 Surveying and Mapping (Journal of the American Congress on Surveying and Mapping) 115
Brown, A.C., *The Last Hero: Wild Bill Donovan* (Vintage Books 1982), ... 2
Brown, *Boundary Control and Legal Principles* (3d Ed. 1986), ... 94
Brown, *Boundary Control and Legal Principles* (John Wiley & Sons, Inc., 2d ed., 1969) 11, 14
Burns, J.M., *The Vineyard of Liberty* (Knopf 1982), ... 4
California State Water Resources Control Board, Water Quality Control Plan for Salinity, San Francisco Bay/Sacramento-San Joaquin Delta Estuary (1988) 197
Carter, Clarence E., *The Territorial Papers of the United States, Vol. II, The Territory Northwest of the River Ohio, 1787–1803* (U.S.G.P.O. 1934) 11
Circular from General Land Office to Surveyor General (1831) as reprinted in Minnick, ed., *A Collection of Original Instructions to Surveyors of the Public Lands, 1815-1881* (Landmark) 296
Clevenger, Shobal V., *A Treatise on the Method of Government Surveying* (1883)(Carben Surveying Reprints) 12, 84
Cobb, "The Great Lakes Troubled Waters," National Geographic (July, 1987) v. 172 282
Colby, "Mining Law in Recent Years" (1945) 33 Calif. Law. Rev. 368 .... 4, 10, 16
Colby, "Mining Law in Recent Years" (1948) 36 Cal. L. Rev. 355 .... 8, 10–11, 13–15
Colby, "Mining Law in Recent Years" (1949) 37 Calif. L. Rev. 592 .... 10, 18, 20
Comment, "Fluctuating Shoreline and Tidal Boundaries: An Unresolved Problem" (1969) 6 San Diego L. Rev. 447 .... 122, 127
Comment, "Land Accretion and Avulsion: The Battle of Blackbird Bend" (1977) 56 Nebr. L. Rev. 814 97
Copleston, Frederick, S.J., *A History of Philosophy, Volume 4* (Image Books 1963) 98
Corker, "Where Does the Beach Begin" (1966) 42 Wash.L.Rev. 33 .... 74, 105, 107, 121

Miscellaneous Texts and Authorities   Page(s)

Dana, Richard Henry, *Two Years Before the Mast* (Bantam Classic 1959) 172
Department of Interior, Bureau of Land Management, "The Meandering Process In The Survey of the Public Lands of the United States (undated) 82
Donaldson, Thomas, *The Public Domain, Its History, with Statistics* (U.S.G.P.O. 1970 reprint) 11
Dostoevsky, F., *Brothers Karamazov* (Penguin 1958) 236
Dwinelle, John W., *The Colonial History of San Francisco* (1877) (reprinted 1978 Ross Valley Book Club) 7
Elgin and Knowles, *Principles of Boundary Location for Arkansas* (Landmark, 1984), ... 85
Ernst, Joseph, *With Compass and Chain, Federal Land Surveyors of the Old Northwest, 1785–1816* (Arno Press 1979) 11
Flushman and Barbieri, "Aboriginal Title: The Special Case of California" (1986) 17 Pac.L.J. 391 57
Frank, "Forever Free, Navigability, Inland Waterways, and the Expanding Public Interest" (1983) 16 U.C. Davis L. Rev. 579 238–240
Gates, *A History of Public Land Law Development* (Pub. Land Law Rev. Comm. 1968), ... 3–8, 10–11, 15–21, 25–26, 203, 329, 337
General Instructions (1834) to Deputy Surveyors in Illinois and Missouri as reprinted in Minnick, ed., *A Collection of Original Instructions to Surveyors of the Public Lands, 1815-1881* (Landmark) 296
General Instructions of 1846 (Wisconsin and Iowa) as reprinted in Minnick, ed., *A Collection of Original Instructions to Surveyors of the Public Lands, 1815-1881* (Landmark) 297
General Instructions to Deputy Surveyors Engaged in Surveying the Finally Confirmed Land Claims in California (January, 1858), reprinted in Uzes, F., *Chaining the Land* (Landmark 1977), Appendix C, pp. 211–214 .... 77, 167
General Instructions to Deputy Surveyors from Office of the Surveyor General, Territory of Florida (1842) as reprinted in Minnick, ed., *A Collection of Original Instructions to Surveyors of the Public Lands, 1815-1881* (Landmark) 297
General Instructions to His Deputies by the Surveyor General of the United States for the States of Ohio, Indiana and Michigan (1850) as reprinted in Minnick, ed., *A Collection of Original Instructions to Surveyors of the Public Lands, 1815-1881* (Landmark) 297
Genesis 1.1 xvii
Gilbert, L. K., *Hydraulic Mining Debris in the Sierra Nevada* USGS Prof. Paper No. 105 (GPO 1917) 142, 171, 184, 196–200, 336
Gore, "The Rising Great Salt Lake: No Way to Run a Desert," National Geographic (June, 1985), V. 167 282
Grimes, Clark on Surveying and Boundaries (4th ed., 1976) 2, 7–8, 10–11, 13, 35–38, 40, 53, 68, 85–87, 89, 92–94, 96–98, 100, 103, 125, 161, 233, 236, 245, 255, 257–258, 263, 266, 271, 285, 287–288, 294, 311, 326–329, 334–335, 337

Miscellaneous Texts and Authorities Page(s)

Hall, *A History of the Foreshore and the Law Relating Thereto* (3d [Stuart Moore] Ed. 1888) 94
Handbook for Title Men (Title Insurance and Trust Co. 1966) 93
Hanna, "Equal Footing in the Admission of States" (1951) 3 Baylor L. Rev. 519 23
Harper, *The Code of Hammurabi King of Babylon* (Univ. of Chicago Press 1904) 189–190
Hawes, *Manual of United States Surveying* (Carben Surveying Reprints 1977), ... 77, 84
Hinds, Norman, *Geomorphology* (Prentice-Hall 1943), ... 142, 330
Hoffman, Ogden, *Report of Land Cases Determined in the United States District Court for the Northern District of California, Vol. I* (1862)(Reprint Yosemite Collections 1975) 7
Hughes, D.E., and Von Geldern, O., "The Determination of the Plane of Ordinary High Tide for Pacific Coast Harbors, with Particular Reference to San Diego Harbor California" (1910) Journal of the Association of Engineering Societies, Vol XLIV, No. 4 .... 105, 119
Hull, W., and Thurlow, C., "Tidal Datums & Mapping Tidal Boundaries (U.S. Dept. of Comm., National Ocean & Atmospheric Administration, National Ocean Survey) 75, 182
III American Law of Property (Casner, ed., 1952) 93
Instructions for Deputy Surveyors (1815), reprinted in Minnick, *A Collection of Original Instructions to Surveyors of the Public Lands, 1815-1881* (Landmark) 240
Instructions from the California Surveyor General to H.A. Higley, Alameda County Surveyor, June 9, 1858 77
Instructions to Surveyor of Public Lands in Oregon (1851) reprinted in Minnick, ed., *A Collection of Original Instructions to Surveyors of the Public Lands, 1815-1881* (Landmark) 297
Instructions to the Surveyors General of the Public Lands of the United States (1855) as reprinted in Minnick, *A Collection of Original Instructions to Surveyors of the Public Lands 1815–1881* (Landmark) 157, 185, 297
Instructions to the Surveyors of General of Public Lands (1855) 240
Irish, John, "The Sacramento Valley" in Muir, J. ed., *West of the Rocky Mountains* (Running Press 1976) 196
James, P.D., *Death of an Expert Witness* (Warner Books 1992) xxiv
Jerome, L.E., "Marsh Restoration," Oceans (Jan.-Feb. 1979), pp. 57 144
Johnson, New England Acadian Shoreline (1925) 118, 183
Jones, Lester E., *Elements of Chart Making* (U.S.C.& G.S. Spec. Pub. No. 38 1916) 118
Josselyn, M.N. and Atwater, B.F., "Physical and Biological Constraints on Man's Use of the Shore Zone of the San Francisco Bay Estuary," *San Francisco Bay: Use and Protection* (AAAS 1982) 142
Keegan, J., *Six Armies in Normandy* (Viking Press 1982), ... 102

Miscellaneous Texts and Authorities   Page(s)

Kennedy, J. Michael, "U.S. Wetlands Swamp by Tide of Neglect" (L.A. Times, Dec. 10, 1988) 144
Kreiger, L., "Location Key Factor in Withstanding Quake" (S.F. Examiner 10/22/89) p. 14 xix
"Landowner Suits Stirring Red River 'Border War'" (February 24, 1985) Los Angeles Times xviii
Leighty, "Public Rights in Navigable State Waters—Some Statutory Approaches" (1971) 6 Land & Water L.Rev. 459 .... 23, 289
Leighty, "The Source and Scope of Public and Private Rights in Navigable Waters" (1970) 5 Land & Water L. Rev. 391 23
Letter, A.F. Rodgers, Assistant, to Pritchett, Superintendent, USC&GS, July 8, 1898 183
Letter from Assistant Henry L. Marinden, U.S. Coast & Geodetic Survey, Appendix No. 16, Report for 1891, Part II, Proceedings of the Topographical Congerence held at Washington, D.C., January 18 to March 7, 1892, pp. 586–587. 109
MacGrady, "The Navigability Concept in the Civil and Common Law" (1975) 3 Fla.St.U.L.Rev. 511 .... 235, 254
Maloney and Ausness, "The Use and Legal Significance of the Mean High-water Line in Coastal Boundary Mapping" (1974) 53 No.Car. L. Rev. 185 .... 107, 115
Maloney, "The Ordinary High-water Mark: Attempts at Settling an Unsettled Boundary Line" (1978) 13 Land and Water Law Review 465 .... 78, 254, 272, 275, 308–309
Manual of Instructions of the Commissioner of the General Land Office to the Surveyors General of the United States Relative to the Survey of the Public Lands and Private Land Claims, May 3, 1881 77
Manual of Surveying Instructions (U.S. Dept. of Int., Bur. of Land Mgmt. 1973) (Landmark Reprint) 4, 10, 13, 77, 82, 84–85, 185, 240, 258, 297, 303
Marmer, H.A., *Tidal Datum Planes* (1951) U.S. Coast and Geodetic Survey, Spec. Pub. 135 Rev. Ed. ... 110–114, 116–118, 326, 328–329, 331–336
McEntyre, John G., *Land Survey Systems* (Landmark 1985) 11
McPhee, John, *Basin and Range* (Farrar, Straus, Giroux 1980) 293
Minnick, *A Collection of Original Instructions to the Surveyors of the Public Lands, 1815–1881* (Landmark) 77, 84, 157, 185, 240, 296–297
Morris, *The Rise of Theodore Roosevelt* (1979) 245
Nermilyea, N., and Moore, J.A., "Ballad of the Republic," The Californians (May/June 1988) 195
Newsweek, "Don't Go Near the Water" (August 1, 1988), p. 47 .... 141, 144
Outline of Land Titles (CLTA) 7–8, 10, 16–17, 167
Parkman, F. *The Oregon Trail* (Feltskog, ed.) (Univ. of Wisconsin Press, 1969) 233
Pattison, William D., *Beginnings of American Rectangular Land Survey System, 1784–1800* (Ohio Historical Society 1957) 11

Miscellaneous Texts and Authorities   Page(s)

Patton, R.S., "Relation of the Tide to Property Boundaries" set out in 2
  Shalowitz, Shore and Sea Boundaries (Dept. Comm. Pub. 10-1 1962)
  Appendix E, pp. 667–679 ...... 111–113, 115–117, 326, 329, 332–333,
                                                                   335–336
Perez, C., ALand Grants in Alta California@ (Landmark 1996) 7
Pestrong, "San Francisco Bay Tidelands" in California Geology (1972) Vol.
  25, No. 2. 170
Pestrong, "The Development of Drainage Patterns on Tidal Marshes"
  (Stanford University Earth Sciences 1965) Vol. 4, No. 2 170
Pestrong, "Tidal-Flat Sedimentation at Cooley Landing, Southwest San
  Francisco Bay" in 8 Sedimentary Geology (Elseweir 1972) 170
Pestrong, R., "Unnatural Shoreline," Environment (Nov. 1974), Vol.16, No.
  9., p.27 144
Porro, "Invisible Boundary—Private and Sovereign Marshland Interests"
  (1970) 3 Natural Resources Lawyer 512 .... 79, 151, 327, 337
Public Lands Surveying: A Casebook (U.S. Dept. of Int., BLM, Cadastral
  Training Staff 1975), ... 15
Reisner, Mark, Cadillac Desert (Penguin 1986) 198
Report of Lt. N.A. McCully, USN, accompanying Descriptive Report
  H-2304 (USC&GS 1897) 173
Restatement Property, § 10 .... 1, 336
Robillard, W., and Bouman, L., Clark on Surveying and Boundaries (7th
  Ed., 1997) 2, 7–8, 10–11, 13, 35–38, 40, 53, 68, 85–87, 89,
      92–94, 96–98, 100, 103, 125, 161, 233, 236, 255, 257–258, 263, 266,
              271, 285, 287–288, 294, 311, 326–329, 334–335, 337
Robinson, *Land in California* (U.C. Press, 1948) 5, 7
San Francisco Chronicle, July 13, 1987 at 1, col. 1 198
Santanaya, G., The Life of Reason (Prometheus Books 1998) 194
Schell, Hal, Dawdling on the Delta (Schell Books 1983) 190
Shabecoff, "How America Is Losing Its Edges," N.Y. Times, May 3, 1987,
  sec. E, p. 5 144
Shinn, C.H., "The Tule Region" in Muir, J. ed., *West of the Rocky
  Mountains* (Running Press 1976) 196
Skelton, *The Legal Elements of Boundaries and Adjacent Properties*
  (1930) .. 35, 39, 41, 53, 68, 93, 125, 233, 236, 257–258, 263, 285, 288
Snell, J., "Search for a Flagpole," Journal of Surveying and Mapping
  (ACSM March, 1967) Vol. 27-1-51 12
Stewart, Lowell O., *Public Land Surveys* (1935) (reprinted Carben
  Surveying Reprints 1976), ... 8, 11
Swainson, O.W., *Topographic Manual* (Dept. of Commerce Special
  Publication No. 144) (1928) 118
T. Sandars, *The Institutes of Justinian* (4th Ed. 1867) 76
Testimony of Augustus Rodgers, George Davidson, James W. Bost, Horace
  A. Higley in *United States v. Peralta* (Nos. 98, 100 U.S. D.C. N.D.
  1871) 181
The American College Dictionary (Random House 1948) 173

Miscellaneous Texts and Authorities   Page(s)

The Times Atlas of World History (Times Books 1979) 189
Theroux, Paul, *The Old Patagonian Express* (Houghton Mifflin 1979) .. 68
Tide and Current Glossary (U.S. Dept. Of Comm., NOAA, NOS
  1984) 106, 110, 113, 141, 212, 329, 332, 336
Tocklin, Adrian M., "*Pennoyer v. Neff*: The Hidden Agenda of Stephen J.
  Field," 28 Seton Hall Law Rev. 75 (1997) 115
Uzes, Francios, *Chaining the Land* (Landmark 1977) 7, 12, 14–15, 77,
                                                              84, 164, 167
Wilson, Donald A., "Early Land Surveyors in New England" (ACSM
  1976) as reprinted in Minnick, Ed., *Plotters and Patterns of American
  Land Surveying* (Landmark 1985), ... 2

Foreign Materials   Page(s)

2 Bl. Comm. 262 (Cooley, 4th Ed., Vol. 1) 95
Attorney General v. Chambers (1854) 4 De G.M. & G., 43 Eng. Reprint
  486. (A reprint of the *Attorney General v. Chambers* opinion can be
  found in 2 Shalowitz, Appendix D, pp. 640–646.) 77, 121
Attorney General v. Parmeter (Ex. 1811) 10 Price 378, 147 Eng.
  Rep. 345 76
Attorney General v. Richards (1795) 2 Anstruther 603, 145 Eng.
  Rep. 980 76
Halsbury's Laws of England, ¶ 297, at 162 (4th Ed. 1984) 315
Las Sieta Partidas (Scott, ed. CCH 1931) Par. III, Tit. XXVIII,
  Law IV. 76, 167
Rex v. Lord Yarborough (1824) 3 B.& C. 91 (K.B.) 95

1

THE BASICS OF LAND TITLE

1.1 INTRODUCTION

Long before the United States became a free and independent nation, six separate prior sovereigns[1] discovered, conquered, governed, and granted portions of the land now constituting the United States' national territory. When the United States gained its independence, these prior conquests and their sometimes peculiar land concessions had to be incorporated into the United States' developing land title[2] system. That system also had to account for and incorporate the results of Manifest Destiny, the United States' own discovery and conquest of the rest of its national territory.

During its formative years, the fledgling federal government added a further complication to the land title system. The United States seemed as anxious to convey to its citizens the land it was acquiring as it was to procure the territory. Piecemeal and episodic divestiture of significant portions of the public domain took place through a series of land disposition programs employed by both federal and state governments. These governments designed those programs to raise capital and to encourage the development of various types of public works and settlement of the nation. And all the various territories acquired and titles recognized or created during the United States' continental expansion had to be located on the earth's face with respect to one another.

---

1. England, France, Spain, Holland, Mexico, and Russia.
2. By the word "title" is meant the means, method, or evidence of one's ownership of land. 63A Am.Jur.2d Property, § 30; Restatement Property, § 10; *Arraington v. Liscom* (1868) 34 Cal. 365, 385.

This conglomeration of different sources of land ownership, episodic acquisitions, and seemingly haphazard dispositions may appear to be of daunting complexity. A stable and secure land title system was, however, a critical necessity to the maturing nation and to that nation's development and progress. Evidence of the importance and challenge of boundary and title questions during the nation's development is that some of the country's most distinguished leaders and thinkers began or punctuated their careers as surveyors.[3]

Fortunately, we may distill some of this complexity into two, seemingly simple, basic principles. First, all land titles, with certain exceptions to be discussed briefly, begin with a land grant, concession, or patent from the sovereign. That grant from the sovereign is known as the original source of title.[4] Second, no matter what the character of the instrument of conveyance, the person who passes title can grant only what that person owns. For example, a patent from the government may be couched in the self-important language adopted by sovereign grants. However, if the government does not own the land that is the subject of the grant, the grant is as worthless as a signed and sealed certificate of ownership for the Brooklyn Bridge purchased from a street vendor.

These two concepts sound simple. In fact, some may wonder what more there is to say on the subject. Rest assured, underlying these simple concepts, like the great bulk of an iceberg lurking below the waterline, are a myriad of detailed principles applicable to property title and boundaries. This chapter is a *basic* primer on land title, with a particular emphasis on titles to lands adjacent to or lying beneath tidal or navigable waters.

This chapter discusses how the United States itself got title to the land area that now constitutes its national territory. Next, the author untangles the means and methods by which prior sovereigns' land grants lying within this territory were recognized and located. This subject in turn leads to an accounting of how the United States established and determined the extent of its public

---

3. For example, it is well known that George Washington began his public career as a surveyor of the Great Dismal Swamp near Norfolk, Virginia. Robillard, W., and Bouman, L., Clark on Surveying and Boundaries (7th Ed., 1997) § 1.03 (hereinafter "Clark 7th"); Grimes, Clark on Surveying and Boundaries (4th Ed., 1976) § 3, p. 7 (hereinafter "Clark 4th"); 19 Collier's Encyclopedia (1961), p. 293. (The author cites both Clark 7th and Clark 4th for ease of reference.) It is not, however, nearly as well known that William J. ("Wild Bill") Donovan, founder of the Office of Strategic Services, the precursor of the modern intelligence apparatus, the Central Intelligence Agency, began his career as a surveyor before becoming a Wall Street lawyer, war hero, and confident to presidents. Brown, A.C., *The Last Hero: Wild Bill Donovan* (Vintage Books 1982), p.18. Nor is it general knowledge that Henry David Thoreau became a surveyor after spending time at Walden Pond. Wilson, Donald A., "Early Land Surveyors in New England" (ACSM 1976) as reprinted in Minnick, Ed., *Plotters and Patterns of American Land Surveying* (Landmark 1985), p. 169, or that Abraham Lincoln ". . . subsisted by working as a surveyor during several years." 14 Encyc. Brit. (14th Ed. 1929), p. 138.
4. This title is sometimes also referred to as "parent" title.

domain and the means of conveyance and disposition of such lands to various persons or entities. A summary of the generous grants made by the United States to individual states follows. This chapter devotes considerable attention to one of those disposition statutes, the United States' grant of swamp and overflowed lands to the states. That grant requires such attention, as it is a potential source of title to lands along, adjacent to, or beneath tidal or navigable water bodies.

The author also examines the source of original title in the states, the constitutional principle of sovereignty, and the Equal Footing Doctrine. Because of strictures imposed on disposition of such sovereign lands, the chapter describes the states' land disposition programs, emphasizing, however, the limitations on state alienation of its sovereign land interests.

The chapter's final topic is other sources of title not involving a grant from the sovereign: adverse possession, agreed boundaries and estoppel. It concludes with a discussion of the importance of "chains of title." A chain of title is the progression and succession of instruments establishing that title to a particular geographic area has devolved from the original title to a specific person or entity.[5]

1.2 UNITED STATES' ACQUISITION OF THE PUBLIC DOMAIN

When European nations discovered the vast and bountiful area that was to become the United States, those nations not only asserted political sovereignty over the territory, but claimed land ownership, title, as well. These exploring sovereign nations founded this claim on what has become known as the doctrine of discovery.[6] The concept that discovering nations gained land title by discovery and conquest has been accepted without question since formation of the United States.

When the United States was founded, it succeeded to all former rights and sovereignty of those nations that preceded it.[7] The United States acquired

---

5. More particularly, chain of title is composed of a chronological list of deeds or other instruments of conveyance originating from the original source of title to the present vested owner. E.g., *Eltman v. Harvey* (N.Y. 1978) 403 N.Y.S.2d 428, 431–432; 1 Ogden's California Real Property Law (1956) (hereinafter "1 Ogden"), p. 709.
6. *Johnson v. McIntosh* (1823) 21 U.S. (8 Wheat.) 543, 573 ("Discovery gives a title which the courts of the Conqueror cannot deny."); *Massachusetts v. New York* (1926) 271 U.S. 65, 85–86; Gates, *A History of Public Land Law Development* (Pub. Land Law Rev. Comm. 1968), p. 1 (hereinafter "Gates").
7. *E.g., Johnson, supra,* 21 U.S. at 584–585; *Weber v. Harbor Commissioners* (1873) 85 U.S. (18 Wall.) 57, 65; *Shively v. Bowlby* (1894) 152 U.S. 1, 57.

lands held by former sovereigns as its public domain[8] by cession, discovery, and conquest.[9]

Figure 1.1 shows how, by virtue of cessions from the original states,[10] the Louisiana Purchase in 1803,[11] the Florida cession by Spain in 1819,[12] the annexation of Texas in 1845,[13] the treaty with England that confirmed the northern boundary (remember "Fifty-four forty, or fight!"),[14] the Treaty of Guadalupe-Hidalgo in 1848,[15] and the Gadsden purchase in 1853,[16] the United States received a vast continental empire of yet untold wealth.[17]

1.3 RECOGNITION OF TITLE GRANTED BY PRIOR SOVEREIGNS

This empire was not entirely unencumbered, however. The discovering nations, often through colonial governments or trading companies chartered by their respective sovereigns, made large-scale grants or concessions of the lands to which they claimed ownership. In part, the purpose of these grants or concessions was to encourage and secure colonization and control of the newly discovered lands.

---

8. "'Public domain' is equivalent to 'public lands' and these words have acquired a settled meaning in the legislation of this country." *Barker v. Harvey* (1901) 181 U.S. 481, 490. These terms are used to describe lands of the United States (or those of a state) subject to sale and disposal under general laws or statutes authorizing disposition and sale of lands. *Ibid.*; 73A C.J.S. Public Lands, § 2; *Newhall v. Sanger* (1875) 92 U.S. 761, 763. Unless otherwise noted, the two terms will be interchangeably used throughout this work.
9. *Johnson, supra*, 21 U.S. at 584–586; *United States v. Curtiss-Wright Corp.* (1936) 299 U.S. 304, 318; *Downes v. Bidwell* (1901) 182 U.S. 244, 322–323 (White, J., concurring); *Mormon Church v. United States* (1890) 136 U.S. 1, 42; 63A Am.Jur.2d Public Lands, § 3.
10. *Howard v. Ingersoll* (1851) 54 U.S. (13 How.) 381, 398–406.
11. Treaty Between the United States and the French Republic, April 30, 1803, 8 Stat. 200; *Chitimacha Tribe of Louisiana v. Harry L. Laws Co.* (5th Cir. 1982) 690 F.2d 1157, 1168, *cert.den.* (1983) 464 U.S. 814.
12. Treaty of Amity, Settlements and Limits, Between the United States of America and His Catholic Majesty, February 22, 1819, 8 Stat. 252; *United States v. Arredondo* (1832) 31 U.S. (6 Pet.) 691, 706.
13. *United States v. Texas* (1950) 339 U.S. 707, 713.
14. Treaty with Great Britain, June 15, 1846, 9 Stat. 869; Gates, *supra*, at 84.
15. Treaty of Peace, Friendship, Limits and Settlement Between the United States of America and the Mexican Republic, February 2, 1848, 9 Stat. 922; Gates, *supra*, at 83.
16. Treaty with Mexico, December 30, 1853, 10 Stat. 103; Gates, *supra*, at 84.
17. Gates, *supra*, at 75–76; Colby, "Mining Law in Recent Years" (1945) 33 Calif. Law. Rev. 368, 370 (hereinafter "Colby I"); Burns, J.M., *The Vineyard of Liberty* (Knopf 1982), pp. 457–464. Alaska was acquired by purchase. Treaty with Russia, March 30, 1867, 15 Stat. 539; Gates, *supra*, at 85. Later acquisitions of Hawaii, Puerto Rico, the Philippines, Guam, Samoa, and the Pacific Trust Territories did not increase the amount of public land. *Ibid.* The genesis of each of the public land states can also be found in *Manual of Surveying Instructions* (U.S. Dept. of Int., Bur. of Land Mgmt 1973) (Landmark Reprint) § 1–23, pp. 10–11.

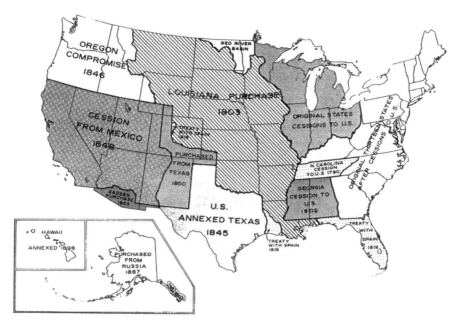

Figure 1.1 Acquisition of the Territory of the United States

For example, to name just some of the states along the East Coast of the United States, such grants included large parts of New Jersey, New York, Virginia, North Carolina, and Georgia.[18] In 19 other states of the United States prior sovereigns also made grants of land to their subjects. These 19 states were as geographically diverse as Michigan and Wisconsin, New Mexico and Arizona, Florida and California.[19] It is estimated that, in total, there were 30,000 to 35,000 claims to title derived from prior sovereigns covering probably 45,000,000 acres. In California alone, there were more than 550 grants totaling almost 8,900,000 acres.[20]

---

18. *Arnold v. Mundy* (1821) 6 N.J.L. 1; *Martin v. Waddell* (1842) 41 U.S. (16 Pet.) 367, 407; *Howard, supra*, 54 U.S. at 398–406; *Gates, supra*, at 34–48; *Massachusetts v. New York, supra*, 271 U.S. at 81; *State v. Pennsylvania Railroad Company* (Del. 1967) 228 A.2d 587, 590.
19. *Gates, supra*, at 89; *United States v. Arredondo, supra*, 31 U.S. at 706; *Chouteau v. Molony* (1853) 57 U.S. (16 How.) 203, 221; *County of St. Clair v. Lovingston* (1874) 90 U.S. (23 Wall.) 46, 46–49. Some of these grants were mammoth. The Arredondo grant in Florida included almost 300,000 acres. But even that grant was dwarfed by the Forbes claim, also in Florida, for more than 2,400,000 acres. *Gates, supra*, at 88.
20. *Gates, supra*, at 86, 88; Robinson, *Land in California* (U.C. Press, 1948), pp. 45–72; Report on the Subject of Land Titles in California (1850) 31st Cong., 1st Sess., Ser. 589, Doc. 18, Ex. 33; *Thompson v. United States* (1959) 8 Ind. Cls. Comm. 1, 7.

1.3.1 Legal Basis for Recognition of Prior Sovereign Grants

It was understood as a matter of high principle that the conquering nation succeeded to the complete sovereignty of its predecessor. Legitimate grants of land, however, made by the former sovereign to its citizens were to be recognized and confirmed by the conquering nation.[21] The foundation for this principle was twofold, one legal, the other practical.

First, based on international law, the conquering nation was bound to respect private titles created by predecessor nations.[22] Second, lands that were the subjects of these grants were never part of the public lands of the United States.[23] Consequently, as a practical matter, until the conquering nation settled and located titles created by the former sovereign, the conquering nation would not know the extent of the lands it owned that it could later grant according to its own laws.[24]

Thus, in all territories the United States acquired as it pushed its borders to the southern and western coasts, Congress provided a process to protect and recognize legitimate titles granted by former sovereigns. Congress also established a system to locate and identify public lands that could later be disposed of or used for public purposes. The procedure and the system worked in concert, though not simultaneously or coincidentally. Our focus first turns to the land grant confirmation process, then to the location and survey of the United States' public lands.

1.3.2 Confirmation of Former Sovereigns' Land Grants—Land Claims Act

By virtue of congressional legislation, the president established and appointed boards of land commissioners to investigate, confirm, or deny the validity of a prior sovereign's land grants.[25] For example, in California the United States created a Board of Land Commissioners; all claims of land title derived from either Spain or Mexico were required to be presented to the Board.[26] Such

---

21. *United States v. Arredondo, supra,* 31 U.S. at 721; *Barker, supra,* 181 U.S. at 483.
22. *Beard v. Federy* (1865) 70 U.S. (3 Wall.) 478, 490; *Botiller v. Dominguez* (1889) 130 U.S. 238, 243–244, 247; *United States v. O'Donnell* (1938) 303 U.S. 501, 510, n. 2.
23. *Botiller, supra,* 130 U.S. at 249.
24. *O'Donnell, supra,* 303 U.S. at 512; *Rodrigues v. United States* (1863) 68 U.S. (1 Wall.) 582, 588.
25. *United States v. Arredondo, supra,* 31 U.S. at 720; *New Orleans v. United States* (1836) 35 U.S. (10 Pet.) 662, 732; *Chitimacha Tribe of Louisiana, supra,* 690 F.2d at 1169. Gates, *supra,* at 89–113, has a fascinating description of development of the process by which these grants came to be recognized and confirmed by the United States.
26. Act of March 3, 1851, 9 Stats. 631 (hereinafter "Land Claims Act"). The Land Claims Act provided that "each and every person claiming lands in California by virtue of any right or title

claims had to be filed within a certain period or were lost forever thereafter.[27] The Board of Land Commissioners had the authority to confirm or reject claims. Decisions of the Board of Land Commissioners were subject to a right of appeal to a United States District Court by either the claimant or the United States. Disgruntled litigants could appeal decisions of the District Court to the United States Supreme Court.[28]

When the United States recognized these grants, it provided a system for surveying them as well. When the Board of Land Commissioners confirmed, or a United States District Court or the United States Supreme Court affirmed, the claim, the United States Surveyor General[29] was duty bound to cause the

---

derived from the Spanish or Mexican government, shall present the same [to the land commissioners]. . . ." Land Claims Act, § 8. If one is interested in a detailed explanation of the operation of the Land Claims Act, one source with which to begin is the *Outline of Land Titles* published by the California Land Title Association. A narrative explanation is found in two eminently readable works: Robinson, *Land in California* (U.C. Press 1948) and Uzes, François, *Chaining the Land* (Landmark 1977), pp. 148–154. There is an interesting wrinkle to the Land Claims Act in California. That Act also authorized appropriate municipal authorities to file claims for organized cities in order to have Spanish or Mexican pueblo grants of 4 square leagues confirmed. *Outline of Land Titles*, *supra*, at 13–15. Titles in the City of San Francisco, for example, are founded on a confirmed pueblo grant. *San Francisco v. Leroy* (1891) 138 U.S. 656, 664–665. For a historical perspective, see Dwinelle, John W., *The Colonial History of San Francisco* (1877)(reprinted 1978 Ross Valley Book Club).

27. Land Claims Act, § 13; *Botiller*, *supra*, 130 U.S. at 244; *Summa Corp. v. California ex rel. Lands Comm'n* (1984) 466 U.S. 198, 203.

28. Land Claims Act, § 8; 1 Ogden, *supra*, § 5.4, p. 134; *Outline of Land Titles*, *supra*, at 6–7. In the case of California land grants, the original records of the confirmation proceedings are still available at selected repositories such as the University of California at Berkeley's Bancroft Library. Several interesting court decisions arising out of the confirmation proceedings concerning the Mexican land grants are published in Hoffman, Ogden, *Report of Land Cases Determined in the United States District Court for the Northern District of California, Vol. I* (1862) (Reprint Yosemite Collections 1975). Ogden Hoffman was the United States District Court judge who decided many of these contests. This collection is also known as "Hoffman's Reports." Another valuable resource is Perez, C., "Land Grants in Alta California" (Landmark 1996). In his book, Mr. Perez lists all private lands claims, identifying those land claims that were confirmed, and provides tables of federal and state court decisions relating to these claims. The book also contains selected congressional and General Land Office documents regarding these claims including, where available, United States surveying instructions to the United States Deputy Surveyors.

29. Prior to 1925, the United States Surveyor General was the officer in charge of survey of the public lands in the General Land Office, an agency of the United States Department of the Interior. Clark 7th, *supra*, § 5.30, pp. 147–148; Clark 4th *supra*, § 67, pp. 76–77. The office was originally created in 1796. Clark 7th, *supra*, at 147; Clark 4th, *supra*, at 126. At one time there were 15 United States Surveyors General. Clark 4th, *supra*, at 127. For example, the United States Surveyor General for California was responsible for public land surveys in California.

Be aware of a potential source of confusion: The title of the state official once charged with the management and disposition of State of California public lands, the Surveyor General of California, is similar. *Boone v. Kingsbury* (1928) 206 C. 148, 184–185, *cert. den.* and *app. dism. sub nom. Workman v. Boone* (1929) 280 U.S. 517; Cal. Pub. Res. Code § 6102. Despite the similarity of nomenclature, the authority and powers of the United States Surveyor General for

land to be accurately surveyed and to prepare and furnish a plat of that survey.[30] The purpose of this survey was twofold. First, the survey determined and located exactly what land the grant included as confirmed by the Board of Land Commissioners. Second, the survey separated lands already granted from those lands remaining as public lands of the United States and that could be disposed of as Congress wished.[31] The surveys formed the basis for a description in a confirmatory patent issued by the United States[32] and were conducted according to United States' law and system of surveys.[33]

Once the commissioner of the General Land Office approved[34] the survey,[35] the United States issued a patent to a claimant.[36] The United States' patent issued to former Mexican citizens (or their successors) confirmed what Mexico had earlier granted. This patent was not a new source of title. It was a recognition, validation, and confirmation of the grant made by the prior sovereign.[37] The title confirmed related back to the date of the original grant.[38]

These confirmatory patents conveyed the entire interest of the United States in the lands encompassed within the patent description.[39] The process was designed to provide repose to titles and ". . . give parties who possess[ed] [the patents] an opportunity of placing them on the records of [the United States], in a manner and form that [would] prevent future controversy."[40] This proved to be the case. Where the government failed to file a timely claim

---

California, a federal official, and the Surveyor General of California, a state official, are entirely separate and distinct.

30. *United States v. Fossatt* (1858) 62 U.S. (21 How.) 446, 450; Land Claims Act, § 13; *Outline of Land Titles, supra*, at 8. Reasonable conformity between the survey and the decree of confirmation was required. *United States v. Halleck* (1863) 68 U.S. (1 Wall.) 439, 455.

31. *Botiller, supra*, 130 U.S. at 249; *Outline of Land Titles, supra*, at 8; Land Claims Act, § 13.

32. *Botiller, supra*, 130 U.S. at 249; *Outline of Land Titles, supra*, at 8; Land Claims Act, § 13.

33. *Fremont v. United States* (1854) 58 U.S. (17 How.) 541, 564–565; *More v. Steinbach* (1888) 127 U.S. 70, 81; *Stewart v. United States* (1942) 316 U.S. 354, 359; *Waterman v. Smith* (1859) 13 Cal. 373, 411.

34. Many surveys approved by the United States Surveyor General for California were subsequently rejected or modified by the General Land Office in Washington. Stewart, Lowell O., *Public Land Surveys* (1935) (Carben Surveying Reprints 1976), pp. 49, 71.

35. The General Land Office was an agency within the Department of the Interior. It was responsible for the execution of the public land laws and for activities concerning management and administration of public lands. It was abolished in 1946 when its functions were combined in the Bureau of Land Management (herein "BLM"). Clark 7th, *supra*, § 5.31, pp. 148–149; Clark 4th, *supra*, at 74–76; Gates, *supra*, at 127–128; Colby, "Mining Law in Recent Years" (1948) 36 Cal. L. Rev. 355, 359–360 (hereinafter "Colby II"). The Commissioner of the General Land Office was head of that office. *Ibid.*

36. *More, supra*, 127 U.S. at 79; *Beard, supra*, 70 U.S. at 492; *Outline of Land Titles, supra*, at 8; Land Claims Act, § 13.

37. *More, supra*, 127 U.S. at 83; *Los Angeles Milling Co. v. Los Angeles* (1910) 217 U.S. 217, 233; *United States v. Coronado Beach Co.* (1921) 255 U.S. 472, 488.

38. *Ibid.*

39. Land Claims Act, § 13; Act of March 3, 1831, § 6, 4 Stat. 492, 494; *United States v. O'Donnell, supra*, 303 U.S. at 509–511.

40. *Fremont, supra*, 58 U.S. at 553–554; *Summa, supra*, 466 U.S. at 206.

under the Land Claims Act, subsequent decisions prevented both federal and state governments from asserting title to areas confirmed in private persons through the confirmatory process.[41]

1.4 THE PUBLIC DOMAIN AND THE UNITED STATES' SURVEY OF THE PUBLIC LANDS

The United States, through the device of the Land Claims Act or similar proceedings, determined that some lands within its new boundaries were not public lands. Most territory acquired as the United States expanded was not affected by this process. One reason, perhaps the most important, was that the conquering nations were not fully aware of the extent of their conquests and thus could not make grants. And for a considerable period enormous areas of the United States were, and for some time remained, unknown, unexplored, remote, unsettled, and inhabited by Native Americans who did not view European encroachment altogether favorably.

Moreover, the entire unoccupied public domain was not subject to disposition or sale under the general land laws.[42] Lands were withdrawn from sale or entry by statute, by the president through executive order, or by the secretary of the interior.[43] The United States has created many military bases and Indian reservations in this fashion. In addition, neither confirmed former sovereign land grants nor lands owned by the states in their sovereign capacities[44] were part of the public domain subject to disposition by the United States.[45]

---

41. *United States v. Coronado Beach Co.*, *supra*, 255 U.S. at 487–488; *Summa*, *supra*, 466 U.S. at 209. *Summa* held that the State of California lost its right to assert a sovereign title claim to an area within the perimeter of a confirmed land grant. The state lost that right because it failed to file a claim in the original confirmation proceedings. *Summa*, *supra*, 466 U.S. at 209. To reach that conclusion, the United States Supreme Court apparently considered the State of California was a "person" within the meaning of the Land Claims Act. This conclusion appeared subject to some doubt; see *Wilson v. Omaha Indian Tribe* (1979) 442 U.S. 653, 667 (states not included as "persons" where statute imposes burden or limitation on state), but the Supreme Court denied a petition for rehearing on that ground. *Rehg. den.*, *Summa Corp. v. California ex rel. Lands Comm'n* (1984) 467 U.S. 1231.
42. *United States v. O'Donnell*, *supra* 303 U.S. at 510; *Union Pacific RR v. Harris* (1910) 215 U.S. 386, 388; *In re Water of Hallett Creek Stream System* (1988) 44 Cal.3d 448, 456, n.3, *cert. den.* (1988) 488 U.S. 824.
43. *In re Water of Hallett Creek Stream System*, *supra*, 44 Cal.3d at 456, n.3; *Federal Power Com. v. Oregon* (1955) 349 U.S. 435, 443–444.
44. How the states received title to such lands will be explained later in this chapter. See notes 111 through 121 and accompanying text in this chapter.
45. *United States v. Stewart* (9th Cir. 1941) 121 F.2d 705, 712, *aff'd, sub nom.*, *Stewart v. United States* (1942) 316 U.S. 354; *United States v. O'Donnell*, *supra*, 303 U.S. at 510; *Borax Consolidated v. City of Los Angeles* (1935) 296 U.S. 10, 17. *Idaho v. United States* (2001) ___ U.S. ___, ___ S. Ct. ___, ___ L.Ed.2d ___, 2001 Lexis 4665 upheld the United States' reservation of public domain lands for an Indian Tribe. The issue arose because the Tribe's reservation encompassed lands underlying a navigable lake and Idaho asserted its sovereign title. The opinion noted that the default rule was that sovereign lands such as the lakebed pass from the United States to

All these lands were not part of the lands surveyed by the United States public land surveys.[46]

The remaining areas were public lands available for sale by the United States. These lands formed the basis of title in substantial portions of the land area of the United States.[47] Indeed, most property holdings in the western United States count as their original title source a United States public domain patent. At first some land holdings based on government patents may have been quite large. Time and necessity, however, ultimately fragmented these holdings into (among other things) individual home sites, shopping centers, office or industrial complexes, and farms. These ownerships include most lands lying adjacent to tidal or navigable waters. Thus, to have a proper foundation for analyzing public land titles and their potential impact on the title to or boundary of lands adjacent to navigable waters, one must acquire at least a working knowledge of procedures leading to conveyance of public lands and the various sources of public land titles.[48]

1.4.1 Background of Public Land Surveying System

The very core of the public domain conveyancing system is the public land survey system. Congress created this system, as might be expected, to resolve a conflict between two competing systems of surveying prevalent in the original thirteen colonies.[49]

Faced with this situation, before the Constitution was a twinkle in James Madison's eye, the Continental Congress fell back upon what may have been

---

the newly admitted State. The sharply divided Supreme Court had different views of the import and effect of the course and history of the United States' relations with the Tribe. The fact critical to the majority was that the public domain reservation by the United States was started before, although not completed until after, the admission of the State of Idaho. Relying on principles of good faith and fair dealing, the majority ruled for the Tribe and denied the state's claim to the bed of the lake. The dissent asserted that Congress' intent to withhold these sovereign lands from the state did "not rise to anywhere near the level of certainty our cases require."

46. E.g., *Manual of Surveying Instructions*, 1973 (Landmark Reprint) § 1-9–1-15, pp. 3–4; Clark 7th, *supra*, § 5.01, p. 132; Clark 4th, *supra*, § 211, p. 249.

47. Gates, *supra*, Appendixes A and B; *United States v. New Mexico* (1978) 438 U.S. 696, 699, n.3 (average of 46 percent).

48. Public land law is an extremely broad field of great complexity. An exhaustive study of this field is considerably beyond the scope of this book. For a good overview, see *Outline of Land Titles*, *supra*, at 22–186. Also, Gates, *A History of Public Land Law Development* (Pub. Land Law Rev. Comm. 1968), provides a very interesting and readable, yet detailed, narrative. Also see Colby I, *supra*, 33 Calif. Law. Rev. 368; Colby II, *supra*, 36 Cal. L. Rev. 355; Colby, "Mining Law in Recent Years" (1949) 37 Calif. L. Rev. 592 ("Colby III").

49. The northern colonies created rectangular townships to divide common or surplus land. A competing system arose in the South. Grants to southern settlers varied in acreage. These settlers could then select such acreage out of the public domain, to be later identified by survey. Irregular parcels of land without any coordinated surveys resulted. Gates, *supra*, at 61–64; Colby II, 36 Cal. L. Rev. 355.

then (and certainly is now) the traditional practice of government. To resolve this incipient dispute, the Continental Congress appointed commissions to study the problem.[50] Unlike its modern counterpart, however, the much maligned Continental Congress actually did something to carry out its commission's recommendations.

In 1785, the Continental Congress enacted the Land Ordinance, which adopted the land surveying system being used in the northern colonies. The Land Ordinance required that the public domain be surveyed and divided into townships 6 miles square.[51] Although adoption of the Constitution nullified this action of the Continental Congress, the Land Ordinance in fact provided the basic method used for subdividing public lands followed to this day.[52]

Later acts fleshed out this system, including the establishment of a General Land Office (later the Bureau of Land Management) to supervise public land surveying.[53] Through the administration of these agencies, what would have been considered by any potentate to be a magnificent empire has been peacefully transferred from government stewardship to private ownership.

1.4.2 Rectangular System of Public Land Surveying

The system of surveying public lands adopted by the United States is known as the sectionalized or rectangular system of surveying.[54] Public land surveys provide a crucial, indeed, sometimes the only, source of boundary informa-

---

50. *Ibid*. The importance to the country of the resolution of this question can be seen by the fact that Thomas Jefferson was the chairman of the first such commission.
51. *Ibid*.
52. *Ibid*. After the adoption of the Constitution, the same heated debate continued about which system of surveying was better. Colby II, *supra*, 36 Cal. L. Rev. at 357. Jefferson's enemy, Alexander Hamilton (then Secretary of the Treasury) prepared a report that ignored the "Land Ordinance" and proposed a modified system based on the practice in the southern colonies. It was not until May 1796 that Congress passed legislation that continued the practice of the "Land Ordinance" in effect and included authorization for appointment of a "Surveyor General" to supervise the making of the surveys and the creation of 6-mile-square townships. *Ibid*.; Act of May 18, 1796, 1 Stat. 464, § § 1 and 2 ("... form townships six miles square. ..."); *United States v. Ashton* (9th Cir. 1909) 170 F. 509, 513, *app. dsm'd., sub nom. Bird et al. v. Ashton* (1911) 220 U.S. 604. *Railroad Company v. Schurmeir* (1868) 74 U.S. (7 Wall.) 272, 285, also contains a brief history of the public land surveying system. Other works include Stewart, L., *Public Land Surveys, History—Instruction—Methods*, *supra*; Pattison, William D., *Beginnings of American Rectangular Land Survey System, 1784–1800* (Ohio Historical Society 1957); Carter, Clarence E., *The Territorial Papers of the United States, Vol. II, The Territory Northwest of the River Ohio, 1787–1803* (U.S.G.P.O. 1934); Donaldson, Thomas, *The Public Domain, Its History, With Statistics* (U.S.G.P.O. 1970 reprint); McEntyre, John G., *Land Survey Systems* (Landmark 1985); Ernst, Joseph, *With Compass and Chain, Federal Land Surveyors of the Old Northwest, 1785–1816* (Arno Press 1979).
53. Gates, *supra*, at 126–128; Colby II, *supra*, 36 Cal.L.Rev. at 358–359.
54. 43 U.S.C. §§ 751, 700; Clark 7th, *supra*, § 5.03, pp. 134–135; Clark 4th, *supra*, at § 63, pp. 71–72; 1 Ogden, *supra*, at 136; Brown, *Boundary Control and Legal Principles* (John Wiley & Sons, Inc., 2d Ed., 1969) § 1.37, p. 62.

tion; an abbreviated description of the procedure used in conducting such surveys should put them in proper perspective.[55] Understanding the inherent limitations and strengths of these early surveys before relying on them is important.

The first decisions in any survey are where and how to begin. When public land surveys commenced, the nation was in its infancy. Consequently, most locales had no surveyed monuments whose geographic positions had been precisely located on the face of the earth. Instead, public land surveyors used virtually the same thing—a prominent mountain or other well-known local landmark or formation. Thus, the rectangular system of surveying began at some substantial landmark that served as a beginning survey point.[56]

From that *initial* point, the surveyor ran a line north and south. This line is the *principal meridian*. The surveyor then ran another line through the initial point east and west. This line was known as the *base line*. In general practice, the surveyor established guide meridians and standard parallels, respectively, 24 miles parallel to and north and south of the baseline and 24 miles east and west of the principal meridian.[57] In California, standard parallels were located 30 miles north and 24 miles south of the base point.[58]

The quadrilaterals created by establishing baselines and guide meridians were further subdivided. From the baseline the surveyor ran lines north and south at 6-mile intervals. These north-south strips of land were known as *ranges*. The east and west ranges were numbered, starting at the initial point with the number 1. For example, "R 1 W" was the first range west of the principal meridian; "R 2 E" was the second range east of the principal meridian.

Next, the surveyor ran lines east and west of the principal meridian at 6-mile intervals. These lines were known as *township lines*. As with ranges, the townships lying north and south of the baseline were numbered, beginning with the number 1 and moving north or south from the baseline. For example,

---

55. The methodology of early surveyors is found in Clevenger, Shobal V., *A Treatise on the Method of Government Surveying* (1883) (Carben Surveying Reprints). The actual practice of rectangular surveying in California is ably described in Uzes, *Chaining the Land, supra,* pp. 159–167.

56. An interesting contrast to this practice was the practice of the United States Coast Survey (USCS) in preparation of nautical charts. In northwestern California the flagpole at Fort Humboldt in Eureka, California, was used as an important base point for construction of coastal nautical charts. Needless to say, in later years, the flagpole disappeared from prominence into oblivion. Its actual location was only discovered by surveying legerdemain. Snell, J., "Search for a Flagpole," Journal of Surveying and Mapping (ACSM March, 1967) Vol. 27-1-51.

57. The earth is round and the meridian lines converge toward the poles. Thus, guide meridians are not quite parallel to the principal meridian, but approach it as they proceed farther to the north. To correct this narrowing, guide meridians are offset at each standard parallel and started anew northward at their original distance apart.

58. Uzes, *Chaining the Land, supra,* at pp. 159–162.

"T 1 N" was the first layer of townships north of the baseline; "T 24 S" was 24 township layers south of the baseline. These 6-mile-square townships were the focal points of the survey.

Each township was described by reference to both the township and range numbers. By using the township and range numbers, the relative location of the township to the base point can readily be determined. Thus, the north line of "T 4 N, R 26 E" is 24 miles (4 townships X 6 miles) north, and the east line is 156 miles (26 ranges X 6 miles) east, of the base point.

Townships were further subdivided into 36 sections. At least theoretically, each section was 1 mile square and contained 640 acres. Sections were numbered consecutively, beginning with section 1. Section 1 was always located in the northeast township corner. Sections were then numbered east to west in the first tier of sections, west to east in the second tier, and so on, until section 36 was reached in the southeast corner of the township. Sections could be further subdivided into quarter sections or quarter-quarter sections, but were not so divided by actual government survey.[59]

Although the standard area of a section is 640 acres, for many reasons there may be a deficiency in actual practice. Fractional sections resulted when, for example, a section encompassed within its perimeter a tidal or navigable water body or a land grant made by one of the nations that preceded United States' sovereignty. If a navigable river ran through a section, the bed of the river was not included in the lands surveyed. The surveyor meandered[60] the course of the river, and the acreage within its banks was excluded from public lands in the section. The irregular units resulting from these deficiencies are called *lots*.[61]

1.4.3 Effect of Public Land Surveys

It is extremely important to remember that when properly completed, these surveys did not merely locate townships and sections, they created them.[62]

---

59. Ogden has an excellent explanation of this system, 1 Ogden, *supra*, at 424–428, as does Colby II, *supra*, 36 Cal. L. Rev. at 361–362. See also 43 U.S.C. § 751, et seq; Clark 7th, *supra*, §§ 5.13–5.24; Clark 4th, *supra*, at §§ 85–99.

60. Meandering refers to a surveying practice that permits the traverse of the margin of a water body to form a meander line. E.g., *Manual of Surveying Instructions*, 1973 (Landmark Reprint), § 3–115. "Meander-lines are run in surveying fractional portions of the public lands bordering upon navigable rivers, not as the boundaries of the tract, but for the purpose of defining the sinuosities of the stream banks, and as a means of ascertaining the quantity of land in the fraction subject to sale, and which is to be paid for by the purchaser." *Railroad Company v. Schurmeir*, *supra*, 74 U.S. at 286–287. For a further explanation of meander lines, see notes 42 through 64 and accompanying text in Chapter 3.

61. Clark 7th, *supra*, § 11.03; Clark 4th, *supra*, at § 213, p. 251; 1 Ogden, *supra*, at pp. 424–426.

62. *Cox v. Hart* (1922) 260 U.S. 427, 436; *Sawyer v. Grey* (1913) 205 F. 160, 163; *Robinson v. Forrest* (1865) 29 Cal. 317, 325.

The township plats prepared as the result of field surveys become part of the instrument of conveyance.[63] In fact, until lands in the public domain were properly surveyed, they were not public lands available for disposition.[64] Unsurveyed lands were not subject to disposition under the public land laws.[65]

Thus, the rectangular system of surveying occupied a quite critical position. Virtually all federal land grants were based on this system of surveying, regardless of the character of the land disposed of by the United States. As discussed later in this chapter, public lands granted to railroads, to the states as swamp and overflowed lands or school lands, or to individuals as, for example, preemption patents, were all surveyed by this methodology and their geographic location and extent established accordingly.

1.4.4 Advantages and Disadvantages of the Public Land Survey System

The public land survey system had certain remarkable advantages. The requirement of government survey before the United States could sell public lands promoted the nation's orderly development. Conflicts of title that seemed invariably to accompany disposition of public lands in a developing country were, in large part, avoided. Archiving, as public records, the official survey notes and plats facilitated identification of land boundaries.[66] The rectangular survey method had its intended consequence of making land descriptions both brief and simple. Finally, reports required of public land surveyors provided an invaluable record of the nature and quality of the largely unspoiled land surveyed.[67]

No system is perfect, however. The rectangular system had its shortcomings. For example, mountaintops, typical of the prominent physical features normally chosen as base points, are not conveniently located at exact multi-

---

63. *Cragin v. Powell* (1888) 128 U.S. 691, 696; *Hardin v. Jordan* (1891) 140 U.S. 371, 380; *United States v. Otley* (9th Cir. 1942) 127 F.2d 988, 993.
64. *United States v. Morrison* (1916) 240 U.S. 192, 210; *Cox, supra*, 260 U.S. at 436; *Sawyer v. Grey, supra*, 205 F. at 163.
65. *Buxton v. Traver* (1889) 130 U.S. 232, 235 ("No portion of the public domain . . . is open to sale until it has been surveyed and an approved plat of the township embracing the land has been returned to the local land office"). There are acts, however, that allow acquisition of rights on unsurveyed lands. Homestead Act, Act of May 20, 1862, 12 Stat. 392; Preemption Act, Act of September 4, 1841, 5 Stat. 453; Act of May 10, 1872, 17 Stat. 94, see *Buxton, supra*, 130 U.S. at 235–237.
66. Hollywood has repeatedly capitalized on this fact as a theme of many cowboy movies. The movie *Silvarado*, one incarnation of this theme, had destruction of local land office records as the principal goal of the villains (cattle ranchers). Destruction of these records would dispossess recent settlers and enable ranchers to run cattle over the settlers' lands without regard for established property boundaries.
67. Colby II, *supra*, 36 Cal. L. Rev. at 364–365; 43 U.S.C. § 751; Brown, *supra*, at 48; Uzes, *Chaining the Land, supra*, pp. 164–167; Id. at Appendix D, p. 217.

ples of 24- or 30-mile intervals. Use of such arbitrarily chosen natural features often meant that surveys beginning at these features would not harmonize where they met.[68] Therefore, in many regions it was necessary to create tiers of fractional townships where surveys met one another as they extended from basic reference points. In addition, the rectangular survey system was not immune from mistakes in measurements due to incompetent surveyors or faulty instruments. Unfortunately, litigation was often required to undo the effects of conflicting surveys and claims.[69]

Moreover, the public land surveying system was, in certain instances, infected with the same malady that plagued the early days of some public disposition programs. Gross survey frauds were perpetrated against the government. For example, in some areas, little, if any, survey work was done in the field or only a few of the most accessible and prominent corners were established. The accuracy of the location of lands described in surveys was often questionable. Unscrupulous or lax surveyors wrote up "paper" field notes in a comfortable office, rather than out in the field. These paper field notes contained gross misrepresentations concerning topography and land character. Public scandals arose, and in some cases criminal proceedings resulted.[70]

Even so, the United States' accomplishment in surveying its vast holdings is genuinely awesome. The law recognized the sanctity of such surveys. Courts established a rule placing a heavy burden on any party seeking to overturn patent boundaries based on such surveys. To succeed, that party must show that the survey was fraudulent or in error.[71]

The following section describes the authority through which such public land patents came to be issued.

1.5 AUTHORITY TO DISPOSE OF PUBLIC LANDS

In discussing the means by which the United States chose to dispose of its public lands, one point should be kept clearly in mind: Congress could dispose

---

68. For example, in California the Bureau of Land Management has noted there is an overlap between the north line of surveys based on the Mt. Diablo Meridian and the south line of surveys based on the Humboldt Meridian. *Public Lands Surveying: A Casebook* (U.S. Dept. of Int., BLM, Cadastral Training Staff 1975), pp. C 3-1–C 3-12.
69. E.g., *French v. Fyan* (1876) 93 U.S. 169; *Smelting Co. v. Kemp* (1881) 104 U.S. 636; *Tubbs v. Wilhoit* (1890) 138 U.S. 134; *Kernan v. Griffiths* (1864) 27 Cal. 87; Colby II, *supra*, 36 Cal. L. Rev. at 363.
70. Gates, *supra*, at 420–422; Colby II, *supra*, 36 Cal. L. Rev. at 363; *United States v. Perrin* (1889) 131 U.S. 55; *United States v. Hall* (1889) 131 U.S. 50; *United States v. Benson* (9th Cir. 1895) 70 F. 591; Uzes, *Chaining the Land*, *supra*, Chapter 6, has a fascinating account of fraudulent public land surveys in California.
71. E.g., *United States v. Otley*, *supra*, 127 F.2d at 995. "The suit here is, in effect, to cancel fourteen . . . patents of the lake lands. In such a case the United States has a more than the ordinary burden of proof." A more detailed discussion of this subject will be found in notes 72 through 79 and accompanying text in Chapter 3.

of and deal with public lands of the United States *in any manner it elected*.[72] Indeed, Congress "paid" for the development of internal improvements such as canals, public roads, and railroads through grants of public lands. One example is particularly illustrative of the breadth and scope of Congress's plenary power to dispose of or otherwise deal with public lands.

During the California gold rush, people from around the world poured into California to seek their fortune in the gold-rich mountains. Many newly arrived prospectors were totally unmindful of, or, more likely, did not a care a whit about what some considered only a minor detail: The gold was located on land held by the United States as part of its public domain. In the absence of congressional action, miners greedily burrowing away at the golden heart of California had no title to these lands, nor to the gold they found.[73]

These miners governed themselves, even to the extent of forming mining districts and adopting rules that regulated how much land each miner could mine and the manner in which the land was to be worked. However, all mining was done without benefit of any legal benediction from the United States. Perhaps comprehending it did not want to fuel secessionist fever a continent away from Washington, and even despite desperate need of the federal government for funds to cover costs incurred in prosecuting the Civil War, the United States acquiesced to and later legislatively confirmed the miners' arrangements.[74] Nonetheless, in most of the public domain, Congress did not usually act on such an *ad hoc* basis.

1.6 PUBLIC LAND DISPOSITIONS—IN GENERAL

Since 1796, Congress has provided for the sale of public lands.[75] The United States has conducted public land sales since at least 1820 under a procedure providing that lands subject to sale be first offered for sale at public auction and only thereafter becoming available for private sale.[76]

---

72. *Alabama v. Texas* (1954) 347 U.S. 272, 273–274; *United States v. San Francisco* (1940) 310 U.S. 16, 29–30.
73. E.g., *Irwin v. Phillips* (1855) 5 Cal. 140, 146:

> Courts are bound to take notice of the political and social condition of the country, which they judicially rule. In this State the larger part of the territory consists of mineral lands, nearly the whole of which are the property of the public. No right or intent of disposition of these lands has been shown either by the United States or the State governments, and with the exception of certain State regulations, very limited in their character, a system has been permitted to grow up by the voluntary actions and assent of the population, whose free and unrestrained occupation of the mineral region has been tacitly assented to by the one government, and heartily encouraged by the expressed legislative policy of the other. Ibid.

74. *Ibid.*; Colby I, *supra*, 33 Calif. L. Rev. at 370–371; *United States v. Gerlach Live Stock Co.* (1950) 339 U.S. 725, 745–748.
75. Act of May 18, 1796, 1 Stat. 464, 467, § 5; Gates, *supra*, at 121–126.
76. E.g., Act of April 24, 1820, 3 Stat. 566, now found in 43 U.S.C. 672; *Outline of Land Titles*, *supra*, at 9; Gates, *supra*, 145–247.

Early statutes authorizing the sale of public lands provided that persons who actually occupied and improved public domain lands could buy such lands at a small price and receive a patent for them. This prior right to buy was known as a "preemption."[77] When public lands in California were opened for entry, Congress made the provisions of the Preemption Act applicable to California.[78]

The Homestead Act changed the theory of public land sales. Before the Homestead Act, lands were sold to settlers; after the Homestead Act, *the government gave lands* to settlers as a bounty to encourage settlement and development.[79]

Congress also made quite liberal grants to states for the accomplishment of some specific purpose.[80] Grants of lands to support public schools, so-called school land grants, began with the Land Ordinance of 1785. This legislation reserved section 16 of each township for the maintenance of public schools within the township.[81] Later admitted states, such as California, received sections 16 and 36 in each township for school purposes.[82] Congress also made generous grants (500,000 acres) to states for internal improvements, for "seminaries of learning," for state buildings, and for "land grant" colleges.[83] Congress designed other granting statutes to encourage private construction of roads, canals, river improvements, and railroads.[84] Congress also enacted various other land disposal statutes to encourage settlement of the nation and development of lands in arid regions.[85]

1.7 SWAMP AND OVERFLOWED LAND GRANT TO THE STATES

The focus of this book is on titles to, and boundaries of, lands along and beneath tidal or navigable waters. Consequently, as far as this book is concerned, the most significant disposition of public lands made by the United States was the grant made to the states for what Congress generically called "swamp and overflowed" lands. As described later in this book, one of the

---

77. Act of September 4, 1841, 5 Stat. 455; Gates, *supra*, at 219–247.
78. Act of March 3, 1853, 10 Stat. 244, 246, § 6.
79. Gates, *supra*, at 387–434; *Outline of Land Titles, supra*, at 96. One politician, 1988 Democratic vice presidential candidate Senator Lloyd Bentsen, described the homesteading of his forbears in South Dakota as a bet of 160 acres with the government that the homesteader would survive the first winter.
80. For a comprehensive narrative of grants to the states, see Gates, *supra*, at 285–339. Gates, in Appendix C, *Id.* at 804–806, enumerates and quantifies these grants by state.
81. *Cooper v. Roberts* (1855) 59 U.S. (18 How.) 173, 177–178.
82. Act of March 3, 1853, 10 Stat. 244, 246, § 6.
83. E.g., Act of September 4, 1841, 5 Stat. 452, 455, § 8, now found at 43 U.S.C. § 857 (500,000-acre grant); Act of March 3, 1853, 10 Stat. 244, 248, § 12 (seminaries of learning); *Id.* at § 13 (public buildings); Act of July 2, 1862, 12 Stat. 503 (land grant colleges).
84. Gates, *supra*, at 341–386; *Outline of Land Titles, supra* at 127–142.
85. Gates, *supra*, at 387–434, 635–698.

principal title and boundary disputes is between titles derived from the swamp and overflowed land grant by the United States and titles derived from or held by a State in its sovereign capacity.[86]

1.7.1 In General

Congress intended the swamp land grants to encourage states with swamp and overflowed lands to drain and reclaim them.[87] Swamp and overflowed grants began with the Act of March 2, 1849, that conveyed such lands to Louisiana.[88] The Louisiana Act was followed by the Arkansas Swamp Lands Act, which granted to the states swamp and overflowed lands within the boundaries of all states then in existence; this included California.[89]

As now contained in the United States Code, the Act simply provides, in pertinent part:

> ". . . [T]he whole of the swamp and overflowed lands, made unfit thereby for cultivation, and remaining unsold on or after the 28th day of September, A.D. 1850, are granted and belong to the several States respectively, in which such lands are situated. . . ."[90]

The Arkansas Swamp Lands Act was a "present grant" of lands that were swamp and overflowed on September 28, 1850.[91] This means that whatever lands were swamp and overflowed in character on September 28, 1850, were conveyed by the United States on that date, subject only to identification as falling within the terms of the grant.

As with so many affairs concerning the government, this granting statute sounds wonderful and perfectly understandable in theory. On closer examination, and in practice, the flaws and ambiguities in the granting legislation became all too apparent. The major problem was basic to the grant itself: Just what exactly were the "swamp and overflowed lands" intended to be conveyed?

---

86. See Chapters 5 and 6.
87. Gates, *supra*, at 321; Colby III, *supra*, 37 Cal. L. Rev. at 592–593; *Wright v. Roseberry* (1887) 121 U.S. 488, 496 ("The object of the grant . . . was to enable the several states to which it was made, to construct the necessary levees and drains to reclaim the lands; and the act required the proceeds from them, whether from their sale or other disposition, to be used, so far as necessary, exclusively for that purpose").
88. Act of March 2, 1849, 9 Stat. 352.
89. Act of September 28, 1850, 9 Stat. 519, found in 43 U.S.C. § 982, et seq. California was admitted on September 9, 1850, just 19 days before the enactment of the Arkansas Swamp Lands Act, Act of September 9, 1850, 9 Stat. 452, and was eligible for the Arkansas Swamp Lands Act grant.
90. 43 U.S.C. § 982.
91. *Wright, supra*, 121 U.S. at 509; *Work v. Louisiana* (1925) 269 U.S. 250, 255.

1.7.2 Definition of Swamp and Overflowed Lands

Putting aside the obscure description found in the granting statutes, supposedly distinctive swamp and overflowed lands have been variously defined. One court's quite general definition described such lands as "wet and unfit for cultivation."[92] Other court definitions, although seemingly more specific in language, in fact provided little help in determining which lands were swamp and overflowed lands and which lands were not. For example:

> ... Swamp lands, as distinguished from overflowed lands, may be considered such as require drainage to fit them for cultivation. Overflowed lands are those which are subject of periodical or frequent overflows as to require levees or embankments to keep out the water, and render them suitable for cultivation. It does not make any difference whether the overflow be by fresh water, as by the rising of rivers or lakes, or by the flow of the tides.[93]

This description could apply equally to lands periodically covered and uncovered by the ebb and flow of the tides. Or it could describe lands within or adjacent to a river that has seasonal stage fluctuations. Such lands could as well be thought of as sovereign lands,[94] not swamp and overflowed lands. This definitional uncertainty was recognized as a problem when Congress enacted the granting legislation. The ambiguity soon caused difficulties in the process by which swamp and overflowed lands were identified and separated from other public lands for conveyance to the states.[95]

1.7.3 Identification of Swamp and Overflowed Lands

Although the swamp and overflowed land grant was an immediate grant of all public lands that were swamp and overflowed in character on September

---

92. *Robinson, supra,* 29 Cal. at 325.
93. *San Francisco Savings Union v. Irwin* (Cir. Ct. D.Cal. 1886) 28 F. 708, 712, *aff'd* (1890) 136 U.S. 578.
94. See notes 111 through 121 and accompanying text in this chapter.
95. Gates, *supra,* at 325–326, describes the effect of this ambiguity as follows:

> Opponents of the Swamp Land Act had warned that the vague and indefinite term "swamp and overflowed lands unfit for cultivation" would make the act difficult to administer. Their forebodings proved more than correct. Almost from the outset the Commissioner of the General Land Office and the surveyors general who were charged with supervising the selections complained about the extreme difficulties involved in determining whether the land was swamp or overflowed on the basis of the surveyors' notes, which proved to be defective, far from complete, and in many instances fraudulent. . . . Butterfield [the commissioner of the General Land Office] had complicated the task of selecting swampland by a series of "unskillful instructions." . . . There was thus added a heavy burden to that already carried by the local land officers. . . . Vacillations from a hard line to a soft line in the instructions did not help matters. Instead they made for uncertainty and confusion.

28, 1850, the lands subject to the grant had to be identified or separated from other public lands. In addition, a patent had to be issued to the state to convert the state's beneficial title into a title the state could freely and confidently dispose of to private persons for purposes of the grant.[96] The Arkansas Swamp Lands Act provided that this identification, also known by the unfortunate term "segregation," was to be accomplished by the Secretary of the Interior.[97] The Secretary was to make accurate lists and plats of swamp and overflowed lands and, on request of the states, issue patents for lands so identified.[98]

Unluckily, the Arkansas Swamp Lands Act failed to describe the manner in which or the procedure by which the Secretary of the Interior was to accomplish these duties.[99] The Act also failed to describe or otherwise specify just what evidence would establish the character of the land as "swamp and overflowed."[100] Instead, the Act left considerable discretion in these delicate matters to the Secretary.

At first, the Secretary of the Interior gave each state a choice between two elections: States could either abide by the showing made in the government surveyor's field notes or could examine the lands themselves and present claims, accompanied by proof, for lands believed to be swamp and overflowed in character.[101] Thus, it was essential for identification by the first method that lands be surveyed and the system of public land surveying be extended as soon as possible to the western states.

The government, however, was not as well staffed in those days as it is today. The nation also seemed much larger and was considerably more difficult in which to get around. Given these two problems and the untimely intervention of the Civil War, the enormous burden of identifying swamp and overflowed lands in the various states is apparent. The Secretary of the Interior was overwhelmed, "swamped" if you will, with applications and demands of the states. His office staff was grossly inadequate to meet those incessant demands for the required segregation of swamp and overflowed lands.[102] Inevitably, confusion resulted, exemplified in the following description:

> ... [A]mong the hundreds of thousands of people rushing westward to find land on which to settle and gain a right of preemption, many squatted upon lands

---

96. *Wright, supra,* 121 U.S. at 509.
97. *Tubbs, supra,* 138 U.S. at 136. The term "segregation" has been defined, in the context of public lands, as referring to the process by which public lands are determined no longer subject to disposal under the public land laws. *Utah Div. of State Lands v. United States* (1987) 482 U.S. 193, 107 S.Ct. 2318, 2325.
98. *Tubbs, supra,* 138 U.S. at 136.
99. Gates, *supra,* at 324–330; Colby III, *supra,* 37 Calif. L. Rev. at 595.
100. *Ibid.*
101. *Ibid.*
102. *Ibid.*

that were later claimed to be swamp. . . . Squatters and indeed, investors in wild land preempted or made cash entries on some lands later to be selected as swamp or overflowed and thought their investments secure, only later to be told that the state was claiming their land. By 1855 many thousands of occupants and purchasers. . . . were in this predicament. Meantime, the states were hurrying to sell their lands, some. . . . selling their swamplands years before patents had been received from the Federal government. There were even instances where two parties each had title, one from the United States and the other from one of the states.[103]

Because of this confusion, the states (prodded by their swamp and overflowed land grantees) went to Congress for help. In vivid testimony to the power of lobbyists even in those days, Congress passed many curative acts to relieve the difficulties of purchasers of swamp and overflowed lands and the embarrassment of both the Land Department and the states.[104] Even with enactment of these so-called curative acts, the process of segregating swamp and overflowed lands remained chaotic and haphazard.[105]

1.7.4 Title Problems in California and Attempts at Their Resolution

Much the same as in other contexts today, California found itself at the forefront of states confronting the vexing problem of segregating swamp and overflowed lands. The gold rush and the subsequent boom (which continues to this day) spawned a great emigration. The crowd of newly arrived Californians in turn created a voracious demand for land, both for living space and as the basis of an agricultural economy. This urgent demand fomented an almost anarchic situation in California concerning disposition of public lands underlying or adjacent to navigable or tidal waters.

California at first accepted applications from prospective purchasers of swamp and overflowed lands *before* the United States even had a chance to survey the townships in which such lands were located.[106] Crazy as it sounds, the courts held that *neither* California *nor* the United States was bound by the action of the other in determining which lands were available for sale as swamp and overflowed.[107]

Because of extraordinarily endemic confusion in California concerning identification and sale of swamp and overflowed lands, interested persons

---

103. Gates, *supra*, at 329–330. The extent of the confusion is evidenced by the fact that in the decades following 1850, contests over more than 3 million acres of land were pending before the General Land Office. *Railroad Company v. Fremont County* (1869) 76 U.S. (9 Wall.) 89, 91.
104. E.g., Act of March 22, 1855, 10 Stat. 634; Act of March 3, 1857, 11 Stat. 251. Gates, *supra*, at 330–332, has a narrative description of this process.
105. *Tubbs*, *supra*, 138 U.S. at 136–138.
106. *Id.* at 138; Gates, *supra*, at 330.
107. Gates, *supra*, at 330; *Kernan*, *supra*, 27 Cal. at 91.

prevailed upon Congress to enact a special statute to solve California's peculiar problems.[108] That act provided four methods for identifying "swamp and overflowed lands" in California.[109] The methods chosen by Congress comprise the entire panoply of options available for identification of swamp and overflowed lands.

First, in all cases in which township surveys had been made under United States' authority and survey plats had been approved, the Commissioner of the General Land Office was to certify to the state as swamp and overflowed lands all lands represented as swamp and overflowed on the approved township plats.

Second, if there were no township plats, the United States Surveyor General for California was to examine segregation plats made by the state.[110] If he found such plats to be in conformity with the rectangular system of surveys, that official was to construct and approve township plats accordingly.

Third, where there was no survey or the state's survey was unacceptable in some regard, the United States Surveyor General was to make segregation surveys on application by the governor.

Finally, in case the state should claim as swamp and overflowed, any land not represented on township plats or in returns of surveyors as swamp and overflowed, then the swamp and overflowed character of the land on the date of the grant and the right of the state to such land were determined by testimony taken before the United States Surveyor General.

California and other states received vast amounts of acreage because of such legislation. In fact, California alone received more than 2 million acres of swamp and overflowed lands from the United States. In turn, California and other states granted these lands to their citizens who wanted to undertake the effort and expense to reclaim the lands. The states then used the proceeds for state purposes. Even before these swamp and overflowed lands were granted to California, however, the state had received title to lands of another character, lands lying beneath tidal or navigable waters. Substantial portions of these lands were similar in physical character and appearance to swamp and overflowed land. The following section discusses title derivation to such lands.

---

108. An Act to Quiet Land Titles in California, Act of July 23, 1866, 14 Stat. 218, now found in 43 U.S.C. § 987.
109. 43 U.S.C. § 987; *Tubbs, supra*, 138 U.S. at 139–140.
110. These plats had been accomplished pursuant to a state statute enacted by legislature to facilitate the process of identifying swamp and overflowed lands. An Act to Provide for the Reclamation and Segregation of Swamp and Overflowed, Salt Marsh and Tide Lands Donated to the State of California by Act of Congress, Act of May 13, 1861, Stats. 1861, Ch. 352, § 19–20. See note 47 and accompanying text in Chapter 5.

1.8 STATE SOVEREIGN LANDS

1.8.1 Equal Footing Doctrine

States also became landowners simply because of their status as sovereign political entities. Each state became an owner of all lands[111] within its boundaries that were waterward or bayward of the ordinary high-water mark of tidal or navigable waters.[112] This title arose as a matter of constitutional principle upon the admission of a state into the Union.[113] Its basis is the Equal Footing Doctrine, a legal doctrine that does not appear in the Constitution but is recognized as constitutionally based.[114] The title received by states to their sovereign lands is absolute[115] and has been ratified and confirmed by Congress.[116]

1.8.2 Public Trust

The states' title to their sovereign lands is unique. Unlike the title that states received to swamp and overflowed lands, States hold sovereign lands in and

---

111. These lands have been known variously as "tidelands," "tide and submerged lands," "sovereign lands" or "sovereign trust lands." For purposes of this chapter, unless the context otherwise indicates, such terms are used interchangeably.
112. E.g., *Pollard's Lessee v. Hagan* (1845) 44 U.S. (3 How.) 212, 223, 230; *Weber, supra*, 85 U.S. at 65–66; *Shively, supra*, 152 U.S. at 57. In the far northeastern states sovereign ownership type rights extend to the great ponds. E.g., Leighty, "Public Rights in Navigable State Waters—Some Statutory Approaches" (1971) 6 Land & Water L.Rev. 459, 471; *Conant v. Jordan* (Me. 1910) 77 A. 938, 940.
113. E.g., *Pollard's Lessee, supra*, 44 U.S. at 230; *State Land Board v. Corvallis Sand & Gravel Co.* (1977) 429 U.S. 363, 374 ("*Corvallis*").
114. The origin of the Equal Footing Doctrine lies in the fact that some of the original 13 states owned land outside their boundaries. Before adoption of the Constitution, during the period of the Articles of Confederation, some of the Original States ceded portions of their lands to the confederation. See note 10 in this chapter. One of the cessions' conditions was that new states created and admitted under the Articles of Confederation should be created with the same rights of sovereignty as the Original States. The Northwest Ordinance of 1787 (as reenacted in 1789 upon the adoption of the Constitution) summarized these conditions with a provision that each state would be admitted into the United States on an equal footing with the Original States in all respects whatever. Act of August 7, 1797, 1 Stat. 50, 52; see generally, Hanna, "Equal Footing in the Admission of States" (1951) 3 Baylor L. Rev. 519, 523. The expression, "equal footing" has been used in all acts of admission commencing with Tennessee in 1796. *Coyle v. Smith* (1911) 221 U.S. 559, 567; Leighty, "The Source and Scope of Public and Private Rights in Navigable Waters" (1970) 5 Land & Water L. Rev. 391, 416, n.4.
115. *State Land Board v. Corvallis, supra*, 429 U.S. at 374.
116. Submerged Lands Act, Act of May 22, 1953, 67 Stat. 29, now found in 43 U.S.C. §§ 1301, et seq., particularly 43 U.S.C. § 1311(a)(1) & (2); *Twombly v. City of Long Beach* (9th Cir. 1964) 333 F.2d 685, 688, cert. den. (1964) 379 U.S. 904; *Furlong Ent. v. Sun Exploration & Prod.* (N.D. 1988) 423 N.W.2d 130, 132.

subject to an inalienable public trust.[117] The public trust is a property interest, an easement the public enjoys over sovereign lands for public purposes and uses.[118] Although the conventional formulation of the public trust is for the purposes of commerce, navigation, and fisheries, the public trust has been much more broadly construed, as graphically described by one court:

> Public trust easements are traditionally defined in terms of navigation, commerce and fisheries. They have been held to include the right to fish, hunt, bathe, swim, to use for boating and general recreation purposes the navigable waters of the state, and to use the bottom of the navigable waters for anchoring, standing or other purposes. . . .
>
> The public uses to which [sovereign lands] are subject are sufficiently flexible to encompass changing public needs. In administering the trust the state is not burdened with outmoded classifications favoring one mode of utilization over another. . . . There is a growing public recognition that one of the most important public uses of the [sovereign lands] . . . is the preservation of those lands in the natural state, so that they may serve as ecological units for scientific study, as open space, and as environments which provide food and habitat for birds and marine life, and which favorably affect the scenery and climate of the area. It is not necessary to define precisely all the public uses which encumber [sovereign lands].[119]

Some view the consequences of land being held subject to the public trust as ravaging.[120] And it is true that some public trust decisions seem harsh to

---

117. E..g., *Illinois Central Railroad v. Illinois* (1892) 146 U.S. 387, 452; *Matthews v. Bay Head Imp. Ass'n* (N.J. 1984) 471 A.2d 355, 358, *cert. den.* (1984) 469 U.S. 821; *City of Berkeley v. Superior Court* (1980) 26 Cal.3d 515, 521, *cert. den. sub nom. Santa Fe Land Improv. Co. v. Berkeley* (1980) 449 U.S. 840; *Boston Waterfront Dev. Corp. v. Com.* (Mass. 1979) 393 N.E.2d 356, 358–359; *White v. Hughes* (Fla. 1939) 190 So. 446, 448.
118. E.g., *Summa, supra,* 466 U.S. at 205; *Marks v. Whitney* (1971) 6 Cal.3d 251, 259–260.
119. *Marks, supra,* 6 Cal.3d at 259–260. For other cases see, e.g., *Just v. Marinette County* (Wisc. 1972) 201 N.W.2d 761, 768; *White v. Hughes, supra,* 190 So. at 449; *Matthews, supra,* 471 A.2d at 363. The public trust encompasses all sovereign lands and has been held to extend to nonnavigable tributaries where actions taken in such tributaries affect public trust interests. *National Audubon Society v. Superior Court* (1983) 33 Cal.3d 419, 435, *cert. den. sub nom. Los Angeles Dept. of Water & Power v. National Audubon Soc.* (1983) 464 U.S. 977. The United States Supreme Court has recognized the broad scope of the public trust as well. *Summa, supra,* 466 U.S. at 205.
120. One industry association group stated in a brief filed with the United States Supreme Court:

> The "public trust" easement, as it has developed in California law, vests most of the attributes of property ownership in the State. The State may "enter upon and possess" trust-encumbered lands to improve them for trust related uses. [Citation omitted.] It has virtually unfettered discretion in dictating the uses to which the lands may be put. . . . Thus, the State, as trustee, may prohibit the fee owner from making any beneficial use whatsoever of trust-encumbered lands. . . .
>
> What is left the landowner whose title is subjected to this permanent public servitude? Not much. . . .

landowners.[121] It is important to keep these matters in mind in considering state land disposition statutes and the effectiveness of the resulting conveyances.

1.9 GRANTS OF STATE LANDS

The states also attempted to promote social policy through land grant legislation.[122] In California, for example, the legislature enacted several statutes supposed to encourage and foster reclamation of swamp and overflowed lands donated to the state by the United States.[123] Such statutes provided for the very favorably priced sale of these lands to purchasers who intended to reclaim them. In addition, these statutes authorized creation of local boards to oversee and approve applications for reclamation of the lands.[124]

The land disposal systems of the states were models of laxness and lack of foresight. Raising money appeared to be the principal, if not the only, goal of the states.[125] In the early years of statehood, this openhandedness often

---

Given the respective rights of the State and the landowner in trust-encumbered lands, it is the State that effectively owns lands that are subject to the public trust easement. The fee owner's "title," such as it is, is a title revocable at the will of the State. He possess what amounts to a mere license to use his lands at the pleasure of the State. Amicus Curiae Brief of the California Land Title Association at 3–5, *Summa Corp. v. California ex rel. State Lands Comm'n* (1984) 466 U.S. 198.

121. In *Colberg, Inc. v. State of California, ex rel. Dept. of Pub. Wks.* (1967) 67 Cal.2d 408, 422–425, *cert. den.* (1968) 390 U.S. 949, construction of highway bridges over navigable waters prevented customers of a riparian landowner's shipbuilding business from navigating their vessels to that business. The state was not required to pay compensation to the riparian landowner. The court held that building the freeway bridges was an exercise of the public trust. Thus, the damage to the shipbuilding business was noncompensable. Ibid. Another case of particular interest is *Lake Mich. Fed. v. U.S. Army Corps of Engineers* (N.D. Ill. 1990) 742 F.Supp. 441. Loyola University of Chicago developed plans to expand its lakeshore campus by filling about 20 acres of lakebed. *Id.* at 443. It sought and received legislative and municipal approval for the project (which included unrestricted access to a seawall area), including a finding by the legislature that since the fill would benefit the public, it had the power to convey the lands to the university. *Ibid.* The university even received a permit to fill the area from the U.S. Army Corps of Engineers under the Federal Clean Water Act and the Federal Rivers and Harbors Act. *Ibid.* The Lake Michigan Federation challenged the issuance of these permits. Despite the legislative finding to the contrary, the district court upheld the challenge on the ground that the conveyance to the university violated the public trust doctrine. The court found the purpose of the conveyance was to benefit a private, not a public, purpose. *Id.* at 444–446. The district court invited an appeal, *id.* at 449–450, but none was filed.
122. Gates, *supra*, at 336–339.
123. E.g., An Act to Provide for the Reclamation and Segregation of Swamp and Overflowed, Salt Marsh and Tide Lands Donated to the State of California by Act of Congress, Act of May 13, 1861, Stats. 1861, Ch. 352.
124. *Ibid.*
125. As one commentator stated:

went unchallenged. In later years, the states themselves, local governments, and ordinary citizens came to rethink and challenge the now-in-hindsight apparently thoughtless grants of lands adjacent to, in, or under tidal or navigable waters. Individual citizens and the government brought lawsuits that established the basic principles that guide us today in analyzing the validity and extent of such titles conveyed by the states pursuant to those early disposition programs.

1.10 LIMITATIONS ON STATE DISPOSAL OF LANDS ADJACENT TO TIDAL AND NAVIGABLE WATERS

Swamp and overflowed lands can be readily disposed of in accordance with the granting and authorizing legislation. Dispositions of state sovereign lands are, on the other hand, subject to very stringent restrictions. The intent of such restraints is to prevent or undo unlawful or improper alienation of the state's property interest, an interest held in public trust. These strictures may be constitutional, statutory, or based on common law as developed by the courts.

The author cannot overemphasize the importance of understanding these restrictions. They limit a state's power to dispose of such lands. They condition or make void or voidable what otherwise might appear to be absolute conveyances from the state to private persons. Although the focus of this discussion is on California, many of these principles are universally applicable. Knowledge of these restrictions will aid surveyors, lawyers, landowners, and others in gauging the strength and security of title provided by state patents and grants.

... With some exceptions it can be said that the early experience of the states in land administration reflects little vision or long range planning, a tendency to rush into leasing or selling without proper consideration of the effects of policies being adopted, and careless management of the funds received from sales and leases. Too frequently, legislatures and public officers appeared to shape policies that would enable them to profit personally.

Few states had any reason to be proud of the record of their land management in the early period—particularly before 1862. While their representatives were complaining that the Federal Land Office was permitting land to become monopolized in many areas by men of capital, the state land offices were letting the lands that came to them slip through their hands and pass to speculators in ways that were rarely tolerated in Washington." Gates, *supra*, at 337.

Courts agree with this assessment. *People v. California Fish Co.* (1913) 166 Cal. 576, 591–592.

126. Tidelands have been defined as lands covered and uncovered by the daily rise and fall of the tide. 2 Shalowitz, *Shore and Sea Boundaries* (Dept. Comm. Pub. 10-1 1962), Appendix A,

1.10.1 Constitutional and Statutory Prohibitions Against Alienation of Tidelands—In General

In California, the state constitution provides that tidelands[126] within 2 miles of incorporated cities or towns are withheld from sale.[127] This is an absolute prohibition on the alienation of tidelands.[128] It does not matter whether the lands no longer have the physical character of tidelands (such as lands that have been filled or reclaimed and are thus no longer subject to overflow by tidal waters). Such lands appear to remain subject to the constitutional prohibition against alienation.[129] No matter in what form the conveyance is couched, or how apparently beneficial to the public, the conveyance may be

---

127. Cal. Const. Art. X, § 3 provides: "All tidelands within two miles of any incorporated city, city and county, or town in this State, and fronting on the water of any harbor, estuary, bay, or inlet used for the purposes of navigation, shall be withheld from grant or sale to private persons, partnerships, or corporations. . . ." Many cases refer to this provision in its former incarnation, Article XV, § 3. The constitutional provision was merely renumbered, not amended or modified. Citation will be made to both provisions as appropriate. In addition, before adoption of the constitution, various legislative acts prohibited sale of tidelands within a certain distance from various governmental subdivisions. E.g., *Dillon v. San Diego Unified Port Dist.* (1972) 27 Cal.App.3d 296, 299.

128. *City of Long Beach v. Mansell* (1970) 3 Cal.3d 462, 482. It is not clear whether this prohibition would forbid a conveyance to the United States of tidelands within the 2-mile constitutional ban. The California Constitution forbids only grants to private parties, partnerships and corporations. Some California cases have held that Article X, § 3 restricts alienation of tidelands to private persons and not to public agencies. *Atwood v. Hammond* (1935) 4 Cal.2d 31, 38; *Cimpher v. City of Oakland* (1912) 162 Cal. 87, 90. Of course, the United States may condemn tidelands. *United States v. Carmack* (1946) 329 U.S. 230, 239; *People of the State of California v. United States* (Ct.Cl. 1955) 132 F.Supp. 208, 211. There is a conflict of authority as to whether the United States, when it condemns tidelands, takes such sovereign lands free of any public trust interest. Compare *United States v. 1.58 Acres of Land* (D. Mass. 1981) 523 F.Supp. 120, 124–125 (take subject to public trust) with *United States v. 11.037 Acres* (N.D. Cal. 1988) No. C-83-4605 JPV (not subject to public trust).

129. *City of Long Beach v. Mansell, supra,* 3 Cal.3d 479:

> . . . It would be contrary to the spirit and purpose of Article XV, section 3 [now Article X, § 3] to conclude that the word "tidelands" as used therein denotes only those public lands which retain the physical characteristics of tidelands at the time of proposed alienation, for such a construction would permit parties to remove public tidelands from the reach of the constitutional provision by simply filling so that such lands were no longer covered and uncovered by the flow and ebb of the tide. It is clear that the framers did not intend to establish a prohibition which could be so easily avoided. We therefore conclude that the word "tidelands" as used in Article XV, section 3, denotes lands which were seaward of the mean high tide line when the provision was adopted in 1879 [or the status of the lands in their last natural condition].

See also *City of Newport Beach v. Fager* (1940) 39 Cal.App.2d 23, 30.

overturned if it runs afoul of this constitutional prohibition.[130] In addition, since 1909, the California State Legislature by statute has prohibited the sale of tidelands whether or not they lay within 2 miles of incorporated cities or towns.[131]

1.10.2 Constitutional Prohibition—Effect on Boundary Line Agreements

An interesting aspect of the constitutional stricture pertains to boundary line agreements sometimes reached by a state and private landowners to resolve title and boundary disputes about lands along or beneath tidal waters. Suppose the area of the boundary line agreement encompasses tidelands within the geographical ambit of the 2-mile prohibition. One may ask whether such an agreement might violate the constitutional prohibition against the sale of tidelands. A California supreme court case provides an answer.

A particular agreement between private and public entities fixed the boundary line between tidelands owned by a public entity and uplands owned by a private entity.[132] The parties accomplished the boundary line through execution of mutual quitclaim deeds. The lands conveyed by the public entities' quitclaim lay within 2 miles of a city. Certain taxpayers claimed that the agreement was an unconstitutional conveyance, violating Article XV, § 3.[133]

The California Supreme Court observed that the purpose of the agreement was not to transfer titles, but ". . . to mark the boundaries of littoral holdings in order to make them certain and permanent and to prevent questions arising in the future concerning the ownership of lands on either side of the line agreed upon."[134] In effect, the court assumed that there was a bona fide boundary dispute and that the boundary line agreement itself established what were and what were not tidelands. Thus, lands on the private party side of the boundary line were not tidelands that were prohibited from being conveyed.[135]

Nonetheless, to ensure that boundary line agreements were not used as a subterfuge to avoid the constitutional prohibition against alienation of tidelands, the court announced certain required specific elements of any valid boundary line agreement:

---

130. In one case, substantial public benefits of harbor improvement would have resulted from a legislatively authorized land exchange agreement. Even so, the court did not approve the agreement because of a potential conflict with the constitutional prohibition against alienation of tidelands. *County of Orange v. Heim* (1973) 30 Cal.App.3d 694, 726.
131. The shore and the bed of the ocean or of any navigable channel or stream or bay or inlet within the State, between ordinary high and low water mark, over which the ordinary tide ebbs and flows is herewith withheld from sale." Cal. Pub. Res. Code § 7991.
132. *Muchenberger v. City of Santa Monica* (1929) 206 Cal. 635, 640.
133. *Id.* at 644–645.
134. *Id.* at 645.
135. *Ibid.*

... When the boundary between public trust tidelands and private uplands is uncertain, and the parties, wishing to fix the boundary in order to prevent future questions of ownership, undertake genuine efforts to determine the true boundary and thereafter agree to a line which fairly represents these efforts, then the subsequent formal "conveyance" in the form of a quitclaim deed ... in furtherance of the boundary agreement does not evidence a "grant or sale" of public ... tidelands with the meaning of article XV, section 3, of the state Constitution.[136]

Courts strictly enforce these requirements (uncertainty as to the location of the boundary, genuine efforts to determine its location, and an agreed boundary line fairly representing those efforts). When it is apparent to the court that a boundary line agreement was merely an effort to draw the boundary to satisfy private owners' desires, no matter where the true boundary might be, a prohibited conveyance may result.[137]

Still, not every state has constitutional or statutory prohibitions on alienation of sovereign lands. Even in California, these constitutional and statutory prohibitions apply only to a certain type of tidelands (tidelands within 2 miles of certain governmental subdivisions). The state may alienate other types of sovereign lands (tidelands not within the 2-mile constitutional prohibition, lands beneath nontidal water bodies, such as lands beneath navigable lakes and rivers) without violating the constitutional or statutory prohibition. Does the nonexistence of statutory or constitutional prohibitions on alienation provide a carte blanche to the state legislatures to permit wholesale disposition of the states' interest in such lands? This question has been the subject of some famous decisions.

1.10.3 Common Law Public Trust—Restriction on Alienation

Recall that state ownership of lands lying beneath tidal or navigable waters is not merely that of a proprietary landowner; the state holds title to such lands as a sovereign in trust for public purposes.[138] That this sovereign trust ownership significantly affects the ability of the state to dispose of such lands

---

136. *City of Long Beach v. Mansell, supra*, 3 Cal.3d at 480.
137. For example, in one case a certain portion of an area included as part of a boundary line agreement resolving some very nasty title and boundary problems had once been part of a navigable channel. The court said: "The indicated portion of the ... agreement certainly does not represent a bona fide effort to establish the *true* boundary between state and proprietary lands; rather it represents an arrangement whereby the boundary is drawn in such a way as to accommodate the homeowners in question *regardless* of where the true boundary line might be." *City of Long Beach v. Mansell, supra*, 3 Cal.3d at 481 (emphasis in original). The implication in the opinion appears to be that since there was no uncertainty as to the location of the former navigable channel, the boundary line was not a good faith effort to figure out location of the true boundary of that channel and adjacent uplands.
138. See notes 111 through 121 and accompanying text in this chapter; *City of Long Beach v. Mansell, supra*, 3 Cal.3d at 482.

is exemplified in a famous United States State Supreme Court case, *Illinois Central Railroad v. Illinois*.[139] In fact, the case provides a direct response to the question of the authority of the legislature to dispose of such sovereign lands without any restriction or inhibition.

The circumstances that produced the case were quite straightforward. The Illinois State Legislature had conveyed the entire waterfront of the City of Chicago to the Illinois Central Railroad Company.[140] The conveyance included nearly the whole of the submerged lands of the harbor lying under the waters of Lake Michigan. Over the years, the railroad proceeded to fill and reclaim such lands.[141]

In an after-the-fact lawsuit by Illinois to undo the conveyance to the railroad,[142] the United States Supreme Court considered whether the Illinois Legislature was ". . . competent to deprive the State of its ownership of the submerged lands in the harbor of Chicago, and of the consequent control of its waters. . . ."[143] Just from the way the judge framed the question, the answer is evident: The legislature could not deprive the state of control of its waters.[144] The Supreme Court's reasoning leading to this holding is critical to one's understanding of the authority of the legislature to dispose of sovereign lands and the continuing authority of the state over such grants or conveyances.

The Supreme Court observed that the state held title to lands underlying waters of Lake Michigan and that this title included control over the waters above such lands.[145] Significantly, the court noted that this title was different in nature and kind from the title by which the state held other lands it intended to sell:

> It is a title held in trust for the people of the State that they may enjoy the navigation of the waters, carry on commerce over them, and have liberty of fishing therein freed from the obstruction or interference of private parties.[146]

That title could not be abdicated or relinquished, except for certain parcels. Such parcels must be used in ". . . promoting the interests of the public therein, or can be disposed of without any substantial impairment of the public

---

139. (1892) 146 U.S. 387.
140. *Id.* at 439–444, 450–451.
141. *Ibid.*
142. *Id.* at 433.
143. *Id.* at 452.
144. *Id.* at 452–453.
145. *Id.* at 452.
146. *Ibid.*

interest in the lands and waters remaining. . . ."[147] In consequence, the Court necessarily held revocable the grant of waterfront lands to the railroad.[148]

As the Supreme Court noted, this was not a unique legal rule. Any novelty arose from the uncommon factual situation rather than from the legal principle being expressed:

> We cannot, it is true, cite any authority where a grant of this kind has been held invalid, for we believe that no instance exists where the harbor of a great city and its commerce have been allowed to pass into the control of any private corporation. But the decisions are numerous which declare that such property is held by the State, by virtue of its sovereignty, in trust for the public. The ownership of the navigable waters of the harbor and of the lands under them is a subject of public concern to the whole people of the State. The trust with which they are held, therefore, is governmental and cannot be alienated, except in those instances mentioned of parcels used in the improvement of the interest thus held, or when parcels can be disposed of without detriment to the public interest in the lands and waters remaining.[149]

The principle stated in this case is known as the common law public trust doctrine. How does this doctrine compare to the constitutional prohibition against alienation of certain tidelands? Does the combination of the common law public trust doctrine and the constitutional limitation completely prevent any alienation of sovereign trust lands? The next section of this chapter discusses these questions.

1.10.4 Common Law Public Trust Compared with Constitutional Prohibition Against Sale of Tidelands

The grant made to the Illinois Central Railroad was voided. This result is much the same as the consequences to grants made in violation of the California constitutional prohibition against the sale of tidelands. In California, however, the state has made grants of tidelands not within the geographic limits of the constitutional prohibition, and courts have upheld such grants.

That courts have upheld these grants accentuates the principal difference between the common law public trust doctrine and the constitutional prohibition of sale of tidelands: The common law public trust does not forbid alienation of tidelands. That doctrine ensures, however, that such lands remain

---

147. *Id.* at 453. ("The State can no more abdicate its trust over property in which the whole people are interested, like navigable waters and the soils under them, so as to leave them entirely under control of private parties . . . than it can abdicate its police powers in the administration of government and the preservation of the peace.")
148. *Id.* at 455–460. This doctrine has continuing vitality in Illinois, as *Lake Mich. Fed. v. U.S. Army Corps of Engineers, supra,* 742 F.Supp. 441, indicates.
149. *Illinois Central, supra,* 146 U.S. at 455–456.

subject to the public trust even after alienation. On the other hand, the constitutional prohibition flatly forbids alienation of tidelands within 2 miles of an incorporated city.[150] Case studies make this difference clear.

One important case applied the common law public trust doctrine to tidelands that were not subject to the constitutional or statutory prohibition against sale. The state sued to quiet California's title to tidally flowed lands.[151] These lands were included within state patents issued under the legislative authority to sell swamp and overflowed lands, salt marsh lands, and tidelands.[152] Although the tidelands concerned were sovereign lands held in trust for the public purposes of navigation and fishery,[153] the California Supreme Court affirmed the power of the state, as part of a plan or system of improvement for the promotion of navigation and commerce, to cut off part of the tidelands and to convey them to private persons or entities, free from public control and use.[154] This was much the same result as obtained in *Illinois Central*.[155]

Notwithstanding that power of disposition, when the court analyzed the statutes authorizing the sale of swamp and overflowed lands, salt marsh lands, and tidelands, it found that those statutes were not at all related to the interests of navigation.[156] In fact, the court found that state approval of the land sales was completely ministerial and the statutes required no determination of usefulness or uselessness of the lands for navigation.[157] These sale statutes applied to immense areas of land, in terms of both acreage and geographic scope.[158] Despite the vastness of the lands affected, the court held that through

---

150. *City of Long Beach v. Mansell, supra*, 3 Cal.3d at 482; *County of Orange v. Heim, supra*, 30 Cal.App.3d at 709.
151. *People v. California Fish Co., supra*, 166 Cal. 576.
152. *Id.* at 582; see Sections 1.7 and 1.9 in this chapter.
153. *People v. California Fish Co., supra*, 166 Cal. at 584. ("A public easement and servitude exists over these lands for those purposes.")
154. *Id.* at 585. ("The most striking instance of the exercise of this power of absolute disposition of such tide or submerged lands by the state of California is found in the laws providing for the improvement of the water-front of San Francisco.")
155. See notes 147 through 149 and accompanying text in this chapter.
156. *People v. California Fish Co., supra*, 166 Cal. at 589–590.
157. *Id.* at 590–591. The court said:

> . . . The [State] surveyor general's approval of the application and survey was necessary. But this requirement applied only to the form of the survey and the application and the qualifications of the applicant. It did not empower him to reject the application on the ground that the land was not suitable for cultivation, or that it was needed for navigation, or that the sale to private use would interfere with or destroy the public easement to which such land is dedicated. The applicant determined what land he desired to buy, he caused it to be surveyed . . . , he made his application and, if he was a person qualified to buy and his proceedings were regular in form, he . . . became entitled to complete the purchase and could compel the officers of the state to execute the title papers necessary to convey it to him, on the payment of the fixed price of one dollar an acre." *Id.* at 591.

158. "The tide lands embraced in these statutes . . . includes the entire sea beach from the Oregon line to Mexico and the shore of every bay, inlet, estuary, and navigable stream as far up as tide water goes and until it meets the lands made swampy by the overflow and seepage of fresh water streams." *Id.* at 591.

such legislation and sale the state did not intend to abdicate its public trust responsibility for such lands.[159]

The court reasoned as follows: The statutes providing for the sale of swamp and overflowed lands, salt marsh, and tidelands for purposes of reclamation did not provide for any means by which the lands fit only for reclamation and not covered by tidal waters could be separated from those on the shores of estuaries, bays, and beaches.[160] Thus, the court recognized that within conveyances made pursuant to those statutes there were lands that the state could have sold free of any public interest and lands that it could not. Should the entire conveyance be revoked, as in *Illinois Central*, or only part of the conveyance, and, if so, how would it be decided as to what part of the conveyance to void? What was the solution?

In this situation the court observed that, as to tidelands, the state had a double right: a public right and a private right.[161] Construing these land disposition statutes,[162] the California Supreme Court found that the Legislature intended the statutes only to dispose of the private right of the state, leaving the public right unimpaired.[163] In effect, the California Supreme Court permitted conveyance of the state's fee title interest, but held that such fee title interest remained subject to the state's public trust easement that the state could not sell.[164]

---

159. It is not to be assumed that the state, which is bound by the public trust to protect and preserve this public easement and use, should have intentionally abdicated the trust . . . and should have directed the sale of any and every other part of the lands along the shores and beaches to exclusive private use, to the destruction of the paramount public easement, which it was its duty to protect, and for the protection and regulation of which it received its title to such lands." *Ibid.*
160. *Id.* at 592. This failure provided evidence that there was no intention to deal with or affect the public interest in commerce, navigation, and fishing.
161. *Id.* at 593. The court quoted another case that had earlier recognized this double right:

> The several states hold and own the lands covered by navigable waters . . . in their sovereign capacity, and primarily for the purpose of preserving and improving the public rights of navigation and fishery. They have in them a double right, a *jus publicum* and a *jus privatum*. The former pertains to their political power—their sovereign dominion, and cannot be irrevocably alienated or materially impaired. The latter is proprietary and the subject of private ownership, but it is alienable only in strict subordination to the former. *Oakland v. Oakland Water Front Co.* (1897) 118 Cal. 160, 183.

162. The court noted the rule of statutory construction, which holds that a court will not construe a statute to impair or limit the sovereign power of the state to act for the public unless such an intent clearly appears. *People v. California Fish Co., supra,* 166 Cal. at 592. Indeed, later in the opinion the court embellished on that principle:

> . . . [S]tatutes purporting to authorize an abandonment of such public use will be carefully scanned to ascertain whether or not such was the legislative intention, and that intent must be clearly expressed or necessarily implied. It will not be implied if any other inference is reasonably possible. And if any interpretation of the statute is reasonably possible which would not involve a destruction of the public use or an intention to terminate it in violation of the trust, the courts will give the statute such interpretation." *Id.* at 597.

163. *Id.* at 593–594; see notes 173 through 174 and accompanying text in this chapter.
164. The court said:

Courts have applied this rule in more contemporary times. The case[165] dealt with the validity of long-ago-authorized sales of lands beneath the waters of San Francisco Bay. The California Supreme Court in a prior decision had upheld the statutes authorizing such sales and the resulting conveyances.[166] Despite vigorous arguments in support of the earlier case, the California Supreme Court overruled it.[167] The landowners also unsuccessfully argued that the court's long recognition of these grants was a "rule of property."[168]

In recognition of the unprecedented effect of its decision,[169] apparently as a matter of judicial policy, the court chose not to give its decision full retroactive effect.[170] Instead, the court that held still-submerged lands and lands that remained subject to tidal action would continue to be subject to the public trust; lands not subject to tidal action and that had been filled, whether or not substantially improved, would be held free of the trust if the fill and improvements were made pursuant to applicable land use regulations.[171] The modern application of the rule has not been limited to tidelands; lands beneath waters of navigable lakes have also been held to be subject to this principle.[172]

1.10.5 Effect of the Existence of the Public Trust Easement

It is important to understand the nature of the public right retained by the state. To the extent a conveyance pursuant to statutes authorizing the sale of

---

... [T]he buyer of land under these statutes receives the title to the soil, the *jus privatum* [the private right], subject to the public right of navigation, and in subordination to the right of the state to take possession and use and improve it for that purpose, as it may deem necessary. In this way the public right will be preserved and the private right of the purchaser will be given as full effect as the public interests will permit." *Id.* at 596.

165. *City of Berkeley, supra,* 26 Cal.3d at 532.
166. *Knudson v. Kearny* (1915) 171 Cal. 250.
167. *City of Berkeley, supra,* 26 Cal.3d at 532.
168. As stated by the court, a "rule of property"

... relates to a "settled rule or principle, resting usually on precedents or a course of decisions, regulating the ownership or devolution of property. . . . [D]ecisions long acquiesced in, which constitute rules of property . . . should not be disturbed. . . ." *Id.* at 532 (citing *Abbott v. City of Los Angeles* (1958) 50 Cal.2d 438, 456–457).

169. After all, the California Supreme Court had to overrule one of its previous decisions, a decision that had been in unquestioned existence for more than 50 years.
170. *City of Berkeley, supra,* 26 Cal.3d at 534. In reaching that determination the court analyzed the alternatives. It could have given its decision full retroactive effect, which would have reduced the value of investments made in reliance on decisions it overruled. Or it could have made the effect of the decision prospective only, which would have made the holding an academic exercise, as the sales were made 100 years before the decision. *Ibid.*
171. *Ibid.* The basis for this decision was that the interests of the public were considered paramount in lands "still physically adaptable for trust uses," while grantees would prevail in cases where tidelands had been "rendered substantially valueless for those purposes." *Ibid.*
172. *State of California v. Superior Court (Lyon)* (1981) 29 Cal.3d 210, 226–232, *cert. den., sub nom. Lyon v. California* (1981) 454 U.S. 865.

swamp and overflowed lands, salt marsh, and tidelands, contains tidelands, the grantee receives fee title to the land. However, that fee title is not absolute. The owner holds the title ". . . subject to the easement of the public for the public uses of navigation and commerce, and to the right of the state, as administrator and controller of these public uses and the public trust therefor, to enter upon and possess the same for the preservation and advancement of the public uses and to make such changes and improvements as may be deemed advisable for those purposes."[173] Consequently, as the United States Supreme Court later noted: "Although the landowner retains legal title to the property, he controls little more than the naked fee, for any proposed private use remains subject to the right of the State or any member of the public to assert the State's public trust easement."[174]

Although the federal and state governments' grand land disposition schemes resulted in passage of title from the sovereigns to vast amounts of acreage in or adjacent to tidal or navigable waters, the extent, nature, and quality of title passing into private ownership may still be at issue. This potential insecurity of title is one of the principal reasons for this book.

1.11 ADVERSE POSSESSION

Not every landowner holds title originally stemming from a government patent. There is another, somewhat unusual, source of title not involving a government patent as parent title: the common law and statutorily acknowledged doctrine of adverse possession. This doctrine recognizes title in someone other than the holder of record title.

Adverse possession is created vis-à-vis the holder of record title when an adverse claimant, for the period required by the applicable statute of limitations, continuously and without interruption, has occupied the land under claim of right or color of title in an open, notorious, and hostile manner and paid the taxes on the land.[175]

---

173. *People v. California Fish Co., supra,* 166 Cal. at 599. The California Supreme Court went on to observe:

> It is to be understood, of course, that if the state has sold tide lands subject to the public easement, prior to making improvements thereon for navigation, as it may do, it cannot, upon putting in force a plan for the subsequent improvement thereof, retake absolute title without compensation. The purchaser will continue to hold [his land], after as well as before the improvement, and if at any time the public easement and use is lawfully and permanently abandoned, the purchaser will then have the absolute and complete estate in the land. And before such improvement is made by the state, he may use the property as he sees fit, subject to the power of the state to abate any nuisance he may create thereon, and to remove any purpresture he may erect thereon. *Ibid.*

174. *Summa, supra,* 466 U.S. at 205.
175. Clark 7th, *supra,* § 22.16; Clark 4th, *supra,* at § 531; Skelton, *The Legal Elements of Boundaries and Adjacent Properties* (1930) § 335 (hereinafter "Skelton"); 1 Bowman, *Ogden's*

This work does not pretend to be a treatise on the doctrine of adverse possession. Commentators such as Clark, Skelton, and Bowman provide detailed discussions and further references that the reader may consult concerning adverse possession. Such sources and each state's statutory and case law should be reviewed to form a complete picture of the elements necessary to establish title by adverse possession. Brief discussions analyzing specific elements of adverse possession, however, are presented in the following sections.

1.11.1 Color of Title

The requirement that adverse possession be under "color of title"[176] raises many interesting possibilities. To state just one example, a person may contend that possession is under color of title derived from a deed from a defunct corporation, but executed by a former officer of that corporation. A corporation can convey title only when authorized by its board of directors to convey property on behalf of the corporation.[177] Even a void deed may, however, constitute color of title if it is accepted in good faith.[178] If the corporation did not authorize the corporate officer who executed the deed for the corporation to do so, or the corporation itself had no title to the land its deed purported to convey, the deed would not affect title, would not provide notice, and would not be evidence of hostile possession required to establish one of the elements of adverse possession.[179] Moreover, because possession under color of title must be in good faith, the element of good faith would surely be lacking where claimants were aware that the person executing the deed had no title to the property.[180]

Revised California Real Property Law (1974) § 4.6, p. 120 (hereinafter "Bowman"). For those who are interested, there are many excellent discussions on the origin of the doctrine. E.g., Clark 7th, *supra*, § 22.02; Clark 4th, *supra*, at § 527.

176. The term "color of title" means a written instrument that attempts to convey, but through some defect does not pass, title. Clark 7th, *supra*, § 22.06; Clark 4th, *supra*, at § 540, p. 709; 1 Bowman, *supra*, § 4.8, pp. 122–123; *Wright v. Mattison* (1855) 59 U.S. (18 How.) 50. On the other hand, the term "claim of right" means the objective acts of the adverse claimant showing an intent to claim the land against all. Clark 7th, *supra*, § 22.07, 727–728; Clark 4th, *supra*, at § 538, pp. 699–700; 6 Miller and Starr, California Real Estate 3d, § 16.3.

177. E.g., Cal. Corp. Code § 1002; 3 Miller and Starr, California Real Estate 3d, § 8.30.

178. Clark 7th, *supra*, § 22.06, p. 725; Clark 4th, *supra*, at § 540, p. 709; *Johns v. Scobie* (1939) 12 Cal.2d 618, 626 (failure to describe property in deed).

179. E.g., Clark 7th, *supra*, § 22.06, p. 726; Clark 4th, *supra*, at § 540, p. 710; *Hager v. Specht* (1878) 52 Cal. 579, 583–584 (deed made by attorney-in-fact who had no authority); *City of Los Angeles v. Morgan* (1951) 105 Cal.App.2d 726.

180. *Yuba River Sand Co. v. Marysville* (1947) 78 Cal.App.2d 421 (claimants who were relying on deed by heirs of one who had no title leased property from true owners).

1.11.2 Character of Possession

If adverse possession is based on color of title, proof of possession is somewhat less difficult. Thus, when adverse possession is based on color of title, the element of occupation is satisfied not only when the claimant has cultivated, enclosed, or improved the land, but also when the claimant has used the land for a pasture, to supply fuel, or for the claimant's ordinary use.[181] On the other hand, when the claimant has not based adverse possession on color of title, but rather on claim of right, in California the adverse claimant must establish that he or she has protected the land by a substantial enclosure or has improved or cultivated the land.[182]

In California and other western states the character of possession necessary to satisfy the element of adverse possession of unoccupied land is an important particular. The question is whether recreational use of lands will satisfy the element of possession. A California court held that use of mountain property on weekends and during the summer for recreational activities and gold mining was sufficient use to establish the element of possession.[183] In other words, to satisfy the requirements of continuous and uninterrupted occupancy, the use and occupancy required need only be appropriate to the location and character of the property.[184] Nevertheless, at least one court found recreational use of property (for picnics, visiting on weekends, sometimes camping overnight) was, as a matter of law, not sufficient to give notice to anyone of a hostile claim.[185]

1.11.3 Payment of Taxes

Another of the elements necessary to establish adverse possession is payment of all taxes that the government has levied and assessed on the property for the required period.[186] In this post-Proposition 13 age, it is difficult to believe that anyone would voluntarily pay such taxes. Some persons have gone so far as having a duplicate assessment established by the local taxing agency in order to pay taxes on land they wanted to possess adversely. Courts take a dim view of such efforts. They hold that one cannot assert adverse possession only on the basis that he or she has paid all taxes levied and assessed for the requisite period.[187] The reasoning behind this rule is that if an adverse

---

181. *Posey v. Bay Point Realty Co.* (1932) 214 Cal. 708, 710–712; 1 Bowman, *supra*, at § 4.8, p. 123.
182. Cal. Code of Civ. Proc. § 324; *Safwenberg v. Marquez* (1975) 50 Cal.App.3d 301, 310.
183. *Newman v. Cornelius* (1970) 3 Cal. App. 3d 279, 290.
184. *Posey, supra*, 214 Cal. at 712; *Kellogg v. Huffman* (1934) 137 Cal. App. 278.
185. *Klein v. Caswell* (1948) 88 Cal. App. 2d 774, 778–779.
186. Clark 7th, *supra*, § 22.16, p. 748; Clark 4th, *supra*, at § 531, p. 686; 1 Bowman, *supra*, at § 4.7, p. 122; 6 Miller and Starr, California Real Estate Law 3d, § 16.23.
187. *Carpenter v. Lewis* (1897) 119 Cal. 18, 24.

claimant can rest on his or her payment of taxes, even though the record owner has already paid taxes, the record owner is powerless to protect him- or herself at all against an adverse claimant, short of bringing a lawsuit to quiet his or her title.[188]

1.12 AGREED BOUNDARIES

Adverse possession is not the only method by which a person can acquire a title that does not originally stem from a grant from the government. Another court-developed source of title is the doctrine of agreed boundaries or acquiescence.

Under this doctrine a property boundary may be located at a different physical location than called for in an owner's vesting documents by virtue of the claim of a different contended physical location by one owner and the acquiescence in that contended location by the adjoining landowner.[189] This doctrine is a venerable one and has even been used to establish the political boundaries of states.[190] In effect, this doctrine establishes a new title.[191] Of course, there must be some notice of the adverse claim and the extent of that claim.[192]

California courts appear reluctant to apply the agreed boundary doctrine. This reluctance apparently stems from considerations of equity and policy.[193]

---

188. *Van Calbergh v. Easton* (1917) 32 Cal.App. 796, 799.
189. Clark 7th, *supra*, § 20.02; Clark 4th, *supra*, at § 491, p. 635; *Bryant v. Blevins* (1994) 9 Cal.4th 47, 54; *Allen v. McMillion* (1978) 82 Cal. App. 3d 211, 214–215 ("The requisites for an agreed boundary are: (1) uncertainty as to the true boundary; (2) an agreement between coterminous owners as to the true boundary; (3) acquiescence to the line so fixed for a period equal to the statute of limitation [for adverse possession]; and (4) the boundary so fixed must be identified on the ground"). *Appel v. Berrman* (1984) 159 Cal.App.3d 1209, 1213; *Mesnick v. Caton* (1986) 183 Cal.App.3d 1248, 1255; *Fogerty v. State of California* (1986) 187 Cal.App.3d 224, 236, *cert. den.* (1987) 484 U.S. 821.
190. *Indiana v. Kentucky* (1890) 136 U.S. 479, 510 ("It is a principle of public laws universally recognized, that long acquiescence in the possession of territory and in the exercise of dominion and sovereignty over it, is conclusive of the nation's title and rightful authority."); *Arkansas v. Tennessee* (1940) 310 U.S. 563, 568. Most recently, the Supreme Court applied the doctrine in the contest over the boundary between New York and New Jersey about Ellis Island. *New Jersey v. New York* (1998) 523 U.S. 767, 140 L.Ed.2d 993, 1014–1015.
191. In contrast to the doctrine of agreed boundaries is that of prescriptive easements. The elements of a prescriptive easement are (1) open and notorious use; (2) continuous and uninterrupted use; (3) hostile to the true owner; (4) under claim of right; and (5) for the period of the statute of limitations, five years. Cal. Civ. Code § 1007; Cal. Code of Civ. Proc. § 321; *Otay Water District v. Beckwith* (Cal. 1991) 1 Cal.App.4th 1041, 1045. The difference between the two doctrines is in the intent of the parties. The agreed boundary doctrine is applicable when there is confusion between the two property owners as to the boundary and the intent by both is to claim only to the true line. In the case of prescription, there is no confusion as to the boundary line; the claiming party intends to claim the property as his or her own. *Otay, supra*, 1 Cal App.4th at 1045; *Mehdizadeh v. Minier* (1996) 46 Cal.App.4th 1296, 1305–1306.
192. *Fogerty, supra*, 187 Cal.App.3d at 236–237.
193. "The common theme of [California case law on the agreed-boundary doctrine] is a deference to the sanctity of true and accurate legal descriptions and a concomitant reluctance to allow

1.13 ESTOPPEL

A similar doctrine creating title without benefit of a government conveyance is the principle of equitable estoppel. Where one has led innocent third parties to believe that one has title with the full power of disposition, an innocent party's title will be upheld against a challenge by the true landowner.[194] In general, the application of the doctrine of equitable estoppel requires the presence of four elements: (1) the party to be estopped is informed of the facts, (2) that party intends that another party will rely on its conduct or acts so the party asserting the estoppel has the right to believe the conduct was so intended, (3) the party asserting estoppel is ignorant of the true state of facts, and (4) that party relies on that conduct to its injury.[195]

In the particular instance of land title, however, courts have been very cautious in applying the doctrine.[196] Consequently, the courts have added a crucial refinement: The doctrine will not be applied to divest title to land in the absence of actual or constructive fraud by the person to be estopped.[197]

1.14 INAPPLICABILITY OF ADVERSE POSSESSION, ESTOPPEL, AND LIKE DOCTRINES TO SOVEREIGN LANDS OR LAND HELD IN TRUST

However beneficial the doctrine of adverse possession is in encouraging landowners to use their land, its scope appears to be limited to lands owned by and property disputes between private parties. The doctrine of adverse possession and like doctrines, such as laches[198] or estoppel, have no application

---

such descriptions to be invalidated by implication, through reliance upon unreliable boundaries created by fences or foliage, or by other inexact means of demarcation." *Bryant, supra,* 9 Cal.4th at 55.

194.   1 Bowman, *supra,* at § 4.19; Skelton, *supra,* § 313 et seq.

195.   *City of Long Beach v. Mansell, supra,* 3 Cal.3d at 489. "Whenever a party has, by his own statement or conduct, intentionally and deliberately led another to believe a particular thing true and to act upon such belief, he is not, in any litigation arising out of such statement or conduct, permitted to contradict it." Cal. Evid. Code § 623.

196.   This caution stems from the rule known as the Statute of Frauds. The Statute of Frauds requires that certain agreements concerning real property be in writing. E.g., Cal. Civ. Code §§ 1091, 1624. Application of the doctrine of estoppel would take title from one person and vest it another without there being any writing evidencing that change of title. *City of Long Beach v. Mansell, supra,* 3 Cal.3d at 489.

197.   *Id.* at 489–490. In construing this refinement the California Supreme Court observed: "It thus appears that the doctrine of equitable estoppel applied to questions of land title . . . differs from that applied to questions involving other matters only in that the culpability of the party to be estopped must be of sufficient dimension that actual or constructive fraud would result if the estoppel were not raised." *Id.* at 491.

198.   Laches is an unreasonable delay in bringing suit plus acquiescence in the act complained about or prejudice to the other party because of the delay. *City of Coachella v. Riverside County Airport Land Use Com.* (1989) 210 Cal.App.3d 1277, 1286.

in the case of sovereign lands and interests held in trust for the public or otherwise devoted to a public use.[199] One particularly interesting situation will illustrate the remarkable force of this exception to the doctrines of adverse possession and estoppel.

In 1945, shortly after the end of the Second World War, the federal government fully recognized the strategic and potential economic value in the oil residing beneath lands underlying the offshore waters of the Pacific Ocean along the coast of California. President Truman declared those offshore oil reserves the property of the United States.[200] This proclamation ran counter to the long-standing belief that those oil-bearing lands belonged to the State of California and not the United States.

The Solicitor General, on behalf of the United States, filed a case in United States Supreme Court against the State of California.[201] The lawsuit sought to quiet California's claims against the United States' proclaimed title in such lands. That litigation has been somewhat misleadingly referred to as the "tidelands cases."[202]

California claimed, among other matters, that it had adversely possessed under color of title the submerged lands lying off its coast. As a consequence, California urged the United States Supreme Court to find that if the United States had any interest in such lands, it had lost its title to California by virtue of California's adverse possession.[203] California recited a long litany of acts to establish this possession: Among other things, California established that the United States had recognized the state's title, even to the extent of acquiring title to portions of the seabed from California and refusing to issue federal oil leases in that area on the ground that California, not the United States, owned the seabed.[204] It was all to no avail. In a passage that encapsulates the reasoning of many courts, the United States Supreme Court said:

---

199. E.g., *United States v. California* (1947) 332 U.S. 19, 39; *In re Water of Hallett Creek Stream System*, supra, 44 Cal.3d at 459; *County of Lake v. Smith* (1991) 238 Cal.App.3d 214, 229; *People v. Shirokow* (1980) 26 Cal.3d 301, 311; *People v. Chambers* (1951) 37 Cal.2d 552, 557; *Patton v. City of Los Angeles* (1915) 169 Cal. 521, 527; Cal. Civ. Code § 1007; Clark 7th, supra, § 22.13; Clark 4th, supra, at § 545, pp. 725–726.
200. *United States v. California*, supra, 332 U.S. at 34, n. 8.
201. It is unusual to begin a case in the Supreme Court, except in certain very narrowly circumscribed instances. U.S. Const., Art. III, § 2 provides for such jurisdiction when a state is a party. Although the Supreme Court has jurisdiction, it must accept that jurisdiction only in cases between two states. 28 U.S.C. § 1251. In cases between the United States and a state, the Supreme Court's jurisdiction is not exclusive, but concurrent; United States District Courts also have jurisdiction regarding such disputes. 28 U.S.C. § 1345.
202. The United States claimed only lands that lay *seaward* of the low-water mark. Thus, tidelands, which lay *landward* of the low water mark, technically were not in issue.
203. *United States v. California, supra*, 332 U.S. at 39.
204. *Ibid*. For a student of California history or for any person interested in how a lawyer ties historical facts into legal argument, California's brief in this case is worthy of exploration. It may be found in any law library that is a repository of United States Supreme Court records. The brief

1.14 INAPPLICABILITY OF ADVERSE POSSESSION, ESTOPPEL, AND LIKE DOCTRINES 41

> ... And even assuming that Government agencies have been negligent in failing to recognize or assert claims of the Government at an earlier date, the great interests of the Government in this ocean area are not to be forfeited as a result. The Government, which holds its interests here as elsewhere in trust for all the people, is not to be deprived of those interests by the ordinary court rules designed particularly for private disputes over individually owned pieces of property and officers who have no authority at all to dispose of Government property cannot by their conduct cause the Government to lose its valuable rights by their acquiescence, laches, or failure to act.[205]

Other courts have subscribed to or followed this rule.[206] The reader should take careful note, however. This exception to the doctrines of adverse possession and estoppel may not apply with equal force against the United States when it asserts title to such sovereign trust lands.

The state and the United States disputed ownership of the bed of a river in North Dakota. The United States claimed it had adversely possessed the bed of the river vis-à-vis the State of North Dakota, and that the State of North Dakota knew of the claim and was now barred from proving its title against the United States by virtue of a federal statute of limitations.[207] Other courts construing that same federal statute had held that under such circumstances a state could have its day in court. Such courts relied on, among other matters, the exception to the doctrine of adverse possession for lands held in public trust.[208] When the case reached the United States Supreme Court, the Court upheld the federal statute (no surprise) and did not allow the State of

---

also recited other examples to support California's claim of adverse possession. The United States' brief was equally well written and obviously was more persuasive.

205. *United States v. California, supra*, 332 U.S. at 39–40. A California court has also discussed the different attributes of public and private land ownership:

> Private and government ownership of land obviously differ in certain fundamental respects. Unlike private property, government land usually cannot be condemned or seized, or obtained by adverse possession. A private owner holds property under local law; the government holds and controls land by virtue "of its own fiat" and may change the rules of ownership at will. [Citation omitted.] In this respect, to speak of government "ownership" is really to describe a variety of governmental *powers* over the regulation and control of the property it holds. *In re Water of Hallett Creek Stream System, supra*, 44 Cal.3d at 459.

206. Skelton, *supra*, § 347; *United States v. Georgia Pac.* (9th Cir. 1970) 421 F.2d 92, 100–101; *Weber, supra*, 85 U.S. at 68–70; *Carr v. Moore* (Iowa 1903) 93 N.W. 52, 54–55. In California, the leading case is *Patton v. City of Los Angeles, supra*, 169 Cal. at 527. It may be that this rule is eroding. In *New Hamsphire v. Maine* (2001)__U.S. __, __ S.Ct. __, __ LE.2d __, 2001 U.S. Lexis 3981, the U.S. Supreme Court applied judicial estoppel against a state asserting title to claimed sovereign trust lands. See note 221 in this chapter.

207. *North Dakota v. Andrus* (D. N. Dak. 1981) 506 F. Supp. 619, 624; *aff'd.* (8th Cir. 1982) 671 F.2d 271; overruled *sub nom., Block v. N.D. ex rel. Bd. of Univ. and Sch. Lands* (1983) 461 U.S. 273.

208. E.g., *State of Cal., etc. v. United States* (N.D. Cal. 1981) 512 F. Supp. 36, 42.

North Dakota to present its title claim against the United States. The Supreme Court acknowledged, however, the force of the exception to the doctrine of adverse possession. It held that the state would not lose title to the lands, only that the State would hold its title claim in abeyance.[209]

Note that the same principles do not apply to private[210] ownerships. Because much of the land in the western United States is federally owned, however, the careful surveyor and lawyer should be aware of this exception.

1.15 EXCEPTION: WHEN ESTOPPEL MAY BE APPLICABLE TO THE SOVEREIGN

There is a further consideration concerning the applicability of the doctrine of estoppel to sovereign lands. In a remarkable case, the state claimed, based on the doctrine of equitable estoppel, that public entities should be prevented from asserting their public trust title claim vis-à-vis many private home owners.[211] The case was unusual because public entities *were seeking to estop themselves*.[212]

The public entities argued that they had filled and improved the lands with the knowledge and agreement of the public entities. In addition, these entities accepted complete authority and jurisdiction over the area, including the granting of various regulatory approvals, constructing and maintaining public improvements (streets), and collecting property taxes. The public entities claimed it would be unjust to persons who had reasonably relied on the public entities' conduct to allow the public entities now to assert a claim of paramount title. These entities insisted that the court should apply the doctrine of equitable estoppel to prevent them from asserting the public title claim.[213]

Finding an estoppel was not difficult in such a case, even including the required element of constructive fraud. The public entities were aware of and had been in a position to resolve the title problems, but had not done so. In fact, the public entities conducted themselves as if no title problems existed.

---

209. *Block v. N.D. ex rel. Bd. of Univ and Sch. Lands* (1983) 461 U.S. 273, 291–292.
210. *United States v. California, supra,* 332 U.S. at 39–40; *In re Water of Hallett Creek Stream System, supra,* 44 Cal.3d at 459.
211. *City of Long Beach v. Mansell, supra,* 3 Cal.3d at 488.
212. *Id.* at 487. In addition, this was an unusual case because the existence and extent of the public interest in the lands were not issue—the case was not a quiet title case. ". . . [I]n such a proceeding arguments of estoppel would be properly advanced only by those who would be injured if paramount title were established." *Ibid.* While not strictly at issue in the case, the court agreed to consider the estoppel issue. The reasons the court gave for its decision to consider the estoppel question are notable: ". . . [T]o perversely turn our backs on equitable considerations which we could not ignore had they been molded in the ordinary matrix of quiet title actions, would be to annihilate the basic concept of [the settlement statute] that some solution other than massive litigation is required for . . . [such] title problems." *Id.* at 488.
213. *Id.* at 487.

By so doing, the public entities misled thousands of home owners. In these circumstances, the court deemed conduct of the public entities "so culpable that fraud would result if an estoppel were not raised."[214]

Even in these extreme circumstances and even considering that all elements of equitable estoppel were present, the California Supreme Court was still acutely aware that it was being asked to apply the doctrine of estoppel to prevent the state from asserting title to lands that it had held in public trust and that were subject to a constitutional prohibition against alienation.[215] The court was also aware that estoppel is not usually available to be applied against a public entity if to do so would negate a policy for the benefit of the public.[216]

The court resolved the tension between these competing doctrines by holding that the government may be bound by an equitable estoppel when the elements requisite to such an estoppel against a private party are present, *and* ". . . the injustice that would result upon a failure to uphold an estoppel [is] of sufficient dimension to justify any effect upon public interest or policy which would result from the raising of an estoppel."[217]

The California Supreme Court had no difficulty in finding that a ". . . manifest injustice[218] would result if the very governmental entities whose conduct has induced . . . reliance were permitted at this late date to assert a successful claim of paramount title and thereby wrest property from the homeowners who [had] settled upon it."[219] The court emphasized the uniqueness of the case, noting the "rare combination of government conduct and extensive reliance . . . [that would] create an extremely narrow precedent."[220] There have been no other California cases with such an unusual combination of circumstances.[221]

---

214. *Id.* at 492. The court also "had no difficulty" in finding that the home owners had no convenient or ready means of ascertaining the true boundaries between the public or private lands. *Ibid.*
215. *Id.* at 493. These lands were tidelands within 2 miles of an incorporated city.
216. *Ibid*; see note 205 and accompanying text in this chapter.
217. *Mansell, supra*, 3 Cal.3d at 496–497. The court quickly disposed of the argument that estoppel could not be applied if the act being permitted was alienation of tidelands in violation of the constitution. The court had held that, in administration of the trust, the state may find it advisable, even necessary, to cut off certain tidelands from the ebb and flow of the tide and by that render them useless for trust purposes. *Id.* at 482. Thus, since the state does have the power to terminate the trust, it was not necessary for the court to decide this issue. *Id.* 499.
218. The supreme court observed that the governmental entities had conducted themselves as if the lands in question were private property free of any public trust claims and ". . . thousands of citizens have settled upon these lands with the same expectation as settling on other lands within in the city." *Id.* at 499.
219. *Ibid.*
220. *Id.* at 499.
221. There have been other unique cases where one may claim a court-applied estoppel against the government, although not in a land title or boundary context. See *Rank v. Krug* (S.D. Cal. N.D. 1950) 90 F.Supp. 773, 804. And one could argue that a later California case applied estoppel against the state in the establishment of the ordinary low-water mark boundary of a navigable

Given the almost infinite variety of ways in which government can interact with lands and owners of lands adjacent to or beneath tidal or navigable waters, it is safe to say that the last word on estoppel has not yet been written.

1.16 CHAINS OF TITLE

Methods and means of acquiring title and derivation of title are important to the surveyor, the lawyer, and the property owner for many reasons. Not only does the source of title limit what may be owned by the grantee and successors thereto, but certain of the instruments transferring title or other interests in land, such as easements, may provide useful information concerning the boundaries of the land. This information can be derived by obtaining or constructing a chain of title.[222]

Examination of the instruments in the chain of title will sometimes provide information not only on what is the original source of title, but also on the extent of title conveyed by each successive instrument that stems from that original source.[223] Especially for land along or adjacent to tidal or navigable water bodies, descriptions contained in such instruments may provide valuable information on the location of particular physical features or objects such as the shore of the ocean or the bank of a river.[224] Further, the type of government patent through which an owner claims title may also have an impact on water rights.[225]

Knowing the source of title is important for yet another reason. The source of title may have a substantial effect on what law will decide any contest over title to or boundaries of lands adjacent to or underlying tidal or navigable waters. This rule, known as the "choice-of-law," will be explored in the next

---

lake. *County of Lake, supra*, 238 Cal.App.3d at 236 (". . . [e]xtensive private reliance on, and general recognition of the [lake level] and its endorsement by the state have only served to strengthen the case for [a particular lake level], whether viewed under the rubrics of relevant precedent, estoppel, or detrimental reliance"). Interestingly, in a case of first impression, the United States Supreme Court applied the doctrine of judicial estoppel against a state claiming title to sovereign land. *New Hampshire v. Maine*, No. 130 Original __ U.S. __, __ S. Ct. __, __ L.Ed.2d __, 01 C.D.O.S. 4303 May 29, 2001. The Court noted that estoppel normally does not apply to states. New Hampshire had, however, persuaded the Court in an earlier original case to accept one interpretation of the location of its river boundary with Maine. It was now urging an inconsistent interpretation to gain "additional advantage at Maine's expense." In this case between two competing states, the Court applied the doctrine of estoppel. "[I]t is a case between two states in which each owes the other a full measure of respect."

222. See note 5 in this chapter.
223. E.g., *Com. v. Morgan* (Va. 1983) 303 S.E.2d 899, 901.
224. For an example of the practical use of documents in a chain of title in an actual case, see note 97 and accompanying text in Chapter 6.
225. *California Oregon Power Co. v. Beaver Portland Cement Co.* (1935) 295 U.S. 142, 162; *Ickes v. Fox* (1937) 300 U.S. 82, 95; *Federal Power Com. v. Oregon, supra*, 349 U.S. at 448 (water rights severed from grants pursuant to the Desert Land Act).

chapter. Fair warning is here given. No matter how technical or boring the topic "choice-of-law" may appear at first blush, the reader will have underestimated. The boredom quotient of the discussion of choice-of-law however, is directly proportional to its impact on a client's land ownership. Forewarned is forearmed.

2

WHAT IS THE CHOICE-OF-LAW?

2.1 INTRODUCTION

Many law schools offer a course with the intriguing name "Conflict of Laws." Prior to reviewing the course description, law students may believe they are about to enter an exciting realm, featuring bitterly fought battles pitting one law against another. Imagine the disappointment when these students peruse a course outline that features some of the more abstruse legal doctrines such as *renvoi* or *lex loci delicti*. Although such concepts may seem either arcane or picayune (or both), one branch of this field should be of particular interest to surveyors and property owners. This branch has particular pertinence to title and boundary disputes concerning lands adjacent to or beneath tidal or navigable water bodies.

This arm of conflict of laws, known as choice-of-law, contains rules the legal system developed to decide which body of law governs a particular boundary or title dispute. To put it in nontechnical terms, this body of law grew out of the legal system's attempt to solve the problem that would arise as each of the 50 states, as well as the United States, developed separate bodies of law to resolve land title and property boundary disputes.[1] When land title or property boundary disputes arise between (1) states, (2) land-

---

1. It was long assumed that application of federal law to such disputes was quite limited. E.g., *State Land Board v. Corvallis Sand & Gravel Co.* (1977) 429 U.S. 363, 375–381. The United States Supreme Court, quoting from an earlier decision, observed:

    "The great body of law in this country which controls acquisition, transmission, and transfer of property, and defines the rights of its owners in relation to the state or to private

owners residing in different states, over lands that had been in one state and now lie in another, or (3) the federal government and a state, courts asked to hear such matters must first decide which body of law governs the dispute. The rules used by courts for making that decision are known as choice-of-law rules. Conflict of laws is a much broader topic, with international law implications,[2] and is beyond the scope of this work.[3]

Choice-of-law questions arise in title or boundary disputes over land along tidal or navigable waterways for two related reasons. First, contesting landowners may derive their titles from different sources. Thus, a choice-of-law issue may be presented to a federal or state court when a boundary dispute arises between states as owners of the beds of tidal or navigable water bodies and the federal government or private persons (holding title by virtue of a federal patent) as owners of upland.[4] The states own the beds of tidal or navigable water bodies by virtue of their sovereignty as states of the Union and claim application of state law to decide title or boundary questions. On the other hand, the federal government or private landowners deriving title from the United States, may insist upon application of federal law to decide these questions. Second, these contests can arise in both state and federal court. Choice-of-law rules decide which law the state[5] or the federal[6] court

---

parties, is found in the statutes and decisions of the state." *Id.* at 378 (quoting *Davis Warehouse Co. v. Bowles* (1944) 321 U.S. 144, 155).

Accord, *State of Cal.ex rel. State Lands Com. v. Superior Court (Lovelace) (1995)* 11 Cal.4th 50, 74.

2. Conflict of laws is defined as ". . . that branch of jurisprudence, arising from the diversity of laws of different nations, states or jurisdictions, in their application to rights and remedies, which reconciles the inconsistency, or decides which law or system is to govern in the particular case, or settles the degree of force to be accorded to the law of another jurisdiction . . . either where it varies from the domestic law, or where the domestic law is silent or not exclusively applicable to the case in point." Black's Law Dictionary (5th ed. 1979), p. 271. Choice-of-law is defined as follows: "In conflicts of law, the question presented in determining what law should govern." *Id.* at 219.

3. For a fascinating discussion of the principles of international law as applied to particular property boundary disputes, see 1 Shalowitz, *Shore and Sea Boundaries* (Dept. Comm. Pub. 10-1 1962) (hereinafter "1 Shalowitz") Chapters 3–5; 3 Shalowitz, Reed, M., *Shore and Sea Boundaries* (Dept. Comm. Pub. NOAA/CSC/20007-PUB 2000) (hereinafter "3 Shalowitz"), Chapters 3–6. 3 Shalowitz also has an excellent discussion about how such cases are litigated.

4. *Cal. ex rel. State Lands Com'n v. United States* (9th Cir. 1986) 805 F.2d 857, 860, *cert. den.* (1987) 484 U.S. 816; *State of Cal.ex rel. State Lands Com. (Lovelace), supra,* 11 Cal.4th at 74.

5. In title and boundary disputes along tidal or navigable waterways, state courts are rarely, if ever, called upon to decide the nature of or to apply federal law. This is because federal law may apply when the federal government is a party; cases in which the federal government is a party are exclusively litigated in federal court. See, e.g., 28 U.S.C. § 1346(f).

6. Federal courts can apply state law. To take just one example, in a case that concerned a dispute over water rights between the United States, the state, and private landowners, the federal court engaged in a lengthy exposition of the California state water law that formed the basis of

applies to decide the effect of changes in geographic location[7] of the boundary water body on littoral or riparian property boundaries.

In an ideal world, such rules would be readily predictable. They would be based on locally or regionally developed and tested standards, founded on long-term experience and proven knowledge of physical characteristics of the particular water body and adjacent landforms. And these rules would take into account local peculiarities of title derivation. Yet, as we will see, this is not an ideal world: Boundary and title decisions by federal courts in Maine may affect boundary and title disputes in California, a continent (some say a world) away. This bizarre result has been occasioned by a series of decisions by the United States Supreme Court and lower federal courts.

This chapter has a very narrow focus, concerned only with the choice-of-law in three specific situations: (1) When an interstate boundary water body changes its geographic location and land that had been geographically in one state ends up in another state by virtue of that change in physical location, (2) when a tidal or navigable boundary water body changes its geographic location and the federal government is the littoral or riparian owner, and (3) when a tidal or navigable boundary water body changes its geographic location and the riparian or littoral owner derives title from either a federal public domain patent or a state patent. As will be seen, choice-of-law decisions in these three situations are not always straightforward or fully explainable based on past decisions.

These are not insignificant issues. From almost the beginning of nationhood, choice-of-law questions such as these have troubled the courts, lawyers, surveyors and property owners.[8] Because the principles in these cases are

---

its decision. *Rank v. (Krug) United States* (S.D. Cal. 1956) 142 F.Supp. 1, 105, *aff'd in part* and *rev'd in part, sub. nom. California v. Rank* (9th Cir. 1961) 293 F.2d 340. The federal court described its function as follows:

> In approaching the problem of the substantive water law . . . , it must be remembered that the primary function of this Court is not to make new law in that field, but to ascertain, from cases decided by the . . . State Courts, what the law *is* on the particular matter, and to accept and apply it. Only if there is a total absence of [state] decisions on the precise point or statute will the Court undertake, as it must, to declare the law. *Ibid.*

7. "Geographic location" and "physical location" are interchangeable terms the author has adopted as a convention in this work. The convention describes the particular latitudinal and longitudinal location of a discrete physical feature that can be definitely marked and measured and later relocated if lost, obliterated, or destroyed. *Kitteridge v. Ritter* (Iowa 1915) 151 N.W. 1097. To illustrate the meaning of this term, consider that part of a river may have been located at a particular latitude and longitude—its original geographic or physical location. As the result of river processes, however, such portion of the river may move to a different geographic location. Whether this change in the physical location of the river effects a change in the property boundary between state-owned riverbed and private uplands is a legal question. See Section 3.10 in Chapter 3. Another concrete physical illustration of this concept is the cutoff of an oxbow bend in a boundary river. One morning the river is located in one's front yard; the following morning it is found in the next state.

8. E.g., *Railroad Co. v. Schurmeier* (1868) 74 U.S. (7 Wall.) 272; *Missouri v. Kentucky* (1870) 78 U.S. (11 Wall.) 395; *Barney v. Keokuk* (1876) 94 U.S. 324.

about the basic relationships of the state and federal legal systems and governments,[9] such cases have importance beyond the seemingly narrow confines of the land title context. Indeed, these cases continue to vex courts, lawyers, surveyors, and property owners.[10]

2.2 INTERSTATE POLITICAL BOUNDARY

Throughout the United States a river often constitutes a boundary between two states. The river is not only a physical barrier; it also separates two political jurisdictions or sovereignties. For example, the Colorado and Missouri Rivers are both physical barriers and political boundaries between California and Arizona and Iowa and Nebraska, respectively.

Suppose, however, that the river forming the political boundary between two states changes geographic location. Lands once in State A on the east side of the river might at a later time lie geographically west of the river. Which law—federal law, State A's law, or State B's law—determines the political boundary between the two states effected by the change in the geographic location of their river boundary? Are the former State A lands, that are geographically on the State B side of the river, now part of State B and subject to its political authority? Are principles applicable to the effect of change in the physical location of the river on the interstate political boundary applicable as well to the effect of such change on the property boundaries of private landholdings?

Over the years the United States Supreme Court has developed a body of federal decisions that supply answers to the first of these questions.[11] Al-

---

9. . . . [T]his is deemed the most important controversy ever brought before this court, either as it respects the amount of property involved, or the principles on which the present judgment proceeds—principles, in my judgment, as applicable to the highlands of the United States as to the low lands and shores." *Pollard's Lessee v. Hagan* (1845) 44 U.S. (3 How.) 212, 235 (Catron, J., dissenting); *California, ex rel. State Lands Comm'n v. United States* (1982) 457 U.S. 273, 277, n.6 (hereinafter "Humboldt Spit").

10. Cases a continent apart in California and Florida illustrate the continuing immediacy of this controversy. In the California case, title was contested to lands uncovered by the recession of a navigable, inland lake (Mono Lake on the eastern slope of California's Sierra Nevada). The lake receded because of human-induced, artificial diversion of the lake's tributary streams to slake the thirst of Los Angeles. The choice-of-law decision was crucial. *Cal. ex rel. State Lands Com'n*, *supra*, 805 F.2d at 860. The State of Florida, by statute, is entitled to all additions to land caused by erection of beach and shore protection structures. A Florida court relied on federal cases to uphold the upland owner's title to these additions to his land. The owner's title was derived from a federal (not a state) patent. *Board of Trustees v. Sand Key Associates* (Fla. 1986) 489 So.2d 34, 36, *cert. den.* (1986) 475 U.S. 1094. Finally, the California Supreme Court applied California law in a case about ownership of land adjacent to a tidal river where the competing parties were the state and an upland owner asserting applicability of federal law. *State of Cal.ex rel. State Lands Com. (Lovelace)*, *supra*, 11 Cal.4th at 74.

11. E.g., *Indiana v. Kentucky* (1890) 136 U.S. 479; *Nebraska v. Iowa* (1892) 143 U.S. 359; *Iowa v. Illinois* (1892) 147 U.S. 1; *Arkansas v. Tennessee* (1918) 246 U.S. 158; *Massachusetts v. New York* (1926) 271 U.S. 65, 93.

though these cases usually turn on their individual facts, two decisions concerning the effect of river movement on political boundaries will be instructive.

Nebraska and Iowa are mutually bounded by the Missouri River, with Nebraska on the east and Iowa on the west.[12] Over a 20-year period, there were marked changes in geographic location of the Missouri River channel near Omaha, Nebraska. Because of these changes, both Nebraska and Iowa claimed jurisdiction over the same area of land.[13]

The Supreme Court decided that the political boundary line would vary as the river moved, despite the rapidity of such river movement.[14] In reaching its conclusion, the Supreme Court noted that certain propositions developed in deciding boundary movement disputes between private parties could also be used in such contests between the states.[15]

A later case, not surprisingly, also concerned the Missouri River.[16] Kansas sued Missouri to locate their common political boundary along 128 miles of the erratic river.[17] The case initially concerned an enormous area of land. As in many interstate boundary conflicts, however, the two states settled most of their disputes. The contest ultimately was reduced in magnitude to the two states' competing claims to political jurisdiction over 2000 acres of land. This land had come to be geographically located on the Missouri side of the Missouri River.[18] The Supreme Court, relying on federal cases, upheld Missouri's claim to the disputed lands.[19]

These two cases involving the Missouri River exemplify disputes between two states. The United States Supreme Court has fashioned a well-developed body of federal law for resolving interstate political boundary disputes.[20] What about property title or property boundary disputes resulting from changes in physical location of a boundary waterway? Are there rules that govern the choice-of-law for such cases?

---

12. *Nebraska v. Iowa, supra,* 143 U.S. at 359–360.
13. *Id.* at 360. Note that political jurisdiction was at stake, not land title.
14. *Id.* at 369–370.
15. *Id.* at 361.
16. The Missouri River is a fecund source of boundary and title disputes. This is partly a result of the value of land adjacent to the river and partly because of the very nature of the river.

   The Missouri River is a vagrant, turbulent stream. Its name reflects this character. The Big Muddy is said to carry more silt than any other river except the Yellow River in China. It is constantly changing its course within the region between its bluffs, shi2fting from side to side as natural forces work upon its flow. *Kansas v. Missouri* (1944) 322 U.S. 213, 216.

17. *Id.* at 213.
18. *Id.* at 214.
19. *Id.* at 228–229. See also *Oklahoma v. Texas* (1923) 260 U.S. 606.
20. *State Land Board v. Corvallis, supra,* 429 U.S. at 375.

2.3 INTERSTATE BOUNDARIES—PRIVATE TITLES AND BOUNDARIES

Some of the most highly visible and well publicized title or boundary disputes arise where riparian landowners, who thought that their lands were in State A, wake up one day to find their land now in State B. A court decision on the political boundary effect of a prior change in the geographic location of an interstate water boundary may cause this upheaval. Questions concerning the title to or boundary of lands that have been geographically relocated to an adjacent state by virtue of a change in location of the political boundary between the two states are resolved most often by virtue of interstate boundary compacts.[21] Wisely, this result is strongly commended by the Supreme Court.[22] For example, by virtue of the interstate boundary compact between California and Arizona, titles good in one state are held good in the other state even though the land may now be geographically located in the other state.[23]

These compacts are reached for good reason. For example, numerous and complicated shifts in the Missouri River boundary between Iowa and Nebraska required the two states to enter into a boundary compact. These shifts resulted not only from actions of the "fickle Missouri River" itself, but also from the sometimes futile attempts of the U.S. Army Corps of Engineers to tame the river. Thus, locating the original interstate boundary line was practically impossible.[24] As a consequence, the boundary compact set the location of the interstate boundary; some of the riparian lands that had formerly been in one state came to be located in the other state by virtue of the compact.[25]

This fact created uncertainty about whether the "new" state would honor titles, mortgages, or other liens that had been created under the laws of the "former" state.[26] The boundary compact attempted to solve this problem by having each state cede to the other and relinquish jurisdiction over lands that,

---

21. The Supreme Court has decided such cases where there is no boundary compact. For example, in *Massachusetts v. New York, supra,* 271 U.S. at 93, the Supreme Court used a state (New York) rule of interpretation of conveyances to decide boundary and title questions between the two states and those private parties claiming by virtue of state grants. The Supreme Court said: ". . . [L]ocal rules for interpreting conveyances should be applied by this Court in the absence of an expression of a different purpose." *Ibid.*
22. *Washington v. Oregon* (1908) 214 U.S. 205, 217, 218.
23. Interstate Compact Defining the Boundary Between the State of Arizona and the State of California (1966) 80 Stat. 340; *California v. Arizona* (1979) 440 U.S. 59, 60. Arizona and California have not always agreed on matters, as witnessed by their monumental confrontation over water resources of the Colorado River. *Arizona v. California* (1963) 373 U.S. 546.
24. *Nebraska v. Iowa* (1972) 406 U.S. 117, 119.
25. *Id.* at 120.
26. *Ibid.*

because of the compact, were now located in the other state.[27] Despite the good intentions of the boundary compact, however, a dispute arose between the two states over land that was, at the time of the lawsuit, indisputably on the Iowa side of the compact boundary.[28]

Iowa's claim to those lands was based on Iowa common law. Nebraska claimed the titles were good in Nebraska. Nebraska asserted that Iowa, by virtue of the boundary compact, was obligated to recognize those titles as good in Iowa as against Iowa's claim.[29] In reviewing the Special Master's Report, the United States Supreme Court applied the laws of both states to issues presented in the case.[30] The Supreme Court decided that which state's law applied would depend on when and in which state the lands finally came to be situated. This result was based on an earlier decision by the Supreme Court. In the earlier case, the Court decided that, with one important condition, how land ending up on one side or the other of an interstate boundary water body is divided between public and private ownership is determined by the law of the state. The condition was that such a decision must not "press back the [interstate] boundary line from where it would otherwise be located."[31]

Unfortunately, private title and boundary disputes along interstate boundary waterways also arise where no interstate boundary compacts exist. One such dispute occurred along the Missouri River boundary between two competing landowners; one landowner claimed the land was in Missouri, the other landowner claimed the land was in Nebraska.[32] The quiet title action began in a Nebraska court.

The Missouri-contending landowner claimed the Nebraska court was without jurisdiction because the land in dispute was in fact located in Missouri.[33] Relying on Nebraska law, the Nebraska court held that the land in dispute was in Nebraska and therefore it had jurisdiction over the dispute.[34] The Missouri property owner did not take this predictable decision lying down; a lawsuit was then filed in Missouri.[35] That lawsuit ended up in federal court[36] and ultimately resulted in a United States Supreme Court opinion upholding the Nebraska court's decision.[37] Believe it or not, all this litigation had nothing

---

27. *Ibid.*
28. *Ibid.*
29. *Id.* at 120–121.
30. *Id.* at 123–126.
31. *Arkansas v. Tennessee, supra,* 246 U.S. at 175–176.
32. *Durfee v. Kieffer* (Neb. 1959) 95 N.W.2d 618; *Durfee v. Duke* (1963) 375 U.S. 106.
33. *Durfee v. Kieffer, supra,* 95 N.W.2d at 621; *Durfee v. Duke, supra,* 375 U.S. at 108.
34. *Durfee v. Kieffer, supra,* 95 N.W.2d at 624; *Durfee v. Duke, supra,* 375 U.S. at 108.
35. *Durfee v. Duke, supra,* 375 U.S. at 108.
36. The reason the suit ended up in federal court was that the two parties were citizens of different states. *Ibid.*; 28 U.S.C. § 1332.
37. *Durfee v. Duke, supra,* 375 U.S. at 115.

at all to do with the location of the interstate boundary. That boundary could be decided only in litigation between the two states.[38]

2.4 LAND TITLE AND BOUNDARY DISPUTES—IN GENERAL

Disputes between two states over their political boundary or the effect of a change in the political boundary on real property titles or boundaries may be somewhat glamorous. Although seemingly more mundane, the choice-of-law decision in land title or boundary cases has become more, not less, unsettled. The long-accepted rule, where interstate political boundary location questions were not concerned and the dispute arose after admission of the state to the Union, was that state law would be the choice-of-law in title or boundary disputes about land along or beneath navigable or tidal waterways.[39] This rule held true even if the source of riparian title was derived from the United States. The Supreme Court had long been content to keep to that well-worn path. Yet in a series of cases, the Supreme Court strayed from this long-accepted course and has misled lower federal courts as well.

The rulings of the United States Supreme Court have veered from one extreme to another. First, the Court decided federal law applied in determining these boundary questions.[40] Then the Court came to the opposite conclusion that, in fact, it was constitutionally required for state law to decide such questions.[41] In the next case, the Court tried to strike a middle ground when

---

38. "It is to be emphasized that all that was ultimately determined in the Nebraska litigation was title to the land in question between the parties to the litigation there. Nothing there decided, and nothing that could be decided in litigation between the same parties or their privies in Missouri, could bind either Missouri or Nebraska with respect to any controversy they might have, now or in the future, as to the location of the boundary between them, or as to their respective sovereignty over the land in question. [Citations omitted.] Either State may at any time protect its interest by initiating independent judicial proceedings here." *Id.* at 115–116. The concurring opinion noted there was an open question of whether the Nebraska judgment would still bind the Missouri landowner if it were later found, either in a case between the two states or as the result of a boundary compact, that the disputed land was in Missouri. *Id.* at 117 (Black, J., concurring).
39. E.g., *Swift v. Tyson* (1842) 41 U.S. (16 Pet.) 1, 17–18; *Barney, supra*, 94 U.S. at 337; *Joy v. St. Louis* (1905) 201 U.S. 332, 342; *Shulthis v. McDougal* (1912) 225 U.S. 561, 570; *Montana v. United States* (1981) 450 U.S. 544, 551. Federal courts had noted, without comment or criticism, the applicability of state law in boundary or title disputes even when the United States was a party. *Philadelphia Co. v. Stimson* (1911) 223 U.S. 605, 632; *United States v. Otley* (9th Cir. 1942) 127 F.2d 988, 993. Clark quoted *Hobart v. Hall* (D. Minn. 1909) 174 F. 433 for the proposition that United States public land grants are construed according to the law of the state in which such lands lay. Robillard, W., and Bouman, L., Clark on Surveying and Boundaries (7th Ed., 1997) § 23.14, p. 801 (hereinafter "Clark 7th"); Grimes, *Clark on Surveying and Boundaries* (4th Ed., 1976) (herein "Clark 4th"), § 565, p. 768. Skelton agreed wholeheartedly. Skelton, *The Legal Elements of Boundaries and Adjacent Properties* (1930) (herein "Skelton") at § 281, pp. 308–309.
40. *Bonelli Cattle Co. v. Arizona* (1973) 414 U.S. 313, 320–321.
41. *State Land Board v. Corvallis, supra*, 429 U.S. 372. *Corvallis* overruled *Bonelli*. *Id.* at 382.

it decided that although federal law would supply the choice-of-law, state law would supply the rule of decision.[42] Finally, the Court wound up where it began when it decided, because of the implications of the longtime federal littoral land ownership and the Submerged Lands Act,[43] that federal law would be the choice-of-law and the rule of decision.[44]

Among other lessons, one critical rule may be learned from these decisions: When the United States, on the one hand, and a private litigant or a state, on the other, are contesting the title to or boundaries of land along a tidal or navigable waterway, the law best protecting federal interests will be applied.[45] Such a result may benefit the federal government. It does little, however, to enhance the certainty of land titles and does great injustice to rules of property that have grown up to reflect local experience and peculiarities. Finally, such a rule also turns the federal judicial system on its head by potentially making a local property boundary dispute a "federal case."

A way to understand the direction of the choice-of-law decisions in the boundary and title area for the last few years is to look at the types of parties who were the disputing litigants. One group of cases decided contests between private litigants or between private litigants and states.[46] A second group of cases concerned disputes between the federal government as one interested party and a state or private landowners as the other.[47] A close look at these groups of cases will readily affirm the underlying reason for this fundamental change in the choice-of-law rule: When federal interests are at stake, federal law will be the choice-of-law.

2.5 LAND TITLE AND BOUNDARY DISPUTES WHEN THE UNITED STATES IS NOT A DISPUTING LANDOWNER

There was a long-held rule in cases of title or boundary disputes along tidal or navigable waterways between competing land owners, neither of which

---

42. *Wilson v. Omaha Indian Tribe* (1979) 442 U.S. 653, 671–676.
43. Submerged Lands Act, Act of May 22, 1953, 67 Stat. 29, set forth in 43 U.S.C. §§ 1301, et seq. The Submerged Lands Act was enacted as a result of the decision in *United States v. California* (1947) 332 U.S. 19. In that case the Supreme Court held that the United States, not the State of California, possessed "paramount rights in and power over" lands off the California coast. *Id.* at 38. This decision shattered the long and staunchly held understanding that the states, not the federal government, were owners of such submerged lands. Congress enacted the Submerged Lands Act to undo the effect of that decision. *United States v. California* (1978) 436 U.S. 32, 37.
44. *Humboldt Spit, supra,* 457 U.S. 273. This case concerned the effect of construction of one of the marvels of coastal engineering at the tip of the North Spit protecting Humboldt Bay near Eureka, California.
45. This has not always been the case. *United States v. Otley, supra,* 127 F.2d at 993.
46. *Hughes v. Washington* (1967) 389 U.S. 290; *Bonelli, supra,* 414 U.S. at 316; *State Land Board v. Corvallis, supra,* 429 U.S. at 365.
47. *Wilson, supra,* 442 U.S. at 657; *Humboldt Spit, supra,* 457 U.S. at 275.

was the federal government. The rule was that the law of the state in which the land was located was the proper choice-of-law.[48] Two cases, however, shook the foundation of that long understanding. The first was the aberrant decision in *Hughes v. Washington*.[49]

Hughes was a contest over artificially caused deposition along the beautiful Washington State Pacific Ocean coast. Stella Hughes, owner of the adjacent upland, claimed the newly precipitated land based on her federal patent. That patent supposedly included the right to all additions to the land patented. On the other hand, the State of Washington claimed title to the additional land as deposits made on tideland and submerged land owned by the state.[50]

Under state law, Mrs. Hughes would have been denied ownership of the land that had grown up adjacent to her land; under "federal law," Mrs. Hughes would win. The Supreme Court decided that determination of who owned the additional land was to be governed by federal law. Justice Hugo Black based this decision on the supposed vital interest of the United States in the international boundary of the United States.[51] The decision contradicted and failed to discuss the long line of Supreme Court cases holding that state law should be applied to decide such boundary disputes.

The next case was *Bonelli Cattle Co. v. Arizona*. This case was a title and boundary contest between the State of Arizona and the riparian landowner over a portion of the former bed of the Colorado River that had become exposed.[52] The Supreme Court first found that the State owned the riverbed by virtue of settled federal law.[53] The Court then framed the question in the case: What was the extent of the state's ownership rights under that law? The Court held that the extent of the state's rights under federal law was a federal question.[54] The effect of this decision was that virtually all title or boundary disputes involving state sovereign land became federal questions.

In the *Corvallis* case, however, the Supreme Court recognized the error of its ways, expressly in the case of *Bonelli* and implicitly in the case of *Hughes*. *Corvallis*, like *Bonelli*, was a dispute between the state as the owner of the riverbed and a riparian landowner.[55] The Supreme Court, in a rare turnabout, reversed *Bonelli*. It held that state law would apply to determine the title and

---

48. See note 39 in this chapter, above.
49. (1967) 389 U.S. 290.
50. *Hughes, supra*, 389 U.S. at 290–291.
51. *Id.* at 292–293. ("The rule deals with waters that lap both the lands of the State and the boundaries of the international sea. This relationship . . . is too close to the vital interests of the . . . Nation in its own boundaries to allow it to be governed by any law but 'the supreme Law of the land.'")
52. *Bonelli, supra*, 414 U.S. at 314–316.
53. *Id.* at 318–319. The settled federal law giving the State of Arizona title was the Equal Footing Doctrine, see notes 111 through 114 and accompanying text in Chapter 1, and the Submerged Lands Act, see note 43 in this chapter.
54. *Id.* at 319–320.
55. *State Land Board v. Corvallis, supra*, 429 U.S. at 365.

boundary questions in this case.⁵⁶ The reason was that the source of the state's ownership to the riverbed, the Equal Footing Doctrine,⁵⁷ did not supply a basis for federal law to be applied; federal law only determined the initial boundaries of and title to the land.⁵⁸ In overruling the *Bonelli* decision, the Supreme Court in *Corvallis* made special note that it was returning to a ". . . system of resolution of property disputes [that had] been adhered to from 1845 until 1973 [the year of the *Bonelli* decision]. . . ."⁵⁹

This was the state of choice-of-law rules until the Supreme Court took its eye off the ball and missed a prime opportunity to finally settle this rule. This missed opportunity arose in a case, however, that was not between private landowners or between private landowners and the state.

2.6 LAND TITLE AND BOUNDARY DISPUTES WHEN THE UNITED STATES IS A DISPUTING LANDOWNER

Because of the enormous extent of federal land ownership in the western United States, title and boundary conflicts along tidal or navigable water bodies between the federal government and competing landowners, either private or governmental, occur with ever increasing frequency. What is the choice-of-law in such a case? Is federal law the inevitable result of the choice-of-law decision? The rule had been that even when the United States was a party, state law was the choice-of-law.⁶⁰ What was once thought to be an immutable standard, however, has been severely eroded in a series of cases.

2.6.1 The Wilson Case

The first of these cases, *Wilson v. Omaha Indian Tribe*,⁶¹ was about a thumb-shaped bend in the Missouri River. This feature, Blackbird Bend, had been gradually washed away and eventually disappeared as the river continued a decades-long westerly movement. The river finally ended up almost 2 miles from its original geographic location at Blackbird Bend.⁶² After many years a title dispute arose concerning ownership of land in the same geographic location as what once had been Blackbird Bend. Confusion over title was only natural. The physical area known as Blackbird Bend, whose geographic

---

56. *Id.* at 370–372.
57. It made no difference to the Supreme Court that the source of the riparian landowner's title was a federal patent. *Id.* at 372.
58. *Ibid.*
59. *Id.* at 382.
60. E.g., *Philadelphia Co. v. Stimson, supra,* 223 U.S. at 632.
61. (1979) 442 U.S. 653.
62. *Id.* at 659.

location once lay west of the Missouri River in Nebraska, was now geographically located east of the river in Iowa.[63]

There were three title contestants in the *Wilson* case: the State of Iowa, owner of a portion of the bed of the river by virtue of the Equal Footing Doctrine; the Omaha Indian Tribe, who originally occupied Blackbird Bend by virtue of a treaty with the United States, and the United States, as trustee on behalf of the Omaha Indian Tribe; and Iowa landowners who occupied and had farmed for years the geographic area that had once been Blackbird Bend.[64] Choice-of-law was crucial to the result; if state law applied, the United States and the Omaha Indians would lose title.[65]

As a predicate to its decision in the *Wilson* case, the Supreme Court noted that the federal government had held the land in trust for the Omaha Tribe since 1869, when the United States had entered into a treaty with the Omaha Tribe. The Court particularly noted that the United States had never yielded title or terminated its interest in the land.[66] The finding of this federal interest was crucial to the case because of the Supreme Court's interpretation of *Corvallis*.[67] Based on this somewhat questionable foundation,[68] the *Wilson* Court decided that federal law[69] would determine the Omaha Indian Tribe's right to the land.[70]

It is important to note that the Court did not find that federal law required the creation of special federal rules to decide the boundary and title question.

---

63. *Ibid.*
64. *Id.* at 656–659.
65. *Id.* at 658.
66. *Id.* at 670. The trust ownership claim of the United States was based on the long-recognized obligation of the United States to protect Indian title. E.g., Flushman and Barbieri, "Aboriginal Title: The Special Case of California" (1986) 17 Pac.L.J. 391, 395–397; *Joint Trib. Coun. of Passamaquoddy Tribe v. Morton* (1st Cir. 1975) 528 F.2d 370, 379.
67. The Supreme Court interpreted *Corvallis* as saying that if there were some applicable federal right, federal law would control. *Wilson, supra*, 442 U.S. at 670. In fact, what *Corvallis* said was that state law should apply, ". . . unless there were present some other principle of federal law requiring state law to be displaced." *State Land Board v. Corvallis, supra*, 429 U.S. at 371.
68. In addition, the Supreme Court assumed the United States had never lost its title to Blackbird Bend. *Wilson, supra*, 442 U.S. at 670. This assumption begs the question. It is based on the conclusion that, as a matter of law, the change in geographic location of the river boundary did not change the property boundary at Blackbird Bend. This was exactly the question in the case.
69. For a period of time in this nation's legal history, some thought federal courts had the power to create law that applied to questions arising in federal court where federal law was silent. The landmark decision of *Erie R. Co. v. Tompkins* (1938) 304 U.S. 64, 78–80, shattered this concept:

> Except in matters governed by the Federal Constitution or by Acts of Congress, the law to be applied in any case is the law of the State. . . . There is no federal general common law. Congress has no power to declare substantive rules of common law applicable in a State. . . . And no clause in the Constitution purports to confer such a power upon the federal courts. . . .

70. *Wilson, supra*, 442 U.S. at 670–671.

Instead, the Supreme Court held that federal law would borrow Nebraska state law as the rule of decision.[71] The *Wilson* Court implicitly recognized the need for each political jurisdiction to establish rules of property that are peculiar to local legal and physical circumstances and equally applicable to all land within the jurisdiction. Thus, even though the *Wilson* property boundary and title dispute was governed by federal law, the Supreme Court found that there was no need to create a uniform federal rule of law[72] applicable to all such controversies, no matter in what context they arose.[73] The laws of the state in which the boundary or title controversy arose would do just fine.

It is significant that the Supreme Court was not compelled to create "federal common law" in the *Wilson* case. *Wilson* concerned a highly favored creature of federal law, Indian trust land.[74] If it was not necessary to create "federal common law" to protect this important federal interest, creation of such rules would appear limited to only a few unique situations. This conclusion was given credence by the reasoning the Supreme Court adopted in deciding whether to create a nationwide federal rule applicable to title and property boundary disputes concerning lands adjacent to or underlying tidal or navigable waters.

The *Wilson* Court noted that creation of a nationwide federal rule or adoption of existing and established state law rules applicable to all other land in the state was a matter of judicial policy. This decision would be "'. . . dependent upon a variety of considerations always relevant to the nature of the specific governmental interests and to the effects upon them of applying state law.'"[75] Among the considerations taken into account in making this judicial policy decision were (1) the need for a nationally uniform body of law to apply in situations comparable to the dispute before the Court, (2) whether application of state law would frustrate federal policy or functions, and (3) the impact a federal rule might have on existing relationships under state law.[76]

After answering these questions, the *Wilson* Supreme Court held that state law, *not* federal common law, should supply the rule of decision to determine ownership of what had once been Blackbird Bend.[77] Because of the impor-

---

71. *Id.* at 678.
72. Interestingly, the Court specifically noted that rules developed in cases about the effect of a change in geographic location of interstate political boundaries did *not* necessarily furnish rules to govern the effect of a change in geographic location in property boundary disputes. *Id.* at 672. As noted in *State Land Board v. Corvallis*, *supra*, 429 U.S. at 375–376, determination of interstate political boundaries was one of four situations in which federal common law had been created in boundary disputes about lands along or beneath navigable or tidal water bodies.
73. *Wilson, supra,* 442 U.S. at 673–674.
74. See note 66 in this chapter.
75. *Id.* at 671–672, quoting *United States v. Kimbell Foods, Inc.* (1979) 440 U.S. 715, 727–728.
76. *Id.* at 672–673.
77. *Id.* at 673.

tance of understanding the reasoning of the Supreme Court, given its later deviations from this once-approved line of thinking, the following explanation is set forth at length:

> We perceive no need for a uniform national rule to determine whether changes in the course of a river affecting riparian land owned or possessed by the United States or by an Indian tribe have been [of a certain character]. For this purpose, *we see little reason why federal interests should not be treated under the same rules of property that apply to private persons holding property in the same area by virtue of state, rather than federal law.* It is true that States may differ among themselves with respect to the rules that will identify and distinguish between [types of boundary movements] but *as long as the applicable standard is applied even-handedly to particular disputes, we discern no imperative need to develop a general body of federal common law to decide cases such as this,* where an interstate boundary is not in dispute. We should not accept "generalized pleas for uniformity as substitutes for concrete evidence that adopting state law would adversely affect [federal interests]."[78]

Thus, in the *Wilson* case, the Supreme Court veered somewhat off its earlier path. Although the Court ultimately returned to the proper course, its question-begging analysis[79] hinted at a disturbingly different route. This route ultimately led into a quagmire of doubt and uncertainty about the proper choice-of-law.

2.6.2 The Humboldt Spit Case

In the *Humboldt Spit* case,[80] the Supreme Court went completely off track. *Humboldt Spit* had its roots in the bumptious town of Eureka, California. Eureka, which fronts on Humboldt Bay, wanted to establish a great port for shipping. There were, however, considerable physical barriers to the use of Humboldt Bay as a port. The bay's entrance migrated because of the effects of the offshore littoral current. In addition, because of a sand bar off the mouth of the bay, great breakers formed. To overcome these natural hazards to navigation, the U.S. Army Corps of Engineers constructed one of the great works of coastal engineering.

The entrance to Humboldt Bay was located between two long sand spits, the North and South Spits. With considerable difficulty and remarkable tenacity, the Corps built two gigantic rubble mound jetties at the tips of these

---

78. *Id.* at 673, quoting *Kimbell Foods, supra,* 440 U.S. at 730 (emphasis supplied). Eventually the Iowa farmers lost in the lower courts. *United States v. Wilson* (8th Cir. 1980) 614 F.2d 1153, *cert. den.* (1980) 449 U.S. 825; *United States v. Wilson* (N.D. Iowa 1981) 523 F.Supp. 874, *rev'd.* (8th Cir. 1982) 707 F.2d 304; *United States v. Wilson* (N.D. Iowa 1984) 578 F.Supp 1191, *cert. den.* (1984) 465 U.S. 1025, 1101.
79. See notes 66 through 68 and accompanying text in this chapter.
80. *California, ex rel. State Lands Comm'n v. United States* (1982) 457 U.S. 273.

spits. The jetties were intended to stabilize the entrance and to scour away the bar. As a result of these coastal works, over the years the Pacific Ocean shoreline along the North Spit shifted seaward. Through the deposition of sand trapped by the north jetty, 184 acres of the former bed of the Pacific Ocean were filled in, covered up, and exposed.[81] These lands, used only for beachcombing and what some used to call "necking,"[82] became the subject of a great constitutional dispute between California and the United States.

The issue was whether federal rules or the rules of real property that had grown up in a state would determine title to or boundaries of the land within its borders, no matter who the present owner of the adjacent land was or what the source of title to such land was. The case became yet another reminder that the South (and claims of state's rights) had lost the Civil War. The Supreme Court, in disregard of its earlier decisions, found that federal, not state, law would control the effect of change in the geographic location of a tidal water body on an adjacent property boundary along the open coast when the federal government was the landowner whose boundary was being defined.[83]

Following the guidance of *Wilson*, the Supreme Court decided that federal law may apply if there is present some other principle requiring state law to be displaced.[84] The Court found there was a dominant federal interest in international relations requiring application of a federal rule to open-coast shoreline changes.[85] The Court relied on *Hughes*[86], a case which was thought to have been overruled.[87]

The Supreme Court did acknowledge, as in *Wilson*, that cases governed by federal law were not inescapably required to be decided by uniform federal rules.[88] Consequently, the Court considered whether federal law should bor-

---

81. The story of how these works came to be and were constructed and the effect of such construction on the Pacific Ocean shoreline is found in U.S. Army Corps of Engineers, San Francisco District, "Survey Report on Humboldt Bay" (Feb. 10, 1950) and earlier reports of the Chief of the Corps of Engineers to Congress. These reports, though written in almost classic bureaucratese, paint a fascinating portrait of how such public works come to be authorized, the tribulations of building them, and the effects of nature's might on humans' works. Reports such as these are an invaluable resource to the careful surveyor or lawyer who takes the time to find and read them.
82. Personal interview with natives of the Eureka/Arcata area.
83. *Humboldt Spit, supra*, 457 U.S. at 282.
84. *Ibid.*
85. *Id.* at 283.
86. (1967) 389 U.S. 290.
87. The belief that *Hughes* had been overruled is founded on Justice Marshall's dissent in the *Corvallis* decision. *State Land Board v. Corvallis, supra*, 429 U.S. at 383, n.1. Justice Marshall completely disposed of the basis of the *Hughes* decision in stating: "It is difficult to take seriously the suggestion that the national interest in international relations justifies applying a different rule to oceanfront land grants than to other grants by the Federal government." *Ibid.* In *Humboldt Spit*, Justice Rehnquist's concurring opinion appears to agree that *Hughes* is no longer sound. *Humboldt Spit, supra*, 457 U.S. at 290 (Rehnquist, J., concurring). Nevertheless, the majority in *Humboldt Spit* and in *Corvallis* refused to disavow the case.
88. *Humboldt Spit, supra*, 457 U.S. at 283.

row state law as the rule of decision. In contrast to *Wilson*, in which the federal and state governments were also competing parties, the Supreme Court found that there was a "dispositive" basis for creating special federal rules—the Submerged Lands Act.[89] It found such "evidence" in one section of the Submerged Lands Act that withheld from the Submerged Lands Act's confirmance of title to the states' land that had been created by "accretion" to lands held or reserved by the United States.[90] Thus, the Court decided that application of state law to resolve the boundary dispute was foreclosed. There was no need to create a federal common law rule, because "under settled federal law"[91] the United States was entitled to the 184 acres that it had created by virtue of building the jetties.[92] Thus, in one instant, without overruling its inconsistent past decisions, the Supreme Court turned its back on long-held understandings pertaining to the choice-of-law decision and, by doing so, created an absolute morass.

Among the many flaws in the *Humboldt Spit* decision were these: First, in *Wilson*, the Supreme Court had found that rules utilized to establish the effect of changes in geographic location of rivers on the political boundary between two states did not necessarily govern the property boundary effect of change in the Missouri River's geographical location.[93] Somehow, in *Humboldt Spit* the Court found that these same rules *were* applicable to the open coast.[94]

Second, the international boundary rationale, the principle basis for the decision, was without foundation.[95] The international boundary was not at all concerned; the coastline, for purposes of the international boundary, had al-

---

89. *Ibid.* The finding that the Submerged Lands Act provided a basis on which to apply federal law completely ignored *Corvallis*, without bothering to distinguish or discredit that case. *Corvallis* specifically noted: ". . . [T]he Submerged Lands Act did not alter the scope or effect of the equal-footing doctrine, nor did it alter state property law regarding riparian ownership. The effect of the Act was merely to confirm the States' title to the beds of navigable waters . . . against any claim of the United States Government. As merely a declaration of the State's pre-existing rights in the riverbeds, *nothing in the Act in any way mandates or even indicates that federal common law should be used to resolve ownership of lands which, by the very terms of the Act, reside in the States.*" State Land Board v. Corvallis, supra, 429 U.S. at 371, n.4 (emphasis supplied).
90. *Humboldt Spit, supra,* 457 U.S. at 283. 43 U.S.C. § 1313(a) was the section of the Submerged Lands Act relied on by the Supreme Court.
91. See notes 93 and 94 in this chapter.
92. *Humboldt Spit, supra,* 457 U.S. at 288.
93. *Wilson, supra,* 442 U.S. at 672.
94. In addition, in *Corvallis* the Supreme Court had found that certain of these cases merely stated rules applicable in certain states, not a "federal common law" rule. *State Land Board v. Corvallis, supra,* 429 U.S. at 380–381, n. 8.
95. Principles of international law set forth in treaties to which the United States is a party are used to decide international boundaries. For example, the United States is a party to the Convention on the Territorial Sea and the Contiguous Zone, March 24, 1961, 15 U.S.T. 1606, T.I.A.S. No. 5639. This convention has been used in determining the offshore boundary between California's submerged lands and those of the United States. *United States v. California* (1965) 381 U.S. 139, 165. That boundary is measured from, as far as concerns the coast line, the ordinary low-water mark. The Convention is not concerned with the dividing line between upland and the bed

ready been established well *seaward* of the lands disputed in *Humboldt Spit*.[96] Thus, the international relations rationale was not supported by the facts mor by the law.

Third, reliance on the Submerged Lands Act to support a choice-of-law rule requiring application of federal law in land title contests between the state and federal governments was truly perverse. The history of the enactment of the Submerged Lands Act established that the Act was never meant to diminish the states' rights in lands they held by virtue of the Equal Footing Doctrine.[97] The purpose of the Submerged Lands Act was to *protect* the states' rights in such lands, a fact that even the Supreme Court had earlier recognized.[98]

2.7 LATER CASES

It is shocking that, notwithstanding these flaws and given the existence of *Corvallis* and *Wilson*, federal courts applied the newly minted and unsettling rule (derived from *Humboldt Spit*) to inland lakes where the United States was a riparian owner.[99] Of course, the international relations "justification" in *Hughes* and *Humboldt Spit* had no bearing on lakes hundreds of miles from the nearest international boundary, and the court did not rely on that slim justification to support its ruling.

---

of the Pacific Ocean, the ordinary high-water mark. There are no international relations in determining the location of the ordinary high-water mark property boundary. *Corvallis*, *supra*, 429 U.S. at 383, n.1 (Marshall, J., dissenting).

96. The coastline for international boundary purposes along this reach of the California coast was established by Supreme Court decree at the seaward end of the North Jetty. *United States v. California* (1977) 432 U.S. 40. The end of the jetty was much farther seaward than the land disputed in the *Humboldt Spit* case. Thus, a decision regarding the location of the upland boundary would not affect the location of the international boundary.

97. The catalyst for the enactment of the Submerged Lands Act was the Supreme Court's validation of the United States' claim. The validation was based on the United States so-called paramount rights to the lands underlying ocean waters beyond the ordinary low-water mark. See note 43 and accompanying text in this chapter. Inland states with significant bodies of nontidal navigable waters were, however, uncertain of the extent of the paramount rights doctrine and joined forces with California and other coastal states to secure passage of the Submerged Lands Act. The Submerged Lands Act was intended to overturn the paramount rights doctrine and restore the states to their former position. *United States v. California*, *supra*, 436 U.S. at 37. As far as concerned lands underlying nontidal navigable waters, such as navigable lakes, the Submerged Lands Act was thought to be merely a confirmation of already vested title. *Bonelli*, *supra*, 414 U.S. at 318; *State Land Board v. Corvallis*, *supra*, 429 U.S. at 371, n. 4; S. Rep. No. 133, 83rd Cong., 1st Sess. (minority views) 10, 15–18, reprinted in 1953 U.S. Code Cong. & Admin. News 1543, 1549–1551.

98. *State Land Board v. Corvallis*, *supra*, 429 U.S. at 371–372, n. 4.

99. *Cal. ex rel. State Lands Com'n*, *supra*, 805 F.2d at 864.

Accepting the other jerry-built argument adopted in *Humboldt Spit*, the federal court found that the Submerged Lands Act supplied a basis on which to apply federal law.[100] The court ignored the certain critical points of similarity in *Wilson*. *Wilson* also concerned inland waters,[101] the United States was also a party, and important federal interests were at stake.[102] *Wilson* had not, however, used the Submerged Lands Act as a basis to support the decision that federal law should be the choice-of-law.[103]

In a particularly off-course fashion, the appellate court distinguished *Wilson* as involving a dispute between riparian owners in which the Submerged Lands Act was not involved.[104] But the Supreme Court in *Wilson* had carefully and specifically noted that the State of Iowa was asserting its sovereign land claims based on the Equal Footing Doctrine and not as riparian owner.[105] Moreover, the federal appellate court ignored three of its own cases that had used state law to decide, without making any reference to the Submerged Lands Act, controversies about the property boundaries of lands along navigable inland waters.[106] Consequently, it appears, regardless of the flimsiness of the rationale in support of the application of federal law, when federal

---

100. *Id.* at 861.
101. Inland or internal waters encompass all waterways within the land territory. Examples of inland waters are rivers, lakes, and bays. These are waters landward of the nation's coastline, the line of ordinary low water in direct contact with the open sea. 1 Shalowitz, § 311, pp. 22–23; *Id.* at Appendix A, p. 283; 3 Shalowitz, Appendix A, p. 387.
102. See notes 66 and 74 in this chapter.
103. The Court also ignored its own statement in *Corvallis* that ". . . nothing in the [Submerged Lands] Act in any way mandates or even indicates that federal common law should be used to resolve ownership of lands which, by the very terms of the Act, reside in the States." *State Land Board v. Corvallis, supra,* 429 U.S. at 371, n.4.
104. *Cal. ex rel. State Lands Com'n, supra,* 805 F.2d at 863.
105. *Wilson, supra,* 442 U.S. at 657, n.2. The State of Iowa's sovereign title formed an important reason supporting the Supreme Court's holding on an additional issue in *Wilson*. In *Wilson*, the Supreme Court refused to apply a statute shifting the burden of proof in Indian title cases to the states asserting Equal Footing Doctrine-based land title. *Id.* at 667.
106. See *Puyallup Indian Tribe v. Port of Tacoma* (9th Cir. 1983) 717 F.2d 1251, *cert. den., sub nom. Trans-Canada Enterprises Ltd. v. Muckleshoot Indian Tribe* (1984) 465 U.S. 1049, *reh. den.* (1984) 466 U.S. 954; *United States v. Aranson* (9th Cir. 1983) 696 F.2d 654, *cert. den. sub nom. Colorado Indian Tribes v. Aranson* (1983) 464 U.S. 982; *United States v. Harvey* (9th Cir. 1981) 661 F.2d 767, *cert. den.* (1982) 459 U.S. 883.

In light of its procedural history, the *Aranson* case is particularly pertinent. *Aranson* concerned the Colorado River Indians' claim to certain portions of the bed of the Colorado River that had become exposed. Initially, relying on federal law, the district court found that the Colorado River Indians owned such land. *Aranson, supra,* 696 F.2d at 657. The district court had applied federal law in reliance on *Bonelli*, a case later overturned by the Supreme Court in *Corvallis*. *Ibid.* Supported by the newly decided *Corvallis*, the other contending parties moved for a new trial on the basis that California, not federal, law should have been applied. *Id.* at 858. Applying *Wilson* (which had been decided in the meantime), the federal appellate court decided that, while federal law controlled the dispute, state law would be used as the federal rule of decision. *Id.* at 658. It is important to note that, although *Humboldt Spit* had been decided almost a year before the

interests are concerned a way will be found to apply federal rules of decision. Of course, this rationale casts considerable doubt on the former choice-of-law rules expressed by the United States Supreme Court and other federal courts, which had been consistently applied when the United States was a disputing party.

These two cases relied on the Submerged Lands Act as requiring a federal rule of decision for coastal lands. Because the Submerged Lands Act could not and did not affect the state's ownership in lands beneath inland navigable waters, the Submerged Lands Act should have provided *no* support for a decision that federal law must also supply the rule of decision in such cases. Read most charitably, *Humboldt Spit* can be supported only as a restricted variant of *Wilson*, distinguished by the fortuity that oceanfront tidelands were concerned, with the Submerged Lands Act supplying the flimsy foundation for the adoption of a uniform federal rule. These elements are absent, however, in the case of nontidal navigable waters. One can only hope that in future decisions federal courts will recognize the true path and return to it.

2.8 EFFECTS OF THESE CHOICE-OF-LAW CASES IN OTHER SITUATIONS

Humboldt Spit appears to have had untoward results in another area where interests of the federal government are not concerned—in boundary contests between private owners or between private owners and the states in which one of the contestants derives its title from a federal public domain patent. Prior to *Humboldt Spit* as reaffirmed by *Corvallis*,[107] it had long been understood that state law determined incidents or rights attached to the ownership of property that bordered on navigable waters and that had been conveyed by the United States into undisputed private ownership.[108] *Corvallis*, in applying state law to decide the contest over ownership of the riverbed between a federal riparian patentee and the state, specifically noted that the fact that private title was derived from a federal patent was not sufficient to require the application of federal law.[109] In litigation in California[110] and Florida[111]

Aranson opinion was announced, the Ninth Circuit did not mention *Humboldt Spit* in its decision. The author is aware that the parties advised the Ninth Circuit of the *Humboldt Spit* decision before the court issued its decision in *Aranson*.

107. See note 56 in this chapter.
108. E.g., *Barney, supra*, 94 U.S. at 337–338; *Joy, supra*, 201 U.S. 342–343; *Borax, Ltd. v. Los Angeles* (1935) 296 U.S. 10, 22.
109. *State Land Board v. Corvallis, supra*, 429 U.S. at 372.
110. In *State of Cal.ex rel. State Lands Com. (Lovelace), supra*, 11 Cal.4th at 74, the contesting private landowner claimed that federal (not state) law determined location of his riparian property boundary. Based on federal law, the property owner asserted that the property boundary continued to change its geographic location whatever the cause of the boundary movement. *Ibid.*
111. *Board of Trustees v. Sand Key Associates, supra*, 489 So.2d at 36.

private landowners whose lands lie along the open coast, and whose titles are derived from federal patents long antedating the enactment of the Submerged Lands Act, are asserting the benefit of the *Humboldt Spit* rule. This approach may have far-reaching effects.

For example, in California and other states, such as Florida, a mix of title derivation is found along the coast. Intermittent Mexican or Spanish land grants are located along the California coast from San Diego to almost 100 miles north of San Francisco. More than 500 miles of this California oceanfront is founded on such land grants. On the other hand, lands not granted by Mexico or Spain came into the public domain by virtue of the cession in the Treaty of Guadalupe Hidalgo. More than 250 miles of oceanfront lands are owned by persons deriving their titles from federal public domain patents, and more than 200 miles of coastline is owned by the United States.

A land title nightmare was immediately created by the *Humboldt Spit* decision. Lands granted by Spain and Mexico were never in federal ownership. Consequently, they are claimed to remain subject to California's property rules.[112] Under the *Humboldt Spit* decision, however, oceanfront lands whose title is derived from federal patents are possibly subject to different boundary rules as created by federal courts all across the country. This may well lead to anomalous and irrational results, especially when the private ownerships having different title derivations lie adjacent to one another. The result may well be a jagged, saw-toothed ownership pattern along the oceanfront, bearing no relationship to any actual configuration of the coast.[113]

The problem is not limited to inequity between private landowners. In a Florida case, the state put in certain beach and shore protection structures that enhanced a particular private landowner's land.[114] A Florida statute reserved to the state enhancement to private lands created by shore and beach protection structures constructed by the state. As in California, where such a rule has been court recognized,[115] the Florida statute fosters the state's policy of

---

112. *Los Angeles Milling Co. v. Los Angeles* (1910) 217 U.S. 217, 225, 227–228; *Carpenter v. City of Santa Monica* (1944) 63 Cal. App.2d 772, 783–787.
113. As the United States Supreme Court observed in a case where the application of state or federal law was in question:

> As Representative Sutherland, later to be a Justice of this Court, succinctly put it, "if the appropriation and use were not under the provisions of the State law the utmost confusion would prevail." [Citation omitted.] Different water rights in the same State would be governed by different laws and would frequently conflict. *California v. United States* (1978) 438 U.S. 645, 667–668.

The Supreme Court acknowledged the legal confusion that would result if federal water law and state water law reigned "side by side" in the same area. *Id.* at 668–669. The same can be said of the confusion if the so-called federal boundary and title law were applied to the same region or area as state boundary and title law.
114. *Board of Trustees v. Sand Key Associates*, *supra*, 489 So.2d at 36. That is, suppose a shore project increased the beach in front of a private individual's property. The statute provides that the increase was owned by the state, not the individual.
115. *Carpenter*, *supra*, 63 Cal.App.2d at 783–787.

maximizing public access to and use of beaches that the state has preserved, protected, or created with taxpayers' funds. Under the *Humboldt Spit* rule, the state's policy would be ignored. These beaches would be private. If the public wanted use of the beaches, application of the power of eminent domain would apparently be required. Use of this power may not be a realistic alternative when consideration is given to the fiscal difficulties in which many local and state governments find themselves.

There is some hope, however, that this confusion will be ended and the long-accepted rule returned: State law is the choice-of-law in riparian or littoral boundary contests even when the federal government or a person deriving title from the federal government is a party. Some federal court cases and a California Supreme Court case provide a hopeful indication of a return to the former choice-of-law rule that state law governs these property boundary and title conflicts.[116] Even the United States Supreme Court seems to be moving back toward the correct path.[117] In an action seeking to quiet the United States' claim of title to a riverbed against the state, the Supreme Court recognized, shortly after the *Humboldt Spit* case, that any claim of adverse possession by the United States against the state would be governed by state law.[118]

Such cases reflect the long-entrenched belief that title and boundary disputes should be decided by local or regionally developed rules and that there should be extreme reluctance to pronounce nationally applicable or "federal common law" rules.[119] Federal courts are not the proper forum for resolution of locally founded real property boundary and title disputes, unless some federal instrumentality should by chance be involved. Even in such a case, federal judges and attorneys representing the United States are required to be members of the bar of the state in which they practice. They are familiar with and can competently argue and apply local real property law. Indeed, where there is a conflict between state or private interests on the one hand and federal interests on the other, a balancing approach has been adopted—federal

---

116. *United States v. Pappas* (9th Cir. 1987) 814 F.2d 1342, 1345, n.8, was a quiet title action between an individual and the United States. The court agreed that construction of federal patents is a question of federal law. The court went on to say that unless Congress expressed a contrary intention, it would construe a federal patent according to the law of the state in which the land lies. Since the federal patent did not declare that federal law governed, the Ninth Circuit construed the patent according to state law. *Ibid.* In *State of Cal.ex rel. State Lands Com. (Lovelace), supra,* 11 Cal.4th at 74–75, the California Supreme Court conceded that federal law "may" apply to oceanfront property owned by or whose title is derived from the United States. The court also declared it is "clear" that state law applies "at least" to a boundary dispute over inland property where the United States has no interest. *Ibid.*
117. See, e.g., *Phillips Petroleum Co. v. Mississippi* (1988) 484 U.S. 469, 484, *rehg. den.* (1988) 486 U.S. 1018 (no reason to depart from "general proposition" that law of real property left to states to develop); *Montana v. United States, supra,* 450 U.S. at 551.
118. *Block v. N.D. ex rel. Bd. of Univ. and Sch. lands* (1983) 461 U.S. 273, 292 n. 28.
119. E.g., *Milwaukee v. Illinois* (1981) 451 U.S. 304, 312; *Erie RR. Co., supra,* 304 U.S. at 78.

interests in carrying out a program are weighed against state control over local interests, with particular weight given to the fact that local real property is concerned.[120]

Finally, there is little need for a uniform nationwide property boundary and title rule, in that federal interests would be little affected by the even-handed application of state law (especially as such rules are applied by federal, not state, judges). Yet the impact of imposing a uniform federal rule on both private and nonfederal public landowners would be particularly detrimental. Landowners acting in reliance on continued application of settled state law would have their titles clouded by decisions of remote federal courts, unfamiliar with local peculiarities, when such courts create and interpret "federal common law." Although such decisions might make eminent good sense in one locale, in another environment their application could contradict physical reality. The Supreme Court has recognized this danger in the water law area[121] and, as exemplified by its reasoning in *Wilson*, should return to the traditional path in the boundary and title area *posthaste*.

Hope springs eternal that choice-of-law decisions concerning title and boundary disputes about lands underlying or adjacent to tidal or navigable waters will return to the well-trodden path. And this path is one where the laws of the states, applied in a nondiscriminatory manner, decide such disputes.

---

120. *United States v. California* (9th Cir. 1980) 655 F.2d 914, 917.
121. *Rank, supra*, 90 F.Supp. at 786; *California v. United States, supra*, 438 U.S. at 653; *United States v. New Mexico* (1978) 438 U.S. 696, 702. The U.S. Supreme Court has observed:

> The very vastness of our territory as a Nation, the different times at which it was acquired and settled, and the varying physiographic and climatic regimes which obtain in its different parts have all but necessitated the recognition of legal distinctions corresponding to these differences. *California v. United States, supra*, 438 U.S. at 648.

3

BASIC LEGAL PRINCIPLES DEFINING PROPERTY BOUNDARY MOVEMENT

3.1 INTRODUCTION

Imagine yourself in a window seat of an airplane on your way across the United States. As the plane glides over the landscape, you may just happen to gaze out the window to take in the view. Almost wherever in the United States you might be, that casual glance could encompass the meandering course of a river, the sprawling outline of a lake, or a majestic ocean coastline. From 28,000 feet (and perhaps even from 28 feet), the definiteness and distinctiveness of the edge of land and the beginning of water seem indelibly fixed at that moment in time. From your lofty perspective, it is easy to agree that use of the apparent permanence and immutability of readily observable, definite physical features such as a river, a lake, or the ocean to mark the boundary dividing competing property ownerships is so wise as to be almost divinely inspired.[1]

---

1. "[A] river is an appropriate frontier. Water is neutral and in its impartial winding makes the national boundary look like an act of God." Theroux, Paul, *The Old Patagonian Express* (Houghton Mifflin 1979), p. 40. Judges and surveyors have also recognized the divine inspiration of this rule. One court opinion characterized such rules as "sacramental." *Swarzwald v. Cooley* (1940) 39 Cal.App.2d 306, 323. The rule of surveying priorities holds that calls to natural monuments in real property instruments have priority over (the more secular) artificial monuments or courses or distances and statements of quantity. Robillard, W., and Bouman, L., Clark on Surveying and Boundaries (7th Ed., 1997) § 14.21 (hereinafter "Clark 7th"); Grimes, *Clark on Surveying and Boundaries* (4th ed., 1976) (herein "Clark") § 308; Skelton, *The Legal Elements of Boundaries and Adjacent Properties* (1930) (herein "Skelton") § 103; *County of St. Clair v. Lovingston* (1874) 90 U.S. (23 Wall.) 46, 62; *Swarzwald, supra,* 39 Cal.App.2d at 321.

How can it be that a large body of law[2] has been and is continuing to develop because of constant and festering property boundary disputes that have germinated from the use of these physical features as riparian[3] or littoral[4] property boundaries? It seems almost heretical. When one closely examines riparian or littoral property boundary disputes,[5] one can see that, fundamentally, their genesis lies in the unpredictable instability and impermanence of the physical land/water boundary. Simply as a matter of physical fact, this natural frontier is constantly and irrevocably in flux. This inherent inconstancy and perpetual oscillation in location clashes with the basic tenets of our faith in using such features as property boundaries—the supposed permanence and indelibility of these physical features.[6]

In fact, no matter how tenaciously we hold to the tenets of permanence and stability, we cannot ignore physical fact: The geographic location of any particular water body in relation to an adjoining landform changes naturally, because of human-made works, or in spite of such works.[7] And it is not only the water body that is changeable in level or in location. The landform itself is mutable. Among other circumstances, plate tectonics, subsidence because of groundwater withdrawal or desiccation of soil, or deposition or erosion of soil by water or wind, change the relationship of the landform to the watercourse. All these elements, and the innately litigious nature of humankind, produce boundary and title disputes along tidal or navigable waterways.

It may be apparent from what has been said, but fair warning should be given anyway. The concepts discussed in this chapter may be difficult for some to picture, much less to understand. With the use of case examples and

---

2. It is perhaps best not to mention too prominently the considerable coterie of surveyors and lawyers that has also flourished in order to assist in the creation and interpretation of this body of law.
3. Riparian land is land along a river or stream. Black's Law Dictionary (5th Ed. 1979), p. 1192; *Alexander Hamilton Life Ins. Co. v. Gov't of V.I.* (3rd Cir. 1985) 757 F.2d 534, 538 n.5.
4. Littoral land is land bordering the ocean, a sea, or a lake. Black's Law Dictionary, *supra*, at 842; *Alexander Hamilton*, *supra*, 757 F.2d at 538.
5. Although the terms "riparian" and "littoral" both denote proximity or nearness to a water body, they are not fungible. Therefore, in the interests of clarity and accuracy, the terms will be used in this work as they are defined in the preceding notes 3 and 4.
6. One court captured this inconstancy with an elegant statement about the ocean shore: "Appearing constantly to change, it remains ever the same." *White v. Hughes* (Fla. 1939) 190 So. 446, 449.
7. For example, one jurist has stated the problem quite plainly:

   Since time out of mind, rivers have been used to mark the boundaries between estates, counties, provinces and nations. This custom has both advantages and disadvantages. The chief disadvantage is that river channels are not static. . . .

   This natural phenomenon has been a continuing source of litigation as to riparian rights. Again and again the common law courts have had to resolve boundary disputes between abutting riparian owners. . . . *United States v. Miller* No. 66-75 (Ct. Cl. 1978), Slip Opinion filed June 14, 1978.

hypothetical physical situations, however, the concepts will be made sufficiently concrete to be understood.

This chapter introduces the term "ordinary high-water mark." It explains that this is a legal term of art created by courts to describe the waterward extent of upland ownership and the landward extent of sovereign land title. After that introduction, the chapter discusses how this legal term has become so baffling and mystifying. To understand why that is so, the reader will be asked to visualize and appreciate a physical demarcation between land and water that is constantly changing its geographic location in virtually every dimension.

The discussion then turns to an explanation, in some detail, of the surveying mechanism designed to cope with this problem, the meander line. Included as part of that discussion will be specific examples of the few situations that are exceptions to the general rule that a meander line is not a property boundary. With this background, the chapter will introduce the reader to legal terms of art developed to describe and define the geographic movement of the physical feature denominated as the property boundary and the property boundary consequences of that geographic movement. Finally, this chapter discusses certain rules intended to aid property owners in establishing the process by which changes occur in the geographic location of the property boundary.

Accordingly, the concepts discussed in this chapter are both theoretically difficult and factually and legally complex. To the property owner, the practicing surveyor, the attorney, and others who wish to be knowledgeable in the principles controlling the property boundary effects of the change in the geographic location of the water body/land boundary, these concepts are of critical importance.

3.2 VARIABLE NATURE OF TITLE AND BOUNDARY DISPUTES

Boundary and title disputes are not merely of academic interest. So that the reader will have a basic understanding of the practical consequences of these disputes (and the value to a surveyor, lawyer, or property owner of understanding the concepts explained here), a short description of the potential nature of title and boundary disputes is in order.

Landowners may have property next to a tidal or navigable river that changes its course. They may need to know where *exactly* on the face of the earth a property boundary description using the "riverbank" as one of its courses should be geographically located when the river has changed its physical location. They may also need to know whether it makes a difference that the change in physical location was caused by natural or human-induced changes in the river regime, such as dams or water diversions.[8]

---

8. See, e.g., *Beaver v. United States* (9th Cir. 1965) 350 F.2d 4, *cert. den.* (1966) 383 U.S. 937; *United States v. Aranson* (9th Cir. 1983) 696 F.2d 654, *cert. den. sub nom. Colorado Indian*

Other property owners may have a lakeside vacation home they wish to renovate and expand toward the lake. Because of setback rules or questions of public access, the owners may request an opinion as to exactly where to geographically locate a waterfront boundary described in a title instrument as the "shore of the lake." When the physical location of the lake boundary fluctuates as a result of climatic precipitation changes or diversions of the lake's tributary water sources, how is this location determined?[9]

Suppose a landowner is a corporation, planning for development of a beachfront property as a casino. The beachfront has grown in extent because of human-induced changes in the littoral regime that conveys the beach, creating and nourishing sand. This additional beach area is intended as the proposed site of the oceanfront pool. Where is the seaward property boundary of the corporation's land ownership?[10]

Finally, suppose the government is condemning a valued client's property. The client asks you to provide support for an appraiser's calculation of the acreage in which the property description has as one of its courses the "ordinary high-water mark of _____"?[11]

These situations are only illustrations. The circumstances in which property boundary or title issues about lands adjacent to navigable water bodies may arise are infinitely varied. Yet it is not only the ever-changing demands and inexorable progress of modern civilization and the capriciousness of nature that provide ready impetus to such disputes. When one couples these factors with the aggressiveness and creativity of littoral or riparian property owners (and their lawyers and surveyors) in asserting title or boundary claims, understanding why there are so many boundary and title disputes is elementary.

3.3 IMPORTANCE OF THE LOCATION OF THE PROPERTY BOUNDARY ALONG TIDAL OR NAVIGABLE WATERWAYS

In all of the preceding examples, the issue was to determine and locate for a particular purpose the property boundary between two competing ownerships along a tidal or navigable waterway.[12] The answer to this question is exceed-

---

Tribes v. Aranson (1983) 464 U.S. 982; *United States v. Gerlach Live Stock Co.* (1950) 339 U.S. 725, 730.
9. See, e.g., *Utah v. United States*, Special Master's Report reproduced in 1976 Utah L. Rev. 1; *State v. Longyear Holding Co.* (Minn. 1947) 29 N.W.2d 657; *Martin v. Busch* (Fla. 1927) 112 So. 274; *Churchill v. Kingsbury* (1918) 178 Cal. 554.
10. See, e.g., *Carpenter v. City of Santa Monica* (1944) 63 Cal.App.2d 772; *California, ex rel. State Lands Comm'n v. U.S.* (1982) 457 U.S. 273 (hereinafter *"Humboldt Spit"*); *Internal Imp. Tr. Fund v. Sand Key Assoc.* (Fla. 1987) 512 So.2d 934.
11. See, e.g., *Borax, Ltd. v. Los Angeles* (1935) 296 U.S. 10, 26; *People v. Wm. Kent Estate Co.* (1966) 242 Cal.App.2d 156, 161.
12. Boundaries other than property boundaries may be determined by this process as well. For example, the location of a riparian or littoral boundary is sometimes critical to the jurisdiction of the Army Corps of Engineers under the Rivers and Harbors Act of 1899, 33 U.S.C. §§ 401 et seq. and the federal Clean Water Act, 33 U.S.C. § 1251, et seq.; *Leslie Salt v. Froehlke* (9th Cir.

ingly important, both to the property owner and to the public, for two basic reasons.

First, tension between the competing ownerships on the two sides of the boundary must be resolved. On the waterward side, one finds the sovereign land ownership of the state or federal government in lands underlying tidal or navigable waters.[13] On the landward side, one finds littoral or riparian ownership of private persons or entities whose title usually derives from the present or the former sovereign. Resolution of the competition between these two ownerships is important because of the application of a strict rule: Except in limited instances, lands owned by the sovereign in its sovereign capacity may not be conveyed into private ownership.[14] For littoral or riparian landowners whose lands lie adjacent to navigable[15] waters, it is crucial to know exactly the geographic location boundary of the sovereign's competing ownership. The physical location of that ordinary high-water mark property boundary determines both the extent of the upland title that has been or can be conveyed and the landward extent to which sovereign public trust title encroaches upon upland ownership. Conceivably, determination of the location of that boundary could result in a *loss* of at least part of the upland owner's title.

Second, recall that even if the lands were conveyed into private ownership, the government did not transfer complete fee title to such sovereign trust lands; the conveyed land remained subject to a reserved and inalienable sovereign property interest, an easement for public trust purposes.[16] This public

---

1978) 578 F.2d 742, 747. In San Francisco Bay, California, the authority of the San Francisco Bay Conservation and Development Commission is governed by the location of such a boundary. Cal. Gov. Code § 66610; *Littoral Development Co. v. San Francisco Bay Conservation etc. Com.* (1994) 24 Cal.App.4th 1050, 1061–1064. Perhaps more important, the water boundary location process may also determine the extent of political jurisdiction between two counties, two states, or two nations. E.g., *Nebraska v. Iowa* (1892) 143 U.S. 359.

13. Before the admission of a territory to statehood, the United States held the lands beneath tidal or navigable waters lying within that territory in trust for the future state. E.g., *Shively v. Bowlby* (1894) 152 U.S. 1, 57. Upon statehood, by virtue of the Equal Footing Doctrine, absolute and complete title to lands underlying the beds of tidal or navigable waters inured to the states. *Id.* at 57 through 58; *Illinois Central Railroad v. Illinois* (1894) 146 U.S. 387, 437. For a more detailed explanation, see notes 111 through 121 and accompanying text in Chapter 1.

14. E.g., *Illinois Central Railroad, supra,* 146 U.S. at 452–454; *City of Berkeley v. Superior Court* (1980) 26 Cal.3d 515, 528–529, *cert. den. sub nom. Santa Fe Land Improve. Co. v. Berkeley* (1980) 449 U.S. 840. The limited circumstances in which such an alienation is possible are described in *City of Berkeley, supra,* 26 Cal.3d at 523–525 and *People v. California Fish Co.* (1913) 166 Cal 576, 597, and notes 126 through 174 and accompanying text in Chapter 1.

15. The principles of navigability will be discussed in later chapters. For now, it is sufficient to say that all tidal waters are navigable for sovereign title purposes, as are all water courses that are susceptible of navigation. E.g., *Phillips Petroleum, supra,* 484 U.S. at 476, 480; see notes 81 through 96 and accompanying text in Chapter 5, notes 9 through 59 and accompanying text in Chapter 7, and notes 47 through 58 and accompanying text in Chapter 8.

16. *City of Berkeley, supra,* 26 Cal.3d at 523–525, and *People v. California Fish, supra,* 166 Cal at 597; see notes 122 through 131 and 138 through 172 and accompanying text in Chapter

trust easement is so encompassing that, according to the United States Supreme Court, the owner is left with only a "naked fee."[17] Obviously, determination of the geographic extent of the retained and inalienable public trust easement is crucial to the property owner. It is just as critical for the public. The sovereign should be aware of the extent of its public trust interest when, among other matters, it acquires shorelands. The extent of the public trust interest is also important when the government attempts to manage that sovereign interest by controlling or allowing access. It is readily apparent that resolution of the location of this property boundary can have consequences that may well be felt for decades.

3.4 ORDINARY HIGH-WATER MARK PROPERTY BOUNDARY—IN GENERAL

The landward property boundary of littoral and riparian lands owned by the sovereign and, conversely, the waterward property boundary of riparian and littoral landowners[18] is the ordinary high water mark.[19] On its face, the term "ordinary high-water mark"[20] has an aura of certainty. On closer inspection, this aura vanishes. Exactly what is meant by "ordinary"? How high is "high" What precisely is a "water mark"? Are there any differences between the physical location of the ordinary high-water mark along tidal waterbodies and

---

1. Only in extremely limited circumstances could complete fee title be transferred and only if certain very stringent conditions were met. See note 14 in this chapter.

17. *Summa Corp. v. California ex rel. Lands Comm'n* (1984) 466 U.S. 198, 205; see notes 173 through 174 and accompanying text in Chapter 1.

18. Certain states have chosen to grant to riparian or littoral owners portions of the states' sovereign interests waterward of the ordinary high-water mark. For example, in California riparian owners along nontidal streams and rivers and littoral owners along lakes have been granted the land between the ordinary high-water mark and the ordinary low-water mark of such water bodies. Cal. Civ. Code. § 830; *State of California v. Superior Court (Lyon)* (1981) 29 Cal.3d 210, 225-226, *cert. den.* (1981) 454 U.S. 865. In other states, owners have been granted the land to the ordinary low-water mark or even the thalweg of rivers. E.g., *United States Gypsum v. Uhlhorn* (E.D. Ark. 1964) 232 F.Supp. 994, 1001, n.2., *aff'd* (8th Cir. 1966) 366 F.2d 211, *cert. den.* (1967) 385 U.S. 1026. At least in California, the lands between the ordinary high-water mark and ordinary low-water mark that have been granted by the state remain and are held by littoral or riparian owners subject to an easement for public trust purposes retained by the state. *Lyon, supra*, 29 Cal.3d at 231.

19. E.g., *United States v. Pacheco* (1864) 69 U.S. (2 Wall.) 578, 595 (estuary); *San Francisco v. Leroy* (1891) 138 U.S. 656, 672 (estuary); *Borax, Ltd., supra*, 296 U.S. at 22–23 (estuary); *Packer v. Bird* (1891) 137 U.S. 661, 666 (tidal river); *Barney v. Keokuk* (1876) 94 U.S. 324, 336 (nontidal river); *Hardin v. Jordan* (1891) 140 U.S. 371, 382 (lake).

20. As we shall see, courts treat as equivalent the terms "line of ordinary high tides" and "high-water mark." The ambiguity in the use of such terms to describe a physical location is apparent. For example, knowledgeable surveyors have considered the word "mark" to connote a boundary. On the other hand, those same surveyors have also pointed out that the word "mark" could mean a physical impression on the shore, such as a debris line.

the physical location of the ordinary high-water mark along nontidal water bodies? And how is one to consistently and predictably locate the ordinary high-water mark when the water body and the landform against which the water body is impressed may be constantly changing their physical relationship to one another? This discussion highlights the need for a consistent, scientifically verifiable location of the ordinary high-water mark property boundary to make that property boundary definite and consistently recognizable or locatable.

About the best one can say is that this legal term of art does not describe a specific geographic or physical location.[21] Thus, the term is part of the nomenclature of the legal system, of judges and lawyers. It has been developed and used to describe and denote the legal property boundary of riparian or littoral lands. Courts have treated the term "ordinary high-water mark" in much the same way as some judges have defined pornography: "I know it when I see it."[22]

We will see that, before development and application of scientific methods to determine the physical location of the ordinary high-water mark (in the case of a particular regime and the accompanying and necessary validation of such methodology by the courts), surveyors and even some courts "knew" an ordinary high-water mark when they saw one.[23] But would another court,

---

21. For example, in the case of tidal regimes one highly regarded commentator defines "ordinary high-water mark" as follows: "Same as Mean High-Water Line. See Ordinary Tides." 2 Shalowitz, *Shore and Sea Boundaries* (Dept. Comm. Pub. 10-1 1962) (herein "2 Shalowitz"), p. 589. "Ordinary Tides" is defined by the same commentator: "This term is not used in a technical sense by the Coast and Geodetic Survey, but the word 'ordinary' when applied to tides may be taken as the equivalent of the word 'mean.' See *Borax Consolidated, Ltd. v. Los Angeles*." The *Borax* opinion referenced by that commentator observed that the property boundary was the high-water mark and, in the situation presented in that case, that legal term meant a line on the ground determined by the course of the tides. *Borax, supra*, 296 U.S. at 22. Because of the oscillation of the tides, the *Borax* Court observed that from the use of the term "ordinary high-water mark" alone it could not state how the ". . . line of 'ordinary' high water [was] to be determined" *Id.* at 23. The same is true for the physical location of the ordinary high-water mark of nontidal water bodies. *Borough of Ford City v. United States* (3rd Cir. 1965) 345 F.2d 645, 649, 650, *cert. den.* (1965) 382 U.S. 902. Thus, commentators and courts alike recognize that it is only after court application of the legal term "ordinary high-water mark" to the physical processes and situations present in a case that geographically locating the ordinary high-water mark property boundary of littoral or riparian lands is possible.
22. *Jacobellis v. Ohio* (1964) 378 U.S. 184, 197 (Stewart, J., concurring).
23. United States Supreme Court opinions validate this point, as do later accounts by commentators. In *United States v. Pacheco, supra*, 69 U.S. at 590, the Supreme Court stated: "By the common law, the shore of the sea, and, of course, of arms of the sea, is the land between ordinary high and low-water mark, land over which the daily tides ebb and flow." Even less helpful was one surveyor's testimony as recounted by a commentator: "Q. What was your answer relative to the historical application of the words "line of ordinary high tide" as meaning? A. Approximately the line of where vegetation ceases to grow." Corker, "Where Does the Beach Begin" (1966) 42 Wash.L.Rev. 33, 53. And in one United States Supreme Court case it was suggested that the United States Coast Survey's rule in the "measurement of waters" determined the location of the ordinary high-water mark. *Knight v. U. S. Land Association* (1891) 142 U.S. 161, 208 (Field, J.,

another lawyer, another surveyor, or another property owner observing identical conditions locate the ordinary high-water mark in the same fashion or at the same geographic location? Probably not. Each person's subjective judgment would most likely be different, even presented with the same facts. Until recently,[24] no "community standards" were available to universally, consistently, and scientifically physically situate the ordinary high-water mark property boundary.

Innate knowledge of the physical location of the ordinary high-water mark property boundary might have been sufficient in the days when land was plentiful, when pressure for development of riparian and littoral property was lacking, and when concern over shoreline and waterway environment was not as passionate. That innate knowledge will not suffice today. There is intense competition for available and developable sites along coasts, bays, rivers, and so forth. Environmental interests in such lands are expressed through a variety of laws, regulations, and cases. As a necessary result, a demand for greater precision in determining the physical location of the littoral or riparian property boundary has evolved. An explanation of the purpose of this property boundary may help in understanding how and where the ordinary high-water mark property boundary may be physically located.

3.5 PURPOSE OF THE ORDINARY HIGH-WATER MARK— TO SEPARATE ARABLE FROM NONARABLE LANDS

The legal term of art "ordinary high-water mark" was developed because of the well-known and readily observed seasonal fluctuations of either a landform itself or an adjacent water body. As far as the author is aware, the vagaries of a particular physical regime, such as the sea or a river, did not inspire the idea. Rather, as the saying goes, necessity was the mother of invention.

A legal concept had to be developed to describe at approximately what stage or height to measure a water body against the land. Creation of this concept was necessary to describe the geographic location of the littoral or riparian property boundary and therefore the geographic extent of lands held by the sovereign and subject to public rights.[25] The line of division was

---

concurring). This approach has been severely criticized by those familiar with Coast Survey practices. 2 Shalowitz, § 11, p. 80; Hull, W., and Thurlow, C., "Tidal Datums and Mapping Tidal Boundaries" (U.S. Dept. of Comm., National Ocean and Atmospheric Administration, National Ocean Survey), pp. 9–10.

24. We will see that in the case of tidal water bodies, a physical location of the ordinary high-water mark has been endorsed and accepted by the courts. See Sections 4.3 through 4.11 in Chapter 4. No such accepted standard exists for nontidal water bodies. See note 172 and accompanying text in Chapter 7 and note 68 and accompanying text in Chapter 8.

25. For an excellent discussion of the development of the concept from Roman times, see Althaus, Helen F., "Public Trust Rights" (U.S. Fish and Wildlife Service 1978), pp. 3–23. All further references to this extremely useful work will be to "Public Trust Rights, at p. ___."

intended to separate lands subject to use for agricultural purposes from lands so frequently overflowed by waters of the adjacent water body that agricultural use was not feasible.[26] The legal boundary was an attempt to separate lands that could be put to private use from lands that should be maintained and preserved for the public.[27]

This property boundary concept is not of recent origin. In fact, every legal system, beginning at least with that of Rome, has produced a rule for the location of property boundaries along waterways. For example, under Roman rule the seashore extended as far as "the greatest winter flood runs up."[28] Land covered by the winter flood run-up was land charged with the right of public use, was owned by the sovereign for the benefit of all people, and was not subject to sale or alienation.[29] The French, Spanish, and Mexican sovereigns also used the "highest high-water mark" or similar descriptive term to define the limit of their sovereign ownership.[30]

The English rule differed somewhat from its continental cousins. In England, the term "ordinary high-water mark" described the landward extent of the king's (or queen's) ownership and, conversely, the seaward or waterward extent of private ownership.[31] This term defined the landward boundary of an area of sovereign ownership that was not the same in geographic scope as the extent of sovereign ownership of other European sovereigns. The English

---

26. E.g., *City of Los Angeles v. Borax Consolidated Limited* (9th Cir. 1935) 74 F.2d 901, 905, *aff'd* (1935) 296 U.S.10; *Howard v. Ingersoll* (1851) 54 U.S. (13 How.) 409, 454–455 (Nelson, J., concurring); *People v. Morrill* (1864) 26 Cal. 336, 356.

27. E.g., *Board of Trustees, etc. v. Mediera Beach Nom., Inc.* (Fla. 1973) 272 So.2d 209, 213. As one court put it:

> These articles [referring to articles of the civil law] establish, in effect, that certain property designated common or public is held by the State and is neither alienable nor susceptible of private ownership, while other property can be owned by anyone and is not subject to restrictions on its alienability. The redactors of our . . . codes were aware, when they included these articles in our law, that in France, public things, such as navigable rivers and their beds, are part of the public domain . . . . Such things are governed by a regime entirely different from that governing private law ownership . . . , for property in the public domain is held by the State not in its proprietary capacity, but for the benefit of all the people. *Gulf Oil Corp. v. State Mineral Bd.* (La. 1975) 317 So.2d 576, 582.

28. Public Trust Rights, at p. 5, citing a translation of the Institutes of Justinian, T. Sandars, *The Institutes of Justinian* (4th Ed. 1867).

29. Public Trust Rights, at p. 4; *Boston Waterfront Dev. Corp. v. Com.* (Mass. 1979) 393 N.E.2d 356, 358–360; *City of Berkeley v. Superior Court*, *supra*, 26 Cal.3d at 521.

30. Public Trust Rights, at p. 15 (under French law the coast extends to the highest regular tide); Las Sieta Partidas (Scott, ed. CCH 1931) Par. III, Tit. XXVIII, Law IV.; *Stewart v. United States* (1942) 316 U.S. 354, 359 (under Spanish and Mexican law the shore included land covered by the tides at their most landward extent whether in winter or in summer).

31. Public Trust Rights, at pp. 23–38; *Attorney General v. Richards* (1795) 2 Anstruther 603, 608, 145 Eng. Rep. 980; *Attorney General v. Parmeter* (Ex. 1811) 10 Price 378, 147 Eng. Rep. 345, 352.

shoreline boundary extended only as far as the reach of ordinary, or what were called "neap," tides.[32]

With the adoption of English common law in this country, the English definition of the shore became ingrained in American jurisprudence.[33] Instructions provided to government surveyors charged with locating the extent of public and private lands required this mythical water mark to be located.[34] Court decisions validated the use of the term, but few have given it any scientific or physical substance.[35] Consequently, although the property boundary of riparian or littoral lands continued to be the ordinary high-water mark,[36] questions remained. What is the ordinary high-water mark property boundary as a physical location? How can it be consistently and precisely located?

---

32. *Attorney General v. Chambers* (1854) 4 De G.M. & G., 43 Eng. Reprint 486, 490. (A reprint of the *Attorney General v. Chambers* opinion can be found in 2 Shalowitz, Appendix D, pp. 640–646.). We will see how use of the term "neap" tides has resulted in some confusion. In Chapter 4 the author will attempt to dispel such confusion. It is enough now to tantalize the reader so that he or she will lust to read ahead.

33. Public Trust Rights, *supra* at 38; *Martin v. Waddell* (1842) 41 U.S. (16 Pet.) 367, 416; *Pollard's Lessee v. Hagan* (1845) 44 U.S. (3 How.) 212, 227; *Jones v. Soulard* (1860) 65 U.S. (24 How.) 41, 65; *Barney*, *supra*, 94 U.S. at 337; *Boston Waterfront Dev. Corp.*, *supra*, 393 N.E.2d at 358–359.

34. E.g., General Instructions to Deputy Surveyors Engaged in Surveying the Finally Confirmed Land Claims in California (January 1858), reprinted in Uzes, F., *Chaining the Land* (Landmark 1977), Appendix C, pp. 211–214; Instructions from the California Surveyor General to H. A. Higley, Alameda County Surveyor, June 9, 1858; Hawes, *Manual of United States Surveying* (Carben Surveying Reprints 1977), p. 46; *Manual of the Instructions for the Survey of the Public Lands of the United States* (U.S. Dept. of Int., Bur. of Land Mgmt. 1973) (Landmark Reprint) § 3-115, pp. 93–94. The Manual of Instructions of the Commissioner of the General Land Office to the Surveyors General of the United States Relative to the Survey of the Public Lands and Private Land Claims, May 3, 1881 provided that the "ordinary low water mark of the actual margin of the river or lakes" was to be located. Minnick, *A Collection of Original Instructions to the Surveyors of the Public Lands, 1815–1881* (Landmark), p. 493. Obviously, there was some confusion about what boundary was to be surveyed.

35. Compare *United States v. Pacheco*, *supra*, 69 U.S. at 590; *DeGuyer v. Banning* (1897) 167 U.S. 723, 735–736; *Coburn v. San Mateo County* (9th Cir. 1896) 75 Fed. 520, 528, with *Borax, Ltd.*, *supra*, 296 U.S. at 22–26.

36. The discussion in this chapter is also applicable to the ordinary low-water mark. The genesis of that boundary is the authority of each state to regulate and control the shores along the coast, on rivers, and on lakes. *Hardin*, *supra*, 140 U.S. at 382. Note 18 in this chapter observes that certain states granted to their citizens littoral or riparian lands to the ordinary low mark, making the property boundary between state sovereign fee ownership and private littoral or riparian ownership the ordinary low-water mark. E.g., Cal. Civ. Code § 830; *State ex rel. Buckson v. Pennsylvania Railroad Co.* (Del. 1969) 267 A.2d 455, 459; *State v. Cain* (Vt. 1967) 236 A.2d 501, 504; *Conran v. Girvin* (Mo. 1960) 341 S.W.2d 75, 80. Also, along the open seacoast, the boundary between lands owned by the state and the submerged lands owned by the United States, but later confirmed and granted to the states, is the ordinary low-water mark. *United States v. California* (1947) 332 U.S. 19, 22; Submerged Lands Act, Act of May 22, 1953, 67 Stat. 29, set forth in 43 U.S.C. §§ 1301, et seq. More detailed discussion of the ordinary low-water mark awaits the reader in following chapters.

3.6 DEFINITION OF THE ORDINARY HIGH-WATER MARK

In later chapters of this work both of the preceding questions are discussed with particular reference to specific physical regimes: open coast tidal shorelines, inland water tidal estuaries and rivers, nontidal navigable rivers, and navigable lakes. For each of those regimes the answers to those questions are different, with the distinction depending on the physical nature of the water body and the surrounding shore and upland. For our present purposes, it is enough to answer these questions quite simply: The "ordinary high-water mark" is a legal term. It describes the property boundary between sovereign lands and uplands. The property boundary should be geographically or physically located by the use of scientific methodology that, if possible, considers and accounts for both the physical character of the shore and the nature of the regime of the boundary watercourse in order to provide predictability and certainty of location.[37]

Even more simply, the "ordinary high-water mark" is not a scientific term, but its geographic location will depend on physical forces or features that may be quantified scientifically—the kind and nature of the shore and its component soils and the type and behavior of the waterway or water body regime by which the shore is bounded. For example, the methodology for determining the physical location of the ordinary high-water mark may well differ when there is a rocky ocean shoreline, as opposed to when there is a sandy beach. And the manner in which the ordinary high-water mark along a river comes to be physically located will differ from how the ordinary high-water mark in a tidal estuary is located in a marsh. The only common factor is that both the shoreline landform and the boundary watercourse regime have to be accounted for and considered. The reason is that there are two physical components of any water boundary determination: the landform's elevation, slope, and composition and the relative elevation of the water level that impresses itself against the landform.[38]

3.7 ILLUSTRATION OF THE COMPONENTS OF THE DYNAMIC ORDINARY HIGH-WATER MARK

Picturing sandy ocean beach, battered constantly by waves of ever varying strength and intensity, is perhaps the best (and easiest) way to visualize the

---

37. On the use of scientific information, methodology, and expert opinion to determine the location of the ordinary high-water mark, see Maloney, "The Ordinary High-water Mark: Attempts at Settling an Unsettled Boundary Line" (1978) 13 Land and Water Law Review 465, 467; *United States v. Cameron* (M.D. Fla. 1978) 466 F.Supp. 1090, 1111–1112; *City of Newark v. Natural Resources Council* (N.J. 1980) 414 A.2d 1304, 1306, *cert. den.* (1980) 449 U.S. 983; *Perry v. State of California* (1956) 139 Cal.App.2d. 379, 396.
38. 1 Shalowitz, *Shore and Sea Boundaries* (Dept. of Comm. Pub. 10-1 1962) (herein "1 Shalowitz") § 64, pp. 89–90.

3.7 THE COMPONENTS OF THE DYNAMIC ORDINARY HIGH-WATER MARK

permanent, yet ever changing components of littoral property boundary. On a grand scale, looking at the reach of an open ocean coastline as a whole, the permanence of the land/sea boundary seems readily observable. The land gradually slopes down (or in some cases precipitously drops) into the sea; the intersection of land and sea is the boundary. Absent some cataclysmic event, such as a great earthquake whereby large sections of land may sink or rise, the location of the convergence of land and sea is, in relative terms, stable. It is when the focus becomes finer and more specific that the ephemeral nature of the open-coast ocean beach littoral property boundary becomes manifest.

One method of refining our focus on the ocean beach littoral boundary is to use our imagination to suspend ourselves over that beach in a hot air balloon on a clear and windless day. Looking down from the still basket of the balloon, we would see below the firm or dry sand, gradually giving way to the boundary water body and its underlying bed. Indeed, assume for the moment that on this hot summer day we could move the surfers, strollers, and sun worshipers for an unobstructed view of the sandy ocean beach and the waves as they roll in, break, and run up onto the beach face. Between the ocean and the upland is the shore, composed, in our example, of the wet sand of a beach. A shore may also be composed of the banks of a river or the mud flats and marsh of a tidal estuary.[39] As each wave inexorably rolls in, breaks, and runs up the beach face, we would see that the wash of each wave does not attain exactly the same location on the beach face. And the season in which we make our observation will color our view of the extent of the beach covered by wave-driven water.[40]

Let us assume we marked the location of the landward extent of a particular wave run-up in the summer with a deeply imbedded, prominent, and stable monument that could be located both horizontally and vertically on the face of the earth with reference to known points. If we could transport ourselves in time to that same beach at the same stage of the tide at the same location in the dead of winter on a windless day, we would see that the location so carefully monumented in the summer was, frequently, if not always, inundated or attained by each winter wave run-up. It would appear, as each wave rolled in and broke, that the winter wave run-up would reach much farther landward and the monument would be covered by the sea to a greater depth than during

---

39. From a slightly different perspective, one commentator has described these components as follows:

> [The composition of a water course] . . . can be described as consisting of: (a) water, (b) a bed, and (c) banks and shore. By way of physical property definition the "bed" is that property which lies beneath the water of the waterway. The "bank and shore" is that property which lies between the high and low water mark. Porro, "Invisible Boundary—Private and Sovereign Marshland Interests" (1970) 3 Natural Resources Lawyer, 512, 516.

Also *Lux v. Haggin* (1884) 69 Cal.255, 417. One case has defined the seabed as the land below the ordinary high-water mark. *United States v. Ray* (5th Cir. 1970) 423 F.2d 16, 20.

40. Bascom, *Waves and Beaches* (Anchor Books 1964), pp. 188 and fig. 58. This little book is considered a classic in the field and an authoritative source.

the summer wave run-up.[41] Which wave run-up, at which season, and at what beach profile, defines the physical location of the ordinary high-water mark property boundary?

Thus, even with our seemingly ideal and enjoyable overhead perspective, the exact geographic location of the littoral property boundary is still not readily apparent. We have only succeeded in affirming that by such observation alone it is not possible to precisely locate the littoral property boundary. Perhaps looking at the beach from a entirely different perspective will allow a precise determination of the physical location of the ordinary high-water mark property boundary. If not, it will at least provide further information to our understanding of the physical processes that underlie property boundary movement. Indeed, from this perspective (described in the following paragraph), each of the basic physical components of that boundary will be presented to us in a more dramatic and even more concrete fashion.

Let us again use our imagination and place ourselves along an ocean beach shoreline. Assume we have the power to take a cross section of the earth at that location at a right angle and perpendicular to the beach, such as is roughly shown in Figure 3.1. Watching this cross section, we would see the ocean surging up against the land. The level to which the ocean surge rises will depend, in very general terms, on the stage of the tide and the slope and elevation of the land relative to the water. Figure 3.1 captures that dynamic situation at one instant in time. It also captures the dynamics of the beach surface, which may change according to the seasons—hence, the summer and winter beach profiles are shown and the relative landward coverage of the same water level may be imagined.

The importance of Figure 3.1 is twofold: First, it graphically shows us that unless some other means that does not refer to the water surface level is

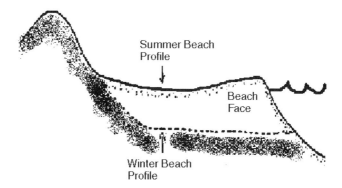

Figure 3.1 Seasonal beach profiles

---

41. *Ibid.*

adopted,[42] in order to physically locate the property boundary, both components forming the boundary should be known: the level of the water and the level of the land. Second, the figure emphasizes the dynamic nature of this boundary; this is a "snapshot" of one instant of time and physical condition, which may be entirely and unpredictably different in the very next instant.

Given the fact of this inconstancy, other than by arbitrarily choosing a particular water level and land elevation as the physical location of the property boundary or adopting some other physical indicia, how can a rational person locate and determine the extent of littoral or riparian property ownership? It may seem that the answer to this question is, as Bob Dylan said, "blowin' in the wind." Fortunately for those of us who live in coastal, lake, or riverine areas, surveyors and lawyers, in forced conjunction with the courts, have, over many years, developed a system and rules that attempt to, and in large part do, provide a framework to resolve this problem.

3.8 SURVEYING TECHNIQUES TO LOCATE THE ORDINARY HIGH-WATER MARK—MEANDER LINES

A discussion of the system and rules used in locating the ordinary high-water mark should begin with the answer to the as-yet-unasked, but what should be the foremost, questions on all readers' minds. Those questions, in one form or another, probably are as follows: What about surveyed property boundary lines along water bodies? Don't such surveyed lines provide the definitive answer to the location of property boundaries along a watercourse? What is the need for any further inquiry once the government has located the extent of its lands and, by virtue of that survey, conveyed them? As the reader will by now have guessed, there is no easy or direct answer to this apparently logical speculation.

Our path to the answers to this series of questions requires us to quickly refresh our recollection about the derivation of land titles. Recall that, with certain exceptions noted elsewhere, land titles are derived from the government.[43] For illustrative purposes and because of the ubiquitous extent of the public domain and public land patents, this chapter focuses on the federal government as the source of title.

---

42. Certain courts have recommended what is known as a vegetation line test. E.g., *Howard, supra*, 54 U.S. at 454–455 (Nelson, J., concurring). In later chapters, the advantages and shortcomings of this test will be examined in some detail. For purposes of this chapter, it need only be noted that the vegetation line test poses many, if not more, of the same questions already raised.
43. The sources of land titles are discussed in Chapter 1.

3.8.1 Public Land Surveys

As part of the disposition process of the public domain, surveys were conducted by United States surveyors to describe and denote the extent of public lands.[44] In conducting those surveys, the surveyors, at some time and probably as a matter of course, would be faced with the decision about where to locate the intersection of public lands with adjacent navigable watercourses. The beds of those watercourses were not part of the public lands of the United States, but were either owned by the states or held in trust for future states by the United States.[45]

The government wished to avoid the obvious confusion that would result if each surveyor individually, and perhaps idiosyncratically, determined just how to survey and geographically locate that boundary. Consequently, the surveyors who conducted those surveys on behalf of the United States were guided by uniform instructions prepared by the United States General Land Office.[46] Over the years these instructions grew more refined and for the last 80 years or so have remained remarkably consistent.[47] Based on these instructions, the initial boundaries of the public lands adjacent to tidal or navigable waterbodies were established by such official surveys and the resultant plats.[48]

Figure 3.2 is an example of such an official plat. It shows a water body, Humboldt Bay, bounded by public land lots, including some swamp and overflowed lands. The United States patent of lands that included or were bounded by a water body incorporated such initial boundaries.[49] Thus, the extent of what was conveyed by the government and what could be conveyed by subsequent owners was established, at least at the time of the government patent.

---

44. The public land surveying process is described in notes 49 through 65 and accompanying text in Chapter 1.
45. See note 13 and accompanying text in this chapter.
46. These instructions are extremely useful in determining what features the United States intended the surveyors to survey in the field. E.g., *Klevin v. Gunderson* (Minn. 1905) 104 N.W. 4, 6; *United States v. Otley* (9th Cir. 1942) 127 F.2d 988, 1000.
47. Department of Interior, Bureau of Land Management, "The Meandering Process In The Survey of the Public Lands of the United States" (undated). For example, the 1902 Manual of Instructions for the Survey of the Public Lands at section 171 requires that in the survey of lands bordering on tidewaters, meander corners are to be set at mean high water. The 1930 Manual of Instructions for the Survey of Public Lands provided at section 231 that "[n]avigable rivers [of a certain width] will be meandered on both banks at the ordinary mean high water mark. . . . Tidewater streams . . . should be meandered at ordinary high-water mark. . . ." The 1973 Manual of Instructions provides that "[a]ll navigable bodies of water . . . are segregated from the public lands at mean high-water elevation." *Manual of Instructions for the Survey of the Public Lands of the United States* (U.S. Dept. of Int. 1973), *supra*, § 3-115, p. 93.
48. *Hardin, supra,* 140 U.S. at 372, and *Foss v. Johnstone* (1910) 158 Cal. 119, 125, both contain a depiction of an official plat that was the result of such surveys.
49. *Railroad Company v. Schurmeir* (1868) 74 U.S. (7 Wall.) 273, 284, 287; *Jefferis v. East Omaha Land Co.* (1890) 134 U.S. 178, 194–195. ("It is a familiar rule of law, that, where a plat is referred in a deed as containing a description of land, the courses, distances, and other particulars appearing upon the plat are to be as much regarded in ascertaining the true description of the land and the intent of the parties as if they had been expressly enumerated in the deed.")

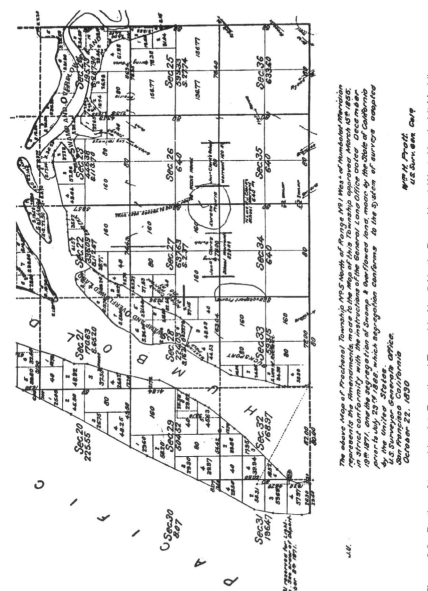

Figure 3.2 Portion of township plat: Fractional Township No. 5 North, Range No. 1 West, Humboldt Meridian

Surveying can be quite precise if instruments and techniques are utilized with that end in mind. Using instruments available in the mid-nineteenth century, surveyors could locate particular geographical points with surprising, even remarkable, precision.[50] With the benefit of laser technology, global positioning systems, and other technical marvels that have been created and developed over the years, the accuracy and precision of modern surveying have truly become awesome. Yet suppose that even the modern-day public land surveyor was instructed as follows:

> Both banks of navigable rivers are to be meandered by taking the courses and distances of their sinuosities. . . . [I]n meandering [the banks of a navigable stream, lake, pond, or bayou] you are to commence at one of those corners on the township line, coursing the banks, and measuring the distance of each course from your commencing corner to the next "meander corner,". . .[51]

Do these surveying instructions precisely define the exact physical location or feature where the surveyor would properly place his rod when he is "coursing the banks," and so forth? What is the "bank"?[52] Assuming one could

---

50. Conversations with William B. Wright, Donald Brittnacher, François Uzes, Roy Minnick, and J. Michael McKown, surveyors extraordinaire; Uzes, François, *Chaining the Land* (Landmark 1977), Chapter 1. This statement, however, assumes that the nineteenth-century surveyors were using the best of instruments. On the other hand, nineteenth-century surveys were often performed with a magnetic compass that was not all that precise and was not intended to be. After all, the government required lengths to be within 100 links per mile, or 1 foot per 80 feet. *Ibid*; Hawes, *supra*, p. 27 (type of compass); Clevenger, Shobal, *A Treatise on the Method of Government Surveying* (1883) (Carben Surveying Reprints), p. 17 (acceptable measurement).
51. Taken from the meander section of the Manual of Instructions for the Survey of the Public Lands 1855 as reprinted in Minnick, *A Collection of Original Instructions to the Surveyors of the Public Lands, 1815–1851*, *supra*, p. 366. In another situation, the government's instructions cryptically stated:

> . . . the boundary lines shall be ascertained by running from the established corners, . . . to the water course. . . *Railroad Company v. Schurmeir*, *supra*, 74 U.S. at 274.

52. One attempt at defining the term "bank" is illustrative of the problem of trying to capture a natural physical feature with mere words and phrases:

> The water-washed and relatively permanent elevation or acclivity at the outer line of the river bed which separates the bed from the adjacent upland . . . and serves to confine the waters within the bed and to preserve the course of the river. 2 Shalowtiz, Appendix A, p. 549.

In contrast to that relatively precise description, some courts believe the term "bank" is the equivalent of the term "shore." For example, one opinion held that "[t]he term 'shore' in its ordinary sense signifies the land that is periodically covered and uncovered by the tide, but it is sometimes applied to a river or a pond as synonymous with 'bank'." *Freeman v. Bellegarde* (1895) 108 Cal.179, 187. Yet the 1973 Manual of Surveying Instructions defines the shore as the space between the margin of the water at its lowest stage and the banks at high water. *Manual of Instructions for the Survey of the Public Lands of the United States*, *supra*, § 3-115, p. 93. This

discern a bank, is the rod placed on the top of the bank, along its slope at the intersection of land and water, or at some other location? Does the placement of the rod at a particular location on the bank forever establish the location of the upland property boundary? Indeed, does this "coursing," no matter how precisely accomplished, make any difference to the upland property boundary as long as the boundary along the waterway is shown or described as the "bank of the river" or "the shore of the lake" or "the beach" or "the ocean"? Does a change in the physical location of the bank affect the location of the property boundary?

3.8.2 Effect of Meander Lines on Property Boundaries

As can be seen, the application of surveying methodology and principles to the location of a shoreline property boundary location appears only to raise more questions. Approaching first the surveying aspects of locating property boundaries along navigable water bodies, a very simple solution has been developed. In public land surveying, where public lands bordered tidal or navigable water bodies, government surveyors were to "meander" the shore of the water body at ordinary high water.[53]

By "meander" is meant, so far as the property boundary is concerned, that the surveyor is, by the ordinary surveying method of measuring and marking courses and distances, to attempt to closely depict the sinuosities of the intersection of the land and the watercourse.[54] The exact location of the intersection of land and water at the time the surveyor happens to be there is unimportant, as the actual property boundary will be the water body.[55]

Meander lines are surveyed because of the difficulty of following the various sinuosities of a water line; by exclusion and inclusion of the irregularities of contour, the meander line produces an average result that closely approx-

---

discussion establishes that there may be as many definitions of the term "bank" as there are persons asked to define the term.

53. *Railroad Co. v. Schurmeir, supra,* 74 U.S. at 286–287; *Hardin, supra,* 140 U.S. at 380–381; *Thomas B. Bishop Co. v. Santa Barbara County* (9th Cir. 1938) 96 F.2d 198, *cert. den.* 305 U.S. 623; Elgin and Knowles, *Principles of Boundary Location for Arkansas* (Landmark, 1984), pp. 158–159; *1973 Manual of Instructions for the Survey of the Public Lands of the United States, supra,* § 3-115, p. 93. There is a lengthy description of the meaning of meander lines in the 1973 Manual, *id.* at §§ 3-115–3-127 and in Clark 7th, *supra,* § 13.01, and Clark 4th, *supra,* § 257.

54. As described by one court, a meander line is ". . . a line . . . described by courses and distances, being a straight line between fixed points or monuments, or a series of connecting straight lines. The line is thus fixed by reason of the difficulty of surveying a course following the sinuosities of the shore; and the impracticability of establishing a fixed boundary along the shifting sands of the ocean." *Den v. Spalding* (1940) 39 Cal.App.2d 623, 627.

55. E.g., *Railroad Co. v. Schurmeir, supra,* 74 U.S. at 286–287; *Hardin, supra,* 140 U.S. 380–381; *Den, supra,* 39 Cal.App.2d at 627; Clark 7th, *supra,* § 13.03, p.342; Clark 4th, *supra,* § 259, p. 312.

imates the quantity of upland contained in the fractional lots bordering on the lake or stream.[56] The official plat made from such a survey shows the meander line, the general form of the water body that has been deduced therefrom, and the adjoining fractional lots bordering on the water body.[57] The patents refer to this plat for identification of the lots conveyed, and the lots in legal effect extend to and are bounded by the water body.[58] But the water body itself ". . . as a natural object, . . . is virtually and truly one of the calls of the description or boundary of the premises conveyed; and all the legal consequences of such a boundary, in the matter of riparian rights and title to land under water, regularly follow." [59]

Thus, the surveyed boundaries of a particular tract adjacent to a watercourse might not describe by course and distance the exact geographic location of the boundary watercourse at the time the surveyor was there. The general physical configuration and geographic location of the boundary watercourse with respect to the land at the time of the survey will, in most cases,[60] be described.[61] A surveyed line is created by this process. That surveyed line may be and often is cartographically depicted on a map, giving the meander line unintended, but significant, weight. For example, the heavy line along the feature in Figure 3.2 denominated as Humboldt Bay is a meander line. Nevertheless, the reader should not be misled by that appearance; *a property boundary line at the geographic location of the cartographically depicted meander line is* not *actually created*. The true property boundary remains the watercourse.[62]

---

56. *Mitchell v. Smale* (1891) 140 U.S. 406, 413; *Railroad Co. v. Schurmeir, supra,* 74 U.S. at 286–287 ("[m]eander lines are run in surveying . . . portions of the public lands bordering on navigable rivers, not as boundaries of the tract, but for the purpose of defining the sinuosities of the banks of the stream and as a means of ascertaining the quantity of [public] land . . . subject to sale. . . .").
57. *Ibid.*
58. *Ibid.*
59. *Ibid.*
60. In certain cases meander lines have been held to be an actual property boundary as a consequence of surveyor error, outright fraud, or judicial fiat. E.g., *Mitchell, supra,* 140 U.S. at 413–414. These instances will be explained later in this chapter. See notes 66 through 83 and accompanying text in this chapter.
61. Although these surveys only describe, in a most general fashion, the sinuosities of the water course, they may be useful in establishing property boundaries where the physical location of such boundary has been obliterated or some significant change has occurred in the physical regime. See, e.g., *State Bd. of Trustees, etc. v. Laney* (Fla. App. 1981) 399 So.2d 408, 410; *Mood v. Banchero* (Wash. 1966) 410 P.2d 776, 779. Later chapters will explain the usefulness of these surveys in determining the physical location of historic property boundaries along particular water bodies.
62. E.g., *St. Louis v. Rutz* (1891) 138 U.S. 226, 243; *Thomas B. Bishop Co., supra,* 96 F.2d at 202; *Los Angeles v. San Pedro etc. R.R. Co.* (1920) 182 Cal. 652, 654, *cert. den.* (1920) 254 U.S. 636; *Den, supra,* 39 Cal.App.2d at 627; Clark 7th, *supra,* §§ 13.01 and 13.03; Clark 4th, §§ 257 and 259. This is also in conformity with the rules establishing priority of calls in instruments of conveyance. Natural boundaries called for in such instruments control over courses and distances

If the rule were otherwise and it was decided that the property boundary of lands adjacent to a watercourse was finally determined when the watercourse was initially surveyed by the public land surveyor, confusion, uncertainty, and unfairness would reign.[63] The littoral or riparian owner would not know from day to day whether a new parcel of land, created by the migrations or ephemeral fluctuations of the boundary watercourse, would prevent the landowner from gaining access to the watercourse. It is that access, the right to be riparian or littoral to a watercourse, that is one of the principle values of such a parcel.[64] These valuable rights could be lost if an intervening ownership was created between the surveyed meander line and the watercourse itself.[65]

3.9 EXCEPTIONS TO RULE—WHEN A MEANDER LINE CAN BE TREATED AS A PROPERTY BOUNDARY

There are certain cases, however, in which a meander line has been considered to be the actual property boundary. In such cases, the result was that an ownership would or could intervene between property claimed to be riparian or littoral, as shown on government plats, and the true geographic location of the adjacent water body.

---

conflicting with the geographic location of such natural boundary. E.g., *County of St. Clair v. Lovingston, supra,* 90 U.S. at 62; *Hunt v. Barker* (1915) 27 Cal.App. 776, 778; Clark 7th, *supra,* § 14.21; Clark 4th, § 277, p. 346.

63. Some of those consequences are set forth in *Mitchell, supra,* 140 U.S. at 412–413.

64. For example, riparian owners in some parts of the country have the almost unfettered right to divert water from the watercourse for use on their land. 1 S. Wiel, "Water Rights in the Western States" (1911) 749, 758; *Lux, supra,* 69 Cal. at 345–374. And it has long been known that one of the most valuable rights held by littoral or riparian owners is the right to "wharf out" and to have access to the water body. *New Orleans v. United States* (1836) 35 U.S. (10 Pet.) 662, 486; *Weber v. Harbor Commissioners* (1873) 85 U.S. 57, 64–65. Also, there is a right to future additions to the land. E.g., *County of St. Clair v. Lovingston, supra,* 90 U.S. at 68. For a brief and detailed discussion of the origin and nature of riparian rights, see *Wernberg v. State* (Alaska 1973) 516 P.2d 1191, 1194. For further descriptions of the riparian or littoral owner's rights see, e.g., *State v. Korrer* (Minn. 1914) 148 N.W. 617, 621–622; *Marks v. Whitney* (1971) 6 Cal.3d 251, 262–263; *Board of Trustees, etc. v. Mediera Beach Nom. Inc., supra,* 272 So.2d at 214. In certain instances, however, the state may interfere with these riparian rights, apparently without there being any right to compensation due the riparian or littoral owner. E.g., *Colberg, Inc. v. State of California, ex rel. Dept. of Pub. Wks.* (1967) 67 Cal.2d 408, 422–425, *cert. den.* (1968) 390 U.S. 949; *Joslin v. Marin Mun. Water Dist.* (1967) 67 Cal.2d 132, 141; *People ex rel. State Water Resources Control Bd. v. Forni* (1976) 54 Cal.App.3d 743, 751–752.

65. *County of St. Clair v. Lovingston, supra,* 90 U.S. at 63; *Coburn, supra,* 75 Fed. at 530. The *Lovingston* case plainly states the rule: "Where a survey and patent show a river to be one of the boundaries of the tract, it is a *legal deduction* that there is no vacant land left for appropriation between the river and the river boundary of such tract." *Lovingston, supra,* 90 U.S. at 63 (emphasis supplied).

For example, in one case the official plat of a survey showed fractional lots supposedly created by a meander line bordering a lake.[66] Party A claimed by virtue of intermediate (known as "mesne") conveyances from an original government grant of those lots.[67] Party A introduced evidence showing the adjacent lake had naturally receded and land that had once been covered by the lake had become uncovered and exposed.[68] As a consequence, Party A claimed, the lake was the boundary of his land and he should own the land exposed as the lake receded.[69] Party B introduced evidence that "tend[ed] to show" that there *never was a lake* in front of the lots, that the lake had never extended to the supposed meander line, and that there had never been any recession of the lake.[70] The jury believed Party B and quieted title in him.

Under this unusual set of facts, the court attempted to rationalize the decision of the jury with the rule that meander lines are not boundaries. The court noted that the government township plats would be conclusive of the existence of the meander line *if* there had been a lake abutting Party A's lots. Thus, Party A would have gained title to the land exposed as the lake receded. Party B could show, contrary to the "evidence" on the township plat, that a lake did not exist and could not have receded.[71]

Many years later a case was brought concerning the very same lake. The earlier opinion was not cited and while the later opinion discussed the heavy burden undertaken when one seeks to set aside such government surveys, the court came to a decidedly opposite result.[72] In this later case[73] the United States itself claimed that the survey by a government surveyor purporting to meander the lake was grossly inaccurate, that patents extending to the supposed shoreline of the lake did not actually extend to the shoreline, and that there was a "buffer zone" of land between the meander line and the shore of the lake.[74] Based on these arguments, the United States attempted to set aside its own conveyances. Because the conveyance had been made based on the United States' survey and the plat created as a result appeared to convey

---

66. *Live Stock v. Springer* (1902) 185 U.S. 47, 49. Although this case concerned a nonnavigable lake, the principles discussed in the opinion are equally attributable to surveys of public lands adjacent to navigable waterways.
67. *Ibid.*
68. *Ibid.*
69. *Id.* at 51.
70. *Id.* at 50.
71. *Id.* at 54.
72. Courts have noted that inconsistent results are not ". . . an unusual result of litigation." *Kitteridge v. Ritter* (Iowa 1915) 151 N.W. 1097, 1099.
73. *United States v. Otley, supra,* 127 F.2d 988.
74. . . .[T]he meander line of the . . . survey did not truly meander the lake at its mean highwater mark . . . nor at the existing shore line thereof, and that the plats of the fractional sections shown as bordering on the shores of the lake . . . had the meander line merely as a boundary thereof and not as the meander of the lake . . ." *Id.* at 992. The maps and plat depicting some of the fractional sections in relation to the lake are depicted in the opinion. *Id.* at 991, 994.

title to the shore of the lake, the United States was held to a very strict and substantial burden of proof,[75] which it could not carry.[76]

Consider just one of those claims as an example: The United States asserted that the survey that led to the patenting of the lands did not truly meander the lake.[77] After describing the regime of the lake and showing that its elevation and surface area coverage were highly variable,[78] the court compared and examined the field notes of the survey with those physical conditions in mind. The court found the United States had not shown that the area between the meander line and the shore was disproportionate. Thus, the court decided, as a matter of fact, that the line surveyed was a true meander of the lake.[79]

On the other hand, when it can be and is shown that a meander line was incorrect or fraudulent, the meander line itself, not the water body, is held to be the actual property boundary.[80] And there are other cases in which a meander line has been treated as the actual property boundary even where there

---

75. The term "burden of proof" is defined in note 130 in this chapter.
76. In such a case the United States has a more than ordinary burden of proof. To avoid such 'solemn evidences of title emanating from the government of the United States under its official seal' requires the observance of the early established rule that it 'cannot be done upon a bare preponderance of evidence which leaves the issue in doubt'. . . ." *United States v. Otley, supra,* 127 F.2d. at 995. Thus, among other matters, the United States failed to establish that it did not intend to exclude the area from the patents, *id.* at 996, that the United States Surveyor General who approved the survey and plat was not aware of the facts, *id.* at 997, or that there was an unusual, or any, amount of land between the meander line established by the patents and the actual location of the shore of the lake, *id.* at 998.
77. *Id.* at 999.
78. The court noted:

   The lake's extreme high water mark is at an elevation of nearly 4,096 feet above sea level. Its lower water mark is between five and six feet less. The difference in feet of elevation is not unusual in any lake, but because of the flat terrain at its highest water level it waters over 80,000 acres of land, while at low water but some 10,000 acres,—that is on an average of over 11,000 acres for a foot rise. . . .

   While from a distance such an area may appear as flat as a billiard table, over its surface are small variations in height which make very difficult the surveying of a meander line of the lake's border. Here are levels where the survey's variation of an inch may mean hundreds of acres in area. *Id.* at 999–1000.

79. *Id.* at 1000. This result is inconsistent with the result in *Live Stock v. Springer.*
80. *Mitchell, supra,* 140 U.S. at 413–414; *Live Stock, supra,* 185 U.S. at 54; *Niles v. Cedar Point Club* (1899) 175 U.S. 300; *Horne v. Smith* (1894) 159 U.S. 40; *United States v. Ruby Co.* (9th Cir. 1978) 588 F.2d 697, 700, *cert. den.* (1975) 423 U.S. 947; *Ritter v. Morton* (9th Cir. 1975) 513 F.2d 942, 947–948; *Smith v. United States* (10th Cir. 1979) 593 F.2d 982, 986–988; *Anderson v. Trotter* (1931) 213 Cal. 414, 420–421, *cert. den.* (1932) 284 U.S. 686; Clark 7th, *supra,* §§ 13.12–13.15; Clark 4th, *supra,* secs. 268–271, pp. 324–332. One policy behind this rule is that the patentee receives all he bargained for in terms of the quantity of land purchased. Also, the fraudulent act of a public official should not hurt the United States. *Mitchell, supra,* 140 U.S. at 413; *United States v. Ruby, supra,* 588 F.2d at 700.

90 BASIC LEGAL PRINCIPLES DEFINING PROPERTY BOUNDARY MOVEMENT

has been no showing that the meander line was incorrectly run or was fraudulent.

In one case, title to land bordering on a river was derived from a confirmed land grant made by the Mexican government prior to the United States' accession of sovereignty.[81] Long after the rancho was confirmed and its boundaries located by survey, the precise location of the river boundary of the land grant came into issue.[82] The physical location of the river boundary of the grant had been obliterated through gold mining activities. There being no other evidence of this riverbank location, the court held that the meander line in the rancho grant survey established the location of the bank of the river so far as the property boundary of the titles derived from that grant.[83]

Notwithstanding that in some unusual situations the geographic location of the ordinary high-water mark can be established through the use of meander line surveys, these surveys in and of themselves are not commonly determinative of the location of the property boundary. The reason they are not determinative is due to the ephemeral and ever changing nature of the riparian and littoral shorelines. The surveyor is present for the survey of the riparian or littoral shoreline for only an instant of time. The riparian or littoral property boundary line should not be frozen in location by the fortuity of the time of year or day the surveyor happens to be there.

3.10 BASIC LEGAL TERMS DESCRIBING THE PROCESS OF PROPERTY BOUNDARY MOVEMENT

As discussed so far, the surveyed boundary line of lands along or adjacent to tidal or navigable water bodies is not usually designed to provide a definite geographic location of the ordinary high-water mark property boundary. The ordinary high-water mark, the legal property boundary, is constantly in motion, not geographically located by the meander line but designed by the law to reflect the vagaries of the boundary watercourse. The boundary may recede farther waterward or encroach more landward, depending on the fluctuations of the watercourse's regime and the configuration and composition of the adjacent landform. The variability of the physical location of this boundary is virtually unlimited, given the eccentricities of nature.

The law, however, abhors this kind of uncertainty.[84] But how does the law resolve the uncertainty? Are there rules or principles delimiting the effect of

---

81. *People ex rel. Dept. Pub. Wks. v. Shasta Pipe Co.* (1968) 264 Cal.App.2d 520, 524.
82. *Id.* at 524–525. The confirmation process of Mexican land grants in California is described in Sections 1.3.1 and 1.3.2 in Chapter 1.
83. *Id.* at 531; *State Bd. of Trustee, etc. v. Laney, supra*, 399 So.2d at 410. We shall see in later chapters of this work how this principle can assist in historic water boundary location.
84. It is "... a principle of public policy that it is in the interest of the community that all land should have an owner...." *Banks v. Ogden* (1864) 69 U.S. (2 Wall.) 57, 67; *Board of Trustees, etc. v. Mediera Beach Nom. Inc., supra*, 272 S.2d at 213. "That is certain which can be made certain." Cal. Civ. Code § 3538.

the movement of the ambulatory littoral or riparian property boundary that might apply to all watercourse regimes—coastal, estuarine, riverine, or lake? Looking at an actual case will place these questions in context.

3.10.1 Study of a Property Boundary Movement Case

Along the Missouri River between Iowa and Nebraska (about which much has already been and is to be said in this work), an 1881 vintage government survey depicted a thumb-shaped area of land forming an oxbow bend in the river.[85] That oxbow bend came to be known as Blackbird Bend. East of the river, Iowa landowners, deriving title from the federal government, had record title extending to the center[86] of the Missouri River.[87] For years these Iowa farmers had cultivated the lands physically adjacent to the river and on its eastern side.[88] To the west, the Missouri River formed the eastern boundary of the reservation of the Omaha Indian Tribe.[89]

Over the course of decades, despite and possibly as the result of the U.S. Army Corps of Engineers' efforts to tame the Missouri River, the geographic location of the river in the Blackbird Bend area changed dramatically.[90] Although some river movement is to be expected, the Missouri River moved west more than 2 miles. The Iowa farmers just kept farming the land along the east bank of the river, probably wondering why they needed so much more seed each year. The farmers contended that the river movement extended their ownership across Blackbird Bend and, like the advance of a particularly voracious plague of insects, the western progress of the river stripped the competing Nebraska landowner, the Omaha Indian Tribe, of its title to Blackbird Bend. After all, the Omaha Tribe's lands were bounded by the Missouri River on the east and the river had moved its geographic location.[91]

The Omaha Tribe asserted ownership to the same land, based on a theory both legally and factually contradictory to the legal and factual theory of the Iowa farmers. The Tribe asserted that a sudden change in the location of the Missouri River had occurred in the past and the mutual eastern and western property boundary of these competing ownerships, the Missouri River, was fixed in geographic location at the time of that event.[92] According to the

---

85. *Omaha Indian Tribe, Treaty of 1854, etc. v. Wilson* (8th Cir. 1978) 575 F.2d 620, 622–627, *vacated and remanded* (1979) 442 U.S. 653 (including portions of government surveys of area); *rev'd*, *Wilson v. Omaha Indian Tribe* (1979) 442 U.S. 653. The survey established a meander line of the shore. *Omaha Indian Tribe, Treaty of 1854, etc., supra*, 575 F.2d at 623, n.4.
86. The center of river is known by a German term, "thalweg." "Thalweg" is defined as the channel continuously used for navigation or the middle of the principal channel of navigation. E.g., *Omaha Indian Tribe, Treaty of 1854, etc., supra*, 575 F.2d at 623, n. 6.
87. *Wilson v. Omaha Indian Tribe, supra*, 442 U.S. at 659–660.
88. *Ibid*.
89. *Id*. at 658.
90. *Id*. at 659.
91. *Id*. at 660.
92. *Ibid*.

Omaha Tribe, the change in the actual geographic location of the river thereafter was irrelevant.

In that case, surveyors and lawyers were required to provide advice and consultation on the availability and likelihood of success of various courses of action to resolve the location of the property boundaries of the disputing ownerships. The result of that process was the two competing theories of river movement, founded on mutually contradictory views of the physical process. This physical process, without question, had resulted in a change in the geographic location of the river. For purposes of this chapter, what is important is not the result in the case.[93] The reader should note that two contenders looking at the same result of a physical process came up with entirely contradictory legal theories. Yet each legal theory urged plausibly supported a competing title and boundary claim. The lesson to be learned is that persons who wish to constructively contribute to resolution of these disputes should obtain a thorough knowledge of the legal basis and physical elements or processes supporting such claims.

3.11 LEGAL TERMS DESCRIBING THE PROCESS OF CHANGE IN PHYSICAL LOCATION

It is imperative the reader become familiar with certain legal terms and expressions that are words of art in property boundary parlance. The reader should have them firmly in mind. Although they are frequently used, they are sometimes imprecisely applied. These terms describe both the physical process by which the geographic location of a boundary waterway changes and the property boundary effect of that movement.[94]

"Accretion"[95] is the legal term used to describe the physical process causing an area of littoral or riparian land to be increased by the gradual and imperceptible deposition of sand, soil, or other solid material on the margin of a watercourse.[96]

---

93. The Blackbird Bend saga has been and will be a sometime focal point in this work. E.g., see notes 61 through 78 and accompanying text in Chapter 2 and notes 68 through 77 and accompanying text in Chapter 7.
94. *Jackson v. Burlington Northern, Inc.* (Mont. 1983) 667 P.2d 406, 407 (accretion describes the process and the rule of law).
95. Sometimes the word "alluvion" is used to describe the deposit itself as well as the act of accretion. For example, a California statute provides: "*Alluvion*. . . . [L]and form[ed] by imperceptible degrees upon the bank of a river or stream, navigable or not navigable, either by accumulation of material or by the recession of the stream. . . ." Cal. Civ. Code § 1014. To avoid such confusion, in this work the author will use the more accepted technical description of the deposit as "deposition" and the legal term "accretion" to describe the result of deposition.
96. E.g., *County of St. Clair v. Lovingston, supra*, 90 U.S. at 66–68; *Jefferis, supra*, 134 U.S. at 192–194; *City of Los Angeles v. Anderson* (1929) 206 Cal. 662, 667; Clark 7th, *supra*, § 24.02; Clark 4th, *supra*, § 567, p. 772; 78 Am. Jur. 2d Waters § 406, p. 851.

"Erosion," the opposite of accretion, is a legal term describing the gradual and imperceptible wearing away or loss of littoral or riparian land by the action of the water.[97]

"Reliction"[98] is the legal term describing the physical process whereby land that was once covered with water becomes exposed or uncovered by the imperceptible recession of the water, usually when the water level is lowered.[99] An element of permanence of the recession of the water is required by some courts.[100]

"Submergence," the reverse of reliction, is the legal term denoting the physical process of the gradual and imperceptible disappearance of land under water and the formation of a navigable water body over it.[101]

Accretion, erosion, reliction, and submergence are all legal terms[102] that attempt to describe a subtle physical process occurring over a period of time. The subtlety and the length of time are signaled by the formula that the change in physical location occur "gradually and imperceptibly." This phrase

---

97. *Philadelphia Co. v. Stimson* (1911) 223 U.S. 605, 624–627; *Miramar Co. v. City of Santa Barbara* (1943) 23 Cal.2d 170, 176; *Coastal Industrial Water Author. v. York* (Texas 1976) 532 S.W.2d 949, 952; *Jackson*, *supra*, 667 P.2d at 407; Clark 7th, *supra*, § 24.02; Clark 4th, *supra*, § 567, p. 773; 78 Am. Jur. 2d Waters § 406, p. 852.

98. Reliction has also been called "dereliction." See *Matter of Ownership of Bed of Devils Lake* (N.D. 1988) 423 N.W.2d 141, 143 n.1. Reliction should not be confused with "reemergence." The term "reemergence" describes the process by which land, once was uncovered by water, becomes covered and then uncovered by water. *Arkansas v. Tennessee* (1918) 246 U.S. 158, 174; *Herron v. Choctaw & Chickasaw Nations* (10th Cir. 1956) 228 F.2d 830, 832; *Horry County v. Woodward* (S.C.App. 1984) 318 S.E.2d 584, 587.

99. Clark 7th, *supra*, § 24.08; Clark 4th, *supra*, § 573, p. 787; *Bear v. United States* (D. Neb. 1985) 611 F.Supp. 589, 593 n.2, *aff'd* (8th Cir. 1987) 810 F.2d 153. Some works improperly equate the term "reliction" with the term "alluvion." This is not correct. "Reliction" is the legal term pertaining to the process by which land is uncovered by exposure through the gradual recession of waters. *Flisrand v. Madson* (S.D. 1915) 152 N.W. 796, 798. Alluvion is land that gradually builds up on the bottom, shore, or bank. *County of St. Clair v. Lovingston*, 90 U.S. at 66; 78 Am. Jur.2d. Waters § 406, p. 851; Clark 7th, *supra*, § 24.02; Clark 4th, *supra*, § 567, p. 773. The terms "reliction" and "accretion," however, are often used interchangeably. *Bear, supra*, 611 F.Supp. at 593 n.2; *Cal. ex rel. State Lands Com'n v. United States* (9th Cir. 1986) 805 F.2d 857, 860, *cert. den.* (1987) 484 U.S. 816.

100. E.g., *Sapp v. Frazier* (La. 1899) 26 So. 378, 380; *Herschman v. State Department of Natural Resources* (Minn. 1975) 225 N.W.2d 841, 843.

101. *Coastal Industrial Water Auth., supra*, 532 S.W.2d at 152–153; *Port Acres Sportsman Club v. Mann* (Texas 1976) 541 S.W.2d 847, 849. See Clark 7th, *supra*, § 24.14; Clark 4th, *supra*, § 579, p. 830. Clark and others seem to equate erosion and submergence. *Ibid.*; 78 Am. Jur. 2d. Waters § 421, p. 868.

102. For those interested in further sources for definitions of these terms, they may be found in various works, some of which follow: *III American Law of Property* (Casner, ed., 1952) § 15.26, pp. 855–856; *Black's Law Dictionary* (Rev. 4th ed., 1968) pp. 36–37, 102, 173–174, 529, 637, 1455, 1594; *Ballentine's Law Dictionary* (3d Ed. 1969) pp. 14, 64, 116, 339, 414, 1085, 1229; 6 Powell, *The Law of Real Property* (3d. Ed. 1939) para. 983, pp. 607–611; 4 Tiffany, *The Law of Real Property* (3d ed. 1939) § 1219, pp. 613–615; *Handbook for Title Men* (Title Insurance and Trust Co. 1966) §§ 1.01, 1.02, 1.04; Skelton, § 292, pp. 331; Id. at § 298, p. 341.

is included as one of the critical elements that all together comprise the physical process described by one of these legal terms. The period of time and the subtlety of the physical process distinguish the physical processes described by those legal terms from the rapid, perceptible, and often violent[103] removal of or addition to land due to the action of water, or the sudden and perceptible change in the physical location of the boundary watercourse. That process is generally known by the legal term *avulsion*,[104] or, by a few, with perhaps more than a little poetic license, as "revulsion."[105]

Over the years, rules have been developed to determine the property boundary effect of changes in geographic location of the ordinary high water mark resulting from the processes of accretion, erosion, reliction, submergence, and avulsion. These rules will be discussed in this chapter only in a general sense in order to supply the reader with a basic understanding of the legal principles of property boundary movement. This preliminary visit to the subject will provide the reader with a firm basis on which nuances to those rules, developed in connection with particular physical regimes, can be discussed in connection with later chapters. In those later chapters, more refined discussions will consider how such general boundary movement principles apply in and to different physical regimes.

3.12 PROPERTY BOUNDARY CONSEQUENCES OF A CHANGE IN GEOGRAPHIC LOCATION

The dynamic state of littoral and riparian property boundaries has long been recognized by the law.[106] This state is reflected in various terms used to describe the characteristic of a property boundary, such as ambulatory or moveable.[107] Countries that are surrounded by water bodies with markedly different physical regimes have developed similar concepts and rules to determine the effect on property boundaries of shoreline movement and change.

---

103. In California, by statute, "avulsion" is defined as "[i]f a river or stream . . . carries away, by sudden violence, a considerable or distinguishable part of a bank. . . ." Cal. Civ. Code § 1015.
104. E.g., *Jefferis, supra*, 134 U.S. at 194; *Nebraska v. Iowa, supra*, 143 U.S. at 366; *Bauman v. Choctaw-Chickasaw Nations* (10th Cir. 1964) 333 F.2d 785, 789, *cert. den.* (1965) 379 U.S. 965; *State v. Gunther & Shirley Company* (Ariz. 1967) 423 P.2d 352, 356; *Nolte v. Sturgeon* (Okla. 1962) 376 P.2d 616, 620; Clark 7th, *supra*, § 24.02; Clark 4th, *supra*, § 567, p. 773; 78 Am. Jur. 2d, Waters § 406, p. 852.
105. Brown, *Boundary Control and Legal Principles* (3d Ed. 1986), p. 199.
106. Angell, *A Treatise on the Right of Property in Tide Waters* (2d Ed., 1847), 249–251; Hall, *A History of the Foreshore and the Law Relating Thereto* (3d [Stuart Moore] Ed. 1888) 395, 785–789; 4 Tiffany, *supra*, § 1220, p. 618. In *County of St. Clair v. Lovingston, supra*, 90 U.S. at 66–67, there is a lengthy explication, harking back to Roman times, of the development of the rules concerning geographic movement of property boundaries along waterways.
107. E.g., *Wright v. Seymour* (1886) 69 Cal. 122, 126; *City of Oakland v. Buteau* (1919) 180 Cal. 83, 87.

3.12.1 Background of Rules

The rugged coastline of the British Isles is pounded and battered by the open ocean, the turbulent waters of the Irish and North Seas and the English Channel. In contrast, the Mediterranean Sea is relatively calm and tideless. Countries bordering the Mediterranean largely followed civil law tradition. England had developed the common law. Remarkably, these countries subscribed to similar rules of law for determining the effect of the dynamics of shoreline movement on adjacent property boundaries.

The reason for such adherence is that the motivations supporting these rules are basically the same: All land should have an owner, access to the boundary watercourse should be preserved for the riparian owner, and, perhaps more important, the growth or loss of land is so slight that the law does not need to recognize it.[108] In England, alluvion formed by the process of accretion belongs to the owner of the adjoining littoral or riparian lands; thus, the physical location of the property boundary moves to account for the accretion.[109] The same is true in civil law countries.[110]

3.12.2 Property Boundary Effect of Accretion and Erosion

In the United States physical regimes and conditions were considerably different from those in the British Isles. In England there were no great nontidal, navigable rivers or lakes as found in the United States.[111] The same rule for

---

108. E.g., 78 Am. Jur. 2d Waters § 412, pp. 859–860; *County of St. Clair v. Lovingston, supra*, 90 U.S. at 68–69; *Jefferis, supra*, 134 U.S. at 191. One basis for the rule was explained long ago by the English jurist, Blackstone:

> And as to lands gained from the sea, either by *alluvion*, by the washing-up of sand and earth, or by *dereliction*, as when the sea shrinks back below the usual watermark; in these cases the law is held to be, that if this gain be by little and little, by small and imperceptible degrees, it shall go to the owner of the land adjoining. [Footnote omitted.] For *de minimis non curat lex* (the law takes not cognizance of small things): and, besides, these owners, being often losers by the breaking in of the sea, or at charges to keep it out, this possible gain is therefore a reciprocal consideration for such possible charge or loss. . ." 2 Bl. Comm. 262 (Cooley, 4th Ed. Vol. 1) (emphasis in original).

In a entirely different context, one author aptly highlighted the basis for the accretion rule. The timid desegregation policy of the early years of the Kennedy administration was designed to avoid arousing the powerful Southern politicians. That policy was described as follows: "It was a policy of accretion, with nothing so sweeping or grand as to touch off a segregationist backlash against them." Branch, Taylor, *Parting the Waters: America in the King Years, 1954–1963* (Touchstone 1988), p. 382.

109. *Rex v. Lord Yarborough* (1824) 3 B.& C. 91 (K.B.); *Jones, supra*, 65 U.S. at 65; *Banks, supra*, 69 U.S. at 67.

110. E.g., *County of St. Clair v. Lovingston, supra*, 90 U.S. at 66–67; *Jefferis, supra*, 134 U.S. at 192.

111. *The Daniel Ball* (1870) 77 U.S. (10 Wall.) 557, 563, and *Phillips Petroleum, supra*, 484 U.S. at 477–479, discuss the difference in such physical conditions.

determining the property boundary effect of the process of accretion and erosion was adopted in the United States as in England and for the same reasons.[112] For example, in a case argued before the Supreme Court by none other than Daniel Webster, the Court restated the time-honored rule:

> ... The question is well settled at common law, that the person whose land is bounded by a stream of water, which changes its course gradually by alluvial formations, shall still hold by the same boundary, including the accumulated soil. No other rule can be applied, on just principles. Every proprietor whose land is thus bounded, is subject to loss, by the same means which may add to his territory; and as he is without remedy for the loss, he cannot be held accountable for his gain.[113]

The rule that the riparian or littoral owner's property boundary continues to change as the geographic location of the shoreline moves by the process of accretion and erosion has been adopted, in one form or another, by many later cases considering the issue.[114]

There are, however, certain situations in which this rule has been held not to apply. Ignoring for the moment the doctrine of avulsion (which will be discussed later), when there is "substantial" accretion, the doctrine does not apply.[115] Nor is the rule concerning the property boundary effect of the processes of accretion or erosion applicable when there is a reemergence[116] or, in some jurisdictions, where the accretion is caused artificially such as by human-made works or causes.[117] Moreover, when the adjoining landowner

---

112. E.g., *Banks, supra*, 69 U.S. at 67; *County of St. Clair v. Lovingston, supra*, 90 U.S. at 68; *Jefferis, supra*, 134 U.S. at 190.
113. *New Orleans v. United States, supra*, 35 U.S. at 486.
114. E.g., *County of St. Clair v. Lovingston, supra*, 90 U.S. at 68; *Jefferis, supra*, 134 U.S. at 189; *Arkansas v. Tennessee, supra*, 246 U.S. at 173; *California, ex rel. State Lands Comm'n, supra*, 457 U.S. at 278; 78 Am. Jur. 2d Waters § 411, p. 856; Clark 7th, *supra*, § 24.02, § 24.09; Clark 4th, *supra*, § 567, p. 773, § 574, p. 791.
115. E.g., *Wittmayer v. United States* (9th Cir. 1941) 118 F.2d 808; *De Boer v. United States* (9th Cir. 1981) 653 F.2d 1313, 1315 (105.22 acres had accreted); *Bear v. United States* (8th Cir. 1987) 810 F.2d 153, 155–156. (The question is whether the accretion is "quantitatively considerable.")
116. See *Arkansas v. Tennessee, supra*, 246 U.S. at 174–75; *Herron v. Choctaw-Chickasaw Nation, supra*, 228 F.2d at 832.
117. E.g., *Carpenter v. City of Santa Monica, supra*, 63 Cal.2d at 787. The California Supreme Court has ameliorated this rule, however. In *State of Cal.ex rel. State Lands Com. V. Superior Court (Lovelace)* (1995) 11 Cal.4th 50, the California Supreme Court upheld the artificial accretion rule, but narrowed the potential scope of its applicability. *Id.* at 73, 76–80. The case is discussed in greater detail in text accompanying notes 144 through 157 in Chapter 5. This is not the rule in many jurisdictions or, in certain instances, in federal court. E.g., *County of St. Clair v. Lovingston, supra*, 90 U.S. at 66; *California, ex rel. State Lands Comm'n, supra*, 457 U.S. at 278; *State v. Sause* (Oregon 1959) 342 P.2d 803, 819–827. In addition, the doctrine appears to apply only for the benefit of the state. *United States v. Aranson, supra*, 696 F.2d at 662; *United States v. Harvey* (9th Cir. 1981) 661 F.2d 767, 772, n.7, *cert. den.* (1982) 459 U.S. 833; *State of Cal.ex rel. State Lands Com. v. Superior Court (Lovelace) supra*, 11 Cal.4th at 73, n.4.

has a hand in the creation of an addition to his land, the rule is also inapplicable.[118] There are other examples as well, in which the accretion rule has been found inapplicable.[119]

3.12.3 Property Boundary Effect of Reliction and Submergence

In regard to the process of reliction, the general rule is the same as in the case of accretion; the riparian or littoral owner's boundary moves its physical location as the boundary watercourse recedes and land is exposed.[120]

The rules concerning submergence are less straightforward. An owner of land that had once been uncovered retained title even though the land had subsided and was thus slowly submerged over decades.[121] The general rule, however, seems to be that the gradual and imperceptible covering, submersion, of land by the incursion of the boundary watercourse causes the movement of the boundary and an ensuing loss of title.[122]

To summarize: With certain limited exceptions, as long as the change in the geographic location of a boundary watercourse is determined to be gradual and imperceptible, the riparian or littoral property boundary continues to move. Thus, such gradual and imperceptible movements can cause the property owner to "gain" or "lose" title as the ambulatory boundary wanders. Will the effect on the property boundary be the same if there is a sudden, perhaps violent, change in geographic location of the boundary watercourse?

3.12.4 Property Boundary Effect of an Avulsion

Where there is an avulsion, a sudden and perceptible change in geographic location of a boundary watercourse, the rule is that the geographic location of the ordinary high-water mark property boundary remains unaltered.[123]

---

118. E.g. *Bonelli Cattle Co. v. Arizona* (1973) 414 U.S. 312, 323, *overld on other grnds*, (1977) 429 U.S. 363; *United States v. Harvey*, *supra*, 661 F.2d at 772, n.7; *Dept. of Natural Resources v. Pankratz* (Alaska 1975) 538 P.2d 984, 989; *State of Cal.ex rel. State Lands Com. v. Superior Court* (*Lovelace*), *supra*, 11 Cal.4th at 79–80.
119. E.g., *Commissioner of Oklahoma Land Office v. United States* (8th Cir. 1920) 270 F.2d 110, 113–114 (main channel exception); *Hudson House, Inc. v. Rozman* (Wash. 1973) 509 P.2d 992, 995 (accretion to tract separated from tract by narrow channel).
120. *Cal. ex rel. State Lands Com'n*, *supra*, 805 F.2d at 860; *Bonelli*, *supra*, 414 U.S. at 325–326; *Honsinger v. State* (Alaska 1982) 642 P.2d 1352, 1354; Clark 7th, *supra*, § 24.08; Clark 4th, *supra*, § 573, p. 787; 78 Am. Jur. 2d § 411, p. 856.
121. 78 Am. Jur. 2d, Waters, § 421, p. 868; *Coastal Industrial Water Author. v. York*, *supra*, 532 S.W.2d at 954.
122. Clark 7th, *supra*, § 24.14; Clark 4th, *supra*, § 579, p. 830. In certain cases the doctrine of reemergence may apply. That doctrine provides that riparian or littoral owners whose land has become submerged may not lose title to such land if the land later reemerges. 78 Am. Jur. 2d § 421, p. 868; Comment, "Land Accretion and Avulsion: The Battle of Blackbird Bend" (1977) 56 Nebr. L. Rev. 814, 817; *Arkansas v. Tennessee*, *supra*, 246 U.S. at 174. For a California case, see *Bohn v. Albertson* (1951) 107 Cal.App.2d 738, 748.
123. *Jefferis*, *supra*, 134 U.S. at 194; *Nebraska v. Iowa*, *supra*, 143 U.S. at 361; *Philadelphia Co. v. Stimson*, *supra*, 223 U.S. at 624; *People v. Ward Redwood Co.* (1964) 225 Cal.App.2d 385,

When stated in such a fashion, the application of this principle seems quite straightforward. Whether the change in the geographic location of the property boundary watercourse has been by the gradual and imperceptible process of accretion and erosion, or by a sudden and perceptible process of avulsion, appears, at first blush, to be quite simple to determine. Needless to say, on closer examination, one discovers that the answer is not quite straightforward.

3.12.5 Accretion or Avulsion? That Is the Question

The characterization of the process of change in the physical location of the boundary watercourse as either accretive or avulsive may not seem unusually difficult. There are many cases, however, in which that characterization may be extremely demanding. In such cases, characterization of the process is not helped by the definition of the terms themselves. After all, how long, exactly, does a "gradual" process take? Is a change in the location of the boundary watercourse that is perceived over a week "imperceptible"? What happens when the change takes place quite rapidly but not immediately? Is such a change "sudden" in the eyes of the law? What if the "sudden" change is not "perceptible"? Does one have to be present at a sudden change for it to be perceptible?

Although these questions may sound like an introduction to a lesson in Descartian philosophy,[124] they are questions that must be asked in order to provide sound advice as to where the property boundary of a littoral or riparian parcel should be located when the boundary water body has changed its geographic location. These specific questions and others will be discussed in later chapters focusing on specific physical regimes.[125]

The author has chosen this method of inquiry because the rules for the characterization of the property boundary effect of such changes seem to have been created with a view to the particular nature of the physical regime being considered.[126] For example, because of the nature of the soils of the banks surrounding the bed of the Missouri River, the courts developed exceptional rules to characterize quite rapid and often readily observable changes in the geographic location of that river.[127] Before we move on to refine our focus,

---

389; *Valder v. Wallace* (Nebraska 1976) 242 N.W.2d 112, 114; *Nolte, supra*, 376 P.2d at 620; *State v. Sause, supra*, 342 P.2d at 818–827; Clark 7th, *supra*, § 24.07; Clark 4th, *supra*, § 572, p. 785; 78 Am. Jur. 2d § 411, p. 858.

124. [A]ll things which I perceive . . . very clearly and distinctly are true." Copleston, Frederick, S. J., *A History of Philosophy, Volume 4* (Image Books 1963), p. 107.

125. See Chapter 4, Section 4.15; Chapter 5, Section 5.8; Chapter 6, Section 6.7; Chapter 7, Section 7.5, and Chapter 8, Section 8.9.

126. The United States Supreme Court has recognized that different physical regimes existing throughout the nation may lead to development of distinct legal doctrines specifically pertaining to peculiarities or distinctive attributes of such regimes. See *California v. United States* (1978) 438 U.S. 645, 648.

127. *Jefferis, supra*, 134 U.S. at 190; *Nebraska v. Iowa, supra*, 143 U.S. at 369–370.

it is appropriate to discuss at this point some general rules that are designed by courts and legislatures to assist in resolving boundary or title disputes.

3.13 PRESUMPTIONS AND BURDEN OF PROOF IN TITLE AND BOUNDARY LITIGATION

The reason for the creation of general rules is based on the kinds of questions asked in the previous section. These questions reflect the difficulty in establishing or proving the nature of a physical event. This event may have occurred at a location remote from ordinary, much less scientific, observation at a time when the event could not be seen because of physical conditions or the cataclysmic nature of the event itself.[128] The rules discussed here were developed by the law to assist littoral and riparian property owners in establishing their ownership or meeting their burden of proof in title or boundary litigation. These rules represent a judgment about the nature of the processes resulting in changes in the geographic location of a water body and, given the regime, the physical likelihood of the particular occurrence of such processes in the absence of or where there is a dearth of evidence.

In very general terms, as a full discussion is beyond the scope of this work, in quiet title actions the plaintiff must recover on the strength of his or her own title and not merely on the weakness of his or her opponent's title.[129] Insofar as that principle concerns the determination of property boundary disputes along tidal or navigable water bodies, some states hold that the burden of proof[130] rests with the party who is not in peaceable possession or who is challenging the existing physical conditions.[131]

To assist in meeting that burden of proof and to aid the trier of fact in title and boundary litigation concerning disputes along tidal or navigable waterways, it is presumed[132] that, as a general matter, a boundary waterway has

---

128. For a court's view of these difficulties see notes 63–66 and accompanying text in Chapter 6.
129. *Ernie v. Trinity Lutheran Church* (1959) 51 Cal.2d 702, 706; *Helvey v. Sax* (1951) 38 Cal.2d 21, 23; *People v. Ward Redwood Co., supra*, 225 Cal.App.2d at 389. In California, the owner of legal title to real property is presumed to be the owner. Cal. Evid. Code § 662; *Rench v. McMullen* (1947) 82 Cal.App.2d 872, 874.
130. To have "burden of proof" means one must prove that his or her position is correct. E.g., Cal. Evid. Code § 115. For a lucid explanation of this concept, see Briscoe, J., *Surveying the Courtroom* (2nd Ed., John Wiley & Sons, New York, 1999), at pp. 65–69.
131. *O'Neill v. State Highway Department* (N.J. 1967) 235 A.2d 1, 11; *Velsicol Chemical v. Environmental Prot. Dept.* (N.J. 1982) 442 A.2d 1051, 1054; *Conran, supra*, 341 S.W.2d at 90 (adverse possession). For example in *Kansas v. Missouri* (1944) 322 U.S. 213, 228, the United States Supreme Court observed that Kansas had the burden of proof. Not only was Kansas the moving party, but the fact was that, at the time of the lawsuit, the land was not in Kansas, but in Missouri. Kansas had the burden of proof to show that the movement of the river was effective to change the political boundary and that the land should be determined to be in Kansas, rather than in Missouri.
132. California defines a presumption as ". . . an assumption of fact that the law requires to be made from another fact or group of facts found or otherwise established in the [lawsuit]." Cal.

moved gradually and imperceptibly and that accretion or erosion has taken place rather than an avulsion.[133] The presumption that accretion or erosion has occurred, rather than an avulsion, favors the riparian or littoral owner's access to the boundary watercourse.[134] The presumption also supports the permanence and stability of water boundaries. Thus, it is presumed that the property boundary of riparian or littoral land remains unchanged unless it is shown that there has been a change in the geographic location of the boundary watercourse.[135] Although these presumptions are valuable, they are not dispositive of all questions.

This chapter has explained how, because of rules developed over hundreds of years,[136] the law accounts for the results of physical processes causing a change in the geographic location of waterways that form property boundaries. Earlier, the chapter briefly explained the process of beach migration—that beaches become wider in summer than in winter. Similar processes occur in tidal estuaries, along rivers and streams, and in lakes. The discussion turns next to the open coast and the problems associated with coastal property boundaries along the ocean shoreline.

Evid. Code § 600(a). There is an excellent discussion about presumptions in Briscoe, J., *Surveying the Courtroom* (2nd Ed., John Wiley & Sons, New York, 1999) Chapter 6.

133. *Mississippi v. Arkansas* (1974) 415 U.S. 289, 296 (Douglas, J., dissenting); see *New York v. New Jersey* (1998) 523 U.S. 767, 140 L.Ed.2d 993, 1016; *Shapleigh v. United Farms Co.* (5th Cir. 1938) 100 F.2d 287, 288; *Hall v. Brannan Sand and Gravel Company* (Colo. 1965) 405 P.2d 749, 750; *Dartmouth College v. Rose* (Iowa. 1965) 133 N.W.2d 687, 689; *Wycoff v. Mayfield* (Ore. 1929) 280 P.340, 342; *Hawkins v. Walters* (Miss. 1981) 402 So.2d 336, 337; Clark 7th, *supra*, § 24.02; Clark 4th, *supra*, § 567, p. 774; 78 Am. Jur. 2d Waters § 427, p. 874.

134. *Jefferis, supra,* 134 U.S. at 197; *Marks, supra,* 6 Cal.3d at 263.

135. *Hunt, supra,* 27 Cal.App. at 780.

136. Gradually and imperceptibly, if you will.

4

PROPERTY BOUNDARY DETERMINATION ALONG THE OPEN OCEAN COAST

4.1 INTRODUCTION

Many of us have traveled along the Pacific Coast Highway for some, if not all, of the more than 1000 miles of California's Pacific Ocean coast. During those journeys some of us may have felt a glimmer of envy directed at those fortunate enough to live or have homes on the ocean side of the highway, on or above the beach and the crashing surf of the Pacific Ocean.[1] The favored geographic position of those homes may give a sense that the beach in front of or below those homes, and even the surf itself, are private preserves of those home owners.[2] Citizen initiative[3] and court enforcement of claimed

---

1. One court quite eloquently described the lure of the ocean beach:

    There is probably no custom more universal, more natural or more ancient, on the sea coasts . . . of the world, than that of bathing in the salt waters of the ocean and the enjoyment of the wholesome recreation incident thereto. The lure of the ocean is universal; to battle with its refreshing breakers a delight. Many are they who have felt the lifegiving touch of its healing waters and its clear dust-free air. Appearing constantly to change, it remains ever essentially the same. This primeval quality appeals to us. "Changeless save to thy wild waves play, time writes no wrinkles on thine azure brow; such as creation's dawn beheld, thou rollest now.'" *White v. Hughes* (Fla. 1939) 190 So. 446, 448–449.

2. See *Nollan v. California Coastal Com'n* (1987) 483 U.S. 825, 827–828. Of course, this situation is not limited to California. The same or very similar circumstances are found along many coastal reaches on all sides of the continent. E.g., *Lusardi v. Curtis Point Prop. Owners Ass'n* (N.J. 1981) 430 A.2d 881, 883.

3. E.g., Cal. Const. Art. X, § 4; California Coastal Act of 1976, Cal. Pub. Res. Code §§ 30000, et seq.; Cal. Gov. Code §§ 53035, 66478.3, 66478.11.

public rights, allegedly created through years of free and unfettered use,[4] have led to the opening of many of these beaches to the public.[5] Yet the success of these claims of public access to the coastline is only part of the story.

It is fair to state that, with some very minor exceptions, there is no such thing as an entirely private beach. This is because that portion of the beach seaward of the ordinary high-water mark is owned by the public and held subject to the public trust.[6] As a general proposition, the public trust itself cannot be conveyed or given away[7] and allows for an extremely broad scope of public use.[8] And land that lies seaward of the ordinary low-water mark is

---

4. The doctrine of implied dedication to the public of a portion of a property owner's land is a topic that is deserving of fuller treatment than possible in this book. Suffice to say that the doctrine of implied dedication has been validated in many cases in California. E.g., *Gion v. City of Santa Cruz* (1970) 2 Cal.3d 29, 39–42; *County of Los Angeles v. Berk* (1980) 26 Cal.3d. 201, 212–215, *cert. den.* (1980) 449 U.S. 836. In brief, if it can be shown that there has been unobjected-to public use of private property for a period of years, the once private property is held to have been dedicated to the public. These implied dedication cases largely concerned beach access, although there are some cases where river access has been at issue. *Bess v. County of Humboldt* (1992) 3 Cal.App.4th 1544, 1549–1551. Other states follow suit or recognize similar doctrines. *City of Daytona Beach v. Tona Rama, Inc.* (Fla. 1973) 294 So.2d 73, 78; *State ex rel. Thornton v. Hay* (Ore. 1969) 462 P.2d 671, 677; *Hay v. Bruno* (D. Ore. 1972) 344 F.Supp. 286, 289; *Lusardi, supra,* 430 A.2d at 888 (the ocean belongs to all citizens); *Smith v. State* (Ga. 1981) 282 S.E.2d 76, 82; *Matcha v. Mattox on behalf of the People* (Tex. 1986) 711 S.W.2d 95, 98, *cert. den.* (1987) 481 U.S. 1024; *Conant v. Jordan* (Me. 1910) 77 A. 938, 943 (great ponds).
5. Some say there is a trend away from increasing public rights along beaches. See *Nollan, supra,* 483 U.S. at 838–842. (requiring that a coastal development permit condition have a "nexus" with a governmental purpose for the permit or restriction), and *Surfside Colony, Ltd. v. California Coastal Commission* (1991) 226 Cal.App.3d 1260 (discussing proof of the nexus (or lack thereof) in one case).
6. As the subject is dealt with elsewhere in this work, see notes 111 through 121 and 138 through 174 and accompanying text in Chapter 1, only a brief recitation is required to refresh the reader's recollection. The public trust doctrine holds that ownership, dominion and sovereignty over land covered by tidal or navigable water to the ordinary high-water mark is vested in the state in inalienable trust for the people. E.g., *Illinois Central Railroad v. Illinois* (1892) 146 U.S. 387, 452; *Matthews v. Bay Head Imp. Ass'n* (N.J. 1984) 471 A.2d 355, 358, *cert. den.* (1984) 469 U.S. 821; *City of Berkeley v. Superior Court* (1980) 26 Cal.3d 515, 521, *cert. den. sub.nom. Santa Fe Land Improv. Co. v. Berkeley* (1980) 449 U.S. 840; *Boston Waterfront Dev. Corp. v. Com.* (Mass. 1979) 393 N.E.2d 356, 358–359; *White, supra,* 190 So. at 448. In some states it has been held that the public trust extends to the dry sand area of beaches. *Matthews, supra,* 471 A.2d at 365–366. New York has a somewhat narrower view. *Tucci v. Salzhauer* (N.Y. 1972) 329 N.Y.S.2d 825, 833.
7. See Section 1.10 in Chapter 1. Recall that in certain limited instances the public trust interest of the state is capable of being alienated for trust-consistent purposes. See notes 147 through 149, 153 through 155, and accompanying text in Chapter 1.
8. See notes 118 through 121 and 173 through 174 and accompanying text in Chapter 1. Even as great a despot as Adolph Hitler recognized this universal truth. General Erwin Rommel, commander of Hitler's vaunted Atlantic Wall, a supposedly impenetrable barrier designed to stop the Allied invasion from England, began the wall at the low-water mark of the English Channel. Keegan, J., *Six Armies in Normandy* (Viking Press 1982), p. 59. This was an obvious recognition that the Allies were exercising their public trust rights seaward of that mark.

usually barred[9] from being conveyed into private ownership.[10] Although these principles give the public a large stake in coastal lands and beaches, just how large a stake and how the extent (or limits) of that stake may be established is the subject of this chapter.

The tension between public and private ownerships along the open coast has given rise to numerous disputes and to many reported cases. In exploring these cases, this chapter will focus on the development and adoption of a legally sanctioned methodology to determine the geographic location of the ordinary high-water mark[11] property boundary. Although not nearly as frequently the point of controversy, the determination of the physical location of the ordinary low-water mark property boundary will also be discussed. The principles concerning determination and physical location of the low-water property boundary have implications beyond the seemingly mundane concerns of where one person's property ends and another's begins.[12]

Once the basics of open coast boundary location are discussed, application of boundary movement principles in the case of changes in geographic location of the seacoast shoreline will be explored. Finally, this chapter makes observations concerning the types of proof and sources of information that will assist in the presentation of such title and boundary claims and how this

---

9. In some of the eastern states, most notably Massachusetts, the area along the shore between the high-water mark and the low-water mark has been granted into private ownership subject to certain public rights, including the rights to fish and to navigate over such lands. Robillard, W., and Bouman, L., Clark on Surveying and Boundaries (7th Ed., 1997) § 25.15 (hereinafter "Clark 7th"); Grimes, *Clark on Surveying and Boundaries* (4th ed., 1976) (herein "Clark 4th") § 596, p. 870; *Boston Waterfront Dev. Corp.*, supra, 393 N.E.2d at 360; *Snow v. Mt. Desert Island Real Estate Co.* (Me. 1891) 24 A.429, 430; *State ex rel. Buckson v. Pennsylvania R.R. Co.* (Del. 1969) 267 A.2d 455, 458; *Com. v. Morgan* (Va. 1983) 303 S.E.2d 899, 901. And in California, a specific type of conveyance that included lands seaward of the low-water mark has been validated by the courts. *City of Berkeley*, supra, 26 Cal.3d at 534 (Board of Tideland Commissioner's patents conveying offshore San Francisco Bay lands including what is now an important part of the business district of San Francisco).

10. *Illinois Central Railroad v. Illinois*, supra, 146 U.S. at 42–460; *People v. California Fish Co.* (1913) 166 Cal. 576, 589–592, 601. But see note 9 in this chapter.

11. Please do not attribute any great measure of significance to the fact that in some reported opinions, different jurists refer to this boundary as the "high-water mark" or "high-tide line." It has long been noted by some federal and state courts that the term "high-water mark" is the equivalent of the term "ordinary high-water mark." *Goodtitle v. Kibbe* (1850) 50 U.S. (9 How.) 471, 477–478; *Borax, Ltd. v. Los Angeles* (1935) 296 U.S. 10, 22–23; *More v. Massini* (1860) 37 Cal. 432, 435.

12. The location of the ordinary low-water mark is the base point from which the territorial or marginal sea of the United States is measured. 1 Shalowitz, Shore and Sea Boundaries (Dept. Comm. Pub. 10-1 1962) (herein "1 Shalowitz") §§ 312, 2211; 3 Shalowitz, Reed, M., Shore and Sea Boundaries (Dept. of Comm. Pub. NOAA/CSC/20007-PUB 2000) (hereinafter "3 Shalowitz"), p. 48. Although the marginal sea is considered part of the national domain of the United States, non-United States flag vessels, including foreign warships, have the right of innocent passage through the marginal sea. 1 Shalowitz, § 312, p. 23.

information has been considered and weighed in the crucible of litigation. Throughout this chapter and those following, the author assumes the reader has a basic understanding of the nature of the littoral property boundary and the terms and principles pertaining to boundary movement as discussed in Chapter 3.[13]

4.2 INITIAL CONSIDERATION OF THE PHYSICAL LOCATION OF THE ORDINARY HIGH-WATER MARK

Early cases discussing the property boundary of lands along the open coast shoreline show a general lack of sophistication concerning that boundary's physical location. Witness a case about a dispute between the Dead Whale Asphaltum Mining Company and a competing miner about the extraction of asphaltum on the seacoast near Santa Barbara, California.[14]

The plaintiffs challenged a state swamp and overflowed land patent issued to their competitor.[15] The swamp and overflowed patent characterized the lands conveyed as tidelands.[16] The court described the physical properties of those lands as follows: The lands lay "immediately" on a rocky and sandy beach, "upon which a heavy surf runs at ordinary high tide; and . . . the remaining portion of the land is covered by the waters of the ocean to a depth . . . sufficient for any class of vessels. . . ."[17]

Perhaps the court believed it was clear that the lands described in the patent were not swamp and overflowed in physical character. In fact, the court did not deign to delineate with any particularity what "usually overflowed" or "neap or ordinary tides" meant and yet still voided the patent.[18] The opinion specifically discussed the legal boundary between swamp and overflowed lands and tidelands. Noticeably, the opinion failed to describe the physical

---

13. See notes 94 through 127 and accompanying text in Chapter 3.
14. *People v. Morrill* (1864) 26 Cal. 336, 337–338. It is believed this same "asphaltum" has made the beaches below the campus of the University of California at Santa Barbara relatively unsuitable for sunbathing; you get tar on your feet. Personal interview, Flushman, M.A., former UCSB student and beach hazard and sunbathing consultant.
15. *People v. Morrill, supra*, 26 Cal. at 352.
16. *Ibid.* The inconsistency between the designation of the patent as one for swamp and overflowed lands and the description of the lands surveyed as tidelands may be argued by some to be indicative of the confused state of the swamp and overflowed land disposition program. See notes 87 through 110 and accompanying text in Chapter 1 and notes 29 through 61 and accompanying text in Chapter 5.
17. *People v. Morrill, supra*, 26 Cal. at 352–353.
18. The court stated:

> The lands [described in the] patent are neither swamp or swampy, nor is there any sensible connection between them and the [planting, growing, or harvesting] "seasons" . . . nor between them and "crops"; and the "average" stated might very well have been spared in view of the fact that a state of "inundation" recurring with the regularity of the tides is the normal condition of those lands. *Id.* at 356.

indicia or the physical feature denoting the geographic location of the ordinary high-water mark property boundary.[19] The court was content to leave as its only guidance as to the physical location of the landward boundary of tidelands and the seaward boundary of uplands, the bald and unhelpful description that tidelands were lands covered by "ordinary tides."

Nineteenth-century cases concerning the seaward property boundary along sandy beaches usually called that boundary the "shore of the ocean." These cases followed the relaxed, but imprecise, standard of judicial description of the littoral property boundary delineation then in vogue. Instead of requiring that the ordinary high-water mark property boundary be physically placed with any precision, courts were satisfied with merely announcing that the "ordinary high tides" covered the lands.[20] This is not a sign of judicial sloppiness. The indefiniteness merely shows that there was a lack of any pressing need for further meticulousness in locating the property boundary.[21] Not only were competing uses practically nonexistent, but the courts did not yet recognize the significance of public interests in the shore.[22]

4.3 INTRODUCTION OF TIDAL MEASUREMENTS TO PHYSICALLY LOCATE THE ORDINARY HIGH-WATER MARK

In the first decade of the twentieth century, interested professionals attempted to provide more precision to determination of the location of the littoral property boundary.[23] These attempts were due, in part, to increasing sophistication

---

19. *Ibid.* Although the Dead Whale Asphaltum Mining Company lost the first skirmish, it ended up winning the war. Its successors successfully continued mining the area, despite protestations of the upland grantees, whose land grants were bounded by the "sea." *More, supra,* 37 Cal. at 433.
20. *Kimball v. Macpherson* (1873) 46 Cal. 103, 107–108. Other cases also broadly and vaguely defined the shore of the sea as ". . . the land between ordinary high- and ordinary low-water mark, the land over which the daily tides ebb and flow." *United States v. Pacheco* (1864) 69 U.S. (2 Wall.) 587, 590. As we shall see, the "daily tides" are of varying heights, largely dependent on astronomic factors. See notes 48 through 65 and accompanying text in this chapter.
21. For a graphic example of lack of precision in locating the ordinary high-water mark property boundary, you may wish to refer to the testimony of one surveyor familiar with early surveying practices along water boundaries:

> . . . [T]hat it was the established practice of the Land Department in surveying areas to establish as high-water mark an actual mark as it appeared on the land where the line could be ascertained or located; that the line had to be somewhat imaginary where there were no physical marks on the ground. . . .

Corker, "Where Does the Beach Begin" (1966) 42 Wash. L. Rev. 33, 59, n. 89 (recounting the testimony of D.E. Hughes, whom we will also hear from later; see notes 23 and 93 in this chapter).
22. See *People v. California Fish Co., supra,* 166 Cal at 595–596; *Illinois Central Railroad v. Illinois, supra,* 146 U.S. at 452–455; *State ex rel. Thornton v. Hay, supra,* 462 P.2d at 674.
23. Hughes, D.E., and Von Geldern, O., "The Determination of the Plane of Ordinary High Tide for Pacific Coast Harbors, with Particular Reference to San Diego Harbor California" (1910) Journal of the Association of Engineering Societies, Vol XLIV, No. 4 (herein "Hughes and Von Geldern"); *Eichelberger v. Mills Land, etc. Co.* (1908) 9 Cal.App. 628.

of littoral and riparian property owners and their lawyers and surveyors, heightened awareness of the value of the public's shore lands and adjoining privately held uplands, and, perhaps even more important, the pressures of ever escalating development and land use. These endeavors gained some recognition in the courts.

In a particularly illustrative case, at issue was the extent of littoral upland conveyed by an instrument containing a legal description calling to and running along the ocean.[24] After an attempted transfer of ownership, party A discovered that the upland tract was not as wide as had been stated in the negotiations. The "shortage" was due to placement of the tract's seaward boundary by survey following those negotiations. Party A wished to avoid the purchase contract because party B had allegedly misrepresented the quantity of upland thought to have been conveyed. Thus, the physical location of the ordinary high-water mark property boundary was critical to the result in the case.[25] The trial court held that the ocean boundary of the upland parcel was "the point of extreme reach of the wash of the waves" or "the highest point of tide level."[26] The appellate court disagreed. That court found the littoral property boundary to be "the line indicated by high-water mark at ordinary or neap tide."[27] Not satisfied with merely reciting the old platitudes culled from past decisions, this time the court tried to provide some certainty to the meaning of "neap tide."

> "The law takes notice of three kinds of tides, viz.: 1. The high spring tides, which are the fluxes of the sea, at those tides which happen at the two equinoctials;[28] 2. The spring tides, which happen twice every month, at the full and change of the moon; 3. The neap, or ordinary tides, which happen at the change and full of the moon, twice in twenty-four hours."[29]

Based on this understanding of the tidal process, the court described the ocean property boundary line of the tract as ". . . the line to which the flow of the water reaches at ordinary or neap tide, unaffected by wind or wave. . . ."[30] Naturally, there is no mention of how exactly one could eliminate the effect of wind and waves. Notwithstanding that slight oversight, this was the first

---

24. *Eichelberger, supra*, 9 Cal.App. at 631 (thence . . . west . . . feet a little more or less to the Pacific Ocean; thence easterly along the Pacific Ocean. . . .")
25. The result of this contest is not important for this work. What is pertinent is the appellate court's discussion of the physical location of the ordinary high-water mark property boundary.
26. *Eichelberger, supra*, 9 Cal.App. at 639.
27. *Id.* at 639–640.
28. Equinoctial tides are those tides occurring near the times of the equinoxes. *Tide and Current Glossary* (U.S. Dept. Of Comm., NOAA, NOS 1984), p. 7. As a source of information and explanation concerning the sometimes arcane terminology of tides, the reader will find the Tide and Current Glossary invaluable.
29. *Eichelberger, supra*, 9 Cal.App. at 639 (citing Angell on Tide Waters, p. 68).
30. *Id.* at 640.

step on the road to a more precise indicator of the physical location of the ordinary high-water mark property boundary, reliance on tidal measurements or observations. Subsequent courts adopted and amplified this step with increasing sophistication.[31]

4.4 USE OF INDICIA OTHER THAN TIDAL MEASUREMENTS— VEGETATION OR EROSION LINES

Courts have not totally accepted determination and location of open coast property boundaries through use of tidal observations.[32] In a case in the State of Washington, the court held that the line of ordinary high tide along the open coast is the line that the water impresses on the land by covering the soil for sufficient periods to deprive the soil of vegetation and destroy its value for agricultural purposes.[33] In Hawaii, the boundary of the state's ownership extends to the upper reaches of the wash of the waves as evidenced by a vegetation line.[34] The use of tidal measurements to set the location of

---

31. E.g., *Forgeus v. County of Santa Cruz* (1914) 24 Cal.App. 193, 195; *F.A. Hihn Co. v. City of Santa Cruz* (1915) 170 Cal. 436, 442; *Carpenter v. City of Santa Monica* (1944) 63 Cal.App.2d 772, 774; *People v. Hecker* (1960) 179 Cal.App.2d 823, 826; *Leabo v. Leninski* (Conn. 1981) 438 A.2d 1153, 1156; *Kruse v. Grocap, Inc.* (Fla. 1977) 349 So.2d 788, 789; *Smith v. State, supra,* 282 S.E.2d at 81; *Hirsch v. Maryland Department of Natural Resources* (Md. 1979) 416 A.2d 10, 12; *Cinque Bambini Partnership v. State* (Miss. 1986) 491 So.2d 508, 514, *aff'd sub nom. Phillips Petroleum Co. v. Mississippi* (1988) 484 U.S. 469, *reh'g. den.* (1988) 486 U.S. 1018; *Matcha, supra,* 711 S.W.2d at 99; *Nollan, supra,* 483 U.S. at 827.
32. Corker, *supra,* 42 Wash. L. Rev. at 43–72; Maloney and Ausness, "The Use and Legal Significance of the Mean High-water Line in Coastal Boundary Mapping" (1974) 53 No.Car. L. Rev. 185, 206.
33. *Harkins v. Del Pozzi* (Wash. 1957) 310 P.2d 532, 534. The authority for this proposition was a case, *Driesbach v. Lynch* (Idaho 1951) 234 P.2d 446. But that case concerned the boundary of a lake in Idaho! Even those with only a dim sense of geography are usually aware that there are no tidelands or tidal waters in Idaho. Later Washington cases are inconsistent with *Harkins*. *Narrows Realty Company v. State* (1958) 329 P.2d 836, 837, n. 2 (using "neap" tides as "ordinary" tides); *Hughes v. State* (1966) 410 P.2d 20, 26, *rev'd on other grnds, Hughes v. Washington* (1967) 389 U.S. 290. And federal courts have held that the vegetation or erosion line tests are applicable only to shoreline property boundaries along nontidal streams or rivers, not to the property boundary of lands along tidal water bodies. *City of Los Angeles v. Borax Consolidated Limited* (9th Cir. 1935) 74 F.2d 901, 904–905, *aff'd* (1936) 296 U.S. 10.
34. *Application of Ashford* (Haw. 1968) 440 P.2d 76, 77; *Tarshis v. Lahaina Investment Corporation* (9th Cir. 1973) 480 F.2d 1019, 1020, n. 1; *Littleton v. State* (Haw. 1982) 656 P.2d 1336, 1343. An extensive dissent in the *Ashford* case forcefully argues against the "uncertain" vegetation line and for the more certain, scientifically determinable line located through use of tidal observations. *Application of Ashford, supra,* 440 P.2d at 80. The dissent's argument is worthy of fuller note:

> ... [W]hat the State is asking this court to do is to declare as *the law* for the determination of the seaward boundaries of private lands in this ... state, which prides itself in being a progressive member of the federal union of states, a practice primitive in concept and haphazard in application and result, which the United States Supreme Court rejects for use

open coast littoral boundaries is the more common and accepted practice.[35]

Despite their somewhat naive unsophisticated initial efforts, courts gradually came to understand and increased their use of scientific principles in reaching decisions. Through continual judicial refinement, court opinions eventually imparted considerable precision to the determination of the physical location of the littoral property boundary. The course of those refinements will next be discussed.

4.5 REFINEMENT OF THE USE OF TIDAL MEASUREMENTS AS THE PHYSICAL INDICIA OF THE ORDINARY HIGH-WATER MARK

Courts in the early years of this century consistently stated that the ordinary high-water mark was the line reached by the "ordinary" or "neap" tides. Unfortunately, these courts had not formulated any physically quantifiable definition of what was meant thereby. In other words, judges would boldly state in their decisions that the location of a boundary was at the "limit reached by the ordinary or neap tides." Pinpointing the physical location of that property boundary merely by using the written description provided by court opinions was not possible for the average person.[36]

A case about the seaward extent of an upland littoral grant exemplifies the problems and confusion resulting from the use of such formulations. In that case, the opinion described the boundary named in the title documents[37] as the "bay."[38] At issue was the physical location of the limit of ordinary high-

---

by the federal government, and to reject for use in this state a practice scientific in concept, uniform in application and precise in result. . . . *Ibid* (emphasis in original).

35. The principal case in support of this rule is *Borax, Ltd. v. Los Angeles* (1935) 296 U.S. 10, 22. Although this case concerned an island in a tidal estuary, its principles are equally applicable to the open coast. Therefore, it will be discussed in this chapter.

36. The evils of vague water boundary description are noted in *Deering v. Martin* (Fla. 1928) 116 So. 54, 63–64. The description in that case referred to water of a certain depth "at high tide." *Ibid.* The court said: "We think this description . . . is too vague and uncertain to constitute a valid conveyance of submerged lands . . . . There is no guide by which a surveyor can take the description in the deed . . . and locate the acreages that are intended to be conveyed." *Ibid.*

37. By now the reader has noticed that many of the boundary disputes discussed in this chapter (as well as the disputes discussed in following chapters) arise out of the interpretation of language in title documents or granting statutes or legislation. Thus, as an important preliminary matter in such boundary contests, it is absolutely crucial to have as complete a chain of title to the property in question as possible; the chain should extend back to the original government grant. E.g., *Com. v. Morgan, supra,* 303 S.E.2d at 900–901. The chain of title enables the surveyor or lawyer to have at his or her disposal the complete series of instruments forming the disputed parcel. The extent of the property or the type of boundary intended to be or capable of being conveyed is ofttimes made plain by an examination of the progression of instruments in the chain. See notes 222 through 225 and accompanying text in Chapter 1.

38. *Forgeus, supra,* 24 Cal.App. 193.

water of that bay.³⁹ The court reconfirmed that the physical location of the property boundary of lands described as bounded by the "bay" was determined by the ordinary or neap tides, as opposed to spring tides.⁴⁰ The court took some pains to point out how that boundary could be physically located.

According to the court, the "United States Geodetic Coast Survey"⁴¹ located the "line of ordinary high-water mark" all along the coast.⁴² As understood by the court, to ensure permanency of that line, that government agency placed bench marks at different locations.⁴³ Consequently, with the glibness born of misunderstanding, the court's opinion noted, by using tide tables published by the government⁴⁴ along with the bench marks, that it was "not disputed" that the ordinary high-water line of that date could have been

---

39. *Id.* at 194.
40. *Id.* at 195.
41. To the author's knowledge there has never been, nor is there now, a United States government agency known by the name "United States Geodetic Coast Survey." The court must have meant to refer to the United States Coast and Geodetic Survey (hereinafter referred to in this work as "USC&GS"). The USC&GS was the successor to the United States Coast Survey (hereinafter referred to in this work as "USCS"). The USC&GS has, in turn, been succeeded by the National Ocean Survey (hereinafter referred to in this work as "NOS") and the United States Geodetic Survey. The USCS, the USC&GS and the NOS were and are charged with, among other things, the measurement of tides and the mapping of the coastal shoreline of the United States in connection with the preparation of charts for navigation. See, generally, 2 Shalowitz, Shore and Sea Boundaries (Dept. Comm. Pub. 10-1 1962) (herein "2 Shalowitz"), Chap. 1; *Smith v. State, supra*, 282 S.E.2d at 79.
42. *Forgeus, supra*, 24 Cal.App. at 195. The USC&GS in its coastal mapping never located such a line. Rather, the USC&GS located the high-water line. This high-water line was the dividing line between land and water for use in navigation charts. 2 Shalowitz, § 441, p. 171. The location of the high-water line was not accomplished by a precise process of leveling. Rather, such a line was determined from observation of the physical characteristics of the beach at the time the coastal surveyor happened to be there. *Id.* at § 4421, p. 174. For a personal statement by a nineteenth-century USC&GS survey assistant as to what constituted the high-water mark in that agency's work, see Letter from Assistant Henry L. Marinden, U.S. Coast & Geodetic Survey, Appendix No. 16, Report for 1891, Part II, Proceedings of the Topographical Conference held at Washington, D.C., January 18 to March 7, 1892, pp. 586–587.
43. According to the opinion, bench marks were marks affixed to a permanent object in tidal observations, or along a line of survey to furnish a datum level. *Forgeus, supra*, 24 Cal.App. at 195. For our purposes, bench marks are more accurately defined thus:

> All along the coast, wherever tides are observed, the Coast Survey establishes tidal bench marks, the elevations of which are determined with reference to . . . tidal planes. . . . The bench marks . . . serve the further purpose of preserving for future use the datum planes that are determined from tidal observations. 2 Shalowitz, § 2314, p. 61; *Id.* at Appendix A, p. 550.

Notice that nowhere is it stated or implied that the bench mark itself is located at the "line of ordinary high-water" or at any specific geographical place, except one arbitrarily chosen for the convenience of and in accordance with the instructions provided to the coast surveyor.
44. These tide tables merely disclose the time and relative elevation of *predicted* high and low waters each day.

(and was) determined and located.[45] The court assumed that bench marks set and located by the USC&GS could be used to set the line reached by the neap tides.

4.6 EXPLANATION OF THE TIDES, TECHNICAL TERMS AND EXPRESSIONS

The courts' opinion in the case described in the preceding section shows a lack of in-depth understanding of the tidal and littoral process.[46] This lack of understanding has continued to plague courts, lawyers, and surveyors in this field.[47] To grasp the court's blunder, the phenomenon of tides and the process of tidal measurements should first be explored. Such knowledge can provide a basis for the understanding of this important aspect of the methodology of tidal littoral property boundary location.

4.6.1 Tide-Producing Forces

In scientific terms, tides are the periodic, usually twice daily, rise and fall of a water surface[48] resulting from the gravitational interactions between the earth, moon, and sun.[49] In a cosmic sense, the major tide-producing forces

---

45. *Forgeus, supra*, 24 Cal.App. at 195–196. Showing its uncertainty and confusion over the use of scientific methodology, the court immediately thereafter attempted to bolster its conclusion by reporting the testimony of an eyewitness who claimed to have seen the waves reach the "ordinary high-tide line" so established. *Ibid.*
46. The court's stated methodology, using published tide tables and USC&GS bench marks, would only provide location of the elevation of the predicted high or low water for any given day. At worst, the result could have been completely misleading and, at best, only a rough approximation of the physical location of the ordinary high-water mark property boundary. For example, the point from which vertical measurements are taken, the reference datum, for the tide tables may not have been the same reference datum used in determining the elevation of the bench marks. For a discussion of the resolution of a datum disparity, see page 225 in this book. Thus, it would be inappropriate to compare the two elevations; you would be mixing apples and oranges.
47. E.g., 1 Shalowitz, § 6412, p. 93.
48. In this book the word "tides" refers to the vertical movement of the surface of the water and is distinguished from the word "current." "Current" refers to the horizontal movement of water. In conformity with the understanding that it is the water surface whose vertical movement is referred to, instead of the term "_____ _____ tide", this work will use the technically more precise term "_____ _____ water". Courts have adopted this convention. *Smith v. State, supra*, 282 S.E.2d at 79.
49. Tide and Current Glossary at p. 21; Marmer, H.A., *Tidal Datum Planes* (1951) U.S. Coast And Geodetic Survey, Spec. Pub. 135 Rev. Ed. (hereinafter referred to as "Marmer"), p. 1. All surveyors and lawyers who are even thinking of becoming involved in the determination of the location of tidal littoral property boundaries should be sure Marmer's work is included in their libraries.

4.6 EXPLANATION OF THE TIDES, TECHNICAL TERMS AND EXPRESSIONS

on earth are the gravitational attractions of the sun and the moon.[50] These two forces operate in conjunction with the hydrographic configuration of the tidal basin to determine the regime of the tides.[51] In other words, the relative positions of the earth, sun and moon and the shape of, for example, the Pacific Ocean basin, produce a unique tidal regime for the Pacific coast. In addition, meteorological conditions, such as wind or barometric pressure, can have short-term, but radical, effects on the tidal regime.[52] A hurricane or a typhoon is an example of what some may euphemistically call "meteorological conditions."

4.6.2 Variability of Tide-Producing Forces

Tide-producing forces are hardly consistent or uniform. The forces vary from locale to locale, from day to day, from month to month, and from year to year; thus, the range of the tides[53] is not constant.[54] There are three principal variables in the range of the tides; these variables are mainly a consequence of the position of the moon and sun in relation to the earth.[55]

4.6.3 Moon's Phase—Spring Tides and Neap Tides

The major variable affecting the tides is related to the phase[56] of the moon. The moon and sun pull together during the times of the new moon and the full moon. During these phases of the moon, the tides rise higher and fall lower and thus have greater range. These tides are known as "spring tides." Tides that rise and fall least and thus have a lesser range occur at about the time of the moon's first and third quarters. These tides are known as "neap tides." Neap tides are produced when the sun and moon are on the opposite side of the earth from each other.[57] In this situation, known as quadrature, the gravitational forces of the sun and the moon on the earth neutralize each

---

50. Marmer, at p. 2; Patton, R.S., *Relation of the Tide to Property Boundaries* set out in 2 Shalowitz, Shore and Sea Boundaries (Dept. Comm. Pub. 10-1 1962) Appendix E, pp. 667–679 (hereinafter referred to and cited as "Patton, at 2 Shalowitz, p. 667–679"). Patton is also an excellent and understandable reference on tides and their relationship to the determination of the physical location of the ordinary high-water mark littoral property boundary.
51. Marmer, at p. 2; Patton at 2 Shalowitz, pp. 668–669.
52. Patton, at 2 Shalowitz, p. 674.
53. The range of tides refers to the vertical magnitude of the rise and fall of the tide. Marmer, at p. 4.
54. Patton, at 2 Shalowitz, p. 668; Marmer, at p. 2.
55. Marmer, at p. 5; Patton, at 2 Shalowitz, p. 669.
56. By "moon's phase" is meant a regularly recurring aspect of the moon with respect to the amount of illumination produced by the moon; i.e., new moon, full moon. 1 Shalowitz, Appendix A, p. 302.
57. Marmer, at p. 5; Patton, at 2 Shalowitz, p. 669.

other. The relative positions of the three heavenly bodies that produce neap and spring tides are shown in Figure 4.1.

4.6.4 Distance from Earth—Perigean and Apogean Tides

The second variable affecting tides is connected with the varying distance of the moon from earth. When the moon is nearest earth, or in perigee, the tidal range is greater; the tide rises higher and falls lower than usual. When the moon is in apogee, or farthest from the earth, the tides rise and fall less than usual. The tides occurring at those times are known, respectively, as perigean and apogean tides.[58]

4.6.5 Declination—Tropic Versus Equatorial Tides

The third variation in the range of the tide is associated with the moon's relation to earth's equator. This is called the moon's declination. When the moon is close to the equator, the two high waters of the day and the two

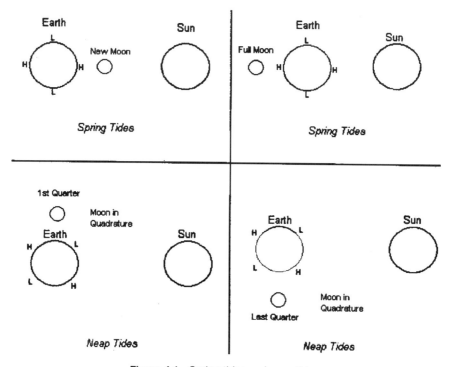

Figure 4.1 Spring tides and neap tides

---

58. Marmer, at p. 5; Patton, at 2 Shalowitz, pp. 669–670.

daily low waters will not differ significantly; that is, the high waters and low waters of the day will resemble each other. As the moon moves away from the equator and its declination increases, there is a difference between the morning and afternoon tides. As expected, when the moon is farthest from the equator—that is, when its declination is greatest—the difference in the two tides is most marked.[59] Tides that occur when the moon is near the equator are called "equatorial tides." Tides that occur when the earth is farthest from the equator at maximum declination are "tropic tides."[60]

4.6.6 Effect of These Variables—18.6-Year Cycle

The three variables affecting the range of the tide do not occur equally in every place in the world. If that were the case, tides the world over would be virtually the same. Thus, it is obvious that one of the variables may affect some locales more than either or both of the other variables. In addition, changes in the tide-producing forces related solely to the gravitational interaction of the sun, moon, and earth vary periodically over time. For example, changes related to the many differences associated with the changing positions of the sun, moon, and earth go through one complete cycle every 18.6 years; changes connected with the phases of the moon go through a complete cycle every 9.5 days.[61] Thus, observations of the tide over a short period at a particular location may not accurately reflect the tidal regime at that location.

4.6.7 Time and Type of Tides

Along with the range of tide, there are two other common features or characteristics of tides at any given place—the time and the type. For purposes of this chapter, the reader can assume that during each slightly more than 24-hour period there will be two high waters and two low waters. Each high and each low water occurs about 50 minutes later than the corresponding high or low water of the preceding day.[62]

Of more importance for purposes of determining the property boundary is the type of tide. By this term is meant the character of the rise and fall of the tide as revealed by an examination of the tide curve.[63] The type of tide is largely governed by the hydrographic characteristics of the tidal basin, such as the Pacific Ocean.[64]

---

59. The difference in the two tides is known as the diurnal inequality. This means the range of one tide is increased at the expense of the range of the other tide. Patton, at 2 Shalowitz, p. 670.
60. Marmer, at p. 5; Patton, at 2 Shalowitz, p. 670.
61. Marmer, at p. 6; Patton, at 2 Shalowitz p. 670.
62. Marmer, at p. 3; Patton, at 2 Shalowitz, p. 669.
63. Marmer, at p. 9. A tide curve is a graphic depiction of the rise and fall of the tide. Time is usually represented by the abscissa and height by the ordinate of the graph. Tide and Current Glossary, p. 22. Examples of tide curves are found in Figure 4.2.
64. Patton, at 2 Shalowitz, pp. 671–672; Marmer, at p. 4.

There are three types of tides. The semidaily, or semidiurnal, type of tide is one in which there are two tidal cycles each day, and such tidal cycles resemble each other. The daily type of tide, as should be readily apparent from its name, is one in which only one high and one low water happens each day. Finally, the mixed type of tide is one in which there are two high and two low waters each day, but there are marked differences in elevation between the two high waters and the two low waters of the day.[65] Figure 4.2. shows examples of tide curves of the three types of tides. The tidal regime in California is a mixed type of tide.

4.7 THE "NEAP TIDES" CONFUSION

Even with the preceeding basic explanation of the tidal process and regime, the reader can readily discern the difficulty in using so-called neap tides as the physical embodiment of the legal term "ordinary tides." The use of the term "neap tides" grew from an early California case, *Teschemacher v.*

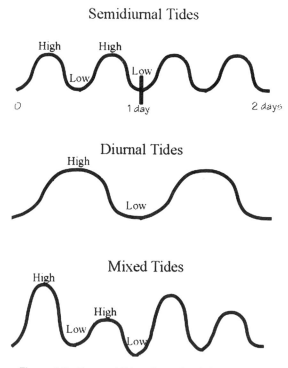

Figure 4.2 Types of tides shown by their tide curves

---

65. Marmer, at p. 9.

Thompson,⁶⁶ about the boundary between a Mexican land grant and a competing state tideland patent. This case concerned location of the ordinary high-water mark in an estuarine marsh. The description of the ordinary high-water mark property boundary created by the court has, as we have already seen, been adopted and applied in cases concerning the open coast shoreline property boundary of uplands and tidelands.

The *Teschemacher* court defined the "usual" or "ordinary" high-water mark as ". . . the limit reached by the neap tides; that is, those tides which happen between the full and the change of the moon, twice in every twenty-four hours."⁶⁷

Nevertheless, we now know that neap tides do not occur as frequently as twice in every 24 hours. Technically, neap tides occur only twice a month when the forces of the sun and moon act to cancel each other.⁶⁸ One can scarcely term this relatively infrequent occurrence "ordinary" in the dictionary sense of the word.⁶⁹ Nor, at least on the Pacific Coast, are the two high waters of the same height; the two high waters each day differ markedly in their height. If the "neap tide" test is used, it is still not possible to decide which of the two daily high tides supplies the limit reached by the neap tides; is it the higher neap tide or the lower neap tide?

As can readily be seen, the use of the term "neap tides" as descriptive of the physical phenomenon defining ordinary tides is fraught with ambiguity and inconsistency.⁷⁰ With help from the government and the scientific community, the courts finally recognized and resolved this uncertainty.

4.8 FOUNDATION OF TIDAL DATUMS

4.8.1 Background

As with almost any question brought to the attention of lawyers and judges, the resolution of how to physically locate the ordinary high-water mark prop-

---

66. (1861) 18 Cal 11. The opinion was written by Justice Field, who later became a justice of the United States Supreme Court. For a fascinating article on the background of Justice Field and his impact on the Supreme Court see Tocklin, Adrian M., "*Pennoyer V. Neff*: The Hidden Agenda of Stephen J. Field," 28 Seton Hall Law. Rev. 75 (1997). On the Supreme Court, Justice Field continued the error of his ways by describing lands below the ordinary high-water mark as those covered by the "neap" or "ordinary tides." *Knight v. United States Land Association* (1891) 142 U.S. 161, 215 (Field, J., concurring); *San Francisco v. Leroy* (1891) 138 U.S. 656, 671.
67. *Teschemacher, supra*, 18 Cal. at 21–22.
68. See notes 56 through 57 and accompanying text in this chapter, and Figure 4.1.
69. In the sense of being "customary or normal", tides that occur only two times a month, 48 out of 365 days, or about 13 percent of the time, are not customary.
70. This ambiguity has been the subject of much critical comment by commentators and courts. E.g., Patton, at p. 668; 1 Shalowitz, at p. 93; Briscoe, J., "The Use of Tidal Datums and the Law" (1983) 43 Surveying and Mapping (Journal of the American Congress on Surveying and Mapping) No. 2, pp. 115, 116; Maloney and Ausness, *supra*, 53 No.Car.L.Rev. at 204; *City of Los Angeles v. Borax Consolidated Limited, supra*, 74 F.2d at 905.

erty boundary along the open coast was a gradual, some may even say glacial, process. The first step in that process was the recognition that there must exist some scientific methodology enabling the littoral property boundary, as a legal concept, to be physically quantified and geographically located with precision.

Fortunately, the United States government was interested in tidal phenomena. This interest stemmed from the fact that considerable resources and commerce may be impacted by or dependent on tidal waters. Safe waterborne commerce has required accurate navigational charts for the coastal and estuarine reaches of the shoreline of the United States. To supply information for these purposes, the USCS and its successors, the USC&GS and the NOS, have operated a system of tide stations at harbors and particular coastal locations. At these places continuous observations of the tide have been made and recorded over a number of years.

4.8.2 Methodology of Tidal Observations

Tidal measurements were either taken and recorded manually by regularly observing the undisturbed water level on a tide staff[71] or by means of an automatic tide gauge.[72] Long-term tide stations, sometimes called control or primary tide stations, have been established at various locations in the United States. Nearly every major port or harbor has one. In addition, short-term tide stations are established from time to time in connection with the preparation of hydrographic charts.[73]

4.8.3 Need for Long-Term Tidal Observations

By virtue of the efforts regarding tidal observations, some basic facts concerning the tides were confirmed, discovered, or refined. Because the rise and fall of the tide varies each day, we may know the characteristics of a particular tidal regime based on only a short series of observations, such as over a day or a week. The results of these observations may differ markedly from characteristics derived from values obtained over a long period and then averaged or meaned. Consequently, in the case of tides long-term averaging is particularly appropriate.[74]

---

71. A tide staff is made from a board 5 to 6 inches wide and approximately 1 inch thick, graduated to feet and tenths with numbers increasing upward. Various methods were used to still the water level. Marmer, at p. 23.
72. For the mechanically inclined reader, automatic tide gauges and their operation are generally described in Marmer, at pp. 26–28.
73. For a more detailed explanation of tide stations and their order, see 2 Shalowitz, §§ 2311, 2312; Marmer, at Chapter 3, pp. 23–31, 67–68.
74. The physical and astronomical variations requiring long-term averaging of tidal values are explained in Marmer, at pp. 6–7; Patton, at 2 Shalowitz, p. 675; and in note 61 and accompanying text in this chapter.

For accurate results, the determination of tidal characteristics should be based on a series of observations or measurements systematically taken over many years.[75] Indeed, an 18.6- (sometimes called a 19-) year period is considered a full tidal cycle. This is because the more important of the periodic tidal variations due to astronomic causes will have gone through complete cycles. In addition, randomly recurring meteorologic variations are assumed to balance out during a period of this length.[76]

These tidal observations conducted by the government, supplied the foundation for scientific development of more precise measurements.

4.9 TIDAL DATUMS

Whatever the character of the tide station or however long the period of observation and recordation of the tides, periodic tidal data has been and is being accumulated through the system of tide stations. One may convert this data, through statistical and mathematical means, into various vertical planes of reference, known as "tidal datums."[77] Tidal datums are quite simply defined and can be readily, accurately, and certainly determined. It is not essential that tidal data be obtained for an entire 19-year tidal cycle. Through statistical and mathematical means, one can derive a 19-year mean from a shorter series of observations.[78]

The basic tidal datums, from the lowest to the highest in terms of relative elevation, are mean lower low water,[79] mean low water,[80] mean sea level,[81]

---

75. Marmer, at pp. 7–8; Patton, at 2 Shalowitz, p. 675.
76. 2 Shalowitz at § 2311, p. 59. One court has required that a 23-year period be used to determine a tidal datum, interpreting quite literally the admonition that the tide be observed for a "long" period. *State v. Pennsylvania Railroad Company* (Del. 1967) 228 A.2d 587, 601. The Delaware Court appears to have lost sight of the reason for the 19-year period—that it accounts for all of the periodic variations in the tide-producing forces—and instead adopted a period that may have coincided with the period of tidal records available. *Borax, supra,* 296 U.S. at 26–27; *City of Los Angeles v. Borax Consolidated Limited, supra,* 74 F.2d at 906.
77. Although the term "tidal datum plane" is sometimes also used, that term is somewhat redundant, as a "datum" is a "plane" of reference. Shalowitz defines "datum plane" as "[a] surface used as a reference from which heights or depths are reckoned. The plane is called a Tidal Datum when defined by a phase of the tide. . . ." 1 Shalowitz, p. 286. In this work, the more technically correct term "tidal datum" will be used. 2 Shalowitz at p. 611.
78. Although the author is certain the reader is lusting for an excruciatingly detailed explanation of just how this is done, that lust will not be satiated in this work. The author would not think of attempting to improve upon Marmer, whose explanation should gratify even the most fervent statophile. Marmer, pp. 87–95.
79. The average height of only the *lower* of the low waters over a 19-year period. Marmer, p. 113; 2 Shalowitz, p. 581. On the Pacific Coast this is the datum for soundings taken for hydrographic charts. 2 Shalowitz, p. 581.
80. The average height of *all the low* waters over a 19-year period. Marmer, p. 104; 2 Shalowitz, p. 581.
81. The mean level of the sea at a particular location determined by averaging the height of the water levels for *all stages* of the tide. This is the primary tidal datum, as all other tidal datums are derived with reference to mean sea level. Marmer, p. 45; 2 Shalowitz, p. 528.

mean tide level,[82] mean high water[83] and mean higher high water.[84] These tidal datums are not magical. They are useful tools for measuring, for different purposes, water levels that are constantly in flux.[85] Certain of these tidal datums have also been used as reference points for the mapping and charting work of the USCS, the USC&GS, and the NOS.

4.9.1 Use of Tidal Datums in Coastal Mapping

The USCS and, later, the USC&GS and the NOS, prepared maps—surveys, really—of shore and coastal areas to construct charts for navigators.[86] These surveys required the features depicted to be referenced to common vertical and horizontal points.[87] Thus, cartographers were able to prepare nautical charts that aided mariners in determining the safety of navigation in a particular area by locating specific objects, features, or structures useful for navigation purposes.

Two types of surveys were prepared: a topographic survey and a hydrographic survey. From these two surveys the coast surveyors constructed a nautical chart. In general terms, the topographic survey is the authority for all upland features; the hydrographic survey is the authority for all submerged features.[88] Courts used these surveys in making property boundary determinations,[89] and their currency in the legal world grew as did that of the methodology utilized to construct the charts.[90]

---

82. This datum is also known as half tide level. It is a tidal datum *midway* between mean high-water and mean low water. Marmer, p. 69; 2 Shalowitz, p. 568. Mean tide may, and usually does, differ from mean sea level.

83. The average height of *all the high-waters* at a location for a period of 19 years. Marmer, p. 86; 2 Shalowitz, p. 581. This is true for semidiurnal or mixed tides. There are some special rules for diurnal tides. *Ibid.*

84. The average of only the *higher of the high* waters at a location over a 19-year period. Marmer, p. 86; 2 Shalowitz, p. 581.

85. For example, mean lower low water is used as the reference datum for nautical charts on the Pacific Coast. Basically, that means the depths shown on such charts are measured from that plane of reference as the 0 or reference datum. The reason for this rule is that it results in shallower charted depths for the benefit of the navigator. 1 Shalowitz, p. 88.

86. 2 Shalowitz, p. 80. Some of the limitations of those surveys are noted. *Ibid.*

87. Swainson, O.W., *Topographic Manual* (Dept. of Commerce Special Publication No. 144) (1928), p. 75. ("The datum plane for elevations adopted by the Coast and Geodetic Survey for topographic work is mean high-water. The mean high-water line which is considered to be the shore line is, therefore, the zero contour.")

88. A much more detailed and authoritative description of these surveys is found in 2 Shalowitz, pp. 79–102, and in Jones, Lester E., *Elements of Chart Making* (USC & GS Spec. Pub. No. 38 1916). These surveys are also discussed in notes 161 through 189 and accompanying text in Chapter 5.

89. *United States v. O'Donnell* (1938) 303 U.S. 501, 508; *City of Oakland v. Wheeler* (1917) 34 Cal.App. 442, 451–452, *error dism'd* (1920) 254 U.S. 659; *Boone v. Kingsbury* (1928) 206 Cal. 148, 186–187, *cert. den.* and *app. dism'd sub nom. Workman v. Boone* (1929) 280 U.S. 517.

90. *Forgeus, supra*, 24 Cal.App. at 195. The bench marks described in the opinion were placed for the use of the USCS and USC&GS in the construction of such surveys. See 2 Shalowitz, pp. 486–494.

4.10 MEAN HIGH-WATER LINE ADOPTED AS THE PHYSICAL LOCATION OF THE ORDINARY HIGH-WATER MARK— THE BORAX CASES

In California, it was during the first decades of the twentieth century that the use of the mean high-water line[91] as the physical location of the legal property boundary between tidelands and coastal uplands first gained currency.[92] The impetus for the adoption of this methodology by the courts was most probably the development and use of tidal datum planes by the United States government and its coastal and shoreline mapping agencies.

During the early twentieth century, a vigorous debate arose between surveyors and coastal engineers over just how the legal term "ordinary high-water mark" was to be translated into a physical location.[93] Legislative enactments and court decisions began using the term "mean high-water line" or "line of mean high-water" to describe the shoreward property boundary of sovereign lands.[94] All these developments led to pressure on the courts to "definitively decide"[95] just how to transform the legal term "ordinary high-water mark" into a physical location.

Since the late nineteenth century, Mormon Island in San Pedro Harbor near Los Angeles was known to be an extremely level, largely barren, mud flat. To the pioneers of Los Angeles it would have seemed absolute lunacy that it took no less than seven separate reported decisions, including two cases in the United States Supreme Court and one in the California Supreme Court, to decide who owned what on this island.[96] The dispute over the boundaries

---

91. The mean high-water line is the intersection of the tidal datum mean high-water with the shore. 2 Shalowitz, p. 581; *Swarzwald v. Cooley* (1940) 39 Cal.App.2d 306, 313; *People v. Wm. Kent Estate Co.* (1966) 242 Cal.App.2d 156, 160; *O'Neill v. State Highway Department* (N.J. 1967) 235 A.2d 1, 9.
92. *Forgeus, supra*, 24 Cal.App. at 195 (apparently misunderstood what the USC&GS had accomplished, but utilized its work); *Strand Improvement Co. v. Long Beach* (1916) 173 Cal. 765, 769 (legislative grant uses term "line of mean high tide").
93. Hughes and Von Geldern, *supra*. Both commentators recognize the deficiency in using the *Teschemacher* formulation of neap tides. Von Geldern contended for a mean of all high waters (mean water), while Hughes sought to convince his audience that the mean of only highest waters (mean higher high water) should be used. The reasoning supporting the use of mean high water was that use of this elevation to determine the ordinary high-water mark would enable the reclamation of marshlands surrounding certain bays and estuaries. These marshes were reportedly just above the elevation of mean high water.
94. *Los Angeles v. San Pedro, etc., RR. Co.* (1920) 182 Cal. 652, 663, cert. den. (1920) 254 U.S. 636; *City of Los Angeles v. Anderson* (1929) 206 Cal. 662, 664.
95. This phrase was first coined by former president Richard M. Nixon to describe the type of Supreme Court decision with which he would comply if ordered to turn over some incriminating tapes to the special prosecutor.
96. *DeGuyer v. Banning* (1891) 91 Cal. 400; *De Guyer v. Banning* (1897) 167 U.S. 723; *City of Los Angeles v. Borax Consol. Ltd.* (S.D. Cal. 1933) 5 F.Supp. 281; *City of Los Angeles v. Borax Consol. Ltd., supra*, 74 F.2d 901; *Borax, supra*, 296 U.S. 10; *City of Los Angeles v. Borax Consol.*

of the land grant encompassing the island, however, resulted in a decision that became a jurisprudential cornerstone in litigation concerning the question of the physical location of the property boundary of lands adjacent to tidal water bodies.

In the famous case of *Borax, Ltd. v. Los Angeles*,[97] the United States Supreme Court decided as a matter of federal law that the ordinary high-water mark, the property boundary between federally patented uplands and tidelands owned by the state, should be physically located by determining the location of the mean high-water line.[98] This holding seems deceptively simple. The reasoning of the case and the preceding decision in the Ninth Circuit Court of Appeals (which the Supreme Court affirmed) is worthy of fuller discussion because of their implications for the location of boundaries in estuaries and tidal rivers.

The basis of the dispute in *Borax* was just how the ordinary high-water mark property boundary of Mormon Island was to be physically located. One party claimed that the property boundary should be located by reference to the mean high-water line; the other party, relying on the *Teschemacher* decision,[99] claimed the boundary was to be located by reference to neap tides.[100] The appellate court fully discussed this claim. The opinion pointed out shortcomings of the "neap tide rule"[101] and the reasons that all high tides must be included.[102] In adopting the mean high-water line as the physical location of the boundary, the court reasoned thus:

Ltd. (S.D. Cal. 1937) 20 F. Supp. 69; *City of Los Angeles v. Borax Consol. Ltd.* (9th Cir. 1939) 102 F.2d 52, *cert. den.* (1939) 307 U.S. 644.
97. (1935) 296 U.S. 10.
98. *Id.* at 26–27.
99. See notes 66 through 67 and accompanying text in this chapter.
100. *City of Los Angeles v. Borax Consolidated Limited, supra,* 74 F.2d at 904–905. Taking a different tack, the city had also contended that the boundary line was determined by " '[a] definite mark upon the ground which has been left by the tide.' " *Id.* at 904. The Ninth Circuit did not agree; it held that such a rule was applicable to the high-water line of streams, not to the location of tidal property boundaries. *Ibid.*
101. ". . . [T]he neap tides as defined [in the *Teschemacher* case] are those which happen between the full and new moon, twice in every 24 hours. As high tides occur twice each day and differ from day to day, the line of neap tides . . . would not establish any definite or fixed boundary." *Id.* at 905.
102. ". . . Justice Field . . . would seem to indicate that the neap tides were all the tides between the full and new moon and inferentially exclude from the definition of neap tides the spring tides occurring approximately at the time of the new and full moon. . . . There is no inherent reason why the highest tide nearest to the full and new moon should be excluded from the average. . . . It is obvious that these spring tides must be included in the tides to be averaged if we are to determine the line usually reached, that is, reached on the average, by the high tide." *Ibid.*

It is evident when we consider the . . . language from Hall on the Sea Shore [on which the *Teschemacher* case relied], and, when we consider the fact that the shore line is the boundary between tillable land or land available for agricultural purposes and land so frequently covered by the sea that it is useless for agricultural purposes, that the average or mean of the high tides would be the appropriate line defining the boundary between the tillable upland and the lands which were ordinarily submerged by the tide.[103]

In affirming this decision, the United States Supreme Court rejected the contention that the location of the ordinary high-water mark littoral property boundary was the physical mark made on the ground by the water. The Supreme Court instead affirmed that in the case of a tidal regime, tidal observations would determine the physical location of the ordinary high-water mark.[104] Adopting the reasoning of the Ninth Circuit, the Supreme Court held that the ordinary high-water mark property boundary of lands adjacent to or along tidal waters would be physically located by use of the mean high-water line.[105] Many states have followed this rule,[106] although there is some confusion in California on its application. In the hope of dispelling and stemming continuation of that confusion, the California aberration, *People v. Wm. Kent Estate*,[107] is next discussed.

---

103. *Ibid.* The court relied on an English case, *Attorney General v. Chambers* (1854) 4 De G.M. & G., 43 Eng. Reprint 486. (A reprint of *Attorney General v. Chambers* is found in 2 Shalowitz, Appendix D, pp. 640–646.) The court also explained how this boundary can be fixed with certainty through long-term tidal observations. *City of Los Angeles v. Borax Consolidated Limited*, *supra*, 74 F.2d at 906.
104. *Borax, Ltd.*, *supra*, 296 U.S. at 22 (". . . [I]t means the line of high water as determined by the course of the tides"). As we shall see in the next chapter, there is an argument that the ordinary high-water mark property boundary may also be geographically located as a matter of law as a consequence of the land grant confirmation process. See Section 5.4.4 in Chapter 5.
105. *Borax*, *supra*, 296 U.S. at 26–27. There has been some criticism of this decision. Corker, *supra*, 42 Wash. L. Rev. at 61–65.
106. *O'Neill*, *supra*, 235 A.2d at 9–10; *Department of Nat. Res. v. Mayor & C. of Ocean City* (Md. 1975) 332 A.2d 630, 632; *State ex rel. Thornton v. Hay*, *supra*, 462 P.2d at 472; *Smith v. State*, *supra*, 282 S.E.2d at 81; *St. Joseph Land, etc. v. Florida State Bd.* (Fla. 1979) 365 So.2d 1084, 1087; *Matcha*, *supra*, 711 S.W.2d at 99.
107. (1966) 242 Cal.App.2d 156. This case ignored prior cases directly on point, which held (1) that the seaward boundary of tidelands was the line of mean high water, *City of Los Angeles v. Duncan* (1933) 130 Cal.App. 11, 14, and (2) that ordinary high water is the average height of all the high waters at a particular locality over a considerable period of time, *Swarzwald*, *supra*, 39 Cal. App.2d at 313. The *Kent Estate* case also ignored a prior California Supreme Court decision that had adopted the mean high-water line as location of the seaward boundary of uplands and the landward boundary of tidelands. *Abbot Kinney Co. v. City of Los Angeles* (1959) 53 Cal.2d 52, 58. Later California cases construed *Kent Estate* as having adopted a mean or average high-tide line. *Lechuza Villas West v. California Coastal Com.* (1997) 60 Cal.App.4th 218, 236–237.

4.11 THE CALIFORNIA ABERRATION DISPELLED—THE KENT ESTATE CASE

In the *Kent Estate* case, the dispute was over the location of a fence on a beach relative to state-owned tidelands. Adjacent property owners intended the fence to exclude the public from an area of a beach that the state contended it owned.[108] The California court purported to follow the reasoning of the *Borax* case. The opinion noted, in dicta,[109] that there was a California rule that used the "average of all high neap tides." As a result, the decision sent the parties back to determine just what that water level was.[110] Subsequent California cases[111] have all but ignored this offhand statement. The argument is that, if adopted, this "California rule" would possibly result in an alienation of sovereign public trust lands not authorized by the legislature.[112]

4.12 THE LOCATION OF THE ORDINARY LOW-WATER MARK

A discussion of the methods of physically locating open coast property boundaries would not be complete without at least a word about the most seaward property boundary—the ordinary low-water mark. The ordinary low-water

---

108. *People v. Kent Estate, supra,* 242 Cal.App.2d at 158.
109. "Dicta" means that part of a decision that is not necessary for the result of the case. 2 Shalowitz, Appendix A, p. 588.
110. *People v. Kent Estate, supra,* 242 Cal.App.2d at 161. Although there was a retrial of the case, the neap tide issue was never resolved in a reported decision.
111. *City of Long Beach v. Mansell* (1970) 3 Cal.3d 462, 478, n. 13.; *Marks v. Whitney* (1971) 6 Cal.3d 251, 257–258; *City of Berkeley, supra,* 26 Cal.3d at 519, n. 1; *Aptos Seascape Corp. v. County of Santa Cruz* (1982) 138 Cal.App.3d 484, 505, *app. dism'd* (1983) 464 U.S. 805. The *Kent Estate* case was roundly criticized in Comment, "Fluctuating Shoreline and Tidal Boundaries: An Unresolved Problem" (1969) 6 San Diego L. Rev. 447.
112. It would be asserted that conveyance of the strip of land between the neap high-tide line and the mean high-water line cannot be accomplished by judicial fiat. Only the legislature, after due consideration of the public trust interest of the state in such lands, can authorize such a conveyance. See, e.g., *City of Berkeley, supra,* 26 Cal.3d at 523–524. As no such determination was made, a judicially caused conveyance would be argued to be in violation of the California Constitution, Art. X, § 3 (if the lands were within 2 miles of a town or city) and California Public Resources Code § 7991, prohibiting alienation of tidelands. In addition, a legislatively unauthorized conveyance would also be claimed to violate the common law public trust limiting alienation of the public trust interest. *Ibid.*; *State of California v. Superior Court (Lyon)* (1981) 29 Cal.3d 211, 226; *cert. den.* (1981) 454 U.S. 865; *rehg. den.* (1981) 454 U.S. 1094; see notes 138 through 174 and accompanying text in Chapter 1.

mark is the seaward boundary of tidelands and the landward boundary of submerged lands.[113] Perhaps because it is not frequently the subject of dispute, due to the strictures on conveyance of tidelands and submerged lands,[114] location of the ordinary low-water mark along the open coast has not been the subject of many court decisions.

However, there is an authority that may guide the surveyor or lawyer in determining how to physically locate the ordinary low-water mark property boundary along tidal water bodies. That authority must be understood in its context to avoid possible misunderstanding and its resulting misapplication.

4.12.1 Use of Submerged Lands Act Definition

In the late 1930s rich and valuable fields of oil were discovered off the southern California coast.[115] Recognizing the value of this resource, California began leasing those oil fields for development and production.[116] Because of California's assumed offshore sovereign land ownership, the federal government at first agreed that California had authority to lease the oil fields. During the Second World War the federal government somewhat belatedly recognized the strategic and financial importance of these resources. The United States then asserted a claim of paramount rights and full dominion over the offshore lands and sued California for their control.[117]

Contrary to the long-held understanding of the states, the United States Supreme Court decided that the federal government, not California, had paramount rights and full dominion over such lands.[118] Texas and Louisiana, other coastal states, were also unsuccessful in attempting to change the Supreme Court's ruling.[119] The United States Supreme Court's *1947 California*

---

113. *E.g., United States v. Pacheco, supra,* 69 U.S. at 590; *San Francisco Savings Union v. Irwin* (9th Cir. 1886) 28 F. 708, 713, *aff'd* (1890) 136 U.S. 578; *People v. California Fish Co., supra,* 166 Cal. at 584.
114. E.g., *People v. California Fish Co., supra,* 166 Cal. at 601.
115. *Boone, supra,* 206 Cal. at 154; 3 Shalowitz, p. 5.
116. *Boone, supra,* 206 Cal. at 194.
117. In the context of these cases the term "tidelands" is not used in its technical sense. Title to tidelands, lands between the ordinary high- and ordinary low-water marks, *was not at issue* in the case. Only title to the lands *seaward* of the low water mark was disputed. *United States v. California* (1947) 332 U.S. 19, 22 (herein referred to as the "*1947 California Decision*"). However, through the modern magic of press-agentry, the offshore boundary cases became known over the years as the "tidelands" controversy.
118. *1947 California Decision, supra,* 332 U.S. at 38–39. A relatively unusual iteration of this doctrine is found in *Native Village of Eyak v. Daley* (1998) 153 F.3d 1090, 1096-1097; *cert. den.* (1999) 527 U.S. 1003 (aboriginal title claim including exclusive right to fish in Prince William Sound, Alaska, denied, based on paramount rights of United States).
119. *United States v. Texas* (1950) 339 U.S. 707; *United States v. Louisiana* (1950) 339 U.S. 699.

Decision and the subsequent *Texas* and *Louisiana* decisions brought nearly universal outcries of "foul" from the states. But after a years-long lobbying effort (tied into conflicting political ideologies),[120] Congress passed and President Eisenhower signed the Submerged Lands Act.[121]

The Submerged Lands Act undid the *1947 California Decision* and the later *Texas* and *Louisiana* decisions as well.[122] The Submerged Lands Act recognized, confirmed, established, and vested in the states the title to and ownership in and the right and power to manage, administer, and so on,[123] the lands[124] the states believed they had lost as a result of the *California* decision.[125] For almost a decade, the matter rested.

In the 1960s offshore oil drilling technology had developed to the point where drilling in ever deeper waters that were ever farther from the coast was possible.[126] The growth of this technology had the consequence of bringing into dispute the seaward boundary and extent of the lands confirmed in the states by the Submerged Lands Act. As far as concerned California, the lands confirmed to the coastal states extended seaward three geographical miles from the "coast line" of each state.[127] The "coast line," the landward point

---

120. As should only have been expected, partisans for California (principally the City of Long Beach) produced a movie supporting the California position. The movie, *Freedom's Shores*, was made during the era of Wisconsin Senator Joseph McCarthy's anti-Communist crusade. In a none-too-subtle attempt to imply that persons favoring the position of the federal government were Communist sympathizers, the film graphically portrayed the paramount rights doctrine as a red stain creeping over the entire country. In debates on measures designed to undo the *1947 California Decision*, supporters were led, in general, by conservative Southern senators. The opponents' main spokesman was Senator Paul Douglas, a northern liberal from Illinois. The dispute even became an issue in the 1952 presidential campaign. The Democratic candidate, Illinois governor Adlai Stevenson, supported the Supreme Court's decision; his Republican opponent, General Dwight Eisenhower, promised to sign legislation overturning the decision. 3 Shalowitz, p. 18, has a brief history of the legislation.
121. Act of May 22, 1953, 67 Stat. 29, set forth at 43 U.S.C. §§ 1301 et. seq. All further citations to the Submerged Lands Act will be referenced to the United States Code.
122. *United States v. California* (1978) 436 U.S. 32, 37.
123. The operative words of the Submerged Lands Act are intended to be words of art. Now that homage has been paid to the artfulness of the legislative craftsmen, in the interest of brevity, the author will substitute the shorthand expression "confirmed" in lieu of the full statutory litany.
124. Those lands were described in the Submerged Lands Act as "lands beneath navigable waters." 43 U.S.C. § 1311(a); 43 U.S.C. § 1301(a). That specific term, so far as it pertains to lands along the open seacoast, was defined by the Submerged Lands Act as ". . . all lands permanently or periodically covered by tidal waters up to but not above the line of mean high tide and seaward to a line three geographical miles from the coast line of each . . . State. . . ." 43 U.S.C. § 1301(a)(2).
125. 43 U.S.C. § 1311(a)(1) and (2); *United States v. California, supra,* 436 U.S. at 37. For a different perspective, see 3 Shalowitz, p. 18.
126. *United States v. California* (1963) 381 U.S. 139, 149 (herein the "*1963 California Decision*").
127. 43 U.S.C. § 1301(a)(2)

or the baseline from which the 3 miles was to be measured, was defined as the "line of ordinary low water along that portion of the coast which is in direct contact with the open sea. . . ."[128] Unfortunately, the Submerged Lands Act did not deign to define the term "line of ordinary low water."

In the *1963 California Decision*, yet another case between the United States and California, the United States Supreme Court determined that the coast line, the line of ordinary low water, would be measured from the mean lower low-water line rather than from the mean low-water line.[129]

This decision is the principal source of confusion in determining how the seaward boundary of tidelands is to be located. Remember, however, that the *1963 California Decision* pertained *only* to the interpretation of the Submerged Lands Act definition of the coast line for its particular purposes. The basis for this decision was the United States' use of mean lower low water as the datum for large-scale coastal charts.[130] Thus, any confusion should be dispelled. This decision had nothing to do with the determination of the location of the seaward property boundary of tidelands.

4.12.2 Mean Low-Water Line Adopted as Physical Location of the Ordinary Low-Water Mark

Cases discussing the ordinary low-water mark property boundary have not relied on the *1963 California Decision* as authority. Neither have these opinions adopted the mean lower low-water line as the physical location of that property boundary. Even considering this decision, California courts defined the location of the seaward boundary of tidelands as the mean low-water line.[131]

In other jurisdictions, early authority adopted a vague, descriptive approach to the physical location of the ordinary low-water mark.[132] Later courts and

---

128. 43 U.S.C. § 1301(c).
129. *United States v. California, supra*, 381 U.S. at 175–176; 3 Shalowitz, p. 43
130. *Ibid.* It is also important to remember that the Supreme Court adopted the definitions of the Geneva Convention (more specifically known as the Convention on the Territorial Sea and the Contiguous Zone, March 24, 1961, 15 U.S.T. 1606, T.I.A.S. No. 5639) for the administration of the Submerged Lands Act. *Id.* at 165. The Geneva Convention deals with rules for *international* boundaries of sovereign nations, *not* property boundaries of land holdings within the national territory of such nations. In the Geneva Convention, the low-water line as marked on large-scale charts of the coastal nation is used a baseline for measurement of the extent of territorial waters. *Id.* at 176. In the case of the United States, the low-water line marked on such coastal charts is the lower low-water line. *Ibid.*
131. *City of Long Beach v. Mansell, supra*, 3 Cal.3d at 478, n. 13; *Marks, supra*, 6 Cal.3d at 257.
132. Clark 7th, *supra*, § 25.15; Clark 4th, *supra*, § 596, p. 870, without pointing out any authority, holds that the low-water mark is the "lowest line made by the receding tide. . . ." Other commentators rely on early Maine or Massachusetts cases. Those cases hold that the low-water mark is the margin of the sea when the tide is out at ordinary tides. Skelton, *The Legal Elements*

legislative bodies in other jurisdictions adopted the convention of using the mean low-water line as the physical location of the ordinary low-water mark.[133] Among those jurisdictions are states that allow private ownership of tidelands down to low-water.[134] Thus, the widely accepted rule appears to be that the seaward boundary of tidelands, the ordinary low-water mark, should be physically located by use of the mean low-water line.[135] We now turn to the practical and legal problems in physically locating the littoral property boundary.

4.13 THE OPEN COAST MEAN HIGH-WATER LINE—FLUCTUATION OF THE LANDFORM

As discussed earlier, the ordinary high-water mark property boundary of open coast tidelands is physically located at the line of mean high water. The plane of mean high water is a fixed plane.[136] The land surface or landform that the water (the datum plane) intersects is, however, constantly in flux.[137] The land surface, whether composed of sand, mud, or cobbles, is ever changing its relative elevation with respect to the adjacent water body. Erosion, deposition, subsidence, upheaval, or any of the many physical phenomena may affect the terrestrial form. The dynamic property of the land surface is most dramatically evidenced in the case of sandy beaches.[138]

The process of beach creation and maintenance is as follows: As the ocean waves meet the sandy shore, they establish a longshore (shoreline) current or littoral drift. When the waves break on the sandy beach, they stir up the sand, putting some sand in suspension in the water. The longshore current picks up

---

of Boundaries and Adjacent Properties (1930) (herein "Skelton") § 284, p. 314. The difficulty in physically locating that boundary as described by the commentators and cases is all too apparent.

133. *DeGuyer v. Banning, supra,* 167 U.S. at 724; *Hynes v. Grimes Packing Co.* (1949) 337 U.S. 86, 90 (used in executive order establishing Indian reservation); *Bradford v. Nature Conservancy* (Va. 1982) 294 S.E.2d 866, 874 (citing Va. Code § 62.1-2.); *State v. Holston Land Co.* (S.C. 1978) 248 S.E.2d 922, 924; *Camping Com'n of Meth. Ch. v. Ocean View Land, Inc.* (Wash. 1966) 421 P.2d 1021, 1022; *Island Harbor Beach Club Ltd. v. Dept. of Nat. Res.* (Fla. 1986) 495 So.2d 209, 214, n.6, *rev. den.* (Fla. 1987) 503 So.2d 327 (seaward limit of statutory beach is mean low-water line); *Opinion of the Justices to the House of Representatives* (Mass. 1974) 313 N.E.2d 561, 565.

134. *State v. Pennsylvania Railroad Company, supra,* 228 A.2d at 601; *Opinion of the Justices to the House of Representatives, supra,* 313 N.E.2d at 565.

135. Insofar as tidal reaches are concerned, including lands encompassed within estuaries and tidal rivers, this is the extent of the author's discussion of the ordinary low-water mark. The same principles that concern the property boundary effect of the change in geographic location of the ordinary high-water mark are thought to be applicable to the ordinary low-water mark.

136. *Swarzwald, supra,* 39 Cal.App.2d at 313; *People v. Wm. Kent Estate Co., supra,* 242 Cal.App.2d at 159–160; *Lechuza Villas West, supra,* 60 Cal.App.4th at 235–237.

137. *Ibid.*; *State v. Pennsylvania Railroad Company* (Del. 1968) 244 A.2d 80, 81.

138. E.g., *Lechuza Villas West, supra,* 60 Cal.App.4th at 235.

this suspended sand and holds it in suspension long enough to deposit it farther along the shore.

As anyone knows who has ever watched or experienced the ocean surf, this process is incessantly repeated. If there is no change in the littoral regime, the suspension and deposition of sand are in equilibrium; new and additional sedimentary materials are seasonally washed down from the upland drainages by storm run off and distributed by watercourses that eventually terminate in the ocean. The ultimate result of this process is the creation and maintenance of sandy beaches fed by the constant erosion of the upland and the deposition of the eroded materials along the coastal reaches.[139]

In most coastal reaches this process creates a regime that fluctuates seasonally. The extent of the beach in summer is measurably greater than its extent in winter. The beach profile itself changes with the seasons.[140] Witness the testimony of an oceanographer:

> . . . Dr. [__] testified that shorelines along Corpus Christi Bay in their natural state oscillate back and forth within a known range. After a hurricane, the oscillation may be as much as thirty or forty feet. In its natural state, [the beach], after damage by a hurricane would be renewed or regenerated by sand moving from one area to another. Dr. [__] termed this process "dynamic equilibrium."[141]

One court expressly considered the effect of this fluctuation on physical location of the upland/tideland property boundary. Prior cases concerning location of the mean high-water line concentrated on determination of the relative height or elevation of the water surface. The *Kent Estate* opinion suggested to the parties that they find out whether the beach movement was substantially the same distance each year in order to "afford[] a basis for fixing an average, mean, or ordinary line of the shore against which the average plane of the water at high tide may be placed to determine a reasonable definite boundary line."[142] There was no supporting legal or technical authority for the court's direction to the parties. What was the legal basis for

---

139. Bascom, Waves and Beaches (Anchor Books 1964), Chapters IX and X, pp. 184–235; *Carpenter, supra,* 63 Cal.App.2d at 776; *Abbot Kinney, supra,* 53 Cal.2d at 55.
140. Bascom, *supra,* at p. 188; *City of Corpus Christi v. Davis* (Tex. 1981) 622 S.W.2d 640, 644.
141. *City of Corpus Christi, supra,* 622 S.W.2d at 644; *Miramar Co. v. City of Santa Barbara* (1943) 23 Cal.2d 170, 175 (relationship between land and water can "shift as easily as the sand"); *Carpenter, supra,* 63 Cal.App.2d at 779–780; *People v. Hecker, supra,* 179 Cal.App.2d at 828–829; *Lechuza Villas West, supra,* 60 Cal.App.4th at 235.
142. In *People v. Wm. Kent Estate Co., supra,* 242 Cal.App.2d at 161, the Court noted there was evidence of beach fluctuation and found that to the extent the land moves, the line of high-water varies. *Id.* at 160. The court sent the case back for retrial for the suggested evidence. *Id.* at 161.

For a further commentary on this aspect of the *Kent* case see Comment, "Fluctuating Shoreline and Tidal Boundaries: An Unresolved Problem" (1969) 6 San Diego L. Rev. 447.

fixing "an average, mean, or ordinary line of the shore"? Is there any scientific or technical justification for so doing? Over what interval shall measurements be taken, and over what period and on what scientific basis?

As a practical matter, these questions were answered in *Lechuza Villas West v. California Coastal Com.*[143] In that case the parties, the state and an adjoining littoral property owner, disputed whether the ordinary high-water mark property boundary line[144] was a moveable boundary line or was fixed in location.[145] The trial court, relying on the *Kent Estate* case, determined that the ordinary high-water mark property boundary line between private property and state tidelands was a fixed line, based on the mathematical average of 37 surveys of the mean high tide line.[146] After closely reviewing *Kent Estate*, the appellate court found that *Kent Estate* had decided that the mean high tide line may vary to the extent the land moves.[147] Thus, the court held that the coastal boundary was an ambulatory boundary.[148]

This holding is recognition of the practical fact that the property boundary is a natural one; its physical location continues to move, to ambulate, unless, through a legal determination, it becomes necessary to fix a location.[149]

---

143. *Lechuza Villas West, supra,* 60 Cal.App.4th at 218.
144. This line was described as ". . . the mark made by the fixed plane of high tide where it touches the land; as the land along a body of water gradually builds up or erodes, the ordinary high-water mark necessarily moves, and thus the mark or line of mean high tide, i.e., the legal boundary, also moves." *Id.* at 235.
145. *Lechuza Villas West, supra,* 60 Cal.App.4th at 232–233.
146. *Id.* at 235.
147. *Id.* at 238. The *Lechuza Villas West* court noted that *Kent Estate* merely suggested, without so holding, that a way to create some certainty as to the boundary might be to calculate an "average or ordinary" line of the shore against which the plane of "ordinary high tide" could be placed to locate a definite boundary line. Neither court discussed what scientific or legal support there was for such averaging.
148. *Id.* at 238–239. The private owners also asserted that this boundary was conclusively established on a recorded subdivision map that had been accepted by the government and recorded. The court easily disposed of this argument. Among its reasons was that a line shown on a map running along the ocean is a meander line, used to ascertain the quantity of land subject to sale and to show the sinuosities of the shore; the mean high-water line is the true legal boundary. *Id.* at 240. See text accompanying note 60 in Chapter 1 and notes 42 through 64 and accompanying text in Chapter 3.
149. In such a case, some contend the winter beach profile is the beach profile against which the elevation of mean high water must be located. Basically, supporters of this contention reason that such a rule provides the public with greatest access to the shoreline because the winter mean high-water line is farther landward. On the other hand, property owners assert the summer beach profile should be adopted. The summer profile provides maximum separation between private property and public tidelands. These contentions have never been litigated to judgment resulting in a reported case, perhaps because of lack of legal support for either position. In *Lechuza Villages West,* the state asserted an analogous position to the winter profile claim. The state asserted that property landward of the ordinary high-water mark was inundated by the higher tides. *Id.* at 243. The state claimed a navigational easement over such lands. *Ibid.* The court found that these tidal waters above the mean high-water line were not navigable and denied the state's claimed easement. *Id.* at 244–245.

4.14 PROPERTY BOUNDARY EFFECT OF THE GEOGRAPHIC MOVEMENT OF THE OPEN COAST SHORELINE

4.14.1 In General

The legal principles concerning property boundary movement are applicable along the open coast.[150] Thus, if land is added through the process of accretion[151] or washed away through the process of erosion,[152] the property boundary follows the changing physical location of the mean high-water line. There are some interesting variables, however, worthy of fuller discussion.

One variable is found in just how legal authorities have defined accretion. One court carefully noted that ". . . addition to land is accretion (and loss of land is deliction [erosion]), only when the changes are gradual and imperceptible."[153] This bland statement gave little hint of what the court would hold when there was an 80-foot horizontal fluctuation in the extent of the beach from summer to winter. The court found that this fluctuation did not meet the definition of gradual and imperceptible and was not within the definition of natural accretion and erosion.[154] Thus, merely restating old bromides does not alleviate the difficulty of establishing by competent evidence that one's situation falls within the rule.

4.14.2 Effect of Artificial or Human-Induced Changes on Coastal Property Boundaries

The uncertainties surrounding the definition of what is and what is not gradual or imperceptible physical movement along the open coast are not the only variable. There is another variable whose effect on the location of the ocean shoreline property boundary seems to be both legally and technically unpredictable. Although we may wish it otherwise, the coastlines along the continent have been substantially modified by human action. Not only have highly visible coastal engineering works been constructed to accomplish particular physical consequences along the coastline, but in many coastal reaches human actions have modified entire coastal regimes.[155]

---

150. *Strand Improvement, supra*, 173 Cal. at 772–773; *Matcha, supra*, 711 S.W.2d at 99.

151. *City of Los Angeles v. Anderson, supra*, 206 Cal. at 666–667; *People v. Hecker, supra*, 179 Cal.App.2d at 835 (for the general rule); *Schafer v. Schnabel* (Alaska 1972) 494 P.2d 802, 806–807; *Honsinger v. State* (Alaska 1982) 642 P.2d 1352, 1353–1354.

152. *Miramar Co. v. City of Santa Barbara* (1943) 23 Cal.2d 170, 175; *Ward Sand & Material Co. v. Palmer* (N.J. 1968) 237 A.2d 619, 622; *Department of Nat. Res. v. Mayor of C. of Ocean City, supra*, 332 A.2d at 638.

153. *People v. Wm. Kent Estate Co., supra*, 242 Cal.App.2d at 160.

154. *Ibid.* The court's novel, but unsupported, attempted resolution of this situation is recounted at note 142 in this chapter.

155. For a discussion of the impacts of a jetty on the coast and the problem of attempting to relate general statements about the effect of human's works on beach processes to a particular beach, see *Surfside Colony, Ltd., supra*, 226 Cal.App.3d at 1263–1264, 1268–1269.

Again focusing on California as an example, piers, jetties,[156] groins,[157] breakwaters, and other coastal works have been put in place. In addition, upland drainage streams no longer flow uncontrolled and free from their sources to the sea; dams and weirs[158] have been built in the stream beds for purposes of hydroelectric power generation, water storage or diversion, or flood control. Such regulation of stream flow affects either the amount or the timing of the sediment load transported by these streams. This sediment load constitutes the upland contribution to the littoral sand transport regime.

Do such changes to the coastal regime, which we can reasonably assume are not visible as they are occurring, affect the location of the property boundary of littoral properties? If a public work increases the extent of the beach in front of a coastal property owner's home, does the coastal property owner or does the state (and the public), which made, caused, allowed, or paid for the coastal work, secure the benefit of the change? What if the result of the coastal work is a reduction in the extent of the beach? What if a private person constructs the work or fills the area in front of his or her property without authority? Does it make any difference whether the upland owner constructs the coastal works pursuant to a government permit? Whose property boundary should bear the consequences? Federal courts and courts in various states have answered such questions in different ways.[159]

In California and in some other states, alluvion deposited gradually and imperceptibly and attributable to the works of humans does not benefit the upland owner; the coastal ordinary high-water mark property boundary is fixed at its location prior to the occurrence of the artificial accretion.[160] One

---

156. A jetty is structure of stones, piles, or other materials projecting into the sea so as to protect a harbor or provide a pier or place for the landing of ships or vessels. See *United States v. California* (1980) 447 U.S. 1 for a lengthy discussion.

157. A groin is a solid structure that generally lies perpendicular to the shoreline and extends out into the foreshore. Its function is to interrupt the littoral drift and produce deposition of sand on the updrift side of the groin in order to widen the beach. E.g., *Muchenberger v. City of Santa Monica* (1929) 206 Cal. 635, 639; *Lummis v. Lilly* (Mass. 1982) 429 N.E.2d 1146, 1148.

158. A weir is a dam or obstruction that backs up or diverts a stream. Ballentine's Law Dictionary (3d Ed. 1949), p. 1364.

159. This difference in result due to the choice-of-law rule is the principal reason that surveyors, lawyers, and property owners should be familiar with the choice-of-law principles discussed in Chapter 2.

160. *Carpenter, supra*, 63 Cal.App.2d at 794; *People v. Hecker, supra*, 179 Cal.App.2d at 837; *Lorino v. Crawford Packing Co.* (Tex. 1943) 175 S.W.2d 410, 414; see *State of Cal. Ex rel. State Lands Com. (Lovelace), supra*, 11 Cal.4th at 76–80 (river; must be direct cause). The upland owner's title in these cases appears to have been based on confirmed Spanish or Mexican land grants, not on federal public land patents. Thus, the choice-of-law controversy resulting from *Hughes v. Washington* (1967) 389 U.S. 290 and *California ex rel. State Lands Comm'n v. U.S.* (1982) 457 U.S. 273 (hereinafter "Humboldt Spit") has not yet arisen. *Carpenter, supra*, 63 Cal.App.2d at 784. But in *State of Cal. ex rel. State Lands Com. (Lovelace), supra*, 11 Cal.4th at 74, the California Supreme Court observed that federal law may apply to oceanfront titles derived from the federal government. The court also noted the choice-of-law was different for inland property in which the federal government had no interest. *Ibid.* Chapter 7, notes 147

state has even enacted a statute proclaiming that land so created does not benefit the upland owner but remains the property of the state.[161] This result is similar to the property boundary consequence of an avulsive change.[162]

When the federal government is the upland owner, however, the highest court in the land has established a different rule. In this case, the federal government undeniably caused a gradual and imperceptible increase in the seaward extent of the Pacific Ocean shoreline through construction of a massive coastal jetty. The Supreme Court determined that, even in California, the United States, as the upland owner, would receive the benefit of the accretion it had caused.[163] The law in this area is in flux,[164] and the careful person should do his or her own research. One can generally state, however, that what some have called the "California rule" is not the rule in the majority of states.[165]

There is claimed to be a substantial policy supporting the California rule. That policy is to protect public ownership of and the public trust interest in tidelands; no matter that the tidelands have become covered up or exposed.[166]

---

through 154, discusses both the choice-of-law and the application of the artificial accretion rule in greater detail.

161. Fla. Stat. (1981) § 161.051. But as we shall soon see, this statute has been seriously undermined by a sharply divided Florida Supreme Court. *Internal Imp. Tr. Fund v. Sand Key Assoc.* (Fla. 1987) 512 So.2d 934, 940–941.

162. See note 123 and accompanying text in Chapter 3.

163. *Humboldt Spit, supra*, 457 U.S. at 284–285. A later federal case has termed this result dicta and criticized the dicta as not fully supported and an "unjustifiably expansive reading of federal common law. . . ." *Alexander Hamilton Life Ins. Co. v. Govt of V.I* (3rd Cir. 1985) 757 F.2d 534, 544–545.

164. Chapter 5, Section 5.8.2, and Chapter 7, Section 7.8, discuss the California Supreme Court's decision on the scope of the application of the artificial accretion rule. One Florida case, *Board of Trustees, etc. v. Medeira Beach Nom., Inc.* (Fla. 1973) 272 So.2d 209, 213–214, discusses reasons that accretion caused by humans should belong to the upland owner and reasons that the land should remain property of the state. The Florida Court favored the rights of the littoral owner. That case was brought before Florida enacted a statute, Fla. Stat. (1981) § 161.051, vesting title to accretion caused by public works in the state. *Id.* at 214. In a later case, this statute was held not to apply to the situation where land was created by construction of a coastal work that was not built by the benefiting littoral owner. *Internal Imp. Tr. Fund v. Sand Key Assoc., supra*, 512 So.2d at 939–941. In the *Sand Key* case, the dissent claimed the state's construction of coastal works cannot cause the state to lose title to its sovereignty lands. *Id.* at 946 (Ehrlich, J., dissenting).

165. E.g., *Schafer, supra*, 494 P.2d at 802; *State v. Sause* (Ore. 1959) 342 P.2d 803, 819. As we note in Chapter 5, Section 5.8.2, and Chapter 7, Section 7.8, the California Supreme Court narrowed the scope of this rule.

166. *State of Cal. ex rel. State Lands Com. (Lovelace), supra*, 11 Cal.4th at 72. ("California's artificial accretion rule was premised on and is consistent with the public trust doctrine and the inalienability of trust lands. . . . The State has no control over nature; allowing private parties to gain by natural accretion does no harm to the public trust doctrine. But to allow accretion caused by artificial means to deprive the State of trust lands would effectively alienate what may not be alienated. . . . This we believe was the driving force behind California's doctrine, and the reason it remains vital today. We thus reaffirm the continuing validity of California's artificial accretion rule.")

There is a competing policy supporting littoral owners: If the geographic location of the property's littoral boundary is changed through no activity of the littoral owner, increasing the area of property, there is no reason for the littoral owner to suffer.[167] This argument concludes that traditional principles of property boundary movement determination should be applied and the increase awarded to the littoral owner.[168]

4.14.3 Effect on the Property Boundary of Filling by the Littoral Owner

The rule is quite different for lands created by the littoral owner. In the case of upland intentionally—non-naturally—created in the geographic location of former tideland or submerged land by actions undertaken by the littoral owner, the majority rule concerning accretion does not apply. It is well established that the upland owner cannot increase the extent of littoral upland by filling in tideland or submerged land to the detriment of the owner of such tideland or submerged land.[169] If the owner of the adjacent tideland or submerged land has permitted or authorized the littoral owner's filling, the artificially induced deposition that increased the extent of littoral upland belongs to littoral owner.[170]

4.14.4 Effect on the Property Boundary of Artificial Erosion

Human-induced erosion is not uncommon along the open coast. The property boundary effect of such erosion may be entirely different from the property boundary effect of other human-induced impacts on the geographic location of the property boundary. Suppose, for example, a coastal work, installed at some distance from a person's property, caused the shoreline along that person's precious beachfront property to erode, the beach itself to be eventually denuded of sand, and the mean high-water line to encroach farther landward onto the person's property. If we follow the rule of artificial erosion, the littoral property boundary would be fixed at its location prior to the downcoast prograding or eroding effect of the coastal work.

In California at least, the opposite result appears to be the case; land that has become submerged as the result of coastal processes generated by human-constructed coastal works is owned by the state.[171] This is an example of the

---

167. *State of Cal. ex rel. State Lands Com. (Lovelace), supra,* 11 Cal.4th at 57.
168. See notes 113 through 114 in Chapter 3.
169. *Alexander Hamilton Life Ins. Co., supra,* 757 F.2d at 538–539. This rule was not applied against the United States in the *Humboldt Spit* case.
170. *Id.* at 545–546. In that case, however, the permit language was held to alter the rule that land artificially reclaimed through lawful means becomes the property of the littoral owner. The permit had an express provision stating that the permit was not intended to alter property rights. *Id.* at 546–547.
171. See *Miramar, supra,* 23 Cal.2d at 175. Some other states appear to agree. *Department of Nat. Res. v. Mayor & C. of Ocean City, supra,* 332 A.2d at 638; *Ward Sand & Materials Co.,*

"heads I win, tails you lose" situation in which many find themselves when litigating with the government on sovereignty issues.[172]

4.14.5 Effect on the Property Boundary of Reemergence and Reliction

On the other hand, sometimes fairness does creep into coastal property boundary cases. Coastal erosion is an endemic problem along many reaches of the nation's open ocean coast. One coastal erosion case should give long-sought comfort to littoral owners. In that case[173] the court held that a nonlittoral owner, whose land eventually became littoral, does not succeed to accretions to that land seaward of the property's original boundary line; the reemerged land belongs to the former and original littoral owner.[174] Would the same rule hold true in the case of a reliction?

The process of reliction is not one associated with lands along the open coast. Indeed, absent a cataclysm such as the rapid onset of a new ice age or an earthquake, any gradual recession of the water or rising of the land is a process we will probably not notice on "our watch."[175]

4.14.6 Effect of Great Storms or Hurricanes on the Property Boundary

Any person who has lived along the ocean coast shore or has seen pictures of great ocean storms knows the destructive force of the sea. This power is the great shaper of the coast. What happens to property boundaries in the wake of such events?

Assume you purchase your dream piece of coastal property. According to the title documents, your seaward property boundary runs along the ocean and the property is about 18 acres in area. After a series of hurricanes and great winter storms, however, you find only 15 acres of that property are

---

supra, 237 A.2d at 622. One should be wary, however, of *State of Cal. ex rel. State Lands Com. (Lovelace)*, *supra*, 11 Cal.4th at 77–80. This case may limit application of this rule to human activities that are too attenuated from the site of the erosion.

172. See *Texas Boundary Case* (1969) 394 U.S. 1, 9 (Black, J., dissenting).
173. *Horry County v. Woodward* (S.C. App. 1984) 318 S.E.2d 584.
174. *Id.* at 589. ("[W]hen riparian land, which was separated from remoter land by a fixed boundary at the time of the original grant, is lost by erosion so that the remoter land becomes riparian, and land is thereafter added by accretion to the land which was originally remote, extending over the location formerly occupied by the original riparian land, the owner of the land which was originally remote has title to the accreted land up to the fixed boundary of his formerly nonriparian tract. All other land so accreted, extending beyond the fixed boundary over the area formerly occupied by the original riparian land, becomes the property of the owner of the original riparian land.")
175. On the other hand, how do we really know that what appears to be the process of accretion is not actually the gradual recession of the ocean caused by uplifting of the tectonic plates? Does it make any difference to the location of the property boundary? Probably not.

landward of the mean high-water line. Seizing the opportunity, your friendly local or state government determines to lease to the Army Corps of Engineers the most seaward 3 acres, now physically located seaward of the mean high-water line. To your consternation, the Corps immediately gives notice that it wishes to fill and reclaim the land for a massive and unsightly seawall, separating your dream land from the ocean. Who owns the 3 acres? The answer to this question will not be very comforting to littoral property owners.

The solution to this problem is dependent on whether the doctrine of avulsion applies to open coast lands. And there is some question as to whether it does. In California there are no reported cases,[176] and one court in Texas has expressed the view that the rule of avulsion may not apply to the tidal coastline.[177]

There are many reasons for not applying the principle of avulsion to the open coast. For example, if the rule of avulsion were applicable to the ocean coastline, private ownership of tidelands could result. As a consequence, the enjoyment of what are now public beaches could be restricted. In addition, application of this rule would make the location of littoral property boundaries uncertain.[178]

One Florida case, however, has applied the doctrine to coastal property.[179] In that case a hurricane filled a tidal channel with sand; the filled-in bed of that tidal channel was retained in State ownership.[180]

4.15 EFFECT OF PRESUMPTIONS ON COASTAL PROPERTY BOUNDARIES

Even if it is assumed that the doctrine of avulsion is applicable to open coast property boundaries, this may be only cold comfort to littoral property owners. The reason is the presumption in favor of accretion and erosion. Littoral

---

176. That may be more a reflection that California truly is God's country or, more likely, that the California coast is subject to a somewhat more benign ocean and meteorological environment than exists along the Gulf or the Atlantic coast. See *Miramar, supra,* 23 Cal.2d at 175.
177. *State v. Balli* (Tex. 1944) 190 S.W.2d 71, 100, *cert. den.* (1946) 328 U.S. 852.
178. *City of Corpus Christi, supra,* 622 S.W.2d at 644.
179. *Siesta Properties, Inc. v. Hart* (Fla. 1960) 122 So.2d 218, 223–224.
180. *Id.* at 223. The source of the sand was adjacent upland. The court said:

> Even if the deposits did come from this source, it is apparent that what occurred on the night of the 1926 hurricane to bring about the condition that existed the following day was the sudden and perceptible deposit of land of one riparian owner upon or against the shore of another riparian owner and upon the bed an intervening tidal [channel], the title to which was in the State of Florida . . . [A] landowner may [not] enlarge his property lines to an extent necessary to take in new lands thus formed, when they lie outside his original boundary lines, or claim title to such lands in their new location. *Ibid.*

owners may be unable to overcome this presumption by proving that the disputed land became submerged as the result of an avulsion.[181] Even where it was undisputed that hurricanes and great winter storms had caused cataclysmic erosion to the coastal property, as in the preceding example, a court has refused to find that an avulsion had occurred.[182] Thus, in the example, the Corps of Engineers could feel perfectly free (as far as the title to the land in concerned) to construct one of its works directly in front of your dream property.

4.16 EXAMPLES OF TYPES OF PROOF IN OPEN COAST PROPERTY TITLE AND BOUNDARY LITIGATION

The application of property boundary movement principles to situations such as those discussed earlier depends on proof—on the lawyer's ability to convince the court that a particular rule is applicable and that the facts support its application. The problems of proof in this area are unusual and require the lawyer or the surveyor to retain and consult with highly skilled specialists in the physical or social sciences.

For example, to establish, as an evidentiary matter, that an avulsion has occurred requires proof of the geographic location of the property boundary before the avulsive event. How is this done when the former location of the boundary has been obliterated?

As a prerequisite, all available sources of surveys, maps, plats, or photographs of the area, cartographically, geographically, or photographically depicting the changing configuration of the littoral shoreline, must be investigated. Sources for such an investigation include the local or regional archives of the USCS, USC&GS, NOS, the U.S. Geological Survey (USGS), the United States General Land Office (now the Bureau of Land Management),

---

181. Because the acreage in question was covered by the sea at the time [of the boundary fixing event], it is presumed that title is in the State." *City of Corpus Christi, supra,* 622 S.W.2d at 644. See also notes 133 through 134 and accompanying text in Chapter 3.

182. . . . [E]rosion can be both sudden and perceptible, and does not have to be always gradual and imperceptible. Losses to the shoreline at [this reach of the Texas shore] to hurricanes and northers are not nearly so sudden as river banks tumbling into the Missouri . . . river[ ]. . . .

There is a further reason. . . . It is undisputed that not all the shoreline loss was attributable to sudden and obvious causes, although it is true that hurricanes and northers have been responsible for a substantial part of the total loss of shoreline. Nevertheless, the evidence is that forces other than hurricanes and northers, such as summertime night winds and quick water action, are at work slowly shifting away the sands of . . . [the b]each. Such forces are classically erosive, not avulsive. *City of Corpus Christi, supra,* 622 S.W.2d at 646.

and the Army Corps of Engineers. Sources for aerial photographs are as numerous and include many of the aforementioned agencies and others, such as the United States Soil Conservation Service, the National Aeronautics and Space Administration, and the Armed Forces. There is a wealth of historic materials at the National Archives in Washington, D.C., and the NOS map archives in Silver Spring, Maryland. Private sources of photographs should not be neglected as they may be more readily obtainable.

In addition, state land management agencies and state archives should be fully explored. Often a property has been the subject of prior court proceedings, so records of the local courts should be searched.[183] Local government records in the files of the city, county or parish clerk, surveyor, or engineer and in the planning, road, water, flood control, and other relevant departments should be explored. Finally,[184] title companies, local historical societies, and local historians and collectors also have much valuable information.

Occasionally, the mere introduction of this documentary evidence will convince the court that a particular coastal process has occurred.[185] In other cases elaborate proof based on time-consuming and expensive technical processes may be necessary. For example, the scale of early USCS topographic and hydrographic surveys is 1:10,000. That scale is equivalent to the scale of 1 in.:833 ft. As can be imagined, when particular areas of such maps are enlarged, the features depicted become distorted and out of scale unless special care is taken. A means of avoiding this problem is computerized digitization of pertinent features cartographically depicted on such maps.[186] This

---

183. For example, in California all of the Mexican and Spanish land grants, many of which included riparian or littoral lands, were the subject of proceedings before the United District Courts. Those records are maintained in the University of California's Bancroft Library on the Berkeley campus. That library also contains extremely valuable maps and plats of the grants. By way of further example, the reader should note that the United States Supreme Court library maintains copies of all the briefs filed in cases which it has considered. On occasion, those materials contain pertinent information concerning the historic physical character of the property that was the subject of the litigation. The National Archives and Records Administration has the patent case files pertaining to individual claims for Mexican and Spanish land grants in California. This holding constitutes records formerly held by the Washington, D.C., office of the U.S. General Land Office.
184. In this case, "finally" means only that the author does not wish to belabor this point. The sources and types of information that may be procurable and valuable in this process are limited only by the imagination and resourcefulness of the person directing or accomplishing the search.
185. In one case, the United States Supreme Court agreed that the process of artificial accretion had occurred, based solely on documents that had been introduced by motion. *Humboldt Spit*, *supra*, 457 U.S. at 275–276 n.1–n.5.
186. By this process the computer is able to recognize particular points on the map that have been electronically captured by various scanning or inputting methods. The computer, with the aid of a plotter, can then reproduce the "digitized" features at the desired scale relatively free of distortion. For example, the shoreline feature cartographically depicted by the USCS surveyors can be digitized and reproduced at a scale appropriate to the base mapping chosen for a particular project. Use of the digitization process assumes the map being digitized itself is not a distorted print of the original. This digitized information may also be utilized in a Geographic Information

is a very expensive process, but in the appropriate dispute may be quite valuable and worth the expense.

Once the relevant information has been obtained, it must be placed in the hands of expert witnesses. These experts can cull through the chaff, find the kernels of fact, and determine the relative usefulness or reliability of those facts to the formation of their conclusions. The experts' conclusions may then be presented to the court as expert testimony or expert-prepared or interpreted exhibits.[187] For example, in one case the state was claiming that the coastal boundary had moved by the process of avulsion rather than accretion, an issue in which the state had the burden of proof.[188] The evidence submitted was described by the court:

System (GIS). A GIS is, in short, a computerized mapping system capable of assembling, storing, manipulating, and displaying geographically referenced information, i.e., data identified according to their locations. A GIS can use information from many different sources, in many different forms. The primary requirement for the source data is that locations for the variables are known. Location may be annotated by x, y, and z coordinates of longitude, latitude, and elevation. Any variable that can be located spatially can be fed into a GIS. Different kinds of data in map form can be entered into a GIS. If the data to be used are not already in digital form, that is, in a form the computer can recognize, various techniques can capture the information. Although maps can be digitized, or hand traced with computer mouse, to collect the coordinates of features, electronic scanning devices will also convert map lines and points to digits. Data capture—putting the information into the system—is the time-consuming component of GIS work. Identities of the objects on the map must be specified, as well as their spatial relationships. Editing of information that is automatically captured can also be difficult. Electronic scanners record blemishes on a map just as faithfully as they record the map features. Extraneous data must be edited or removed from the digital data file. The USGS Web site, http://infor.er.usgs.gov/research/gis/title.html, has a more detailed explanation of the GIS.

187. Some, but not all, jurors and even some judges may choose to believe an expert witness just because the witness proclaims him- or herself to be an expert. This natural reaction has even been noted by a well-known mystery writer. See note 16, in the Preface. The job of the lawyer and the expert consultant, depending on whether one is a proponent or opponent of such testimony, is to either enhance or defeat that natural reaction.

The court also plays a significant role. While a detailed discussion regarding the admissibility of expert testimony is beyond the scope of this work, a short explanation will be valuable. The U.S. Supreme Court has set standards for determining whether scientific expert testimony may be admitted as evidence in trials in federal court. *Daubert v. Merrill Dow Pharmaceuticals* (1993) 509 U.S. 579, 597, called on trial judges to act as "gatekeepers" to ensure that scientific experts' opinions are both reliable and relevant. The "*Daubert* test" includes testing, peer review, error rates, and acceptability in the relevant scientific community. *Id.* at 592–595. *Daubert* applies not only to scientific experts but to other technical experts who are not scientists. *Kumho Tire Company, Ltd. v. Carmichael* (1999) 526 U.S. 137, 143 L.Ed.2d 238, 251–253.

While *Daubert* changed the standard for admissible expert scientific testimony in federal courts, it did not change the law in California. *People v. Leahy* (1994) 8 Cal. 4th 587, 612. In California, in order for scientific methods to be deemed reliable, they "must be sufficiently established to have gained general acceptance in the particular field in which [they belong]. General acceptance means a consensus drawn from a typical cross-section of the relevant, qualified scientific community." *Leahy, supra,* 8 Cal. 4th at 612. The burden of showing general acceptance is with the proponent of the evidence to show a scientific consensus. *Id.* at 611.

188. See notes 133 through 134 and accompanying text in Chapter 3 and note 181 in this chapter; *Hous. Auth. of City of Atlantic City v. State* (N.J. 1984) 472 A.2d 612, 613.

The State relied for its proofs of coastal shifts more than 110 years ago on the contemporaneous reports and logs of the keeper of the . . . Lighthouse . . . ; on a topographical survey by the [USC&GS] in 1869; on a succession of survey maps showing the high-water line at the site in the 1860s and 1870s by engineers of the . . . Lighthouse District; and on expert opinion testimony by Dr. [___], a coastal geologist. The State introduced in evidence a clipping from the New York Times of March 31, 1870, reporting a gale in the area. . . ."[189]

Although such proof may establish that a boundary movement occurred over a short period, this does not guarantee the result. In the case described, the court found that a two-month migration of the boundary did not constitute an avulsion.[190]

Moreover, use of historic maps to establish the location of the physical features with respect to current topography may be difficult. Those maps were usually, but not always, prepared with great accuracy.[191] Surveyors sometimes made blunders in the location of physical features they were cartographically depicting. Further, the methods used by surveyors to place themselves and the features they were to depict precisely on the surface of the earth were just not as accurate or precise in historic times as they are now.[192] Whether a shoreline feature was 30 feet either way did not appear to matter very much when the only people for miles around were the USCS surveyor and his rodman. Yet today a 60-foot range in location can affect millions of dollars worth of property and improvements. In sum, although important, these historic sources require close inspection and careful and painstaking interpretation.[193]

4.17 EFFECT OF THE BURDEN OF PROOF ON THE OUTCOME OF COASTAL TITLE AND PROPERTY BOUNDARY LITIGATION

There is little dispute that, among other matters, in-depth investigations of all reasonable sources of information, application of the expertise of highly

---

189. *Ibid.* The few reported cases concerning open coast littoral lands show similarity in the types of proof and experts used. E.g., *City of Corpus Christi, supra,* 622 S.W.2d at 644; *People v. Hecker, supra,* 179 Cal.App.2d at 827–832; see *Lechuza Villas West, supra,* 60 Cal.App.4th at 236.

190. *Ibid.* Later in this work, in connection with chapters on property boundary location along rivers and lakes, the standards, or lack of standards, for determining how rapid or sudden a shoreline movement must be to be termed avulsive as opposed to accretive, will be discussed. See Section 7.7 in Chapter 7 and Section 8.9 in Chapter 8.

191.   2 Shalowitz, § 4422, p. 175

192. *Ibid.* For example, the high-water line feature on such maps could be within 10 meters of its actual location. *Ibid.* This is a 60-foot range of error.

193.   It is important to be familiar with the methodology used in preparing these surveys and the instructions under which the surveyors were operating. In particular cases, it may also be important to be aware of what was actually happening in the field when the surveys were prepared. The National Archives contain reams and volumes of correspondence of the USCS and the USC&GS surveyors. In some instances, these documents contain interesting and fascinating details about exactly what these surveyors were doing and what features were being mapped.

skilled consultants to the results of those investigations, and formation and presentation of well-supported conclusions are necessary elements of case preparation and presentation. They are, however, not determinative. The result in a case may still turn on who has the burden of proof. Witness one case in which the factual issue was whether a certain coastal salt pond had been open to the sea during a particular period and, if so, how such an opening had been closed.[194] Maps, aerial photographs, and lay witnesses all provided conflicting facts.[195] Because the party in possession established that it had good title, the burden was on the claiming party to prove its case; the claiming party had failed to do so.[196] On the other hand, the claiming party argued that all it had to do was to dispute the evidence of good title and that the burden of proof remained with the party in possession.[197]

In discussing these cross-claims, the federal court noted bleakly that the burden of proof in quiet title actions is not much discussed by the courts, but that there were some general rules.[198] The burden of proof on all essential issues usually rests with the party who initiates the lawsuit by the filing of a complaint.[199] On the other hand, if the defendant makes an affirmative claim, that party has the burden of proving its claim.[200] In this case the court held that both parties had the burden of persuasion to establish the strength of their respective titles.[201] Although the party in possession had shown its title, the claiming party convinced the court that the salt pond had once been connected to the ocean, had been cut off by artificial fill, and properly belonged to it (the claimant), rather than the party in possession.[202]

It should be noted that in littoral or riparian property boundary or title disputes requiring re-creation of historical conditions and events, courts have not required mathematical certainty of proof. Indeed, even highly skilled technical experts even make mistakes and sometime admit them.[203] Courts recognize this fact:

---

194. *Alexander Hamilton Life Ins. Co., supra,* 757 F.2d at 539.
195. *Id.* at 539–540.
196. *Id.* at 541.
197. *Ibid.*
198. See notes 132 through 135 and accompanying text in Chapter 3.
199. *New Jersey v. New York* (1998) 523 U.S. 767, 140 L.Ed.2d 993, 1015; *Alexander Hamilton Life Ins. Co., supra,* 757 F.2d at 541; 65 Am.Jur.2d Quieting Title and Determination of Adverse Claims, § 78, p. 207; see note 129 in Chapter 3.
200. *Alexander Hamilton Life Ins. Co., supra,* 757 F.2d at 541; 65 Am.Jur.2d Quieting Title and Determination of Adverse Claims, § 79, p. 209; see note 129 in Chapter 3.
201. *Alexander Hamilton Life Ins. Co., supra,* 757 F.2d at 541. This was because the claiming party had affirmatively stated it had title to the property. *Ibid.*
202. *Ibid.* On appeal, the trial court's findings cannot be overturned unless clearly erroneous. *Ibid.* Of course, it is the extremely rare and unusual case where an appellate court will substitute its judgment for that of a trial judge and overturn the trial court's factual findings. After all, the trial judge heard all the witnesses, observed their demeanor, their confidence in their knowledge of the facts, and the basis for their conclusions and their apparent truthfulness. The rule is recognition that the cold record on appeal, consisting of pleadings and transcripts of testimony, does not exactly reflect what occurred in the crucible of trial and the trial judge's rulings should not be changed except in circumstances out of the ordinary.
203. *Id.* at 542–543.

> We recognize that Dr. []'s opinion is not free from doubt, but there are many cases in which certainty is unobtainable. No closed-circuit television camera keeps sentinel over the weathered shores to record whether indigenous materials are washed up by the waves or deposited by human beings. Dr. [] has the status of an expert because he has knowledge, training, and experience in his calling, and he is thereby privileged to express an opinion. . . . This opinion need not be categorical in order to merit reliance; rather, in the context of a civil case, it simply must be sufficiently persuasive to convince a trier of fact that the expert's opinion as to what occurred is more likely correct than not.[204]

This commonsense statement and the principle of reasonableness it embodies should be kept constantly in mind as we turn our focus to estuarine areas in the following chapter. The property boundary movement principles may largely be the same in estuarine areas as on the open coast, but the problems of proof dramatically escalate.

---

204. *Id.* at 543.

5

PROPERTY BOUNDARY DETERMINATION IN ESTUARINE AREAS

5.1 INTRODUCTION

By looking at the progression of maps depicting the United States as it expanded over the continent, an observer will readily conclude that initial settlement, as well as later population centers and development, clustered around many of the great bays and estuaries[1] indenting the nation's coastlines.[2] Beginning with the ill-fated Roanoke Colony in Pamlico Sound, North Carolina, and Jamestown in the Chesapeake Bay, Virginia, pilgrims and colonizers sought, among other matters, a protected harbor from which to engage in waterborne commerce and navigation and a nearby area to cultivate for sustenance.

The great cities of the East (Boston, Baltimore, Philadelphia, and New York), the South (Charleston, Mobile, and New Orleans), and the West (San Diego, San Francisco, and Seattle) were established largely as a consequence of favored locations beside tidal bays and estuaries. These estuaries and bays were freely connected to the ocean and the ocean tides. Their physical configuration also protected the harbors from storms and swells of the open sea, providing suitable and safe anchorages for the mariner. In addition, freshwater

---

1. The term "estuary" means a semi-enclosed coastal body of water that is freely connected to the open sea and within which ocean water is diluted by fresh water from upland river and stream drainages. See Tide and Current Glossary (U.S. Dept. of Comm., NOAA, NOS 1984), p. 7.
2. This development has progressed to the point where at least 70 percent of the population of the United States is clustered within 50 miles of one coastline or another. Newsweek, "Don't Go Near the Water" (August 1, 1988), p. 47. Further support for this statement is found in the 2000 presidential election. In that election, the eastern and western seaboard states' electoral votes alone nearly decided the election for one candidate.

runoff from river or stream drainages contributed to the estuary's unique nature as a productive area of both plant and animal life. Chesapeake Bay and San Francisco Bay are only two important examples of this distinctive feature.

Almost completely surrounding[3] the tidal bays and estuaries were extensive tidal marshes or meadowlands.[4] An example of this geomorphic[5] feature, is the tidal marsh system that once stretched from San Francisco Bay into the Sacramento-San Joaquin Delta. The extent of the pre-European-settlement marshes in the San Francisco Bay system[6] was about 500 square miles.[7] Figure 5.1 [illustration from Gilbert p.76.] on the following page is a graphic illustration of the extent of those tidal marshes.

The geographic reach of tidal marshes once surrounding San Francisco Bay is much less today. Large portions of these marshes have long since been reclaimed; only some isolated pockets of tidal marsh now exist along the fringes of the almost fully developed estuary.[8] For example, among other developments now existing in the place of these tidal marshes adjacent to the San Francisco and San Pablo Bays are portions of the Port of Oakland's busy and vital harbor, single-family residential homes, industrial parks, portions of

---

3. As along the open ocean coast, there are areas in these estuaries where sand or cobble beaches may have formed due to a particular upland regime. As far as riparian property boundary determination is concerned, the property boundary determination principles and their application to this feature are much the same as in the case of open coast beaches discussed in Chapter 4 and will not be discussed again in this chapter.
4. Courts have treated the terms "meadowlands" and "marshes" as equivalent terms. *Bradford v. Nature Conservancy* (Va. 1982) 294 S.E.2d 866, 870, 877.
5. If one is going to be comfortable in this field, one might as well get used to scientific terms such as "geomorphic." Geomorphology is the study of the characteristics, origin, and development of landforms. Hinds, Norman, *Geomorphology* (Prentice-Hall 1943), pp. 12–13. Especially in estuarine environments, the complexity of legal, technical, and practical problems of boundary determination are pronounced. Consequently, it is important to have a solid grasp of the physical features and processes pertaining to the estuarine regime, the scientific disciplines that study them, and the technical nomenclature used to denote and describe such features and processes.
6. This system includes San Pablo and Suisun Bays and portions of the Sacramento-San Joaquin Delta (known herein as the "Delta"). Because of certain unique features of the Delta and its regime, however, Delta property boundaries will be discussed separately in Chapter 6.
7. Gilbert, L. K., *Hydraulic Mining Debris in the Sierra Nevada* USGS Prof. Paper No. 105 (GPO 1917), p. 78, Table 17 (hereinafter "Gilbert"); Josselyn, M. N. and Atwater, B. F., "Physical and Biological Constraints on Man's Use of the Shore Zone of the San Francisco Bay Estuary," *San Francisco Bay: Use and Protection* (AAAS 1982), pp. 57, 58 (herein "Josselyn: Constraints on Shorezone Use"); Atwater, B. F, et al., "History, Landforms, and Vegetation of the Estuary's Tidal Marshes," *San Francisco Bay: The Urbanized Estuary* (AAAS 1979), pp. 347, 352 (herein "Atwater: Tidal Marshes"). Property boundary determination is a technical, specialty field. Consequently, the surveyor's or lawyer's library should contain significant scientific papers or treatises discussing particular and unique physical features of the geographic area in which the surveyor or lawyer may be practicing. Some of these papers may be publically available on the internet.
8. Atwater: Tidal Marshes, *supra*, at p. 353, fig. 3.

5.1 INTRODUCTION 143

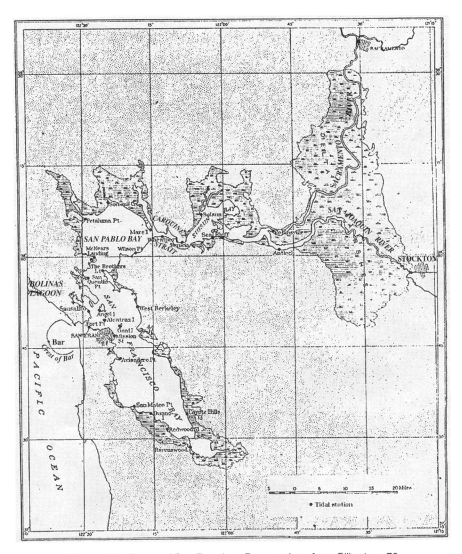

Figure 5.1 Scope of San Francisco Bay marshes, from Gilbert, p. 76

a now-abandoned Air Force base, and the solar salt production works of the Cargill (formerly Leslie) Salt Company. Each of these developments was or is in some way critical to the regional economy.[9]

From these few examples alone, one can see that over the years, population and development pressure on lands surrounding estuaries has not decreased. If anything, these pressures have risen exponentially. The result is that the press of civilization has modified the estuarine regime.[10] Tidal marsh areas immediately adjacent to such water bodies are still ardently and persistently sought for development to relieve these pressures. Consequently, title and boundary disputes between public and private interests[11] concerning these former tidal marshlands have arisen and will most certainly continue to arise.

In this country's legal history, some of the earliest contests leading to reported cases were about sovereign rights in and authority over lands underlying, or tidal marshes adjacent to, bays and tidal estuaries.[12] Such lands provided a venue for these boundary and title disputes for two simple reasons.

First, such controversies continually arose in the discrete geographic areas where, because of the growth of population and commerce, conflict between competing public and private interests was most likely to occur. Second, some

---

9. A final example (that could not be resisted): Former tidal marshes have been replaced by a football stadium, the Meadowlands in New Jersey (home of the 1986 and 1990 Super Bowl Champion New York Giants). *New Jersey Sports & Exposition Authority v. McCrane* (N.J. 1971) 292 A.2d 580, 585–586, *app. dismd* (1972) 409 U.S. 943. Joe Montana is probably still hearing Jim Burt's squishy footsteps swiftly and relentlessly approaching over the Astroturf-covered former marsh.

10. Some contend this modification has not been for the better. E.g., Kennedy, J. Michael, "U.S. Wetlands Swamp by Tide of Neglect" (L.A. Times, Dec. 10, 1988), pp. 1, 26; Newsweek, "Don't Go Near the Water (August 1, 1988) p. 46; Shabecoff, "How America Is Losing Its Edges," N.Y. Times, May 3, 1987, sec. E, p. 5; Pestrong, R., "Unnatural Shoreline," Environment (Nov. 1974), Vol.16, No. 9., p.27; Jerome, L. E., "Marsh Restoration," Oceans (Jan.-Feb. 1979), p. 57.

11. The dispute over those portions of these marshes that may fall within the definition of "wetlands" (see 33 C.F.R. §328.3(7)(b), the filling of which is regulated by section 404 of the federal Clean Water Act (33 U.S.C. § 1344), is only tangentially discussed. That subject is important (and complex) enough for its own work. The U.S. Supreme Court's decision in *Solid Waste Agency of Northern Cook County v. U.S. Army Corps of Engineers* ("SWANCC") (2001) ___ U.S. ___, 121 S. Ct. 675, 148 L.E.2d 576 is evidence of the importance of these issues. The U.S. Supreme Court ruled that the CWA does not authorize the Corps to regulate so-called "isolated" waters and wetlands, which are located away from rivers, lakes, and other readily recognizable waters. 148 L.Ed. 2d at 587. The Corps had asserted that the use of such waters and wetlands by migratory birds sufficed to invoke Congress' constitutional power to regulate interstate commerce. Id. at 584. In the 5-4 decision, the Court questioned whether the commerce power extended as far as the Corps supposed, but offered no answer, holding that in any event Congress never intended to regulate isolated waters and wetlands when it enacted the CWA in 1972. Id. at 589.

12. *Arnold v. Mundy* (N.J. 1821) 6 N.J.L. 1 (Raritan Bay); *Martin v. Waddell* (1842) 41 U.S.(16 Pet.) 366 (Raritan Bay); *Pollard's Lessee v. Hagan* (1845) 44 U.S.(3 How.) 212 (Mobile Bay); *Weber v. Harbor Commissioners* (1873) 85 U.S.(18 Wall.) 798 (San Francisco Bay).

persons thought that these tidal lands were worthless breeding grounds for mosquitoes and disease. Others recognized their value to the public for such vital uses as fishing and fowling. Thus, definition of the nature, limits, and extent of public versus private rights and authority came to have singular importance to the developing country. These property boundary and title disputes necessarily arose at the very inception of the American legal system's consideration of the nature and extent of public rights and powers. Merely because there was early judicial consideration of title and boundary questions in regard to bays and estuaries, however, does not mean that the courts have resolved all title and boundary questions. Far from it.

Such title and boundary disputes have occurred, and can still occur in many unusual contexts.[13] Today such disputes spring from condemnation proceedings,[14] from the development permit or development financing process,[15] or from basic title and boundary controversies between adjoining landowners.[16] In all such cases the determination of the location of the landward[17] boundary of tidelands[18] and, as a consequence, the landward extent, nature, and limit of the public trust interest of the state, have remained the crucial question.[19]

---

13. In one case, the United States' title to an abandoned United States Air Force base built on alleged tidelands was challenged by the state when the United States attempted to sell the base as surplus property free of any sovereign public trust title interest. *State of Cal., etc. v. United States* (N.D.Cal. 1981) 512 F.Supp. 36, 37–38.
14. *O'Neill v. State Highway Department* (N.J. 1967) 235 A.2d 1.
15. Most of such disputes are, for a number of reasons, ultimately resolved by agreement short of court decision. Consequently, there are not as many reported cases as one might expect. Some cases are: *City of Berkeley v. Superior Court* (1980) 26 Cal.3d 515, cert. den., sub. nom. *Santa Fe Land Improv. Co. v. Berkeley* (1980) 449 U.S. 840; *Board of Public Works v. Larmar Corporation* (Md. 1971) 277 A.2d 427; *Warren S. & G. Co., Inc. v. Commonwealth, Dept. of E.R.* (Pa. 1975) 341 A.2d 556.
16. *Velsicol Chemical v. Environmental Prot. Dept.* (N.J. 1982) 442 A.2d 1051.
17. The author conceives of the physical location of the ordinary high-water mark property boundary in a marsh environment as if one were looking down on the marsh from directly overhead and could place the boundary line through and over the tide marsh much like a road divider. Thus, the convention of this chapter will be to refer to the location of lands relative to the ordinary high-water mark property boundary in the horizontal rather than the vertical aspect. Use of the term "landward" or "bayward" rather than the terms "above" or "below" will be this book's convention when discussing the ordinary high-water mark in a marsh environment. The reason for this convention is the physical nature of the marsh itself. The surface of the marsh is extremely flat, with little or no slope. Conceiving of the physical location of the property boundary location in a vertical aspect is not only difficult, but is misleading. The relative physical location is more precisely and accurately expressed in horizontal rather than in vertical terms.
18. For purposes of this chapter, unless otherwise specified or apparent from the context of the term, the term "tidelands" encompasses submerged lands as well.
19. Briefly, as this issue is more fully explored in Chapters 3 and 4, lands bayward of the ordinary high-water mark of tidal waters, tidelands, are subject to the public trust. See notes 111 through 121 and accompanying text in Chapter 1, notes 13 through 15 and accompanying text in Chapter 3, and notes 6 through 10 and accompanying text in Chapter 4. Marshlands landward of the ordinary high water mark are not subject to the public trust. These lands may however, be

146   PROPERTY BOUNDARY DETERMINATION IN ESTUARINE AREAS

The issue is significant. Should such marshland be found subject to the public trust, that finding would drastically affect, for example, the availability of compensation for the condemned land or the possibility of freely developing one's land.[20] Thus, this chapter intends to assist the surveyor, lawyer, or other professional whose advice is sought in answering one of the most challenging of title and boundary questions: where and how to establish the physical location of the ordinary high-water mark property boundary in a tidal marsh.

In answering that question, this chapter[21] will discuss why, due to the derivation of title to tidal marshlands and the very physical features and properties of the tidal marsh environment, property boundary determination in tidal marshes is so difficult and confusing. In an attempt to dispel the confusion, the chapter will explain the title derivation of tidal marshlands and will analyze the legal effect of the government patenting process. That discussion will be followed by an analysis of various means, methods, and contentions conceived to resolve some of this uncertainty: the legal doctrine of estoppel, the enactment of "curative" acts by Congress and state legislatures, and the promotion of an argument that would severely limit the geographic extent of tidelands.

After such an analysis, this chapter will discuss various methods of physically locating the ordinary high-water mark property boundary in tidal marshes. It will explore the circumstances in which one may establish the geographic location of the ordinary high-water mark property boundary as a matter of law. Then the author will address the issue of the physical indicia of the ordinary high-water mark property boundary and the applicability of boundary movement principles to tidal marshes. The remainder of the chapter will feature some legal and practical problems and suggested approaches in determining and proving the physical location of the ordinary high-water mark property boundary in a tidal marsh environment.

---

subject to various land use and other development constraints including the requirements of action 404 of the Clean Water Act administered by the U.S. Army Corps of Engineers. *SWANCC, supra*, 148 L.Ed 2d at 584.

20.   E.g., *Summa Corp. v. California ex rel. State Lands Com'n* (1984) 466 U.S. 198, 205; see notes 120 through 121 in Chapter 1.

21.   Chapter 4 discussed the progress of the development and adoption of the mean high-water line as the court-approved physical location of the ordinary high-water mark, the legal property boundary of tidelands and public lands along tidal waters. That discussion need and will not be repeated in this chapter in connection with location of the property boundary between tidelands and public lands in tidal marshes. Some may claim a recent United States Supreme Court case appears to reinforce this conclusion. This case apparently rejected other suggested means of locating the ordinary high-water mark property boundary in estuaries and instead accepted the mean high-water line as the physical location of that property boundary. See *Phillips Petroleum Co. v. Mississippi* (1988) 484 U.S. 469, 481, n.9, *rehg. den.* (1988) 486 U.S. 1018; note 125 and accompanying text in this chapter.

5.2 DIFFICULTY AND CONFUSION IN TITLE AND BOUNDARY DETERMINATION IN TIDAL MARSHES

In the author's opinion, property boundary determination in tidal marshes is confusing and difficult for three principal reasons. Understanding these reasons, the reader will be better able to focus on the true issue, the geographic or physical location of the ordinary high-water mark property boundary.

The first reason for confusion is the undeniable fact that courts, commentators, lawyers, surveyors, and property owners have muddled the legal character[22] and physical character of tidal marshlands. For example, the terms "tidelands," "salt marsh,"[23] and "swamp and overflowed lands" could each describe the physical character or attributes of tidal marshland. On the other hand, two of the same terms, "tidelands" and "swamp and overflowed lands," could as well describe tidal marshland's legal character. The consequences of those designations, insofar as public and private rights are concerned, are vastly different.[24] Thus, one must keep clearly in mind the distinction between the physical character and the legal character of tidal marshland when reading or examining any discussion or explanation of tidal marshland titles or boundaries.

The second reason for misunderstanding is that the public holds a dual interest in lands of the legal character of tidelands—the *jus privatum* and the *jus publicum*.[25] Although the *jus privatum* could be alienated, the *jus publicum* could not.[26] This dichotomy leads to an apparently perplexing result. The government may have regularly issued a patent to a private purchaser purporting on its face to convey the entire title interest of the state in lands characterized as tidelands. However, only the state's proprietary rights—the *jus privatum*—could be lawfully sold and conveyed. Unless certain conditions were met, the state's public trust property interest—the *jus publicum*—could

---

22. The phrase "legal character" is a term of art. It means the type or nature of legal title to land as reflected by its source of title. To avoid confusion, it is the author's intent specifically and clearly to distinguish the "legal character" of marshland from its physical features and attributes, the "physical character," of marshland.
23. There also appears to be considerable confusion over whether the term "salt marsh" is intended to denote a particular legal character of tidal marshlands. One California court first equated this term with lands that were of the legal character of "swamp and overflowed lands." *Rondell v. Fay* (1867) 32 Cal. 354, 364. In the very next case in the California official reporter, the same California court equated the term "salt marsh" with lands that were of the legal character of "tidelands." *Ward v. Mulford* (1867) 32 Cal. 365, 373. Although later cases also equate salt marsh with tidelands, *White v. State of California* (1971) 21 Cal.App.3d 738, 743 ("saltmarsh tidelands"), no firm conclusions regarding the legal character of the land can be derived solely from the description of the physical character of tidal marsh as "salt marsh."
24. See notes 19 and 20 in this chapter.
25. *Oakland v. Oakland Water Front Co.* (1897) 118 Cal. 160, 183.
26. *Ibid.*; *People v. California Fish Co.* (1913) 166 Cal. 576, 592–594.

not be conveyed.[27] Therefore, despite the seemingly absolute language of such government patents ("all right, title and interest"), title conveyed by the patent would remain subject to the public trust easement held by the state—an easement quite broad in scope.[28]

The third and final reason that there is much confusion and difficulty in tidal marsh title and boundary determination is found in the very nature of estuarine tidal marshes. Far from providing a ready resource for the definitive and final answer, the physical character of tidal marshes only enhances the uncertainty. Obtaining meaningful or accurate physical measurements in a tidal marsh is quite difficult. Such marshes have little or no topographic relief. Vegetation cover also largely obscures their surfaces. Further, the marsh itself is dynamic; it changes according to the fluctuations of the elements of the estuarine regime. Finally, humans have altered or modified the physical regimes of most estuaries (and even the tidal marshes themselves). Thus, no matter how interesting the measurement of the physical features of currently existing tidal marshes may be, any such measurement must be carefully planned to be useful and convincing. What is more important, one should thoroughly vet and understand the purpose of the exercise and its intended result.

It is hoped that the reader will keep the preceding explanations in mind as this chapter progresses. The author intends each topic in this chapter to give the reader the tools with which to hold at bay further confusion and uncertainty in title and boundary determination about land that once was (and perhaps still is) a tidal marsh. The first such topic focuses on the title derivation of tidal marshlands—their legal character.

5.3 TITLE DERIVATION OF TIDAL MARSHLANDS

5.3.1 Ordinary High-Water Mark—Boundary of Tidelands and Swamp and Overflowed Lands

Tidal marshlands are generally comprised of lands that are either swamp and overflowed or tidelands in legal character. Swamp and overflowed lands are a type of United States' public lands granted to the states by the United States.[29]

The waterward boundary of swamp and overflowed lands is the ordinary high-water mark.[30] By definition, all swamp and overflowed lands lie land-

---

27. *Ibid.*; *Illinois Central Railroad v. Illinois* (1892) 146 U.S. 387, 453; *City of Berkeley v. Superior Court, supra,* 26 Cal.3d 515, 524.
28. See notes 19 and 20 in this chapter.
29. See text accompanying notes 87 through 91 in Chapter 1.
30. *People v. Morrill* (1864) 26 Cal. 336, 356; *United States v. O'Donnell* (1938) 303 U.S. 501, 519; *Work v. United States* (9th Cir. 1927) 23 F.2d 136; see *Martin v. Busch* (Fla. 1927) 112 So. 274, 284 (swamp and overflowed lands adjacent to a lake).

ward of the ordinary high-water mark.[31] Lands lying waterward of the ordinary high-water mark of tidal or navigable waters are tidelands[32] owned by the states as a consequence of the Equal Footing Doctrine as recognized and confirmed by the Submerged Lands Act.[33] The boundary question appears deceptively simple.

All one has to do is to locate the ordinary high-water mark boundary. Then, like Solomon, one may pronounce lands landward of that line as swamp and overflowed lands and lands bayward of that line as tidelands. One may wonder what all the fuss is about. As with most matters associated with property boundary and title disputes about lands along, adjacent to, or underlying a water body, little should be taken for granted.

5.3.2 Segregation of Swamp and Overflowed Lands from Tidelands by Visual Observation of Their Physical Character

In tidal estuaries, such as San Francisco Bay, distinguishing lands of the legal character of swamp and overflowed lands from lands of the legal character of tidelands has long been a problem. Succinctly stated, the problem is that, as a physical matter, swamp and overflowed lands immediately abutted tidelands. One court pithily described the situation, saying that tidelands "merged into them [swamp and overflowed lands] imperceptibly, making it difficult to distinguish between them."[34] Swamp and overflowed lands were similar in physical character and appearance to the tidelands to which they were contiguous and it was ". . . often difficult or impossible to accurately locate the lines of separation, and, indeed, such line might change temporarily in times of flood."[35] In other words, merely by looking at a tidal marsh, an observer, even a trained surveyor, could not visually differentiate the tidal marsh into lands of the legal character of swamp and overflowed lands and lands of the legal character of tidelands based solely on the purported "distinctive" physical characteristics of either character of lands. Courts have not made the problem of identification any less difficult or complex.

5.3.3 Judicial Guidance for Segregation of Swamp and Overflowed Lands

As should be expected, courts have not provided unequivocal standards or tests by which one could judge, purely by the content of the court's written

---

31. *Ibid.*
32. E.g., *Martin v. Waddell, supra,* 41 U.S. at 410; *Pollard's Lessee, supra,* 44 U.S. at 229; *Weber, supra,* 85 U.S. at 65–66; *Shively v. Bowlby* (1894) 152 U.S. 1, 26; *Borax, Ltd. v. Los Angeles* (1935) 296 U.S. 10, 15; *State Land Board v. Corvallis Sand & Gravel Co.* (1977) 429 U.S. 363, 374.
33. See notes 111 through 121 and accompanying text in Chapter 1.
34. *People v. California Fish Co., supra,* 166 Cal. at 591.
35. *Forestier v. Johnson* (1912) 164 Cal. 24, 32.

description of the physical character of a tidal marsh, whether the tidal marshlands at issue in the case were swamp and overflowed lands or tidelands in legal character. For example, one court described the physical character of tidelands as follows:

> . . . To render lands tide-lands, which the State by virtue of her sovereignty could claim, there must have been such a continuity of the flow of tidewater over them, or such regularity of the flow within twenty-four hours, as to render them unfit for cultivation, the growth of grasses or other uses to which upland is applied.[36]

In a purported contradistinction, a court described the physical character of swamp and overflowed lands as follows:

> . . . Swamp lands, as distinguished from overflowed lands, may be considered such as require drainage to fit them for cultivation. Overflowed lands are those which are subject to periodical or frequent overflows as to require levees or embankments to keep out the water and render them suitable for cultivation. It does not make any difference whether the overflow be by fresh water, as by the rising of rivers or lakes, or by the flow of the tides.[37]

Thus, in their early benchmark decisions, courts provided very little guidance to surveyors, lawyers, government officials, or property owners that would help them in physically distinguishing, in tidal marshlands, lands of the legal character of tidelands from lands that were swamp and overflowed lands in legal character.[38]

---

36. *San Francisco v. Leroy* (1891) 138 U.S. 655, 671.
37. *San Francisco Savings Union v. Irwin* (9th Cir 1886) 28 F. 708, 712, *aff'd*, (1890) 136 U.S. 578.
38. In the 1860s, two courts separately described the physical character of an undisturbed tidal marsh that lay on the east side of south San Francisco Bay. One court described a portion of the marsh as follows:

> . . . Between [the uplands] and the water there extends a large tract of level land, traversed by sloughs or estuaries in every direction and producing salt grasses and other distinctive vegetation. . . .
> . . . Whether the marsh land around the bay is regularly overflowed every year I am not informed; but from the distinctive character of the vegetation it may be inferred, that it is occasionally covered with salt or brackish water.
> Throughout by far the greater part of the year the marsh land is considerably above the ordinary limit of high water. . . . *United States v. Estudillo* (1862) No. 234 "San Leandro."

Just two years later, in an opinion written by Justice Field, a former justice of the California Supreme Court, the United States Supreme Court described virtually the same tidal marshlands as follows:

> . . . On the side of the bay there are about two leagues of salt or marsh land. The whole of this land is covered by the monthly tides at the new and full moon, and a part of the

Some may see lack of judicial guidance as evidence that it is not possible, purely by visual examination of the physical features or attributes of a tidal marsh, to determine the precise physical location of the boundary between lands of the legal character of tidelands and lands of the legal character of swamp and overflowed lands.[39] This difficulty was not eased by the manner in which the United States attempted to identify which tidal marsh lands were swamp and overflowed in legal character.

5.3.4 Legal Effect of the United States' Segregation and Patent of Tidal Marshlands as Swamp and Overflowed Lands

The chain of title to tidal marshland will most likely show that title to such land is derived from a state patent to an individual. That patent will identify the authority by which it is issued and the legal character of the land conveyed by the patent. Figure 5.2 and Figure 5.3 (shown on the following pages) are examples of such patents (although they have been retyped for legibility). Figure 5.2 is an example of a tideland patent, while Figure 5.3 conveys swamp and overflowed land. For our purposes, it may be assumed that the patent will identify the tidal marshland conveyed as either swamp and overflowed land or tideland in legal character.

In the case of the state swamp and overflowed land patents, understanding the process leading to the state's issuance of such patents to individuals is important. With this knowledge, the value of the government's determination of the legal character of the land conveyed[40] may be explored and understood.

Swamp and overflowed lands[41] were a species of public land granted by the United States to the states. The United States' swamp lands grant was

---

land is covered by the daily tides. *United States v. Pacheco* (1864) 69 U.S. (2 Wall.) 587, 589.

39. One commentator is even more graphic in his description of this boundary, calling it "invisible."

> Throughout the United States exist untold acres of marsh, meadow and estuarine areas. This immense boundless real estate commonly borders rivers, streams, creeks, bays, oceans and similar bodies of water and waterways. In many areas, such tracts are inundated and overflowed periodically by the tides, especially by the Spring and extraordinary tides. Symbolic of these unmeasured acres are the characteristics of the merging of water and land beyond the shore and the high water mark, of the watercourse across acres yielding miles of grasses . . . of many varieties. In such instances, the line of watercourse often disappears and becomes an invisible boundary. Porro, "Invisible Boundary—Private and Sovereign Marshland Interests" (1970) 3 Natural Resources Journal 512.

40. Swamp and overflowed lands adjacent to navigable rivers and lakes were also identified and sold to private persons. E.g., *Churchill Co. v. Kingsbury* (1918) 178 Cal. 554, 557; *Martin v. Busch*, *supra*, 112 So. at 279. The principles discussed in this section are equally applicable to swamp and overflowed lands adjacent to rivers and lakes.

41. A more complete explanation of the derivation and identification of swamp and overflowed lands is found in notes 87 through 105 and accompanying text in Chapter 1.

U.S. of America, State of Cal.)
    ---- To ---)
<u>*Elsa A. Oliver*</u>)

    United States of America, State of California. To whom these presents shall come, greeting. Whereas, the Legislature of the State of California has provided for the sale and conveyance of the Tide Lands belonging to the State by virtue of her sovereignty by Statutes, enacted from time to time, And whereas it appears by the certificate of the Register of the State Lands office No. 217 bearing date July 22nd A.D. One Thousand, Eight Hundred and Ninety-Nine and issued in accordance with the provisions of law that the tracts of Tide Land hereinafter described have been duly and properly conveyed in accordance with the law, that full payment has been made to the State for the same, and that Elsa A. Oliver is entitled to receive a Patent therefore; said lands being situated in Alameda County and described as follows, to wit; Location No. 200-State Tide Lands, Alameda County Township No. 4 south range and No. 2 West Mount Diablo Meridian; Sections No. 6 Fraction in SE one quarter of NW one quarter and Fraction in SW one quarter of NE one quarter of Section 6 more particularly described in the field notes of said locations as follows;

    <u>*Beginning at a point*</u> *20 chains east from the one quarter section corner dividing section One (1) of Township Four (4) south range Three (3) west from sections Six (6) Township Four (4) south range Two (2) west and running thence east 40.00 chains to swamp and overflowed land survey No. 291 thence along same north 12.65 chains to the south boundary of swamp and overflowed land survey No. 279. Thence along said boundary of swamp and overflowed land survey No. 279 west 40.00 chains thence south 12.65 chains to the place of beginning.*

    Run by the true meridian magnetic Variation 17 degrees 10 minutes east containing Fifty $^{60}/_{100}$ (50 $^{60}/_{100}$) acres. Now therefore, All of the requirements of the acts of the State Legislature in relation to Tide Lands, having been fully complied with, I Henry T. Gage, governor of the State of California by virtue of authority in me vested, have granted, bargained, sold, and conveyed, and by these presents do grant, bargain, sell and convey unto the said Elsa A. Oliver all the above described lands with the appurtenances thereto belonging to have and to hold unto her, the said Elsa A. Oliver, her heirs and assigns forever.

    <u>*In testimony*</u> *whereof I, Henry T. Gage, governor of the State of California have caused these letters to be made Patent, and the seal of the State of California to be hereunto affixed given under my hand at the City of Sacramento this the 19th day of August in the year of our Lord one thousand, eight hundred and ninety nine.*

State Seal	*Henry T. Gage*
Attest	*Governor of the State of California*
By: J. Roesch, Dep.	*Secretary of State*
Countersigned	*Wm. J. Wright*
	Register of the State Land Office

Figure 5.2 Tideland Patent, State of California to Oliver

termed a "present" one. This meant that the public lands granted were deemed to have been granted at the time of congressional authorization. There was, however, no actual instrument of conveyance particularly describing the public lands encompassed in the United States' swamp land grant. In other words, there was no instrument or conveyance from the United States recorded in local property records that described such lands. The swamp land grant was most often perfected by the United States' later identification (known by the name "segregation") from the public domain of public lands swamp and overflowed in character and by the subsequent patent of such segregated lands to the state.[42]

---

42. E.g., *Wright v. Roseberry* (1887) 121 U.S. 488, 509; *Tubbs v. Wilhoit* (1891) 138 U.S. 134, 136.

UNITED STATES OF AMERICA --- STATE OF CALIFORNIA

TO ALL TO WHOM THESE PRESENTS SHALL COME, GREETING:

WHEREAS, under the provisions of an Act of Congress of the United States, approved the twenty-eighth day of September, A.D. one thousand eight hundred and fifty, entitled "An Act to enable the State of Arkansas, and other States, to reclaim swamp land within their limits," in which Act the manner of selecting and setting apart swamp and overflowed lands is fully set forth: AND WHEREAS, in accordance with the provisions of the Act of Congress, the title of which is above recited, the Legislature of the State of California, on the twenty-eighth day of April, A.D. one thousand eight hundred and fifty-five, passed an Act entitled "An Act to provide for the sale of swamp and overflowed lands belonging to this State," and on the twenty-first day of April, A.D. one thousand eight hundred and fifty-eight, passed an Act entitled "An Act to provide for the sale and reclamation of the swamp and overflowed lands of this State," and on the eighteenth day of April, A.D. one thousand eight hundred and fifty-nine, an Act amendatory thereof, and on the twenty-seventh day of April, A.D. one thousand eight hundred and sixty-three, passed an Act entitled "An Act to provide for the sale of certain lands belonging to the State," which Acts authorize the location and disposal of a portion of the swamp and overflowed lands donated to the State of California by the Act of Congress, the title of which is above recited; AND WHEREAS, it appears by the certificate of the Register of the State Land Office, Number 684, bearing date April 4th, A.D. one thousand eight hundred and sixty-eight, and issued in accordance with the provisions of said last named Act; that he has been duly notified by the County Treasurer, that James Crooks, has paid Five hundred and forty seven 88/100 dollars to the State of California, in full payment for Two-hundred and forty seven 88/100 ACRES OF SWAMP AND OVERFLOWED LANDS, situated in San Mateo County, and described as follows, to wit: SURVEY NO. 95 SWAMP AND OVERFLOWED LANDS San Mateo County – TOWNSHIP NO. Five (5) South RANGE NO. four (4) West of Mount Diablo MERIDIAN: SECTIONS No. 12 of 13 being the fractional South half (1/2) of Section 12; and the fractional North East quarter (1/4) of Section 13. Surveyed in accordance with the provisions of the Act of April 18th, 1859, and more particularly described in the field notes of said survey as follows: Beginning at the corner of Sections Seven (7) Twelve (12) Thirteen (13) and Eighteen (18) on line between Ranges Three (3) and Four (4) West, in Township No. Five (5) South, Base of Meridian of Mount Diablo; thence North forty (40.00) chains to quarter section corner, between Section 7 of 12; thence West, seventy (70.00) chains to the Eastern boundary of the "Bulgas Rancho"; thence (with said boundary South 49°30' East, thirty five (35.00) chains; South 32°15' East twenty eight (28.00) chains; South 50°00' East, thirty seven 14/100 (37.14) chains to line between Ranges Three (3) and Four (4) West; thence North, (on said Range line) between section, 13 of 18, thirty 28/100 (30.28) chains to the place of beginning, containing two hundred and forty seven 88/100 (247 88/100) Acres, lines run by the true Meridian Magnetic Variation 15°45' East.

NOW, THEREFORE, all the requirements of the Act of Congress, as well as the Acts of the State Legislature, in relation to swamp and overflowed lands, having been fully

Figure 5.3 Swamp and Overflowed Land Patent, State of California to C. M. Hitchcock

complied with, I, Henry H. Haight, Governor of the State of California, by virtue of authority in me vested, have granted, bargained, sold, and conveyed, and by these presents do grant, bargain, sell, and convey unto C.M. Hitchcock assignee of James Crooks, all the above described lands with the appurtenances thereunto belonging, to have and to hold, unto him the said C.M. Hitchcock his heirs and assigns forever.

IN TESTIMONY WHEREOF, I, H. H. Haight, Governor of the State of California, have caused these letters to be made patent, and the seal of the State of California to be hereunto affixed.

GIVEN under my hand, at the City of Sacramento, this, the twentieth day of May in the year of our Lord one thousand eight hundred and sixty-eight.

H.H. Haight,

Governor of State.

ATTEST:

H.L. Nichols,

Secretary of State.

COUNTERSIGNED:

John W. Bost,

Register of State Land Office.

Figure 5.3 (*Continued*)

Whatever the intent of the swamp lands grant, the process by which lands ultimately came to be granted by the United States to the states and by the states to private persons was hardly orderly or systematic. Almost immediately after the swamp and overflowed grant was made by the United States and even *before* the lands granted had been identified as swamp and overflowed by the United States, the states, through legislative authorization, sold swamp and overflowed lands to private persons.[43] Nevertheless, it was only those lands, swamp and overflowed in physical character on the date of the Swamp Lands Act,[44] that were conveyed to the states by virtue of the Swamp Lands Act.[45]

How were such swamp and overflowed lands identified, and what was the effect of that identification? Shouldn't the existence of patents regularly issued

---

43. E.g., *People v. California Fish Co.*, *supra*, 166 Cal. at 589–590.
44. For example, the Arkansas Swamp Lands Act was made applicable to California on September 28, 1850. Act of September 28, 1850, 9 Stat. 519 (now 43 U.S.C. secs. 981, et seq.).
45. *Ord Land Co. v. Alamitos Land Co.* (1926) 199 Cal. 380, 385.

to private persons by or through governmental land management authorities, patents that characterized and conveyed the tidal marshlands as either swamp and overflowed lands or tidelands,[46] end once and for all any question about the legal character of the tidal marshlands? Shouldn't issuance of a swamp and overflowed patent from the United States to the state on the initiative of the United States, because of legal requirements, or on the request of the state,[47] end inquiry as to the legal character of the land conveyed? The reader will not be shocked to learn that there are arguments made to the contrary.

Suppose United States land management officials in the General Land Office (now the Bureau of Land Management)[48] made an administrative determination that lands were swamp and overflowed in legal character. Further suppose that the United States subsequently issued a swamp and overflowed land patent to the state. Finally, suppose that under authority of state legislation, state authorities patented tidal marshlands to a private party as swamp and overflowed lands. Even with the concatenation of these acts by the government, some claim with ardent conviction that the legal character of tidal

---

46. If the state patent characterizes tidal marshlands as tidelands in legal character, there is an argument that there is no need for further inquiry concerning the legal character of the land described. It can be asserted that the tideland patentee and his or her successors are precluded from introducing physical evidence that would repudiate the characterization of the lands conveyed by the patent as tidelands in character. In effect, so the argument goes, one who holds title by virtue of such a patent cannot challenge his or her own source of title. E.g., *Newcomb v. Newport Beach* (1943) 7 Cal.2d 393, 398; *Steel v. Smelting Co.* (1892) 106 U.S. 447; *French v. Fyan* (1876) 93 U.S. 169. On the other hand, this argument seems *not* to apply to the state, which appears always able to invalidate its patents. E.g., *People v. California Fish*, supra, 166 Cal. at 611–612. Thus, the state is presumably not bound by a state patent's characterization of the lands as swamp and overflowed in legal character. See notes 59 through 60 and accompanying text in this chapter. This is yet another example of the "heads I win, tails you lose" syndrome of litigating sovereign title with the state.

47. See text accompanying notes 97 through 110 in Chapter 1. For example, the State of California adopted "An Act to Provide for the Reclamation and Segregation of Swamp and Overflowed, Salt Marsh and Tide Lands Donated to the State of California by Act of Congress." Act of May 13, 1861, Stats. 1861, Ch. 352 ("1861 Act"). While Congress did not "donate" tidelands to the State, the 1861 Act was intended to spur the United States to take action to identify and patent swamp and overflowed lands to California. The 1861 Act required county surveyors to perform "segregation" surveys. The state then sent these "segregation" survey maps to the U.S. General Land Office to assist in preparation of township plats. The township plats led to issuance of swamp and overflowed patents to the states by the United States. It has been argued that the 1861 Act required county surveyors to segregate only swamp and overflowed lands from the "high lands." *Id.* at § 19. And it has also been claimed the 1861 Act recognized that the state owned the tidal marshlands as either swamp and overflowed lands or tidelands no matter how their legal character was perfected. Thus, the argument concludes that the request to the United States for issuance of a township plat prepared in conformity to those segregation maps was merely intended as a request for what amounted to a quitclaim deed for whatever interests the United States had in lands waterward of the segregation line between the high lands on the one hand and the swamp and overflowed, salt marsh, and tidelands on the other. There has been no trial or appellate court determination of the validity of these claims.

48. See note 35 in Chapter 1 for a fuller explanation of the role and history of these agencies.

marshlands as swamp and overflowed lands has *not* been conclusively and finally determined.[49] This argument is based on a deceptively simple premise.

Tidelands are owned by the states, not the United States. Consequently, the United States would act beyond its authority in segregating and patenting tidelands as swamp and overflowed lands; the United States did not own tidelands, and tidelands were not part of the public lands of the United States after the United States' disposition.[50] Thus, to the extent that tidelands are conveyed, segregation and patent of such lands by the United States to the state as swamp and overflowed lands is not a conclusive determination of the legal character of the lands conveyed. According to this argument, such patents remain subject to challenge by the states or by someone claiming through title derived from a state tideland conveyance.[51]

It is also asserted that the state stamp of approval, established by issuing a swamp and overflowed patent to the purchaser, makes no difference. In some states, when the authority and ability to immediately sell acres and acres of swamp and overflowed lands was allied with the state's need for money to pay for badly needed internal improvements, land sale programs were established that have been characterized by some as "fire sales."[52] The purchasers themselves decided the legal character of the land that they requested to be patented.[53] In addition, courts added to the confusion by holding that neither the United States nor a state was bound by the actions of the other in determining which lands were available for sale as swamp and overflowed lands.[54]

Thus, some argued that the necessity of working through the fine distinctions required to differentiate tidelands from swamp and overflowed lands was lost in the state's efforts to sell these lands swiftly.[55] Attempts to direct

---

49. The California legislature enacted a series of acts to try to cure the claimed flaws in these titles. See *People v. California Fish Co., supra*, 166 Cal. at 590. Essentially, those acts attempted to legalize patents in the situation in which a person had received a patent from the state but the applicant had made a mistake in classification of the legal character of conveyed land. *Ibid.* As we will see in notes 73 through 79 and accompanying text in this chapter, some contend these curative acts did not result in ratification of conveyance of the state's public trust interest, the *jus publicum.*
50. Recall the discussion in Chapter 1, Section 1.7.4, that provides a graphic example of the adage that one cannot sell what one does not own.
51. *Smelting Co. v. Kemp* (1881) 104 U.S. 636; *Steel v. Smelting Co., supra*, 106 U.S. 447; *Borax, supra*, 296 U.S. at 18–19; *United States v. O'Donnell, supra*, 303 U.S. at 508; *Dreyfuss v. Badger* (1895) 108 Cal. 58, 66–67; *Klauber v. Higgins* (1897) 117 Cal. 451, 458; *Williams v. City of San Pedro, etc. Co.* (1908) 153 Cal. 44, 47.
52. For example, the State of California at first accepted applications to purchase swamp and overflowed lands before approval by the United States of the township plat for the township in which the lands for sale were located. *Tubbs, supra*, 138 U.S. at 138. Some state officials decried these practices. E.g., Annual Report of the State Surveyor General for the Year 1855.
53. *People v. California Fish Co., supra*, 166 Cal. at 591.
54. *Kernan v. Griffith* (1864) 27 Cal. 87, 91.
55. Long after the land sale programs were largely completed and governmental officials and a few citizens raised questions of just what had been sold, this fact appears to have been recognized by the courts:

state surveyors by issuing surveying instructions were also seen to be ineffective.[56] Although federal surveyors operated under similar instructions,[57] it made little difference. Some claim the federal government has no authority to determine the extent of the state's tidelands; federal authority extends only to the public lands of the United States, such as swamp and overflowed lands.[58] As a consequence, the United States' segregation of swamp and overflowed land merely separated swamp and overflowed land from other species of public lands, not from tidelands.[59] As a consequence, it is frequently asserted that even those persons whose title appears to have been (1) derived from a state swamp and overflowed patent regularly issued by the state and (2) founded on a United States swamp and overflowed patent regularly issued to the state must be prepared for a title dispute.[60]

> [T]he idea that the state held a double right, public and private, in . . . tidelands and that the [state's] private right in [tidelands] was proprietary, as were the swamplands with which the tidelands were classed, and subject to a similar disposition, and this disposition [of tidelands] might be subject to the public easement, without injury to the public right, does not seem to have been suggested or have been considered [in early cases concerning the sale of tidelands and swamp and overflowed lands]. No mention is made of the practical difficulty arising from the doctrine declared; that of finding a means of determining where the line should be drawn between the lands authorized to be sold [swamp and overflowed lands] and those not within the legislative intent [tidelands]. The statutes afford no means for such differentiation [referring to acts providing for the sale of swamp and overflowed, salt marsh and tidelands] . . . . [T]here are vast bodies of land bordering on the bay of San Francisco and over navigable bays and waters of the state with respect to which the doctrine would produce the utmost confusion and uncertainty. . . . There are all conceivable varieties of conditions. No line of demarcation or classification is given. No distinction can be made which can be applied with certainty to all locations. *People v. California Fish Co., supra,* 166 Cal. at 592.

56. For example, under the 1861 Act, *supra,* the State of California instructed county surveyors to segregate the swamp and overflowed lands from the "high land." The state instructed these surveyors that in accomplishing the segregation, they were to run only the line of division between the swamp and "United States [public] lands." Report of Board of Swampland Commissioners (1861). Thus, it is contended the legislature intended *only* the differentiation between "high land" on the one hand and "swamp and overflowed, salt marsh and tideland" on the other. Neither this Act nor the instructions issued to the county surveyors are claimed to require that any distinction be made between lands that were of the physical character of tidelands and those that were of the physical character of swamp and overflowed lands. Act of 1861, *supra,* at secs. 19–20; *Forestier, supra,* 164 Cal. at 33; *People v. California Fish Co., supra,* 166 Cal. at 592.
57. Instructions to the Surveyors General of the Public Lands of the United States (1855) as reprinted in Minnick, *A Collection of Original Instructions to Surveyors of the Public Lands 1815–1881* (Landmark), pp. 371–372.
58. *Borax, supra,* 296 U.S. at 15–19.
59. One trial court has so held. *State of California, ex rel. Public Works Board v. Southern Pacific Transportation Company* (1983) Sacramento County Superior Court No. 277312, Statement of Decision, para. A. 6–7, p. 6 (sometimes referred to herein as the "*State v. Southern Pacific (Delta Meadows Decision)*"). This decision is the only court determination of which the author is aware where such issues have been tried to judgment. The decision is reprinted in Appendix A. All further references to the decision in this chapter will be to Appendix A.
60. There is great value, however, in the regular issuance of such patents by the government. For example, the burden of proof would be on the party challenging the patents. See notes 129

Courts have recognized this uncertainty of title.[61] In response, astute surveyors and lawyers representing competing swamp and overflowed or tideland title claimants have conjured various stratagems that attempt to cure or avoid the uncertainty. It is to these arguments that this chapter's attention now turns.

5.4 MEANS TO REMEDY UNCERTAINTIES OF TITLE TO TIDAL MARSHLANDS

5.4.1 Doctrines of Estoppel and Adverse Possession

Those whose long-held tidal marshland titles have been challenged have been understandably upset. After all, the state had already regularly patented tidal marshlands as swamp and overflowed lands. Patent holders or their successors had peaceably occupied and typically improved such lands and faithfully paid property taxes for many years. As a consequence, these landowners have argued that the state should be prevented, "estopped" in legal terminology, from asserting its sovereign public trust title claim to such tidal marshlands.[62] The argument is quite straightforward and on its face appears to be fair and equitable. The courts have not supported it, however, except in extremely limited circumstances.[63]

Some have claimed (and some courts have held) that when the government is asserting public trust rights to these claimed sovereign lands, statutes of limitation recognizing long-unchallenged private possession do not apply nor can the government be prevented, or estopped, from asserting such public trust title claims.[64] Nor has the government's long delay in asserting its claim to these sovereign rights been held as bar to their later assertion.[65] While these rules may appear inequitable to some, their expressed purpose is to

---

through 131 and accompanying text in Chapter 3. As we will see later in this chapter, establishing the physical character of the land as a matter of fact in order to establish its legal charactar as other than swamp and overflowed land is quite difficult. Thus, allocation of the burden of proof is significant.

61. E.g., *Forestier, supra*, 164 Cal. at 32; *People v. California Fish Co., supra*, 166 Cal. at 596.

62. E.g., *City of Long Beach v. Mansell* (1970) 3 Cal.3d 462, 487; *People v. California Fish Co., supra*, 166 Cal. at 602–603.

63. *City of Long Beach v. Mansell, supra*, 3 Cal.3d at 496–497; see notes 211 through 221 and accompanying text in Chapter 1.

64. E.g., *United States v. California* (1947) 332 U.S. 19, 40; *State of California v. United States, supra*, 512 F.Supp. at 40–43; *State v. Superior Court (Fogerty)* (1981) 29 Cal.3d 240, 244–246, cert. den., sub nom. *Lyon v. California* (1981) 454 U.S. 865. A fuller discussion of both the applicability to the government of statutes of limitation and the doctrine of estoppel may be found in notes 199 through 209 and accompanying text in Chapter 1.

65. E.g., *United States v. California, supra*, 332 U.S. at 40; *State of Cal., etc. v. United States, supra*, 512 F.Supp. at 40; *People v. Department of Housing and Community Development* (1975) 45 Cal.App.3d 185, 195; *Cinque Bambini Partnership v. State* (Miss. 1986) 491 So.2d 508, 521, aff'd, subnom. *Phillips Petroleum Co. v. Mississippi* (1987) 484 U.S. 469.

prevent the loss of the public's rights in these valuable lands because of the negligence or sloth of the government and its officials or employees.[66]

There is one type of dispute, however, in which a statute of limitation has been held to prevent the assertion of a sovereign title claim by the states. In the specific case of title and boundary disputes with the United States, Congress has conditioned the availability of federal court jurisdiction for such suits on compliance with a statute of limitations.[67] This statute applies to all those who would bring quiet title actions against the United States, including sovereign states.[68] But although such state claims may be barred by the passage of time, state title is not lost.[69] Title would remain unresolved unless a litigant could induce the United States to file its own action or, possibly, if the United States transferred title to a nongovernmental entity or person.[70] In a very interesting twist, the United States Supreme Court noted that should the United States wish to obtain good title, application of state law would resolve that question.[71] As noted earlier, state adverse possession statutes of limitations do not ordinarily apply when the state is asserting title to public trust lands.[72]

The assertion of the application of the legal rules of estoppel or adverse possession, however, was not the only method by which purchasers of swamp

---

66. *Ibid*. With the exception noted earlier, see notes 196 through 210 in Chapter 1, this policy seems deeply ingrained in American jurisprudence. Cases holding otherwise, *Bryant v. Peppe* (Fla. 1969) 226 So.2d 357, 358, have been overturned, *Bryant v. Peppe* (Fla. 1970) 238 So. 2d. 836, 838.

67. 28 U.S.C. § 2409a. This is part of a waiver of sovereign immunity by the United States under the Quiet Title Act. The Quiet Title Act permits suits against the United States in federal court to hear title disputes in land "in which the United States claims an interest." 28 U.S.C. § 2409a(a). For an interesting case construing this provision, see *State of Alaska v. Babbitt* (9th Cir. 2000) 201 F.3d 1154. In that case, the United States was accused of playing "dog in the manger" by not "claim[ing] an interest" in land where it was apparent if it did the United States would lose title.

68. *Block v. N.D. ex rel. Bd. of University & Sch. Lands* (1983) 461 U.S. 273, 288.

69. The U.S. Supreme Court stated:

[This s]ection . . . does not purport to strip any State, or anyone else for that matter, of any property rights. The statute limits the time in which a quiet title suit against the United States can be filed; but, unlike an adverse possession provision, [the section] does not purport to effectuate a transfer of title. If a claimant has title to a disputed tract of land, he retains title even if his suit to quiet his title is deemed time-barred under [the section].

Block, supra, 461 U.S. at 291.

70. "A dismissal pursuant to [the section] does not quiet title to the property in the United States. The title dispute remains unresolved." *Ibid*.

71. *Id*. at 298, n. 28. ("In many instances, the United States would presumably eventually take the land by adverse possession, but, if so, it would be purely by virtue of State law.")

72. *Ibid*. ("[In the *Block* case the State] asserts that the disputed land is public trust land that cannot ever be taken by adverse possession under [State] law."); see notes 199 through 209 and accompanying text in Chapter 1.

and overflowed lands or their successors in interest attempted to solve title or property boundary questions.

5.4.2 Curative Acts: An End the Uncertainty of Title to Tidal Marshlands?

At least in California, after the land sales programs had largely been completed, the state legislature finally recognized the potential problems it had created. The legislature attempted to correct the claimed deficiencies in the land sales legislation and procedure, after the fact, through a series of what were known as "curative acts." In these acts, the legislature purported to ratify and confirm—"cure"—the supposedly contaminated conveyances resulting from the disposition program.[73]

No attempt will be made to analyze each of the many curative acts. Because all the curative acts are susceptible to the same flaw, to a greater or lesser degree, one example should suffice.

By the terms of one such statute,[74] the federal government's characterization of lands the United States conveyed to the state as swamp and overflowed lands is made conclusive and binding on the state. According to this statute, all that need be shown is that the United States returned[75] the lands as swamp and overflowed and the state issued a patent for the lands.[76] If both prerequisites are met, the lands would be "held to be of the character so returned."[77]

The statute, however, has been claimed to be ambiguous. According to this argument, the term "returned" is left undefined.[78] In addition, the statute does not specify what kind of patent the state must issue. Thus, it is asserted that if the United States returned the lands as swamp and overflowed lands and the state issued a *tideland* patent for such lands, the statute would hold the lands to be swamp and overflowed in legal character, not tidelands.

Moreover, some have asserted that neither the language nor the legislative history of this legislation gave any consideration whatever to the public trust interest in tidelands. If the statute is interpreted as has been suggested, some claim it could result in alienation of tidelands erroneously or improperly char-

---

73. Some of the early curative acts are noted in *People v. California Fish Co.*, supra, 166 Cal. at 590. For a description of some such of acts, see note 49 in this chapter.
74. California Public Resources Code § 7552 provides in pertinent part:

> Lands within this State which are returned by the United States as swamp and overflowed lands, and shown as such on approved township plats, shall, as soon as patents are issued therefor by this State, be held to be of the character so returned.

75. The term "returned" is not defined by the statute. Returned, however, is understood to mean the identification or segregation as public lands of the legal character of swamp and overflowed lands.
76. Cal. Pub. Res. Code § 7552.
77. *Ibid.*
78. See note 75 in this chapter.

acterized as swamp and overflowed lands.[79] Consequently, some contend that these curative acts do not "cure" "infected" conveyances or indisputably extinguish the state's sovereign public trust interest in lands that were tidelands in legal or physical character.[80]

The possible failure of absolute legislative solutions has not daunted riparian property owners. They have taken another tack and developed a legal argument to attempt to restrict the geographic scope of tidelands. The method used has been to narrowly define tidelands as lands underlying *navigable*[81] waters despite the tidality of such waters.

5.4.3 Navigability Versus Tidality—Must the Waters Overlying Tidelands Be Navigable, Not Merely Tidal Waters?

Tidelands are defined, as a matter of federal law,[82] as lands beneath tidal or navigable waters whose landward boundary is the ordinary high-water mark.[83] Some riparian land-owners, however, have argued that no matter the unquestioned tidality of waters overflowing tidal marshes, those tidal waters must be navigable for the underlying marshlands to be held as tidelands.[84] Consequently, followers of this school of thought have contended that there can be no lands of legal character of tidelands within a tidal marsh. They reason, simply as a matter of physical fact, that such a tidal marsh and the tidal distributary waterways within the marsh are not part of navigable waters, even though indisputably overflowed by tides.[85] This contention seeks to limit the

---

79. One trial court has found that this section cannot be so construed. *State of California v. Southern Pacific (Delta Meadows Decision)*, Appendix A, at para. A. 12.–19., pp. 7–9. Recall the rule of statutory construction that holds a statute will not be construed to impair or limit the sovereign power of the State to act for the public unless such an intent clearly appears. E.g., *People v. California Fish Co., supra*, 166 Cal. at 592, 597. See note 162 and accompanying text in Chapter 1.
80. E.g., *Illinois Central Railroad, supra*, 146 U.S. at 453; *People v. California Fish Co., supra*, 166 Cal. at 590–599; *City of Berkeley, supra*, 26 Cal.3d at 525.
81. The concept of navigability is discussed in great detail in Section 7.2 in Chapter 7. It is sufficient for now to say that waters are navigable as a matter of law when such waters are navigable in fact.
82. Determination of the extent of the tidelands obtained by the states as a result of the Equal Footing Doctrine, including the tests for determining what are "navigable waters," is a federal question determined by federal law. *California, ex rel State Lands Comm'n v. U.S.* (1982) 457 U.S. 273, 279; *Borax, supra*, 296 U.S. at 22; *United States v. Utah* (1931) 283 U.S. 64, 75; *United States v. Holt State Bank* (1926) 270 U.S. 49, 55–56.
83. E.g., *Weber, supra*, 85 U.S. at 65; *Knight v. United Land Association* (1891) 142 U.S. 161, 183; *Borax, supra*, 296 U.S. at 15.
84. There is some California authority for this argument. In brief, two lower courts have stated that the tidal test of navigability was not adopted. Rather, these cases state that the test of navigability adopted was whether the waters were navigable in fact. *People ex rel. Baker v. Mack* (1971) 19 Cal.App.3d 1040, 1045, 1048; *Bohn v. Albertson* (1951) 107 Cal.App.2d 738, 747.
85. E.g., Robillard, W., and Bouman, L., Clark on Surveying and Boundaries (7th Ed., 1997), § 25.15 (hereinafter "Clark 7th"); Grimes, *Clark on Surveying and Boundaries* (4th ed., 1976)

geographic extent of sovereign lands to lands underlying the actual navigable channel of the water body, whatever the water body's tidal character.[86] In other words, virtually all tidally flowed marshes and nonnavigable tidal distributaries waterward of the mean high-water line would be disqualified as tidelands in legal character.

A number of opinions from state courts across the country rejected this argument.[87] In fact, the Supreme Court opinions do not appear to limit the definition of navigable waters solely to waters that are navigable in fact. Just the opposite is the case. Supreme Court opinions concerning nontidal but navigable water bodies appear to expand the definition of navigable waters.[88] This apparent expansion of the definition of navigable waters has been reaffirmed by a United States Supreme Court case that arose in Mississippi's Mobile Bay, *Phillips Petroleum Co. v. Mississippi*.[89]

At stake in the *Phillips Petroleum* case were lands underlying a branch of a bayou and 11 small drainage streams.[90] These streams were tributaries and distributaries of a tidal river.[91] No matter that such waters were concededly

---

(herein "Clark 4th") § 596, p. 869; *Baer v. Moran Bros. Co.* (1894) 153 U.S. 287, 288; *Cinque Bambini, supra*, 491 So.2d at 513. In the *Cinque Bambini* case it was argued tidelands were lands that underlay waters that were navigable-in-fact and tidal influence was essentially irrelevant. *Ibid.* This argument was repeated in the United States Supreme Court. *Phillips Petroleum, supra*, 484 U.S. at 477–478.

86. *Phillips Petroleum, supra*, 464 U.S..at 479–480; *Cinque Bambini, supra*, 491 So.2d at 514–515; *City of Newark v. Natural Resource Council* (N.J. 1980) 414 A.2d 1304, 1310, *cert. den.* (1980) 449 U.S. 983.

87. *Wright v. Seymour* (1886) 69 Cal. 122, 124; *Los Angeles v. San Pedro R.R. Co.* (1920) 182 Cal. 652, 658, *cert. den.* (1920) 254 U.S. 636; *White v. State of California, supra*, 21 Cal.App.3d at 742; *City of Newark, supra*, 414 A.2d at 1310–1311; *Trustees of Freeholders, etc. v. Helner* (N.Y. 1975) 375 N.Y.S.2d 761, 770; *Cinque Bambini, supra*, 491 So.2d at 513–516. One of the clearest statements is from New Jersey:

> . . . [T]he navigability and tidal tests are clearly separate criteria. In England all the "land below high-water mark belonged to the British nation, and was vested in the king . . . in trust for the public. . . . Sovereign ownership was based on tidal flow and the sovereign had no title to lands beneath non-tidal rivers. In England, non-tidal watercourses were, with a few small exceptions, non-navigable and the public interest in unobstructed commerce was never threatened by the limiting character of a tidal test. [citations omitted].
>
> In this country, however, many interior rivers are navigable but non-tidal, and the interest in unobstructed commerce was threatened by the possibility of private ownership of these navigable waters. Thus, in the nineteenth century the United States Supreme Court saw fit to create a navigability basis for sovereign ownership *separate from the tidal test* [citation omitted]. In so doing the Court did *not* reject the tidal standard as the basis for sovereign ownership of land beneath tidally-flowed, but non-navigable waters. *City of Newark, supra*, 414 A.2d at 1310–1311 (emphasis supplied).

88. *Shively, supra*, 152 U.S. at 34, 43; *Montana v. United States* (1981) 450 U.S. 554, 551; *Cinque Bambini, supra*, 491 So.2d at 514.
89. (1988) 484 U.S. 469.
90. *Id*. at 472.
91. *Ibid*.

not navigable, the State of Mississippi claimed these tracts as lands underlying tidal waters and issued oil and gas leases for the lands.[92] In upholding Mississippi's title claim, the Supreme Court made three crucial points: (1) The Court reaffirmed its past decisions that land beneath tidal waters came to the states upon their admission.[93] (2) The Supreme Court's earlier decisions upholding state ownership of lands beneath nontidal, but navigable, fresh waters did not have the effect of withdrawing from public trust coverage lands beneath tidal waters.[94] (3) The concession that the states owned tidelands bordering on oceans, bays, and estuaries, even when these tideland areas were hardly navigable, was a recognition that these waters were part of the sea and was the reason the tidal test had long been the measure of sovereign title.[95] Consequently, since tidal marshlands were subject to the ebb and flow of the tide and lie waterward of the ordinary high-water mark, they were tidelands in legal character regardless of whether it was physically possible to navigate in the waters covering them.[96]

Thus, remedies by landowners attempting to cure or avoid the potential uncertainty of title to or the boundaries of tidal marshes have not been accepted without serious challenge. Some of these remedies have not survived the challenge. One final method of attempting to end title uncertainties has not yet been tested in the crucible of litigation, but is worthy of discussion.

5.4.4 The Impact of Spanish/Mexican Land Grants on the Location of the Ordinary High-Water Mark[97]

In California, as in most of the South and the Southwest, French, Spanish, and Mexican sovereigns preceded the United States.[98] These sovereigns all

---

92. *Ibid.*
93. *Ibid.*
94. *Id.* at 479–480. In fact, after the Supreme Court listed some cases in which it had recognized the public trust doctrine and the diverse nature of public trust uses, it stated: "It would be odd to acknowledge such diverse uses of public trust tidelands, and then suggest that the sole measure of the expanse of such lands is the navigability of the waters over them." *Id.* at 476.
95. *Id.* at 480.
96. *Id.* at 476. The dissent argued, in substance, that the public trust should extend only to navigable waterways because ". . . its fundamental purpose is to preserve them for common use for transportation." *Id.* at 488 (O'Connor, J., dissenting).
97. Land title and property boundary determination are not solely technical matters requiring rigid and reflexive application of surveying methods to physically locate monuments or features. The following is exemplary of the kind of argument that one may develop if one understands the interrelationship of law, land title, history, and surveying. An understanding of each or all may be necessary to construct a title or boundary opinion. Using historic factual, surveying, and legal authority could obviate the need to retain expert witnesses or undertake the physical testing necessary to reestablish the physical character of tidal marsh in the mid-nineteenth century. Exploration of such authority may not always be determinative. Such exploration will usually, if not in all cases, provide sound guidance in constructing and defending an opinion on the geographic location of the property boundary or the extent and character of title to property. In addition, the discussion that follows also reemphasizes the impact of the land title system on property boundary location.
98. See notes 18 through 20 and accompanying text in Chapter 1.

made grants to their subjects to reward service or to encourage settlement. When the United States gained additional territory settled and occupied under the authority of prior sovereigns, international law or treaty required the United States to recognize and confirm lawful grants made by those former sovereigns to their former subjects.[99] These confirmed grants were then located according to the laws of the United States and thereby separated from the public domain available for disposition.[100]

An argument has been advanced that relies on the intent and effect of this process. It is contended that in certain instances, land grant confirmation proceedings concerning Mexican and Spanish grants bounded by a navigable water body[101] determined, as a matter of law, the geographic location of the ordinary high-water mark and, therefore, the waterward boundary of any potential swamp and overflowed lands.[102] The argument is as follows:

Many Spanish and Mexican land grants in California were geographically located along the seacoast, estuaries, or rivers.[103] Restricting our focus to estuaries and bays, such land grants usually described the grant's bayward boundary as "the bay" without any further particularized description. The confirmation decree issued by the United States, because of the land grant confirmation proceeding, adopted the description of the boundaries as stated in the original grant. In other words, if the original grant described its seaward boundary as the "bay," that bay boundary was confirmed by decree of the United States.

The boundaries of the confirmed grant were required to be located in conformity with that decree.[104] By following the decree's mandate, the United

---

99. For a detailed explanation of the land grant confirmation process, see text accompanying notes 21 through 41 in Chapter 1.
100. E.g., *Fremont v. United States* (1851) 58 U.S. (17 How.) 541, 564–565; *Waterman v. Smith* (1859) 13 Cal. 373, 410–411.
101. The descriptions of the geographic boundaries of the area conveyed by these land grants were often imprecise. *Fremont, supra,* 58 U.S. at 558. Under Mexican law there was a procedure, juridical possession, to make the land grant's boundaries more precise and to locate these boundaries geographically by reference to distinct and agreed-upon physical features. E.g., *Waterman, supra,* 13 Cal. at 410–411. For an example of an actual Mexican land grant and the proceedings surrounding its confirmation, one ready source of reference is Uzes, François, *Chaining the Land* (Landmark, 1977), Appendix K. The juridical possession proceedings are described at *id.* at 283–286. The bayward boundary of the confirmed grant described by Uzes was the "Bay of San Francisco." *Id.* at 286.
102. This argument could have substantial practical consequences. At least in California, Spanish and Mexican land grants form the basis of title for most lands surrounding San Francisco Bay, the southern California coastline, and many other equally valuable properties along the California river system. Proof of such titles may be made through authenticated copies of the grant. Cal. Evid. Code section 1605.
103. E.g., *City of Los Angeles v. Venice Peninsula Properties* (1982) 31 Cal.3d 288, 294, *ovrld on other grnds, sub. nom., Summa Corp. v. California ex rel. State Lands Com'n* (1984) 466 U.S. 198 (seacoast); *United States v. Pacheco, supra,* 69 U.S. at 589 (estuary); *Packer v. Bird* (1891) 137 U.S. 661, 663 (river).
104. E.g., *United States v. Halleck* (1863) 68 U.S. (1 Wall.) 439, 455; *Teschemacher v. Thompson* (1861) 18 Cal. 11, 25.

States honored its obligation to recognize prior sovereign grants. In addition, the United States was required to accurately locate those grants for purely selfish motives. The grants had to be precisely located in order to reliably separate lands conveyed by the prior sovereign's land grants from the public domain remaining available for disposition by the United States. This public domain included those portions of the public domain that were swamp and overflowed in character and had been granted to the states.[105]

The argument goes on to state that cases examining Mexican land grant boundaries in California have affirmed the rule that when the land grant decree of confirmation described the waterward property boundary as the "bay" or used a term describing a tidal or navigable watercouse, under the common law, the property boundary was to be located at the ordinary high-water mark of that bay or watercourse.[106] One case discussed proper location of a bay property boundary when the confirmed land grant was bounded by tidal marshlands immediately adjacent to that bay. The opinion also discussed whether the United States' surveyed "bay" property boundary was properly located. The physical location of the land grant's bay boundary excluded such tidal marshlands from incorporation into the United States' confirmatory patent for the land grant.

In that case, the confirmation decree established the bayward boundary of the rancho grant as the "Bay of San Francisco."[107] An objection to the later United States' survey of the land described in the decree claimed that adjacent tidal marshland should have been, but was not, included by the survey within the boundaries of the rancho.[108] That is, the rancho boundary should have been physically located at "the line of low water mark" to include such lands.[109] The Court rejected this contention:

> By the common law, the shore of sea, and, of course, of arms of the sea, is the land between ordinary high and low-water mark, the land over which the daily tides ebb and flow. When, therefore, the sea, or a bay, is named as a boundary,

---

105. *United States v. O'Donnell, supra,* 303 U.S. at 512. In *O'Donnell,* issuance of a United States' swamp and overflowed patent to the state had been confirmed as a result of federal administrative hearings and federal court decisions. *Work v. United States, supra,* 23 F.2d 136. The Supreme Court held that a competing Mexican land grant, confirmed for the same lands, took precedence over the later United States' swamp and overflowed patent to the state, although the United States swamp and overflowed grant had been court-affirmed. *United States v. O'Donnell, supra,* 303 U.S. at 513.
106. E.g., *Teschemacher, supra,* 18 Cal. at 20–22; *DeGuyer v. Banning* (1897) 167 U.S. 723, 735–736; *Los Angeles v. San Pedro etc. R.R. Co., supra,* 182 Cal. at 654–655; *Stewart v. United States* (1942) 316 U.S. 354, 359; *United States v. O'Donnell, supra,* 303 U.S. at 509; *Peterson v. United States* (9th Cir. 1964) 327 F.2d 219, 221.
107. *United States v. Pacheco, supra,* 69 U.S. at 588. Mexican or Spanish land grants were commonly known as "rancho" grants.
108. The lands the survey omitted from the confirmed grant were described as tidal marshlands covered "by the monthly tides at the new and full moon" and partly "covered by the daily tides." *Id.* at 588–589.
109. *Id.* at 590.

the line of ordinary high water mark is always intended where the common law prevails.[110]

Thus, it is claimed that the United States Supreme Court appeared to approve a survey of a Mexican land grant's Bay of San Francisco boundary that had physically located the ordinary high-water mark property boundary landward of tidal marshlands, which were only "partly covered by the daily tides." The asserted importance of this case is the holding that the bayward property boundary of Mexican land grants adjacent to a tidal water body was located as matter of the common law at the ordinary high-water mark. Consequently, by definition, all swamp and overflowed lands, if any, would have been included within the perimeter of the land grant; all lands lying bayward of that surveyed ordinary high-water mark boundary were tidelands in character as a matter of law.[111]

On the other hand, some may assert that Mexican land grants located as the result of confirmation proceedings, were to be located according to *Mexican* law, *not* under the law of the United States. According to this argument, in the case of a grant with a confirmed boundary calling to the "bay," that boundary should have been located by Mexican law at the line of highest tides rather than at the ordinary high-water mark.[112] The consequence of this argument is that swamp and overflowed lands would not have been included within the confirmed land grants, as bay property boundaries of such land grants would have been located at the line of highest high tides. This dispute remains unresolved.

There are those who contend that if the United States placed the land grant's bay boundary in accordance with Mexican law, at the line of highest high tides the land grant, whose confirmed bay or seaward boundary was the "bay" or the "sea" or other similar term denoting a shoreline boundary, would have lost its riparian character.[113] A strip of public domain land would

---

110. *Ibid.*
111. Further, even if the land grant's waterward property boundary was incorrectly geographically located so that, for example, the grant included tidelands, once the United States approved the survey and the patent was issued, it was too late to challenge the resulting conveyance.

> The question whether there was such a prior grant and what were its boundaries were questions that had to be decided in the proceedings for confirmation and there was jurisdiction to decide them as well if the decision was wrong as if it was right. The title of California [to the tidelands] was in abeyance until those issues were determined. . . . [A]lthough it well may be that . . . the grant could have been construed more narrowly, that was a matter to be passed upon and when the decree and patent went in favor of the grantee it is too late to argue that they are not conclusive. . . . *United States v. Coronado Beach Co.* (1921) 255 U.S. 472, 487–488

112. Cf. *Stewart, supra,* 316 U.S. at 359.
113. Under Mexican law, as under the common law, the shore was held in trust for the benefit of the public and burdened with public uses. See notes 19 through 20 and accompanying text in this chapter; *United States v. Stewart* (9th Cir. 1941) 121 F.2d 705, 710, *aff'd, sub nom. Stewart*

5.4 MEANS TO REMEDY UNCERTAINTIES OF TITLE TO TIDAL MARSHLANDS 167

be interposed bayward of the line of highest high tides, the Mexican law shoreline property boundary, and landward of the line of ordinary high tides, the common law shoreline property boundary. This strip of public land would have been available for ultimate disposition to private persons as either swamp and overflowed lands or under another public land disposition program. In such a case, in contravention of the international law and the treaty obligations of the United States, an intervening private title would have destroyed the confirmed Mexican grant's riparian character,[114] an attribute that international law and the Treaty of Guadalupe Hidalgo constrained the United States to protect.[115]

It is argued that the United States seems to have recognized the importance of maintaining the riparian character of these land grants.[116] According to the argument, the United States' location of the property boundary of a rancho grant calling to the "bay" at the ordinary high-water mark under the common law rule accomplished two important goals: It prevented any intervening private title between the riparian land grant and the boundary watercourse. And it did not interfere with, but instead continued, the protection of the riparian character of the grants to the same extent as under Mexican law.[117] Therefore, it may be asserted, as a matter of law, that no swamp and overflowed lands existed between the confirmed land grants and the adjacent tidal water bodies. If this argument were adopted, any tidal marshlands bayward of the ordinary high-water mark property boundary would be tidelands in character as matter of law.[118]

---

v. United States (1942) 316 U.S. 354; *Lux v. Haggin* (1886) 69 Cal. 255, 316; Las Sieta Partidas (Scott, ed. CCH 1931), Partidas 3, Title 28, Law 3; Althaus, Helen F., "Public Trust Rights" (U.S. Fish and Wildlife Service 1978), pp. 16–23. Thus, grants calling to the bay as a boundary were assured of their riparian rights under Mexican law. *Stewart v. United States, supra,* 316 U.S. at 359; see *Lux, supra,* 69 Cal. at 313–314.

114. See *Aptos Seascape Corp. v. County of Santa Cruz* (1982) 138 Cal.App.3d 484, 503, *app. dismd* (1983) 464 U.S. 805; 2 Ogden's Revised California Real Property Law (Cont. Ed. Bar 1975) §§ 26.2–26.3, pp. 1238–1239.

115. E.g., *Leese v. Clark* (1862) 18 Cal. 535, 572–573; *Teschemacher, supra,* 18 Cal. at 23–24; *Beard v. Federy* (1865) 70 U.S. (3 Wall.) 479, 491–492; *Soulard v. United States* (1830) 30 U.S. (4 Pet.) 511, 511–512.

116. For example, United States Deputy Surveyors charged with surveying confirmed land grants were instructed that when the surveyor came upon swamp and overflowed lands adjacent to tidewaters, the surveyor was to locate the line of ordinary high tides. General Instructions from the U.S. Surveyor General for California to the Deputy Surveyors Engaged in Surveying the Finally Confirmed Land Claims in This State (1858) as reprinted in Uzes, *supra,* at Appendix C, p. 212.

117. "This [the argument as set forth in notes 107 through 111 and accompanying text in this chapter] would appear to be a complete answer to those who claim that as Mexican grants bordering on the sea extended only to the highest high tide line there must be a strip of public land between that line and the line of ordinary high water which could be open to entry." Outline of Land Titles (CLTA) § 15, p. 10; 2 Ogden, *supra,* § 26.3, p. 1239.

118. As earlier noted, see note 106 in this chapter, the Mexican land grant confirmatory proceedings are now final and cannot be reopened. Courts have applied this rule evenhandedly. Lands,

This argument has not been accepted by any court at either trial or appellate level. In any event, the argument is inapplicable where confirmed land grants do not bound tidal marsh areas. Thus, the question remains, how can the ordinary high water mark boundary between tidelands and swamp and overflowed lands be located? A beginning point is to figure out what are the physical indicia of the ordinary high-water mark property boundary in tidal marshes. Are they any different from that along the open coast?

5.5 MEAN HIGH-WATER LINE—THE PHYSICAL LOCATION OF THE ORDINARY HIGH-WATER MARK PROPERTY BOUNDARY IN TIDAL MARSHES

It has already been established that courts have adopted the mean high-water line as the physical location of the ordinary high-water mark property boundary in tidal areas.[119] Indeed, the case that is the main authority for that conclusion concerned an estuary in tidal mud flats.[120]

The reader must be wondering whether the factual investigation and development necessary to establish the location of a historic mean high-water line is not just another technical and sophisticated (and somewhat glorified) version of Monday morning quarterbacking. After all, the adoption of the mean high-water line as the physical location of the ordinary high-water mark was only completely endorsed in the mid-1930s. For the majority of properties, its adoption was decades after the property boundary surveys had been undertaken that led to issuance of the questioned conveyances of tidal marshlands. Simply put, who cares where the mean high-water line is currently located or was located in the 1850s? Isn't the location of the mean high-water line simply irrelevant? Don't the swamp and overflowed or tideland surveys by state-authorized surveyors tell us all we need to know about the location of the property boundary?

It has been argued that approving the surveying methods and practices of the mid-nineteenth century, accomplished for purely private interests during a period that many have claimed was rife with shameful fraud and corruption, is in contravention of public policy.[121] A half century after the issuance of

---

indisputably tidelands in character, have been included within confirmed land grants and their inclusion within those grants has survived later challenges by both federal and state officials. E.g., *United States v. Coronado Beach, supra*, 255 U.S. at 487–488; *Summa, supra*, 466 U.S. at 208–209 (failure of state to file timely claim in confirmation proceeding bars later claim of public trust interest in tidelands asserted to be included within boundaries of confirmed land grant).

119. How the mean high-water line was developed and came to be accepted as the physical location of the ordinary high-water mark in tidal regimes is found in Sections 4.2 through 4.11 in Chapter 4.

120. *Borax, supra*, 296 U.S. at 12; *City of Los Angeles v. Borax Consolidated Limited* (9th Cir. 1935) 74 F.2d 901, 903, *aff'd* (1935) 296 U.S. 10.

121. *City of Berkeley, supra*, 26 Cal.3d at 522; *Hellman v. City of Los Angeles* (1899) 125 Cal. 383, 387 (judicial notice of the inaccuracy of early surveys).

conveyances based on such surveys, the California Supreme Court found that the surveying methods, procedures, and practices used to describe and patent swamp and overflowed lands and tidelands had not considered the public's interest in tidelands contained in tidal marshes.[122] At least in California, when sovereign trust resources are concerned, there appears to be continuing authority to reexamine past activities in light of modern practices.[123] Moreover, some contend that the United States Supreme Court rejected other suggested tests[124] and reaffirmed use of the mean high-water line as the measurement of the landward boundary of tidelands.[125]

Thus, there is a strong argument supporting the use of the mean high-water line as the physical location of the ordinary high-water mark property boundary along tidal water bodies.

Merely agreeing on the mean high-water line as evidence of the physical location of the ordinary high-water mark property boundary, however, does

---

122. *Forestier, supra,* 164 Cal. at 33; *People v. California Fish Co., supra,* 166 Cal. at 591–592; see notes 49, 53, and 55 and accompanying text—in this chapter.

123. E.g., *National Audubon Society v. Superior Court* (1983) 33 Cal.3d 419, 425-426, *cert. den., sub nom. Los Angeles Dept. of Water & Power v. National Audubon Soc.* (1983) 464 U.S. 977. In that case the California Supreme Court held that a water diversion permit could be reexamined for its consistency with present public trust values 40 years after its issuance and the expenditure of hundreds of millions of dollars for the permitted water diversion facilities. The court made this decision even in light of an administrative finding, made at the time the state issued the permit, that there was *no* authority to consider public trust values as part of the permit process. Given the *National Audubon* and other opinions, it is little wonder private property owners along the riparian or littoral shorelines are extremely wary of the existence of questions concerning their property boundaries and title and the potential applicability of the public trust easement to their property.

124. In *Phillips Petroleum* one party argued that adoption of the mean high-water line as the physical location of the ordinary high-water mark was to be ignored. *Id.* at 481, n.9. Instead, this party suggested that the headland-to-headland rule should be used to divide lands subject to the public trust from those lands not so subject relying on *Knight v. United Land Association* (1891) 142 U.S. 161, 207–208 (Field, J., concurring). The "headland-to-headland" rule provides that when geographically delimiting the extent of a bay, a line was to be drawn across the mouths of distributary or tributary creeks, sloughs, or rivers. In *Knight,* the United States Supreme Court approved a survey of a Mexican grant for the pueblo of San Francisco; the survey followed the ordinary high-water mark of San Francisco Bay, but crossed and did not enter a tidal distributary creek of that bay. *Knight, supra,* 142 U.S. at 164–165. Thus, application of this rule would exclude tidal marshes and the beds of tidal distributary or tributary creeks as lands of the legal character of tidelands. The Brief of the American Land Title Association as Amicus Curiae in Support of Petitioners stated: "We might add that the difficulties of distinguishing between non-navigable "edges" of a navigable waterway and wholly non-navigable "separate" waterbody are much the same in each case and afford no argument for ignoring the difference when tidal waters are involved. Moreover, the method for drawing the dividing line between two distinct bodies of tidewater is well established."

125. *Phillips Petroleum Co., supra,* 484 U.S. at 481. On the other hand, some contend that Justice O'Connor, in her dissent in the *Phillips Petroleum* case, recognized that the mean high-water line is the appropriate measure of the extent of tidelands. *Phillips Petroleum, supra,* 484 U.S. at 489 (O'Connor, J., dissenting). Other claimed difficulties in the headland-to-headland rule are described in notes 69 through 71 and accompanying text in Chapter 6.

not tell us where that property boundary should be located in a tidal marsh regime. To guide our determination of that location, we first need to explore the application of the principles of property boundary movement to the tidal marsh regime.

5.6 LEGAL CHARACTER OF TIDAL MARSHLANDS DETERMINED AT THE TIME OF THE SWAMP LANDS ACT GRANT; CUSTOMARY PROPERTY BOUNDARY PRINCIPLES DETERMINE EFFECT OF CHANGE IN PHYSICAL LOCATION OF THE ORDINARY HIGH-WATER MARK

Most swamp land grants were made in the mid-nineteenth century; in California, the grant was made on September 28, 1850, 19 days after California was admitted into the Union. Because the character of the land encompassed within the swamp land grant must be shown at the time of that grant, it is necessary to prove the physical and legal character of the tidal marshlands in the 1850s.[126] This is a daunting task, especially when coupled with the technical difficulties in physically locating the mean high-water line boundary between tidelands and swamp and overflowed lands in a tidal marsh.

Notwithstanding that the physical and legal character of the tidal marshland is required to be established at the date of the Swamp Lands Act, the property boundary is not fixed in that geographic location. Later changes in the tidal marsh regime may alter the physical location of the ordinary high-water mark property boundary. Such a change in physical location could affect both the property boundary location and how one may prove property boundary location. Consequently, even assuming that the physical location of the ordinary high-water mark in the 1850s (or at another appropriate or relevant historical date) can be established, that physical location is dynamic. The question is: Does the after-1850 movement in physical location affect the property boundary? To answer that question, this chapter briefly explores the nature of tidal marsh regimes.

5.7 THE TIDAL MARSH REGIME AND CHANGES TO THAT REGIME

Tidal marsh origin, formation, and maintenance are closely related to the tidal regime of the estuary and the amount of sediment contributed to and transported and later deposited by the tidal waters.[127] Focusing first on the tidal

---

126. See notes 44 and 45 and accompanying text in this chapter.
127. E.g., Atwater: Tidal Marshes, *supra*, at 348–349, 357–362; Pestrong, "Tidal-Flat Sedimentation at Cooley Landing, Southwest San Francisco Bay" in 8 Sedimentary Geology (Elseweir 1972) p. 251; Pestrong, *The Development of Drainage Patterns on Tidal Marshes* (Stanford University Earth Sciences 1965) Vol. 4, No. 2.; Pestrong, "San Francisco Bay Tidelands" in California Geology (1972) Vol. 25, No. 2, p. 27.

regime, in many of today's estuaries there were once considerably larger areas of tidal marshlands and many more tidal distributary channels.[128] Before the settlement and development of such areas, these marshes and tidal channels had been available to receive and be nourished by the tidal waters of the bay. Through a system of dikes, drains, levees, weirs, dams, or other flood control and reclamation works, these tidal marshes were reclaimed and tidal channels cut off and removed from tide flow.[129] Reclamation affected the tidal prism.[130] This in turn affected the range of tides.[131] The tidal marsh regime and the physical character of remaining tidal marshes, particularly their relationship to mean high water, may well have been affected by these changes.[132]

Another type of change that may have affected location of the mean high-water line in a tidal marsh regime concerns the second element of the tidal marsh regime: sedimentation. To take just one example, in the case of the San Francisco Bay system, there was a marked increase in sedimentation because of hydraulic mining.[133] From the inception of hydraulic gold mining activities in the Sierra Nevada until they were stopped by court order in the 1880s, this method of mining resulted in the deposit of an almost incomprehensible amount of debris into the San Francisco and San Pablo Bays.[134] These activities were human-caused, and their effects on the tidal marsh regime were interrelated.[135] What effect, if any, did these activities have on the physical location of the ordinary high-water mark property boundary in the tidal marshes of the San Francisco Bay system?

---

128. See Figure 5.1.
129. Atwater: Tidal Marshes, *supra,* at 352–354; Gilbert, *supra,* at 85–88.
130. The tidal prism is the measure of volume of the tide. It can be conceived of as the volume of water between the plane of high tide and the plane of low tide within the bay. Gilbert, *supra,* at 71.
131. Gilbert, *supra,* at 85, 88.
132. For a discussion of the effects of some of these human-induced activities on marsh formation, see Atwater: Tidal Marshes, *supra,* at 352–359.
133. Hydraulic mining is defined and described in *Woodruff v. North Bloomfield Gravel Min. Co.* (9th Cir. 1884) 18 F. 753, 756; see note 28 and accompanying text in Chapter 6.
134. An account of this environmental tragedy is masterfully chronicled from a scientific standpoint in Gilbert. Then-contemporary cases describing the impacts of this process on the streams and rivers of Northern California are *Woodruff, supra,* 18 F. at 756–768; *People v. Gold Run D. & M. Co.* (1884) 66 Cal. 138, 144–145. See note 29 and accompanying text in Chapter 6. A more current discussion is found in *State of Cal. ex rel. State Lands Com. v. Superior Court (Lovelace)* (1995) 11 Cal.4th 50, 76–77.
135. Gilbert wrote in his monumental study:

> Through the interlocking of subjects my attention has been drawn to matters apparently remote from the problems of mining debris. Mining debris merged, both bodily and in its effects, with debris sent to the streams by agriculture and other industries; the aggravation of valley floods due to the clogging of channels by debris was inseparable from the aggravation due to the exclusions of floods from lands reclaimed for agriculture; the weakening of tidal currents at the Golden Gate by the deposition of debris in the bays is inseparable from the weakening by the reclamation of tide lands. . . . Gilbert, *supra,* at p. 7.

5.8 PROPERTY BOUNDARY EFFECT OF CHANGES IN THE TIDE MARSH REGIME

Assume that the tidal marshland and the physical regime that created and maintained the tidal marsh has not been disturbed or altered by human activity. In that case, under accepted boundary determination principles, the ordinary high-water mark property boundary would be physically located at the currently existing line of mean high water as it has been modified by the processes of accretion, erosion, reliction, and submergence.[136] Thus, by way of example, a tidal marsh determined to be swamp and overflowed in legal character in 1850 because of the marsh's then-extant physical position landward of the geographic location of the ordinary high-water mark could later be found to be tidelands in fact and in legal character by virtue of the physical movement of that property boundary.

5.8.1 Effect on the Property Boundary If the Activities Took Place on the Tidal Marsh

As we have seen, in the real world, finding a tidal marsh unaltered by humans is difficult.[137] Do human-instigated changes to the tidal marsh regime and to the physical location of the mean high-water line affect the geographic location of the legal property boundary between swamp and overflowed lands and tidelands?

If one makes certain assumptions, this question does not seem as difficult to answer as it may first appear. By way of example, assume that a party can establish that human activities were undertaken or accomplished on the tidal marshland itself.[138] In such a case, the answer to the question of the physical location of the property boundary appears to be relatively straightforward.

---

136. See notes 106 through 122 and accompanying text in Chapter 3; Cal. Civ. Code § 1014; *City of Oakland v. Buteau* (1919) 180 Cal. 83, 87. Proof of the location of the mean high-water line in a tidal marsh regime is generally discussed in Section 5.9 and in greater technical detail in Sections 6.8 through 6.9 in Chapter 6.

137. *State of Cal.ex rel. State Lands Com. v. Superior Court (Lovelace)*, *supra*, 11 Cal.4th at 78 ("[N]umerous human activities have indirectly as well as directly caused a change in water flow and the accumulation of sediment in virtually every river and tideland in the state").

138. Such a showing will require the testimony of expert witnesses, such as historians, geographers, or historical geographers. Such experts may have examined documents, records, or other available information that recount when and by what means, method, or activity humans changed the landform's physical character. This information may not be available in all cases. Even if available, the information may not be specifically focused on the land in dispute. Except in extremely remote areas, however, some information is usually available and in some detail. For example, then-contemporaneous accounts of the historical physical characteristics or attributes of an area are even found in great literature. Richard Henry Dana's *Two Years Before the Mast* has detailed accounts of the flora and fauna of the unspoiled California coast at San Diego, Santa

5.8 PROPERTY BOUNDARY EFFECT OF CHANGES IN THE TIDE MARSH REGIME 173

Thus, if the tidal marsh itself was diked, leveed, or warped,[139] there would appear to be a persuasive argument that an avulsive change had occurred at the time humans accomplished such activity.[140] For example, when finished, a dike would suddenly exclude the tidal waters. By way of further example, in the process of warping, sediment-carrying fresh water would flood the marsh. Both human-instigated actions would perceptibly change the marsh's character and relative elevation. Thus, the physical character of the tidal marsh would arguably have been immediately and perceptibly changed.[141] As a consequence of such a change, one might reasonably conclude that the physical location of the boundary between tidelands and swamp and overflowed lands should be fixed in its immediately preoccurrence physical location.[142]

Suppose, for example, that in conjunction with the construction of a regional harbor and transportation center, a mole[143] was built to accommodate

---

Barbara, Monterey, and San Francisco. For examples of the type of materials that have been used in a litigated case, see Chapter 6, Section 6.9.8.

139. "Warping" is a term of art. It does not mean the tidal marsh' psyche was altered for the worse. Technically, the term "warping" describes the method of reclaiming tidal marshlands by leveeing the marsh, then directing sediment-laden freshwater sources over the land, allowing sediment to be deposited and salts to be leached out. See *Peabody v. City of Vallejo* (1935) 2 Cal.2d 351, 362, 369. For example, a then-contemporary account of the process stated:

> Besides the natural causes tending to make the stream change its course, various other causes as damming or ditching, either openly or covertly by people, some of whom wish the stream to overflow the marshes, freshening them and depositing sediment on which farm produce is raised. Report of Lt. N.A. McCully, USN, accompanying Descriptive Report H-2304 (USC&GS 1897).

140. An avulsive change is often defined as one that is sudden and perceptible. See notes 103 through 105 and accompanying text in Chapter 3.
141. *Garrett v. State* (N.J. 1972) 289 A.2d 542, 547 (artificial avulsion of tidal creek); *Cinque Bambini, supra,* 491 So.2d at 520 (artificial change in boundary by dredging treated as an avulsion); *Georgia v. South Carolina* (1990) 497 U.S. 376, 404 (construction of training wall by U.S. Army Corps of Engineers found to be primarily avulsive in character because of rapidity of change).
142. See note 141 in this chapter; *Arkansas v. Tennessee* (1918) 246 U.S. 158, 173; *City of Long Beach v. Mansell, supra,* 3 Cal.3d at 469, n.4; see *Beach Colony II v. California Coastal Com.* (1986) 151 Cal.App.3d 1107, 1116; cf. Cal. Civ. Code § 1015. The result may be the same if courts treat such activity under California law as an artificial change. See *State of Cal.ex rel. State Lands Com. v. Superior Court (Lovelace), supra,* 11 Cal.4th at 79–80; see notes 143 through 157 in this chapter. The property boundary effect of these human activities may depend on the identity of the parties to the subsequent boundary dispute. If the contest over accretions to a tidal marsh is between two nonsovereign (private) parties, there may be a different result even when the accretion is artificially caused. E.g., *Forgeus v. County of Santa Cruz* (1914) 24 Cal.App. 193, 199–200; *United States v. Harvey* (9th Cir. 1981) 661 F.2d 767, 772, n.7, *cert. den.* (1982) 459 U.S. 833. Courts have not applied the artificial accretion rule in favor of private litigants. *State of Cal. ex rel. State Lands Com. (Lovelace), supra,* 11 Cal.4th at 73, n.4. California's rule is not followed by a majority of states nor by the federal courts. See note 117 and accompanying text in Chapter 3.
143. A mole is a massive structure, usually of stone, that is set up in the water to act as a breakwater or pier. The American College Dictionary (Random House 1948), p. 782. Such a mole

railroad and ferry transportation. Such a large work could well change the marsh/tide/sediment regime. For example, by changing the tidal circulation pattern, the mole might cause either a gradual increase or decrease in the accumulation of sediment to marshes both close to the project and some distance away. What is the property boundary effect of such human activities that were part of a regionwide series of changes, remote in location from the tidal salt marsh?

5.8.2 Effect on the Physical Location of the Property Boundary of Activities Not Directly on the Tidal Marsh

Continuing to focus on California, the usual rule is that the processes of accretion and reliction serve to move the physical location of the riparian or littoral property boundary only if the change in physical location of that boundary is the result of natural causes.[144] That is, if the mean high-water line in the tidal marsh changes its physical location by the process of gradual and imperceptible deposition or erosion of alluvion (accretion or reliction) in the natural course of events, the physical location of the ordinary high-water mark property boundary would be the present line of mean high water.[145] The same result does not follow, however, for artificially induced erosion or submergence.

California Civil Code section 1014 addresses only the natural processes of accretion and reliction. A strict reading of this statute may lead one to the conclusion that the processes of erosion, and submergence are not encompassed within its scope. Therefore, in California, as in the majority of states, when the property boundary changes its physical location through the process of erosion or submergence, no matter what the cause of the change in physical location, there is support for the argument that the property boundary continues to move landward; the property boundary's physical location is not fixed as in the case of "artificial" accretion or reliction.[146]

---

was constructed in the Oakland, California, harbor as the terminus for railroad traffic bound for and the embarkation point for ferry service to San Francisco.

144. E.g., Cal. Civ. Code § 1014; *State of Cal. ex rel. State Lands Com. v. Superior Court (Lovelace), supra,* 11 Cal.4th at 66, *Carpenter v. City of Santa Monica* (1944) 63 Cal.App.2d 772, 794.

145. Such rules apply no matter what the regime in which the marsh may be found. *Honsinger v. State* (Ala. 1982) 642 P .2d 1352, discussed the property boundary effect of the process of glacio-isostatic uplift. Galcio-isostatic uplift is the gradual rise of the earth's crust that occurs when the downward pressure exerted by a glacial ice mass lessens. This process occurred and exposed some tidal marshland. Both littoral owners and the state claimed the land. *Id.* at 1353. The state insisted that the process of glacio-isostatic uplift was not a reliction. It argued that the legal concept of reliction referred only to a situation where the water itself receded. *Id.* at 1354. The court held that reliction encompassed the emergence of existing soil either through recession of water or through a rise of the bed. *Ibid.* Therefore, the recession caused by glacio-isostatic uplift was a reliction and the littoral owners owned the exposed land. *Ibid.*

146. See, e.g., *State v. Superior Court (Fogerty), supra,* 29 Cal.3d at 247–249 (lake); *Miramar Co. v. City of Santa Barbara* (1943) 23 Cal.2d 170, 175 (ocean beach); *County of St.Clair v.*

In contradistinction, if gradual and imperceptible deposition or erosion of alluvium results directly from human-induced events or human-created works, California law treats the property boundary effect as if it were an avulsion; the ordinary high-water mark property boundary is fixed in location.[147] The question is, What does "directly" mean in the case of human-instigated activity not occurring on the property itself?

Application of this rule in the case of regionwide changes, some of which are natural and others of which are human-induced, has been simplified.[148] The California Supreme Court limited the scope of California's artificial accretion rule.[149] In fact, the issue arose in connection with claims regarding the effect of hydraulic mining on the ordinary high-water mark property boundary of a parcel along a river. The court found that the connection between hydraulic mining in the mountains and the change in physical location of the river parcel's property boundary many miles away ("alluvium placed in the river by hydraulic mining which eventually collects downstream") was too "attenuated to render the accretion artificial under California's rule."[150] The court expressed two concerns with the application of the artificial accretion rule in these circumstances: First, given the impact of human activity in California, this exception would swallow the rule that accretions go to the upland owner[151] and, second, there would be significant problems of

---

Lovingston (1874) 90 U.S. (23 Wall.) 46, 68 (river); 78 Am.Jur.2d Waters § 410 (1975); Annot. (1975) 63 A.L.R.3d 249. The *Fogerty* case concerned a lake regulated by dams. A substantial reason for adoption of the rule that the current physical location is the location of the property boundary was what was characterized by the court as a "monumental evidentiary problem" in reconstructing former lake levels. *State v. Superior Court (Fogerty), supra,* 29 Cal.3d at 248. This problem of proof provided a "convincing justification" for accepting the current physical location as the property boundary. *Ibid.* Such difficulties in proof would not appear to be any easier in a tidal estuary where many interrelated changes occurred over a long period throughout the region and in contributory, but distant, drainages. See Section 5.7.

147. *State of Cal. ex rel. State Lands Com. (Lovelace), supra,* 11 Cal.4th at 66–73 (discussion of cases); *Carpenter, supra,* 63 Cal.App.2d at 794; Cal. Civ. Code § 1014; see note 142 in this chapter, above.

148. Some parties had suggested that the courts should first determine whether the artificial or natural causes were controlling in each particular instance. *Alexander Hamilton Life Ins. Co. v. Govt of V.I.* (3rd Cir. 1985) 757 F.2d 534, 547 n.13 ("The . . . court might find it impossible to subdivide the [land in dispute] according to how various parts of it came to be uplands because of the scarcity of available evidence. If this is the case, the court should determine the *primary* cause for the plot's existence—natural accretion or artificial fill—and quiet title accordingly.")(Emphasis in original.) See *Georgia v. South Carolina, supra,* 497 U.S. at 404 (changes primarily avulsive in character).

149. *State of Cal.ex rel. State Lands Com. v. Superior Court (Lovelace), supra,* 11 Cal.4th at 76–80.

150. *Id.* at 77–78. Another case has noted this same distinction. *Brainard v. State of Texas* (1999) 12 S.W.3d 6, 22 ("The distinction between wholly natural and artificially-influenced changes in the shoreline is unworkable as the sole criterion to determine the ownership of land along the shore.")

151. "Over the years, numerous human activities have indirectly as well as directly caused a change in water flow and the accumulation of sediment in virtually every river and tideland in

proof.[152] In consequence, the court adopted a practical rule that, because of its importance, is set forth at length:

> We hold . . . that accretion is artificial if directly caused by human activities . . . that occurred in the immediate vicinity of the accreted land. Accretion is not artificial merely because human activities far away contributed to it. The dividing line between what is and is not in the immediate vicinity will have to be decided on a case-by-case basis, keeping in mind that the artificial activity must have been the *direct* cause of the accretion before it can be deemed artificial. The larger the structure or the scope of human activity . . . the farther away it can be and still be a direct cause of the accretion, although it must always be in the general location of the accreted property to come within the artificial accretion rule.[153]

What seems clear, given this case-by-case, flexible approach, is that proof of the cause of the accretion will require introduction of expert testimony.[154]

There may, however, be some relief of that burden. Some courts have held that if a watercourse, once altered by artificial conditions, continues in that changed condition for an extended period, the property boundary of the land adjacent to that watercourse is located in a "new" natural condition; in effect, the artificial condition becomes a natural one.[155] Courts have applied this rule, however, only in cases of property boundaries in which the water body submerged adjacent lands and increased, not decreased, the water body's geographic extent.[156]

In sum, under the Supreme Court formulation, evidence that changes in the physical location of the mean high-water line resulted from (1) a regional pattern of human actions, major in scope, (2) occurring over time, (3) at some distance from the tidal marsh, and (4) without any particular or immediately discernable impact on the specific tidal marsh in which the property boundary

---

the state. To view all of this as artifical accretion would effectively eviscerate the general rule." *State of Cal.ex rel. State Lands Com. v. Superior Court* (*Lovelace*), *supra*, 11 Cal.4th at 78.

152. The supreme court observed there would be "difficult evidentiary problems" if the upland owner must trace accretions back to their source, especially given the period of time since hydraulic mining activities had taken place. The court then noted that problems of proof as to the direct cause of the accretion "will be minimized if the artificial accretion rule is limited to human activities in the immediate vicinity of the accreted area." *Ibid.*

153. *Id.* at 79–80 (italics in original).

154. Such testimony could include that of coastal oceanographers, fluvial hydrologists, historians, historical geographers, civil engineers, plant ecologists, and soils scientists. Such expert opinion, as one court has stated, ". . . has been developed to a high point of accuracy." *Perry v. State of California* (1956) 139 Cal.App.2d 379, 396. But see notes 203 through 204 and accompanying text in Chapter 4 and Section 6.9.2 in Chapter 6.

155. *Chowchilla Farms, Inc. v. Martin* (1933) 219 Cal. 1, 18; *State v. Superior Court* (*Fogerty*), *supra*, 29 Cal.3d at 248–249.

156. *State v. Superior Court* (*Fogerty*), *supra*, 29 Cal.3d at 248–249.

is to be located, would not support application of the artificial accretion rule.[157] Thus, to answer the rhetorical question that generated this discussion, the mole probably would not affect the location of the ordinary high-water mark in the region's marshes, except in those marshes in the immediate vicinity of the mole where the evidence established physical changes linked to the mole. In such a case, the ordinary high-water mark property boundary would be fixed at its location before construction of the mole.

Even with this guidance from the Supreme Court, each boundary dispute in which impact of human activity is an issue should be determined on its own facts. This places a heavy burden on the litigants and their surveyors, consultants, and lawyers to develop as complete a factual record as possible. It is to the development and proof of such facts that we next turn.

5.9 PROOF OF THE LEGAL CHARACTER OF TIDAL MARSHES AND THE PHYSICAL LOCATION OF THE HISTORIC ORDINARY HIGH-WATER MARK

Development of the information and evidence needed to assist the court in its determination of the location of the historic ordinary high-water mark is concentrated in two broad areas: historical character, condition, and attributes of the land in the 1850s and how, when, and why such character, condition, and attributes were or may have been changed.[158] These are intimidating tasks.

For example, among other matters, two considerable hurdles must be overcome. First, massive change of the tidal marsh regime over time has frequently left the remaining tidal marshes so altered or obliterated that present-day physical conditions may bear little relation, if any at all, to historic tidal marsh conditions. Second, there is the sheer technical difficulty in physically measuring and marking in a tidal marsh the ordinary high-water mark boundary between tidelands and swamp and overflowed lands.

Although difficult, these tasks are possible. An attempt to recount each source or method of proof is bound to fall short of completeness. Instead,

---

157. One trial court held that the ordinary high-water mark property boundary was fixed at its location prior to the onset of such events. *State v. Southern Pacific* (*Delta Meadows Decision*), Appendix A, para. C.5., pp. 14–15. This decision was prior to the decisions in *Fogerty* and *Lovelace*. Compare *Delta Meadows Decision*, Appendix A, para. C.5., pp. 14–15, and *Littlefield v. Nelson* (10th Cir. 1957) 246 F.2d 956, 958–959.

158. The rationale for this focus is that no matter what has happened with the tidal marsh regime, the physical character of the land as swamp and overflowed and, as a consequence, the physical location of the ordinary high-water mark property boundary must be established as of the date of the Swamp Lands Act grant. See notes 44 through 45 and 126 and accompanying text in this chapter. The location of the property boundary thereafter is subject to the rules and quandaries briefly outlined in notes 127 through 154 and accompanying text in this chapter.

178 PROPERTY BOUNDARY DETERMINATION IN ESTUARINE AREAS

this chapter will explore in some detail certain of the more readily available and frequently relied upon types of proofs and their limitations. The chapter will then make general comments on the advantages and disadvantages of some frequently suggested types of physical investigations.

5.9.1 Means to Establish the Physical Character of Tidal Marsh in Historic Times

Turning to the proof of the physical character of the tidal marsh in 1850, it is important to know what sources of information are available.

We have already discussed[159] the fact that there are historical accounts or descriptions of many estuarine areas. In edition to written accounts, historic mapping by various governmental authorities provide other sources of information about physical character. One readily available and quite often used source of such cartographic information is the historic topographic and hydrographic mapping undertaken by the United States Coast Survey (USCS) and the United States Coast and Geodedic Survey (USC&GS).[160] Because of the widespread acceptance of the practice of relying on such maps for boundary determination purposes, a detailed exploration of their creation and purpose is worthwhile.

5.9.2 USCS and USC&GS Mapping

Although other governmental authorities, such as the United States Army Corps of Engineers and the State Swampland or local Reclamation Districts, also mapped tidal marsh areas in the nineteenth and early twentieth centuries, courts have used USCS and USC&GS topographic mapping in a way that has endorsed their veracity and accuracy.[161] In addition, these maps narrowly focus on coastal and shoreline regions and are relatively precise and accurate. Further, verification of their accuracy is possible, as such maps are regularly and continually updated through later mapping projects of the same coastal or shoreline areas. This updating enables one to compare two maps, providing a snapshot of distinct time periods. This comparison may be used to establish the consistency of portrayal of the physical features and to determine the cultural or physical changes that may have occurred in the tidal marsh and surrounding area in the period between publication of the two maps. Moreover, the USCS and USC&GS have had a definite interest in the accuracy of the cartography because of the reliance on that cartography in constructing navigation charts.

---

159. See note 138 in this chapter.
160. See notes 86 through 90 and accompanying text in Chapter 4.
161. See note 89 in Chapter 4.

These facts all confirm the signal importance of such maps to those interested in boundary determination. Given the probable paucity of other sources of historic topographic information, detailed examination of the USCS and USC&GS topographic mapping of tidal marshes accomplished in the nineteenth century, before reclamation of those marshes, should be of great assistance in ascertaining their prereclamation physical character. These maps, however, are not talismanic; one should understand the limitations of such mapping, and any conclusions concerning the location of the property boundary should be drawn with those limitations in mind.[162]

Figure 5.4 on the following page is a portion of a USCS Topographic Map[163] showing a tidal marsh in south San Francisco Bay as USCS surveyors surveyed that marsh in the late 1850s.[164] Even to an untrained observer, this map depicts a fibrous land area interlaced with many sinuous watercourses. Neither vertical elevations were shown for the marsh features nor contour lines placed in the marsh area. One may assume, therefore, that the surface of the marsh was devoid of any notable topographic relief. These obvious matters aside, as with any third-party evidence, one should be aware of the limitations and value of these maps for property boundary purposes. Thus, the method of preparation of these maps, including pertinent instructions to the surveyor, should be examined.

5.9.3 Method of Preparation of USCS and USC&GS Maps

Historic USCS and USC&GS topographic maps were prepared with the use of a planetable.[165] Basically, this means that a particular type of surveying instrument, the planetable, was used to accomplish the survey. There were no written descriptive field notes of the physical features that the surveyor could use for plotting the survey in the comfort of an office; the coast surveyors actually created the map manuscript as the USCS or USC&GS topographer/cartographer progressed through the marshland.[166]

---

162. We will see, among other limitations, that these surveys were not prepared for property boundary purposes. See note 180 and accompanying text in this chapter.
163. "T" is the indicator that the map is a topographic survey. "H" is the indicator that the map is a hydrographic survey.
164. Portions of the marsh were described in two cases. See note 38 in this chapter.
165. In very basic terms, the planetable consisted of a drawing board mounted on a tripod. Set on the drawing board was an instrument called an "alidade." An alidade was simply a ruler on which a telescope was mounted. 2 Shalowitz, *Shore and Sea Boundaries* (Dept. Comm. Pub. 10-1 1962) § 411, p. 160. A picture of a planetable is found in the Report of the Superintendent of the U.S. Coast Survey, 39th Cong., 1st Sess., House Ex. Doc. No. 75, Appendix 22 (1865), fig. 30., following p. 231 (hereinafter referred to as "1865 Planetable Manual").
166. "Delineating the shoreline, sketching the contours, mapping the roads and other topographic features are accomplished in the field while the terrain is in full view of the topographer." 2 Shalowitz, § 41, pp. 159–160. Shalowitz has a valuable description of just how the planetable is used. *Id.* at 160–164.

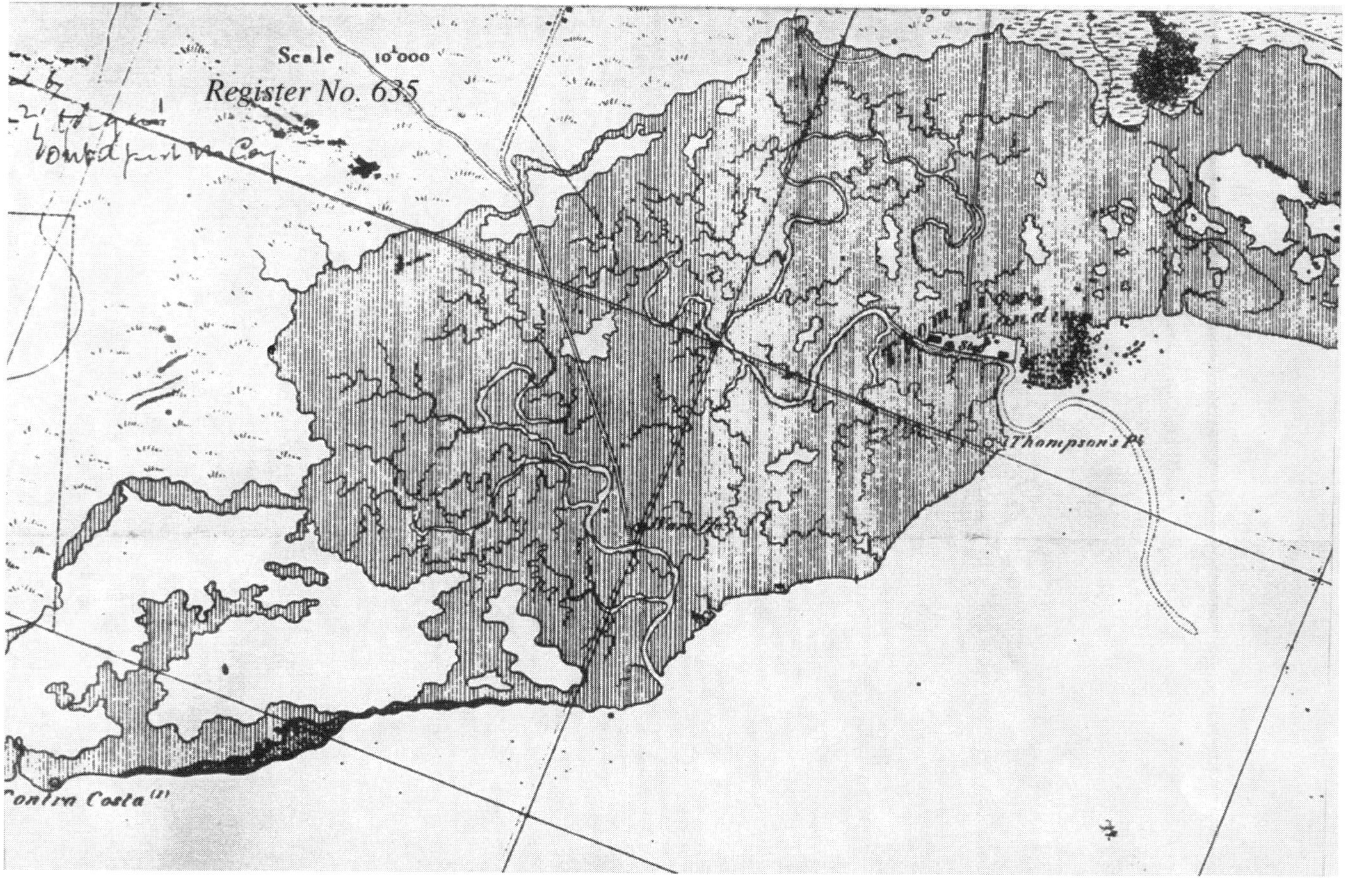

Figure 5.4 Portion of USCS topographic map

The USCS carefully instructed surveyors who accomplished those surveys about how features they observed in the field were to be shown cartographically.[167] Insofar as pertinent here, the high-water line, the line dividing land and water, was the most important feature that was to be depicted on the USCS maps.[168] Coast surveyors were instructed that along the shore the high- and low-water lines were to be marked accurately and determined, if possible, at both spring and neap tides.[169] According to the instructions to the cartographer, the cartographic symbolization of the high-water or shoreline feature was to be "full and black, the heaviest lines on the sheet. . . ."[170]

5.9.4 Interpretation of a USCS Map

Figure 5.4 appears to show that the cartographic feature indicated with arrow A, the shoreline feature, is depicted as called for in the instructions. It is one of the heaviest lines on the sheet. Figure 5.4 also confirms that the area was a tidal marsh, as shown by the cartographic representation indicated with Arrow B. The symbol at Arrow B, chosen by the field surveyor to depict the type of vegetation found by the coast surveyor, is representative of salt marsh vegetation.[171]

Relying on instructions requiring the shoreline feature be the average or ordinary high-water line,[172] then-contemporaneous testimony concerning USCS topographic mapping practice and maps,[173] and descriptive reports[174]

---

167. In this chapter we are concerned with the location of the ordinary high-water mark. Thus, a detailed review of all instructions relating to the use of the planetable is beyond the scope of this work. If you are interested in a broader perspective about the preparation of these maps, there is an overview of the instructions given in 2 Shalowitz, §§ 412, 42, pp. 164–168. The surveyor or lawyer wishing to understand how planetable surveying was accomplished would do well to become familiar with these instructions.
168. 2 Shalowitz, § 441, p. 171.
169. 2 Shalowitz, § 42, p. 166, citing handwritten instructions from Ferdinand Hassler, the first superintendent of the USCS; 1865 Planetable Manual, p. 219.
170. 1865 Planetable Manual, p. 230.
171. One may readily accomplish interpretation of the cartographic symbolization of physical features on these USCS maps. For our purposes, the Rosetta stone of USCS and USC&GS topographic map symbolization is found in Shalowitz. 2 Shalowitz, Chapter 4. See especially pp. 189–192 and figure 50, p. 201. One should consult persons familiar with the type of symbolization utilized by the USCS and USC&GS. In an appropriate case, these persons might testify about their considered opinion and interpretation of such maps and any particular questioned cartographic symbolization.
172. 1865 Planetable Manual, p. 219.
173. Testimony of Augustus Rodgers, George Davidson, James W. Bost, Horace A. Higley in *United States v. Peralta* (Nos. 98, 100 U.S. D.C. N.D. 1871). Rodgers and Davidson were members of the USCS who directed or actually prepared the topographic surveys such as T-635. Bost was then the surveyor general of California. Higley was then the Alameda County surveyor, the county in which the land depicted in T-635 is (Figure 5.4) located.
174. Descriptive reports are written reports that sometimes accompany topographic and hydrographic surveys. The field engineer or his or her associates prepare the descriptive reports to supplement the survey with information that they cannot graphically depict and to note important

from the USC&GS, some have suggested that the most waterward edge of the marsh depicted on USCS and USC&GS maps as a heavily inked line was the physical location of the ordinary high-water mark. Like many contentions concerning the interpretation of littoral title or boundary documents, this contention appears to have some support.[175] However, it is not universally accepted.

5.9.5 Cautions on the Use of and the Value of USCS and USC&GS Maps for Property Boundary Purposes

According to a recognized authority in USCS and USC&GS topographic map interpretation, in a tidal marsh area it is assumed that the high-water line has not been determined unless there is some evidence to the contrary on the face of the topographic survey.[176] Thus, some claim that USCS and USC&GS maps such as shown in Figure 5.4 depict only a tidal marsh interlaced by sloughs and ponds and populated by salt marsh vegetation. In other words, these persons contend such maps depict no property boundaries.[177]

First, the purpose of the USCS and USC&GS topographic surveys had nothing to do with the location of a property boundary, much less with location of the elusive ordinary high-water mark property boundary. The primary purpose of USCS and USC&GS topographic and hydrographic surveys was to provide data for the construction of nautical charts.[178]

Second, neither the scale of the topographic map nor the use of the planetable as the method would have been adopted by the USCS or the USC&GS to demarcate property boundaries based on tidal definitions.[179]

Third, according to some USC&GS officials, the high-water mark property boundary, as a physical location, is not shown on these maps.[180]

---

results. They may also throw important light on a particular area of shoreline. 2 Shalowitz, § 1242, pp. 85–86. The descriptive reports relied on were those accompanying USC&GS Topographic Surveys T-2252, T-2312, and Hydrographic Survey H-2304. Descriptive reports may be available from the National Ocean Survey and the National Archives.

175. Note that a similar contention was rejected by the United States Supreme Court. See notes 124 through 125 and accompanying text in this chapter.
176. 2 Shalowitz, § 4431, p. 176.
177. *Id*. at § 451.
178. *Id*. at § 11, p. 80.
179. *Ibid*. For example, when asked to perform a survey of a political boundary between the District of Columbia and the state of Virginia, the USC&GS tied the mean high water line horizontally to triangulation markers and vertically to tidal bench marks. This kind of accuracy was not needed (or available) for the early USCS surveys.
180. *Id*. at § 4431, p. 176; Hull, W., and Thurlow, C., "Tidal Datums & Mapping Tidal Boundaries" (U.S. Dept. of Comm., National Ocean & Atmospheric Admin., NOS), pp. 9–10 ("Hull"). Commander R.S. Patton, the late director of the USC&GS, described the practice of the USC&GS as follows:

5.9 PROOF OF THE LEGAL CHARACTER OF TIDAL MARSHES

Fourth, considering the physical character of the tidal marsh, identifying the high-water line in such tidal marsh areas would be extremely difficult.[181] Therefore, in interpreting such mapping, one must assume that unless there is some evidence on the survey itself, the high-water line has not been determined.[182]

Finally, the descriptive reports relied on in support of the argument that the shoreline feature depicted on USCS or USC&GS maps such as Figure 5.4 established the location of the high-water line may be held by some to be equivocal and by others to be contradictory. This is because some USC&GS surveyors who wrote those descriptive reports assumed that the level of the tidal marsh in its natural condition was coincident with the plane of ordinary or mean high water.[183]

Thus, merely by examining the face of USCS or USC&GS topographic maps, determining the physical location of the ordinary high-water mark property boundary between tidelands and swamp and overflowed lands may not be possible. In other words, other than the lands depicted by such maps as the beds of the tidal creeks and sloughs that traverse and interlace the marsh,

---

As a rule, in the case of marsh lands, the line of high water is not definitely located on the charts. In general one may assume that it lies somewhere between the lines limiting the seaward and landward edges of the marsh. One cannot, however, be sure that the outer line, or seaward limit of the marsh, is below high water, because ultimately salt marsh grows upward to a level approximately even with, or slightly above, high water. All one can say is that in general the higher tides penetrate to a line somewhere between the inner and outer limits of marsh [as depicted on USC&GS topographic maps]. Johnson, New England Acadian Shoreline (1925) Cap. XII, p. 424 (personal communication with Patton); *Best Renting Co. v. City of New York* (1928) 162 N.E. 497, 501 (where Patton testified that the shoreline feature on such charts represented only the extreme limit of the marsh on the shore).

181. 2 Shalowitz, § 4432, p. 176. Shalowitz states:

In surveying such areas, the [USCS and USC&GS] [have] not deemed it necessary to determine the actual high-water line but rather the outer or seaward edge of the marsh, which to the navigator would be the dividing line between land and water. [Footnote omitted.] Therefore, from the topographic survey alone, and in the absence of any corroborating collateral information, no conclusion could be drawn as to the exact location of the high-water line, nor as to the condition of the marsh area with reference to the tidal plane of high water; that is whether the ground itself was above water, or whether only the marsh grass was above water, and ground below water at the time of high tide. *Id.* at p. 177.

182. *Ibid.* A trial court in a case involving a tidal marsh rejected this argument: "... [USC&GS] charts do not locate the high water line in tidal marshes; only the apparent dividing line between the marsh land and water is located." *State v. Southern Pacific* (*Delta Meadows Decision*), Appendix A, para. B.2., p. 10.

183. Letter, A.F. Rodgers, Assistant, to Pritchett, Superintendent, USC&GS, July 8, 1898. The National Archives in Washington, D.C. is the repository of such correspondence. Investigation of this archival resource is worthwhile in the appropriate case.

there is no way of determining, without other information and purely from the cartographic depiction of the marsh, the physical location of the ordinary high-water mark property boundary between lands of the legal character of swamp and overflowed lands and tidelands.

That is not to say that these maps are useless. USCS and USC&GS topographic maps are valuable in helping to determine the physical character of the tidal marshlands. From the interpretation of the cartographic symbolization of physical features and attributes, such as are shown in Figure 5.4, one may conclude that the tidal marsh was extensive,[184] extremely flat,[185] vegetated by a species of salt marsh vegetation,[186] and "covered" at some frequency by tidal waters of the bay.[187] This information, while not determinative of the location of the property boundary, will at least provide experts with a vital historical snapshot of the physical attributes or character[188] of the marsh. This snapshot will be extremely useful in forming their conclusions.[189]

---

184. The scale of the topographic maps was $1'' = 833'$. Thus, for example, the area of the tidal marsh encompassed in Figure 5.4 was almost mile in length and width.

185. The lack of relief can be determined by lack of contour lines or absence or infrequency of any elevations on the map relative to the vertical reference datum.

186. 2 Shalowitz at 188–192.

187. See note 38 in this chapter. Other helpful conclusions may also be drawn from a careful examination of these USCS or USC&GS topographic maps. The author is aware of one case in which a highly knowledgeable surveyor testified that it was possible to examine the original manuscript maps and observe that there are "pin prick points," tiny, pin-sized holes in the original manuscript map. That surveyor claimed that these "pin prick points" were evidence of the exact location the USCS or USC&GS cartographer actually set up the planetable in the tidal marsh. Based on the observation that "pin prick points" were found within the marsh, the surveyor concluded the marsh was firm and not subject to frequent tidal inundation. Because he observed no "pin prick points" in the manuscript offshore, the surveyor concluded that the offshore lands were too wet and soft for the USCS surveyor to have set up the planetable. A more detailed examination by another knowledgeable surveyor found pin prick points scattered not only throughout the marsh, but also in the mud flats bayward of the shoreline. This discovery called into some question the conclusions drawn by the first surveyor, as it led to the contradictory conclusions that the USCS or USC&GS surveyor either set up his instrument in the mud or that the offshore lands were not too wet and soft for planetable setup.

188. For example, the graphic portrayal of vegetation as salt marsh vegetation will assist a botanist, plant taxonomist, or plant ecologist in forming a conclusion as to the species of plant that probably inhabited the tidal marsh. Based on such knowledge, it is possible such experts may have an opinion as to the frequency of tidal inundation required to support such a stand of vegetation. For example, in Gilbert it is stated:

> The higher levels of all the [San Francisco Bay system] marshes are marked by samphire (*Salicornia*) and alkali grass (*Distichlis*), but the flora of the lower and broader parts varies with the salinity of the water. About San Francisco Bay the dominant plant is creek sedge (*Spartina*) . . . . Gilbert, *supra*, at p. 77.

189. A detailed discussion of the types of proof and the expert testimony that can be used to establish the historical physical character of tidal marshes is found in Sections 6.8 through 6.9 in Chapter 6.

5.9.6 Other Types of Mapping

One further example of the types of mapping available to establish historic physical character and features will be helpful. Topographic and hydrographic mapping were not the only types of mapping accomplished during the mid-nineteenth century. Coincidentally, land surveying by federal, state, and local government surveyors was being concluded.[190] Unlike that of the USCS and USC&GS topographic and hydrographic surveys, the purpose of the surveys that led to the preparation of maps or plats was to locate property boundaries. Earlier in this chapter we have seen how certain of those surveys could be used to support an argument that would have geographically located the ordinary high-water mark property boundary as a matter of law.[191] These surveys and the field notes created during the surveys provide useful information on the physical character of the area being surveyed.[192] In reviewing these surveys, it is important to keep in mind the instructions provided to the surveyors.

In surveying the extent of public lands, federal surveyors necessarily encountered navigable watercourses whose sinuosities had to be established in order to estimate the amount of public land available for disposition. Tracing the instructions given to the surveyors about the types of features they should survey may assist in forming a conclusion about the features actually surveyed and those features that the surveyors did not survey but may have existed at the time of the survey.[193] Given these limitations and caveats, these surveys

---

190. These surveys should be used in property boundary determination only after careful analysis and study of all available boundary information. Some of these surveys were not based on actual field work, and many were outright shams. E.g, *Hellman, supra,* 125 Cal. at 387; see note 70 and accompanying text in Chapter 1.
191. In addition, in certain instances meander line surveys accomplished by the United States in conjunction with defining the extent of public lands may be used to define and locate the true property boundary. See notes 66 through 83 and accompanying text in Chapter 3.
192. Another source where field notes can be found is the original peg books or "tablets," as they were referred to in the BLM surveying manual. These are the actual volumes in which the U.S. Deputy Surveyors kept the actual field record and sketches of the measurements. Note that these tablets are not the same as "field notes" in Bureau of Land Management (BLM) parlance.
193. The 1855 Instructions to the Surveyors General of Public Lands of the United States as contained in Minnick, *A Collection of Original Instructions to Surveyors of the Public Lands 1815–1881* (Landmark) at pp. 366–367, contained no specific instructions for meandering along tidal watercourses. The instructions required the surveyor to meander not only navigable rivers, but all lakes and "deep ponds" greater than 25 acres and navigable bayous. It was not until 1890 that the General Land Office expressly instructed surveyors in the surveying of lands adjacent to tidewater; they were told to meander at the "high-water line." 1890 Manual of Instructions. These instructions have been substantially continued to the present. Manual of Instructions for the Survey of the Public Lands of the United States 1973 (Landmark Reprint), p. 96. The surveys accomplished pursuant to those instructions may be the only information available. Therefore, it would be very helpful for the surveyor and the lawyer to have a complete library of the instructions

and their accompanying field notes and plats can provide much useful information.[194]

5.9.7 Other Sources of Factual Information

Since other sources of factual information about the historic character of the land are many and varied, examining each of them in detail would be repetitive. Pointing out some of them, however, should suffice to give the surveyor, lawyer, and property owner a starting point for further and more creative factual investigation: court records in the local courthouse; university libraries and rare document repositories; museums; local historical societies; state archives; state land management agencies; city or county surveyors, engineers, recorders, treasurers, and auditors; public works and road departments; local surveyors; documents in the chain of title, and so forth. Only the investigator's imagination and creativity limit the type and number of sources of information available.[195]

5.9.8 Suggestions for Exhibits

Each source of information should provide a snapshot in time about the physical character, features, or attributes of the tidal marsh. From this information a party may reconstruct a chronological history of the location of various features or the change in physical attributes of the tidal marsh.

One method to effect this reconstruction is to prepare graphic representations of the pertinent information about physical character, features, or attributes in relation to the property in question. This can be accomplished through use of the products of a GIS system[196] or with overlays justified to same-scale base exhibit maps. Pertinent narrative information should also be geo-

---

issued to governmental surveyors, including any special instructions that concerned a specific type of survey or region.

194. The author recalls that in examining the field notes of one such township survey, the government surveyor wrote that he was unable to set a meander corner. This statement alone is unremarkable. However, the field notes supplied the reason: The location of that corner was under 4 feet of navigable water. This information called into considerable question a later "survey" by the local county government surveyor that characterized those same lands as swamp and overflowed lands.

195. For example, the cultural use of tidal marshland may be important in supporting a particular physical location of the ordinary high-water mark. Recall that the ordinary high-water mark property boundary is asserted to separate those lands suitable for agriculture from those lands so frequently affected by tidal waters that the lands are not suitable for agriculture. ". . . [T]he shore line is the boundary between tillable land or land available for agricultural purposes and land so frequently covered by the sea that it is useless for agricultural purposes. . . ." *City of Los Angeles v. Borax Consolidated Limited*, *supra*, 74 F.2d at 905; see note 26 in Chapter 3. Thus, if the cultural evidence establishes that the tidal marshland was vacant and unsuitable for agricultural uses, such evidence could support an argument that the physical location of the ordinary high-water mark is landward of the tidal marsh.

196. See note 186 in Chapter 4 and notes 12 and 97 in Chapter 6.

graphically located with respect to the land in dispute. This kind of graphic presentation is easy to understand and interesting to the court and other persons to whom it is presented. Such information may also be valuable for and relevant to any physical investigations that may be required.

5.9.9 Types of Physical Investigations

Physical investigations of tidal marshlands are difficult and expensive. Sometimes, however, a physical investigation may be needed and should be accomplished. Such scientific analyses may be needed to establish the current location of various historic, now obliterated, physical features or attributes or to examine the strata of tidal marshland that has been reclaimed. Before any physical investigation is undertaken, its purposes, limitations, and expected uses should be clearly understood, planned, and explained to all concerned.

For example, suppose one was attempting to prove the physical character of a completely reclaimed tidal marsh in which all evidence of its prereclamation physical character had been obliterated or severely disturbed. Suppose, further, that it was suggested that a topographic survey of a presently existing, apparently undisturbed, tidal marsh be undertaken in order to determine the vertical relationship of that marsh to various tidal datums. Assuming such a topographic survey could be precisely accomplished, what would one have achieved except the possibility of unjust enrichment of the lawyer who suggested it and the surveyor who accomplished it?

Among other matters, picking out just any undisturbed marsh will not be satisfactory. The relative elevation of the tidal marsh to the tides and the consequent location of the mean high-water line in the marsh is ofttimes determined purely by local conditions.[197] Moreover, failure to properly instruct the placement of the surveying rod on the marsh surface could well affect the result.[198] Suffice it to say, before any such surveying is undertaken, a no-holds-barred conversation exploring every angle and pitfall should take place with those recommending the survey.

Other types of physical investigations are subject to the same infirmities. Some believe that coring the soil of the tidal marsh is valuable. These samples contain soil strata and vegetative remains, and their examination by soils scientists and botanical taxonomists may be fruitful.[199] On the other hand, some courts have held that botanical "surveys"[200] do not in and of themselves

---

197. Atwater, Tidal Marshes, *supra*, at 359.
198. Due to the softness of or the vegetation coverage on the tidal marsh surface, rod placement is critical. The surveyor should instruct the rodmen to place the rod on, not through, the marsh floor and beside, not on, marsh vegetation.
199. One court has apparently endorsed geological examination of core drilling through existing fill to the elevation of prefill mean low water to establish the location of the prefill mean low-water line. *State ex rel. Buckson v. Pennsylvania Railroad Co.* (Del. 1971) 273 A.2d 268, 270.
200. In general terms, specialists can undertake such a survey through the interpretation of plant signatures on true color or false color (infrared) aerial photography. See *City of Newark, supra*, 414 A.2d at 1306. Botanists, plant ecologists, and photointerpretation specialists are called for in

provide sufficient proof of the location of the mean high-water line.[201] The results of carefully accomplished physical investigations, coupled with other information, can be highly persuasive of the location of the mean high-water line.[202]

This chapter has only touched on some of the many complex factual and legal issues that can arise, and have arisen, in boundary and title disputes over tidal marshlands. The next chapter will examine some of the same issues concerning estuarine marshes in a more concrete setting—an actual case that was tried to a final judgment. A unique component of this case was that the geographic location of the land in dispute was in an area where tidal ocean waters and freshwater river flows met and mixed.

---

such work. Other methods include on-site vegetation population identification and survey in relation to known vertical and horizontal locations and the preparation of vegetation cover maps. If such "surveys" are intended to be introduced, strict adherence to the requirements for their admissibility appears to be required. *Velsicol Chemical, supra,* 442 A.2d at 1053–1054 (state fails to prove basis supporting preparation of maps of biological delineation of mean high tide).
201. *Velsicol Chemical, supra,* 442 A.2d at 1054.
202. See *State v. Southern Pacific* (*Delta Meadows Decision*), Appendix A, para. B. at pp. 10–13.

6

PROPERTY BOUNDARY DETERMINATION ALONG AND IN TIDAL RIVER REGIMES

6.1 INTRODUCTION

Even from the technologically sophisticated perspective of the later part of the twentieth century, the United States appears similar to ancient Mesopotamia in at least one respect. Mesopotamia, the Cradle of Civilization, was situated in the delta of the Tigris and the Euphrates Rivers, the Fertile Crescent. Similarly, within the United States are two such great river deltas, the Sacramento-San Joaquin Delta in California and the Mississippi Delta in Louisiana. Great centers of civilization also grew up around them.

All of these deltas have at least two features in common: highly fertile land and abundant nearby water to irrigate that land. In Mesopotamia, as in the Egyptian Nile delta, these physical attributes were the lodestone that attracted and eventually led to the formation of a highly complex, organized civilization, complete with an intricate legal system.[1]

The author does not intend to discourse at length on the development of the Indo-European legal system, a riveting topic to be sure. One salient fact, however, should not be ignored. A purpose of both ancient and modern legal systems was to to assure delta landowners of relatively secure property boundaries in order to instill the confidence necessary for those owners to develop, bring, and keep their land in fruitful production.[2] Even considering the wealth

---

1. E.g., The Times Atlas of World History (Times Books 1979), pp. 52–55.
2. For example, the Code of Hammurabi, created by one of the ancient rulers of Mesopotamia, contains laws that were devoted to ensuring that landowners would maintain dikes to prevent flooding of neighboring parcels. Harper, *The Code of Hammurabi King of Babylon* (Univ. of Chicago Press 1904) §§ 53–56, pp. 29–31. "If a man neglect to strengthen his dyke and do not strengthen it, and a break be made in his dyke and the water carry away the farm land, the man

189

of boundary rules developed over the course of centuries, not every vagary of nature or humans that affected the geographic location of property boundaries could be foreseen. Thus, the great river deltas have persisted in providing title and boundary disputes requiring riparian property boundary rules to be continually refashioned to account for this unique regime.[3]

The ancient river deltas would have only a passing interest for us today, save for the fact that the Fertile Crescent has been the site of not only the now-concluded Iran-Iraq War, but of what is known as Desert Storm, the miraculously easy victory of a coalition of nations over Iraq. The Iraqis tried to settle their boundary (as well as other) disputes in one fashion. This chapter will focus, however, on the current, nonviolent methodology of resolving property boundary location and title disputes. The methodology will be applied in a fascinating physical environment and region much closer to home, the Sacramento-San Joaquin River Delta.[4] The location of the Delta in relation to San Francisco Bay and Pacific Ocean is shown in Figure 6.1 (found on the following page).

This chapter takes a different approach from that in other chapters of this work. This chapter describes an actual title and boundary dispute over particular delta lands tried to a final decision, the *Delta Meadows Case*.[5] That case concerned an area of the Delta commonly known as "Delta Meadows." Delta Meadows is thought to be one of the last remaining, relatively unaltered examples of the physical conditions of the Delta prior to the onset of European civilization.[6]

Figures 6.1 and 6.2 (see page 192) shows the geographic location of Delta Meadows within the Delta.[7] The *Delta Meadows Case* concerned only a portion of the Delta Meadows area—a strip of land on which a railroad roadbed and trestle had been constructed by a predecessor of the once-goliath Southern Pacific Railroad Company.[8]

---

in whose dyke the break has been made shall restore the grain which he had damaged." *Id.* at § 53, p. 29. In modern times, a California court has also had occasion to discuss the effect of a Delta levee failure on the property owner's right to reclaim the land. *Bohn v. Albertson* (1951) 107 Cal.App.2d 738, 748–751.
3. *Ibid.*
4. For ease of reference and to adopt the vernacular term by which the Sacramento-San Joaquin Delta is generally known, this chapter will refer to that region as the "Delta." The geographic limits of the Delta are defined in California Water Code § 12220.
5. *State of California, ex rel. Public Works Board v. Southern Pacific Transportation Company* (1983) Sacramento County Superior Court No. 277312. For ease of reference, the case is not referred to by its technically correct name. Instead, the convention of this chapter is to adopt the name of the case as it was known by the litigants and the court, the *Delta Meadows Case*. The Statement of Decision resulting from the *Delta Meadows Case* is reprinted in Appendix A hereto and is cited in this chapter as *Delta Meadows Decision*, Appendix A.
6. For a down-to-earth description of the attributes of Delta Meadows, physical and otherwise, see Schell, Hal, *Dawdling on the Delta* (Schell Books 1983), pp. 65–72.
7. Figure 6.2 also shows the many reclamation and levee maintenance districts that have grown up in the course of reclamation of the Delta.
8. That area will be referred to in this chapter as the "railroad strip." Reference to the Delta Meadows area includes the railroad strip. The reverse is not intended.

6.1 INTRODUCTION 191

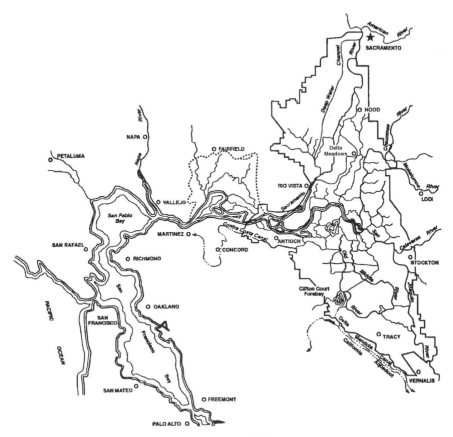

Figure 6.1 General Delta map

Figure 6.3, a smaller-scale map, shows Delta Meadows and adjacent areas in greater detail. The discrete portion of the Delta Meadows area that was in dispute in the *Delta Meadows Case*, the railroad strip, and the other pertinent geographic features, such as the Sacramento River and the Snodgrass Slough, are all plainly shown. The location of the Delta Cross Channel and its relation proximity to Delta Meadows is shown in Figure 6.2.

In question in the *Delta Meadows Case*[9] was the extent of land, if any, in the railroad strip that was subject to the public trust easement of the state.[10]

---

9. The title issue, extensively analyzed in notes 29 through 61 and accompanying text in Chapter 5, will be only briefly discussed. That issue was whether an unbroken chain of title to the railroad originating from state swamp and overflowed patents established fee simple absolute title in the railroad free of any state retained public trust easement, or whether it could still be established that such lands, or portions thereof, were tidelands in which the state retained and which remained subject to an inalienable public trust interest.

10. This information was not required merely for academic reasons. The title and boundary dispute arose in a condemnation proceeding undertaken by the State of California to acquire the railroad strip. The quantity of area of the railroad strip in which title was held by the state,

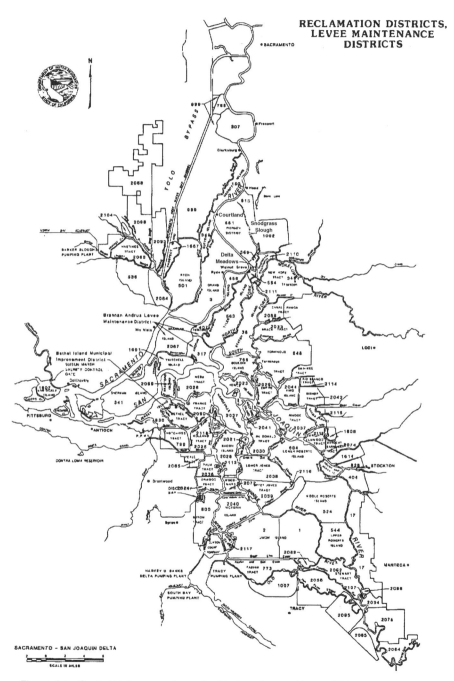

Figure 6.2 Central Delta map—Including Reclamation and Levee Maintenance Districts

6.1 INTRODUCTION 193

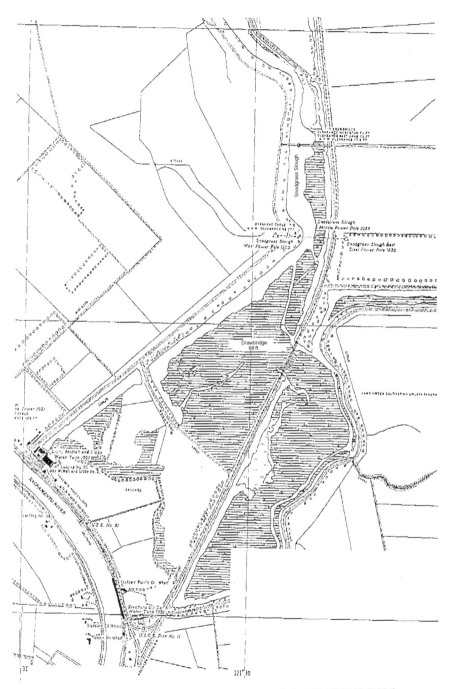

Figure 6.3 Portion of USC&GS airphoto compilation Plat T-5019 (1931)

There was no question of the invalidity of the original state swamp and overflowed land patents, only the extent of the title those patents attempted to convey. That is, within the perimeter boundaries of the swamp and overflowed patents, were there any tidelands in which the state retained a public trust easement, and, if so, what was the geographic extent of those tidelands?

A necessary predicate to answering such questions was the decision on how and where the boundary between swamp and overflowed lands and tidelands was to be physically located within the railroad strip where the physical regime and the topography of Delta Meadows had been modified. The very fact that such questions have arisen and were tried to a decision in relatively recent times recommends the value of the practical approach of this chapter. To paraphrase George Santanyana, those who ignore past cases are doomed to repeat them.[11]

Land title and property boundary disputes over a particular piece of property cannot be divorced from the larger physical framework of which such property is usually only a relatively small part. To place this specific land title and property boundary dispute in context, the pre-European-settlement character of the Delta and its regime and humans' substantial modifications to the regime are described. The stage having been set, certain preliminary legal/factual issues peculiar to the Delta will be briefly addressed. The title question, the issue of the tidality of the adjacent watercourse, the application of boundary movement principles to the Delta in light of the many modifications to its regime and the continuing dispute over how the physical location of the ordinary high-water mark property boundary is to be established will be reviewed as well. These issues, however, will not be the main focus of this chapter.

The main burden of the chapter is a discussion of the types of inquiries and expert investigation undertaken in the *Delta Meadows Case*. Such undertakings founded the re-creation and reestablishment of the court-accepted physical location of the ordinary high-water mark property boundary between lands of the legal character of tidelands and lands of the legal character of swamp and overflowed lands. In the course of that discussion this chapter will explain how the historic physical location of the ordinary high-water mark and the changes in that location over the course of years can be reconstructed and cartographically depicted. That explanation will include, among other matters, the kinds of investigations undertaken by the experts and the methods used to introduce and present the evidence to the court.[12]

---

whether in the form of fee title or in the form of a public trust easement, would affect the extent and the value of the land being condemned. See Cal. Code of Civ. Proc. §§ 1250.310(c), 1260.220. The state would not pay for land it already owned. Nor would the state pay fair market value for any land in which it already had a substantial property interest, viz., the public trust easement.
11. Santanaya wrote: "Those who cannot remember the past are condemned to repeat it." Santanaya, G., *The Life of Reason* (Prometheus Books 1998), p. 284.
12. Changes in technology have occurred since trial of the *Delta Meadows Case*. These technological advances do not alter proof required to establish the legal position. Technology, among

Finally, the accumulated facts obtained by the parties and the parties' experts after their investigations, and the conclusions based on the facts reached by those experts, will be melded with the applicable legal standards and rules to explain the basis for the court's decision. That court decided as follows: Within the perimeter boundaries of swamp and overflowed patents issued by the state in the Delta Meadows area, patents that on their face purported to convey an absolute fee title, there were tidelands that remained subject to a public trust easement reserved by the state for public trust purposes. Before the specific details of the *Delta Meadows Case* are reviewed, the physical character and regime of the Delta and Delta Meadows and the impact of human activity thereon will be recounted.

6.2 THE DELTA REGIME AND DELTA MEADOWS

Today's Delta bears little resemblance to the pre-European-settlement Delta. It is crisscrossed by highways, railroad roadbeds, tracks, and trestles and populated with thriving small and medium-size towns and cities.[13] Moreover, great systems of water diversion and flood control works and levees have changed the Delta regime. These works protect and nourish cultivated fields, orchards, and rice paddies that were once swamps.

But a little more than 100 years ago the Delta was almost totally wild, virtually undeveloped and untouched by humans. The geological setting and physical process that formed that estuarine wilderness supplies a crucial link in understanding the methodology of physically locating the ordinary high-water mark property boundary.

The essential features of the Delta's geology have been explained in these lyric terms:

---

other matters, eases and makes more affordable and flexible the way in which such proof is presented. For example, use of Global Positioning System (GPS) technology aids surveyors in precisely locating boundary descriptions in relation to known monuments or physical features. The technology has even been incorporated into golf carts to enable the Sunday duffer to determine just how far he or she is from the green. And the Geographic Information System (GIS) system in widespread use allows the graphic depiction of topography, boundary locations, and historic land use relative to one another with a remarkable facility. See note 186 in Chapter 4. Further explanation of these technological advances is beyond the scope of this book. Suffice it to say that expert consultants should be involved in deciding whether, when, and how best to employ this technology.

13. One of those towns, Stockton, California, is thought by some to be the "Mudville" of Ernest Lawrence Thayer's classic poem about baseball, "Casey at the Bat." Nermilyea, N. and Moore, J.A., "Ballad of the Republic," The Californians (May/June 1988), p. 43. Some say that before the development and construction of flood control works, there was "no joy in Mudville" during the rainy season either. Stockton is at the confluence of two rivers, the Calaveras and the San Joaquin, both of which frequently flooded during the rainy season before they were "controlled" by dams and diversions.

> ... San Francisco Bay, that greatest of our inland seas, is locally divided into a half dozen bays, but it is all one water, and the evidences of geology are that it once filled the valley whose streams now enter it by a common delta. In that far time picturesque California was like another world. The peaks of the Sierra ... were active volcanoes. ... Those great volcanic torches lighted that age to its close, and where that sea had been their mighty mirror there is now a valley. This great trough is elevated at the ends, and sags in the middle where the two streams that flow from either extreme and give a name to each have escaped into the bay.[14]

Within that valley the early European explorers found seemingly endless fen[15] lands that reminded some of Holland or Belgium.[16] These basic features have remained unchanged for more than 15,000 years.[17]

The following account of the geology and geomorphology of the Delta is based on the undisputed testimony of expert witnesses at trial. The three principal physical science expert witnesses were a historical geographer with training in geology, a coastal oceanographer who was also a registered geologist, and a civil engineer experienced in delta and river hydraulics. These witnesses had wide experience in estuarine environments such as the Delta, San Francisco Bay, and the Mississippi River Delta.[18]

---

14. Irish, John, "The Sacramento Valley" in Muir, J., ed., *West of the Rocky Mountains* (Running Press 1976), pp. 364–366.
15. Fens are marshes. Fenway Park in Boston, the home of baseball's Boston Red Sox and the famous Green Monster left field wall, is built on or nearby former marshlands. That the Red Sox are thinking of building a new stadium and destroying this jewel of a park is unspeakable.
16. One commentator noted:

> The topographical engineers in their surveys of California find that one of the most remarkable natural features of the state is the "hollow land," the "heart of its tule region," that lies between the great Sacramento and San Joaquin valleys, belonging in some degree to both, and yet in many respects different from either. Here is the waste and surplus of the two vast drainage basins from Shasta to Tehachapi; here in the midst of broad sloughs, are the beginnings of another Belgium. . . .

> The "Holland of California" embraces the greater part of a district about fifty miles long and from twelve to thirty miles wide. . . . If one counts all the "swamp and overflowed" lands along the axis of the valley, the total area will not be less than 1600 square miles. The most of it lies in either fresh or brackish water; it is between river and tide, and when properly dyked can be irrigated at will . . . . Shinn, C.H., "The Tule Region" in Muir, J., ed., *West of the Rocky Mountains* (Running Press 1976) at pp. 380–382.

17. An excellent beginning for the discussion of the geology of the Delta is Gilbert, L. K., *Hydraulic Mining Debris in the Sierra Nevada* USGS Prof. Paper No. 105 (GPO 1917) (herein "Gilbert"). A more technical paper is Atwater, B.F., "Ancient Processes at the Site of Southern San Francisco Bay: Movement of the Crust and Changes in Sea Level," San Francisco Bay: The Urbanized Estuary (AAAS 1979), p. 31 et seq. (herein "Atwater: Ancient Processes").
18. Other sources of information for this account are Gilbert, Chapter III; Atwater, Ancient Processes; Atwater, B.F., "Geologic Maps of the Sacramento—San Joaquin Delta, California"

Twenty-five thousand years ago the Delta was a river valley through which the Sacramento and San Joaquin Rivers flowed. At that time, the ocean was at a low level as a result of the Ice Age. The Pacific Ocean shoreline in the vicinity of what is now San Francisco was at about the Farrallon Islands (which lie some 20 or so miles offshore). As the ice melted and retreated, the level of the sea rose. (Sea level is believed to be continuing to rise today.) As sea level rose, it flooded progressively greater areas, finally intruding through the Golden Gate from its lowest level. After the ocean invaded what is now San Francisco and San Pablo Bays, the rise of the sea poured over the "sill" at the Carquinez Strait and infiltrated Suisun Bay and the Delta.

Rather than becoming a bay, which would have continued to fill as the ocean rose, the surface of the Delta began to be builtup. This buildup was a result of the establishment of an aquatic plant community, aided by the contribution of sediments transported by the Sacramento and San Joaquin Rivers and their tributaries.[19] As sea level continued to rise to its present level, the land surface growth of the Delta kept pace. Facilitated by both the flourishing organic plant community and the sediment transported from the Sierras by the rivers, the land surface of the Delta has maintained a close relationship with ocean sea level for the past 15,000 years or so.

In modern times, at least in a geological sense, the greatest topographic relief in the pre-European-settlement Delta was found in the natural levees that were formed along the Sacramento and San Joaquin Rivers.[20] Natural levees also lined other tributary and distributary streams, but were considerably more modest in elevation than the levees along the two major rivers.[21] Behind the natural river levees, there were overflow basins. These overflow basins were depressions into which overbank flows from the rivers were held until the overflow waters could be drained by distributary sloughs.[22] In these

---

(USGPO 1982); State Water Resources Control Board, Water Quality Control Plan for Salinity, San Francisco Bay/Sacramento-San Joaquin Delta Estuary (1988); *Gray v. Reclamation District No. 1500* (1917) 174 Cal. 622. The author also relied, in general, on studies concerning various aspects of the Delta regime produced by the California Department of Water Resources. These studies are too numerous to list, but all contain valuable information on various aspects of the Delta and its regime.

19. This process resulted in the Central Delta soils being composed mainly of organic plant material, or peat. As one moves away from the Central Delta, the soils have greater mineral content. As we will see, the soil type makes a considerable difference in maintaining the relative elevation of the surface of the land after reclamation.

20. Natural levees were formed as overbank (flood) flows caused the river to deposit its sediment load due to the decreased flow velocity. E.g., Gilbert, p. 14; *Gray, supra*, 174 Cal. 626–627.

21. *Gray, supra*, 174 Cal. at 626–627.

22. Gilbert describes the function of these back swamp areas as follows:

In times of flood the waters regularly over-topped the banks and filled the adjacent basins, through which they moved slowly and from which they drained gradually back to the main channel as the flood subsided. The detrital load of the flood was spread by them over the

basins were found the fen, marsh, or low lands, which were flat and splintered by sloughs and covered with tules, cattails, reeds, and rushes or full of thick canebrakes.[23] The back swamp areas, as these basins were also known, were classified as swamp and overflowed lands in legal character by federal and state surveyors. Delta Meadows lies at the lower end of such an overflow basin away from the Central Delta.

This Delta regime was not bound to last. Much the same as the relentless intrusion of the rising ocean waters into the Delta, although at a much, much faster pace, European settlement thrust its way into the Delta. The result of this invasion has been that over the years, the surface of the Delta itself and the Delta regime have been substantially modified. It is to these human-made or human-caused modifications to the Delta that we next turn.

6.3 THE IMPACT OF HUMAN ACTIVITIES ON THE DELTA

Almost as reflected in *Chinatown*, Roman Polanski's classic cinematic account of just how and for what purpose Los Angeles obtained water from the Owens Valley,[24] there are those who see the Delta as simply a holding pond, a reservoir. In that "reservoir" Northern California water is expected to patiently await its eventual, but inevitable, dispersion to the lawns, swimming pools, and car washes of Southern California.[25] The great water projects of

---

whole inundated tract, including the delta marshes, which acted as a system of settling basins. Gilbert, p. 15.

Gray, supra, 174 Cal. at 628, provides a more detailed description of the function of the back swamps:

The river poured its excess waters into them in low places and breaks through the banks, known as sloughs, and over the banks themselves in sheets. These sloughs had well-defined channels as they cut through the higher lands of the river banks, which channels feathered out into nothingness as the sloughs poured their waters into the vast flat area of the basin's low lands. Here and there along the lines of these basins were such sloughs, and always at the lower end was one through which, as the river waters receded until they were below the level of the plane of the submerged area, the waters of these basins were gradually drawn back into the stream.

23. The author's personal experience in the Delta confirms the overgrown and visually obscured nature of the pristine Delta. While following one intrepid expert during a field investigation, the author stepped onto the edge of the hidden bank, lost his balance, and fell backward into a small, but deep, marsh slough totally obscured by thick vegetation. The author sadly reports that this event was and has been the occasion for a great deal of entirely unwarranted merriment.
24. A very much more refined and accurate account can be found in Reisner, Mark, *Cadillac Desert* (Penguin 1986), Chapter 2.
25. The defeated Peripheral Canal initiative was a prominent, but not the most recent, vestige of that type of thinking. San Francisco Chronicle, July 13, 1987 at 1, col. 1. According to press reports, two-thirds of all drinking water in the state passes through the Delta. Although this water is what has made the San Francisco Bay system a great and unique natural resource, there are those who persist in thinking that this water is "wasted." *Ibid*.

California, the federal Central Valley Project (CVP) and the monumental State Water Project (SWP), are some of the fruits of that vision. These water diversion projects drain immense quantities of fresh water from the Delta and "ship" the water through project canal systems to Central and Southern California.[26]

Purely from a physical standpoint, the Delta from which this water is drawn is considerably different from the pre-European-settlement Delta. The principal event that caused the physical alteration of the Delta and its regime was the California gold rush. Although the surface of the Delta was not actually mined for gold, the forty-niners and their successors discovered in the Delta another precious commodity: enormously fertile land. On this land a wide variety of food crops could be grown in great abundance and with relative ease. At first crops were cultivated to feed the rapidly expanding population of California that had been attracted by the gold rush. Later, the world at large benefited from the Delta's rich soil. This "gold rush" instigated and promoted the reclamation of the Delta's marshes by the construction of levees and drains. Through this time-tried system, the Delta's great marsh islands were removed from tidal and river flow.[27]

Reclamation of the Delta was not the only effect of the gold rush on the Delta's physical regime. In the early days of the gold rush, the forty-niners worked the mountains with hand implements. But as the deposits on the surface and in the streams were played out, "modern" technology reared its head in the form of hydraulic mining.[28] The consequences of hydraulic mining operations were profound.[29] The debris temporarily came to rest in the can-

---

26. The Central Valley Project of the federal government has been discussed in many cases. E.g., *United States v. Gerlach Live Stock Co.* (1950) 339 U.S. 725, 728–729; *Rank v. (Krug) United States* (S.D. Cal. 1956) 142 F.Supp. 1, 39–41, *aff'd in part* and *rev'd in part, sub. nom. California v. Rank* (9th Cir. 1961) 293 F.2d 340; *United States v. State Water Resources Control Bd.* (1986) 182 Cal.App.3d 82, 98–100. Their effect on the Delta and San Francisco Bay is a continuing source of controversy. *United States v. State Water Resources Control Bd., supra*, 182 Cal.App.3d at 110–111.
27. Gilbert, pp. 78–79, 87–88.
28. Hydraulic mining . . . is the process by which a bank of gold-bearing earth and rock is excavated by a jet of water, discharged through the converging nozzle of a pipe, under great pressure, the earth and *debris* being carried away by the same water, through sluices, and discharged on lower levels into the natural streams and water-courses below." *Woodruff v. North Bloomfield Gravel Min. Co.* (9th Cir. 1884) 18 F. 753, 756.
29. Numerous sources describe these consequences. E.g., Gilbert; *Woodruff, supra*, 18 F. at 756–768; *People v. Gold Run D. & M. Co.* (1884) 66 Cal. 138, 144–145; *Gray, supra*, 174 Cal. at 629. These sources bring to the reader's attention a fascinating era of California and civil engineering history. Perhaps the most telling description of the effects of hydraulic mining is found in an almost offhand statement buried in the monumental work by G. K. Gilbert. After summarizing the amount of mining debris caused by hydraulic mining operations at some 1665 million cubic yards, Gilbert noted:

> To most laymen, and possibly to some engineers not concerned with great movements of earth, the term 1,000,000 cubic yards conveys no definite meaning. It helps us to a conception of the actual magnitude of the hydraulic-mining operations of the Sierra to know

yons and valleys of the Sierras and in the channels of the rivers, such as the Sacramento River. Permanent deposits of the "slickens," the finer sediments, were also made in the back swamps and delta marshes of the Sacramento and San Joaquin Rivers and in the Suisun and San Francisco Bays.[30]

The gold rush-instigated hydraulic mining had other effects on the Delta as well. Great floods of the mid and later nineteenth century were increased in likelihood and magnitude because of the clogging of river channels with mining debris.[31] Further, navigation of the Delta's watercourses was also substantially impaired.[32] As a result, a channel was dredged in and levees, weirs, and bypasses were constructed along the Sacramento River to remedy the debris problem. Great dams were built across the rivers in the mountains to tame, in part, the flows of both the Sacramento and San Joaquin Rivers. As part of the flood control works, the Delta Cross Channel was created in the early 1950s to encourage better water distribution in the Delta.[33]

Thus, events and circumstances both in and many miles away from the Delta in geographic distance made substantial modifications to the pre-European-settlement Delta regime. These changes were not limited merely to physical alteration of the immediate land surface and topography of the Delta—the entire physical regime of the Delta was modified. Delta Meadows, considered by many to be the last vestige of the pre-European-settlement physical condition of the Delta, was not spared from the impact of these modifications.

6.4 THE IMPACT OF HUMAN ACTIVITIES ON DELTA MEADOWS

Inspection of aerial photographs, review of historic accounts, and on-the-ground investigation revealed that the Delta Meadows area had been modified by human activity. Levees had been constructed to assist in the reclamation of nearby lands and new channels dredged to assist in the construction of those levees. Further, on Delta Meadows itself two sets of low-lying levees were discovered, the results of long-ago failed attempts at reclamation. The most significant work in Delta Meadows was a massive roadbed, trestle, and

---

that the volume of earth thus moved was nearly eight times as great as the volume moved in making the Panama Canal. Gilbert, p. 43.

30. *Id.* at p. 46. A table summarizes and totals the deposits. *Id.* at p. 50. It was estimated that fenlands and marshes were the recipients of about one-quarter of the fine silt that was created as the result of the hydraulic mining operations. *Ibid.*; *Gray, supra,* 174 Cal. at 629.
31. *Woodruff, supra,* 18 F. at 761–762; *People v. Gold Run D. & M. Co., supra,* 66 Cal. at 145; *Gray, supra,* 174 Cal. at 629.
32. *Ibid.*
33. In general terms, the Delta Cross Channel permitted water from the Sacramento River system to be distributed by the San Joaquin River system during certain flow conditions. *United States v. State Water Resources Control Bd., supra,* 182 Cal.App.3d at 99.

turntable, all of which had been constructed between 1905 and 1907 by a forerunner of the Southern Pacific Railroad. Figure 6.3 shows the location of this roadbed and the dredger cut that was accomplished as part of the roadbed construction.

What is manifest, even from this relatively succinct and very general description of the physical character and setting of the Delta, and of Delta Meadows in particular, is the considerable prospect that both title and boundary questions would arise as the result of the "civilization" of the Delta. The *Delta Meadows Case* is a prominent example of how such questions can arise and have been answered.

6.5 TITLE TO DELTA MARSHLANDS

Turning first (and briefly) to the title question, the very physical character of the Delta made it difficult, if not impossible, to even gain access to many areas of Delta Meadows in order to make or conclude the survey of the public lands.[34] Overlooking such obstacles, federal surveyors in their "survey" of the Delta townships, sometimes assisted by information from state surveyors, described the entire Delta as swamp and overflowed in legal character. Figure 6.4 (on the following page) is an official plat that represented the entire Delta as notoriously swamp and overflowed land.

In the *Delta Meadows Case* no township plat was introduced, as none had been constructed.[35] Instead, the federal township survey field notes were introduced. Those field notes were prepared by the federal government for the purpose of depicting the extent and location of public lands available for disposition and the identification of swamp and overflowed lands already granted to the state.[36] Despite the fact that the United States deputy surveyor surveying the federal townships could not physically explore the terrain in the vicinity of Delta Meadows, virtually the entire area was characterized by that surveyor as swamp and overflowed land in the federal township survey. A federal swamp and overflowed patent was issued as well to the State of California by the United States.

"Swamp and overflowed surveys" were accomplished by local surveyors retained by private purchasers. These surveys, which led to the issuance of state patents to these persons, were sometimes not much better than the federal township surveys that had been accomplished in the Delta. In some cases

---

34. In the *Delta Meadows Case*, a surveyor, expert in federal and state land surveying practices, retraced the federal township survey. That expert surveyor observed that the federal surveyor has been unable to conclude the township survey because of the physical impenetrability of the terrain in the vicinity of Delta Meadows.
35. The entire Delta was described as notoriously swamp and overflowed by the United States.
36. See notes 96 through 110 and accompanying text in Chapter 1 and note 41 and accompanying text in Chapter 5; *Delta Meadows Decision*, Appendix A, para. A.4.

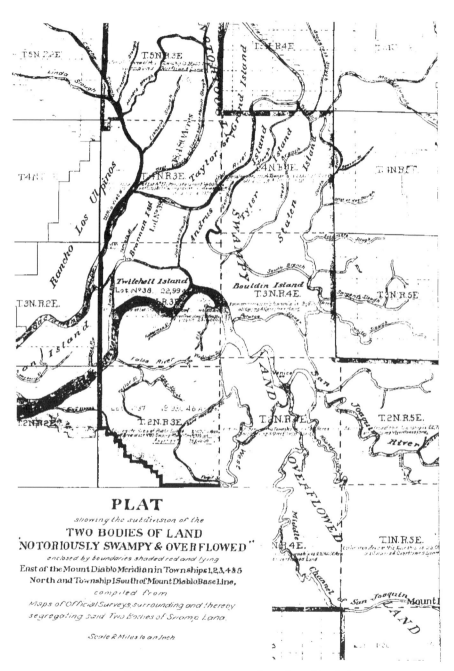

Figure 6.4 Notoriously swamp and overflowed land plat

these state surveys were inaccurate, and accepted surveying practices were often ignored. Indeed, the requirement that there be an actual field survey was neglected by some surveyors.[37] Thus, it has been asserted that the existence of these surveys and the government patents resulting therefrom do not end the inquiry concerning the true character of the land.

In Chapter 5, the conclusivity of the determination of the legal character of the lands was discussed. The argument was made that the federal government's identification of and patent to the state of land as swamp and overflowed in legal character, and confirmation, ratification, or adoption of that characterization and patent by the state, were not conclusive or binding on the state. Thus, it was claimed that the state was not precluded from seeking, at a later time, to protect and assert its competing tideland title by disputing the swamp and overflowed characterization.[38]

In the *Delta Meadows Case* the court found that the *only* boundary determined by the federal surveyors in accomplishing the federal township surveys was the boundary between two characters of public lands of the United States, public lands that were swamp and overflowed in legal character and public lands that were uplands available for disposition under public land laws other than the Arkansas Swamp Lands Act.[39] Federal surveyors did not separate

---

37. An 1874 report found

[a] mighty mass of evidence . . . to show the utter looseness and extremely wanton manner in which the swamp lands . . . had been managed in this States. . . .

That, through the connivance of parties, surveyors were appointed who segregated land as swamp, which were not so in fact. The corruption existing in the land department of the General Government has aided this system of fraud.

Again, the loose laws of the State, governing all classes of State lands, has enabled wealthy parties to obtain much of it under circumstances which, in some countries where laws were more rigid and terms less refined would be termed fraudulent; but we can only designate it as keen foresight and wise (for land grabbers) construction of loose, unwholesome laws. . . .

. . . [T]here are many acres of land held in this State which are not in fact swamp land, but they have been segregated, listed, and patented as such.

Your committee is satisfied, from evidence, that the grossest frauds have been committed in swamp land matters in this state. . . . Report of the Joint Committees on Swamp and Overflowed Lands and Land Monopoly (1874).

Courts and commentators concur generally in this conclusion. E.g., *City of Berkeley v. Superior Court* (1980) 26 Cal.3d 515, 522–523, *cert. den.*, *sub. nom. Santa Fe Land Improv. Co. v. Berkeley* (1980) 449 U.S. 840; Gates, *History of Public Land Law Development* (Pub. Land Law Rev. Comm. 1968) at pp. 321–335; see notes 95 through 107 and accompanying text in Chapter 1 and notes 44 through 57 and accompanying text in Chapter 5.

38. See notes 41 through 61 and accompanying text in Chapter 5.

39. Appendix A, para. A., second subpara. 4. This conclusion was supported by the unanimous opinion of expert surveyors retained by both sides to testify concerning federal and state surveying practices. One may argue that the court's finding was judicial recognition of what the legislature appeared to acknowledge in "An Act to Provide for the Reclamation and Segregation of Swamp and Overflowed, Salt Marsh and Tide Lands Donated to the State of California by Act of Con-

tidelands from swamp and overflowed land, nor did they have the authority to do so.[40] And although the state also issued swamp and overflowed patents encompassing Delta Meadows, the court found the state's issuance of those patents did not enhance the federal determination that the lands were swamp and overflowed in legal character.[41]

Thus, the issuance of both federal and state patents denominating the lands conveyed as being swamp and overflowed in legal character did not preemptively solve the boundary issue. The physical location of the ordinary high-water mark boundary between swamp and overflowed lands and any tidelands possibly included within the boundaries of such patents remained to be determined.[42] In the *Delta Meadows Case*, proof of the physical location of this ordinary high-water mark boundary in 1850 was required in order to establish that, at the time of the Arkansas Swamp Lands Act grant to the State of California, tidelands existed within Delta Meadows.[43] It is to one of the predicates necessary to establishing that physical location that this chapter now turns its attention.

6.6 TIDALITY OF THE ADJACENT WATERCOURSE—CONSEQUENCES AND PROOF

The proof of the geographic location of the ordinary high-water mark property boundary consumed most of the effort and attention of the parties at the trial of the *Delta Meadows Case*. Because of the mixed tidal and river regime, however, part of the proof of the geographic location of that boundary involved a showing of the tidality of the boundary watercourse, Snodgrass Slough. Its tidality was important for two reasons.

First, for title purposes, if the tidality of Snodgrass Slough could be shown, navigability was argued to be irrelevant.[44] Second, depending on whether the

---

gress," Act of May 13, 1861, Stats. 1861, Ch. 352 ("1861 Act"). A purpose of that Act was to have the state assist the United States in quickly separating lands that the United States should have conveyed as swamp and overflowed land to the state. To further that purpose, the 1861 Act required the county surveyors to locate only the line between "high lands" that would or could be held by the United States as public lands and swamp and overflowed lands, salt marsh lands, and tidelands, all of which would come to the state in some fashion. *Id.* at § 19.

40. Appendix A, para. A.7.; *Borax, Ltd. v. Los Angeles* (1935) 296 U.S. 10, 17–19.
41. Appendix A, para. A.11–12. According to the argument, state patents were issued based on the prospective purchasers' applications. The purchasers' applications characterized the land as swamp and overflowed. If the applications were in the proper form, state officials had no authority to deny the patent, nor to determine whether the lands applied for were tidelands or swamp and overflowed lands or even lands that were not subject to sale. *Ibid.*; *People v. California Fish Co.* (1913) 166 Cal. 576, 590–591.
42. Appendix A, para. A.
43. See notes 44 through 45 and accompanying text in Chapter 5.
44. See notes 81 through 96 and accompanying text in Chapter 5. Factual evidence of Snodgrass Slough's navigability was also introduced out of an abundance of caution. Such evidence included

boundary watercourse was tidal or not, a different property boundary may have had to be located.[45] But in the context of title and boundary disputes concerning riparian lands in an estuarine environment, where tidal currents mix with river flows, what constitutes tidality as a matter of law? How can tidality be proven as a matter of fact?

Despite the fact that the question of tidality of a water body arises in many contexts other than land title or boundary disputes,[46] a direct, an unequivocal answer to these questions has not yet been provided in any judicial opinion. Courts have adopted various formulations in addressing tidality, such as "ebb and flow of the tide,"[47] "tidal action,"[48] or "tidal influence."[49] Court opinions also seem compelled to describe the physical characteristic of the watercourse constituting "ebb and flow" of the tide, "tidal action," or "tidal influence," each in a slightly different manner.[50] Given the creative energy of lawyers and surveyors, these different formulations may give rise to at least some debate over whether or not, at a particular geographic location, a specific

---

pictures of steamers, barges, and skiffs on the slough at or farther upstream of Delta Meadows. On the final day of trial, the court took a field trip by boat to and around the Delta Meadows area.

45. In some states, such as California, the determination of whether a water body is tidal is quite important. This is because, by statute, the owner of upland bordering on tidewater takes to the ordinary high-water mark, while the owner of upland bordering on a navigable lake or navigable, but nontidal, stream takes to (owns to) the ordinary low-water mark of the lake or stream. E.g., Cal. Civil Code § 830. For some other states that have made grants or allowed ownership to low water, see *Shively v. Bowlby* (1894) 152 U.S. 1, 18–26.

46. For example, in determining the jurisdiction of the U.S. Army Corps of Engineers under section 10 of the Rivers and Harbors Act, 33 U.S.C. § 403, and section 404 of the Federal Water Pollution Control Act, 33 U.S.C. § 1344, it is necessary to find out what constitutes "navigable waters." Under both such acts "navigable waters" include tidal water bodies. 33 C.F.R. §§ 329.1, 329.4; 40 C.F.R. § 230.3(s)(1). "Tidal waters" are defined by the Corps as those waters that "... rise and fall in a predictable and measurable rhythm or cycle due to the gravitational pulls of the moon and sun." 33 C.F.R. § 328.3(f). Also, in determining the extent of admiralty jurisdiction, the tidality of the waters may be significant. See, e.g., *Executive Jet Aviation, Inc. v. City of Cleveland* (1972) 409 U.S. 249, 253; *Hassinger v. Tideland Electric Membership Corporation* (4th Circ. 1986) 781 F.2d 1022, 1025–1027, cert. den., sub. nom. *Coast Catamaran Corp. v. Hassinger* (1986) 478 U.S. 1004.

47. E.g., *Martin v. Waddell* (1842) 41 U.S. (16 Pet.) 367, 407 ("navigable waters of the Raritan River and bay, where the tide ebbs and flows"); *United States v. Pacheco* (1864) 69 U.S. (2 Wall.) 587, 590 ("daily tides ebb and flow").

48. *Sawyer v. Osterhaus* (9th Cir. 1914) 212 F. 765, 775 ("subject to the tidal action of the waters of the bay"); *An-Son Corporation v. Holland-America Insurance Company* (10th Cir. 1985) 767 F.2d 700, 792.

49. *Mobile Transportation Company v. Mobile* (1903) 187 U.S. 479, 487 ("lands were situated upon a navigable river far above the tidal influence").

50. For example, in the *Mobile Transportation Co.* case, "tidal influence" appeared to constitute the rise and fall of waters as a daily occurrence. *Id.* at 486. In another case, a lake separated from the ocean was not found subject to tidal action even in light of expert testimony that changes in water level of the ocean were immediately transmitted from the ocean to the lake and influenced the lake's water level. *An-son Corporation, supra,* 767 F.2d at 702.

watercourse is tidal. Fortunately, there appear to be means at hand to resolve the question.

First, courts have judicially noticed[51] the tidality of water bodies. For example, in an 1832 case in which the issue was the extent of admiralty jurisdiction, the Supreme Court took judicial notice of the tidality of the Mississippi River at New Orleans.[52] In taking such judicial notice, the court was careful to state that it was not crucial whether tidal influence was either great or little.[53]

Second, tidality may be proven by evidence or testimony concerning the physical character or regime of the water body. Various types of physical criteria have been considered sufficient to establish tidality. One judge set forth what he termed the "ordinary idea of the ebb and flow of the tide: the presence of salt water and if the water comes back from the ocean or sets upward in a current."[54] That same judge also conceded that it was settled that the tide is considered to ebb and flow

> ... in any place where it affects the water daily and regularly, by making it higher or lower in consequence of its pulsations, though no current back be caused by it. [Citations omitted.] Yet this of course must be a visible, distinct rise and fall, and one daily caused by the tides, by being regular, periodical, and corresponding with their movements.[55]

---

51. Judicial notice is an accepted legal device that allows courts to accept certain facts as true without the necessity of any proof. Courts have called such facts ". . . facts, particularly with respect to geographical positions, of . . . public notoriety. . ." *The Planter* (1833) 32 U.S. (7 Pet.) 324, 342. According to a more modern formulation:

> A judicially noticed fact must be one not subject to reasonable dispute in that it is either (1) generally known within the territorial jurisdiction of the trial court or (2) capable of accurate and ready determination by resort to sources whose accuracy cannot be questioned. Fed. Rules of Evid. § 201 (b).

Each state also has its own rules of when and of what judicial notice may be taken. E.g., Cal. Evid. Code § 450–460. For a more detailed discussion of this concept, see Briscoe, J., *Surveying the Courtroom* (2nd Ed., John Wiley & Sons, New York, 1999), Chapter 8 and references cited therein. For an example of a judicially noticeable fact, see, e.g., *Fullerton v. State Water Resources Control Bd.* (1979) 90 Cal.App.3d 590, 604; *Rutten v. State* (N.D. 1958) 93 N.W.2d 796, 798 (judicial notice of drought).
52. The Supreme Court stated:

> We think, that although the current in the Mississippi, at New Orleans, may be so strong as not to be turned backwards by the tide, yet if the effect of the tide upon the current is so great as to occasion a regular rise and fall of the water, it may properly be said to be within the ebb and flow of the tide. *The Planter, supra,* 32 U.S. at 343.

53. *Ibid.* There is a location where tidal influence ends, however. The Corps of Engineers defines that location as ". . . where the rise and fall of the water surface can no longer be practically measured in a predictable rhythm due to masking by hydrologic, wind, or other effects." 33 C.F.R. § 328.3(f).
54. *Waring et al. v. Clarke* (1847) 46 U.S. (5 How.) 440, 496 (Woodbury, J., dissenting).
55. *Ibid.* Among the salient facts in that case that argued against tidality were that the locale in dispute was 203 miles from the sea; the current of the Mississippi River, the watercourse in

In the *Delta Meadows Case* the issue of the tidality of Snodgrass Slough, the boundary watercourse, arose because of the mixed river and tidal regime. Snodgrass Slough at Delta Meadows was 84 miles from the Golden Gate ocean tides. And these were not highway miles; the 84 miles were filled with the bends and turns created by a river and tidal regime. In addition, there had been major modifications to the Delta regime. Thus, even if Snodgrass Slough was currently tidal, there was some doubt of its tidality in 1850.

Expert witnesses[56] concluded, based on descriptive documentary evidence created in or close to 1850,[57] that Snodgrass Slough was tidal in 1850 in the Delta Meadows reach. Physical evidence, including the current existence of tides, also established tidality.[58] Based on such evidence, the court found that Snodgrass Slough was tidal in the Delta Meadows reach and that its water level was controlled by the water levels in the ocean.[59]

Proof of the tidality of the boundary watercourse in 1850 is only the preliminary step in physically locating the ordinary high-water mark property boundary at the line of mean high water.[60] The next step is to decide the date

---

question, was so great as to be seen and felt 40 miles out to sea; the tides at the mouth of the river were but 18 to 20 inches high; and there were many bends in the 203 miles of river between the locale in dispute and the ocean. *Ibid.*

56. The experts were a historical geographer, considered the preeminent authority on the historical geography of the Delta, a surveyor long experienced in the determination of the location of boundaries and titles along tidal or navigable water courses and in the Delta in particular, and a coastal oceanographer knowledgeable in estuarine and riverine regimes and in tidal mechanics and measurements and their interpretation in such regimes.

57. The types of documentary evidence offered and considered were descriptions contained in the field notes of a surveyor of a nearby local reclamation district, a map that showed reclamation of lands above and below Delta Meadows including the installation of tide gates, and the field notes of state engineers who had studied the nearby area. All such documents were created in the period of 1860–1880. Certified copies were admitted into evidence without an objection.

58. For example, to show that ocean tides some 84 miles away controlled the water levels at Snodgrass Slough in the Delta Meadows reach, the oceanographer examined 15 NOS tide stations in the path tides would take to reach Delta Meadows. The oceanographer's conclusion was graphically presented by two tide staffs that represented tidal datums at the Golden Gate and at Snodgrass Slough, respectively. The comparison was astounding. Mean high water at Snodgrass Slough in the Delta Meadows reach differed by only 4 inches from mean high water at the Golden Gate, some 84 miles away. This comparison established the ocean's control of the water level at Delta Meadows and the unlikelihood that recent (geologically speaking) modification of the Delta regime had precipitated the tidality.

59. *Delta Meadows Decision*, Appendix A., para. B.7. Other types of proof may also establish tidality, such as the presence of vegetation adapted to areas subject to tidal flow, the existence or remnants of insects or invertebrates that live in tidally flowed locales, the type of soil that indicates a fully saturated, anaerobic condition, and tidal drainage patterns. See e.g., *United States v. Ciampitti* (D. N.J. 1984) 583 F.Supp. 483, 489.

60. The process of the development and adoption of the mean high-water line as the physical indicia of the ordinary high water mark along tidal waters is set forth in notes 91 through 106 and accompanying text in Chapter 4. In the *Delta Meadows Case*, the court adopted the mean high-water line as the physical location of the ordinary high-water mark. *Delta Meadows Decision*, Appendix A, para. B.5.

in time at which this boundary must be located after 1850. To take that step requires an understanding of property boundary movement principles.[61]

6.7 THE IMPACT OF PROPERTY BOUNDARY MOVEMENT PRINCIPLES

The issue in the *Delta Meadows Case* was, What effect, if any, did the regionwide, human-induced modification of the Delta regime have on the geographic location of the ordinary high-water mark property boundary between swamp and overflowed lands and tidelands in the railroad strip? Subsidiary questions were, What was the rule when movement of the boundary was produced by a mixture of forces: some human-induced, others natural, some gradual and imperceptible, others perceptible and sudden? Did it make a difference that some of these events or forces occurred outside the Delta region, that some took place within the Delta region but not specifically at Delta Meadows, or that some took place directly on the railroad strip? Was the ordinary high-water mark property boundary fixed in geographic location at a particular time as the result of these modifications? At what time? Or did the property boundary continue to move, regardless of the modifications to the regime?

Of course, the parties in the *Delta Meadows Case* presented conflicting answers to these questions.[62] The court rejected the parties' submissions.[63]

---

61. Those principles are summarized in notes 94 through 127 and accompanying text in Chapter 3. A more detailed discussion relating these boundary movement principles to estuaries is contained in notes 136 through 154 and accompanying text in Chapter 5.

62. Note: The parties, positions were taken long before the decisions in *State of Cal.ex rel. State Lands Com. v. Superior Court (Lovelace)* (1995) 11 Cal.4th 50 and *State v. Superior Court (Fogerty)* (1981) 29 Cal.3d 240. The state, recognizing the modifications to the Delta regime, claimed that many gradual and abrupt forces had changed the geographic location of the ordinary high-water mark property boundary within Delta Meadows after 1850. Admitting there was little guidance in case authority, the state asked the court to determine which type of force was the dominant or controlling force. See *Alexander Hamilton Life Ins. Co. v. Govt of V.I.* (3rd Cir. 1985) 757 F.2d 534, 547, n.13. *Delta Meadows Decision*, Appendix A, para. C.2–3. According to the state, the changes to the geographic location of the ordinary high-water mark property boundary were mainly gradual between 1850 and the time the railroad began construction of the roadbed in 1907. Thus, the state's position was that the ordinary high-water mark boundary continued to move until the time of the railroad construction, when a sudden, avulsive change occurred, freezing the boundary as of that time (1907). This position is largely consistent with *State of Cal.ex rel. State Lands Com. V. Superior Court (Lovelace), supra,* 11 Cal.4th at 79–80. See notes 146 through 152 and accompanying text in Chapter 5. On the other hand, the railroad urged a rule of convenience—reference to present-day conditions should establish the ordinary high-water mark property boundary. *Delta Meadows Decision*, Appendix A, para. C.2–3. The railroad's position could be claimed consistent with the rule in *State v. Superior Court (Fogerty), supra,* 29 Cal.3d at 248.

63. The court made this ruling after the case was tried. In consonance with its legal position regarding the time when the boundary should be located, the state submitted proof of the location of the property boundary for both 1850 and 1907.

The court could not determine with "the necessary degree of precision" whether the boundary changes that had occurred over the years were gradual or avulsive.[64] In light of the lack of authority[65] to support the "controlling or determining influence" theory, the court ruled that the property boundary was fixed in location in 1850.[66] Even though this ruling has, as a result of later decided cases, been undermined, examination of the trial of the case is instructive about how one might prove the physical location of the ordinary high-water mark property boundary in 1850.

6.8 PROOF OF THE PHYSICAL LOCATION OF THE HISTORIC ORDINARY HIGH-WATER MARK—USE OF USGS MAPS

Recall that in determining the physical location of the ordinary high-water mark property boundary between swamp and overflowed lands and tidelands, both of the elements of the boundary must be known: the water level and the topography and relative elevation of that topography.[67] In the *Delta Meadows Case* these two elements had to be established as of 1850, long before there were any local measurements of the water level or any detailed survey of the topography survey and subsequent mapping of the land. In addition, Delta Meadows itself, the area surrounding Delta Meadows, and the Delta regime as a whole had been modified by both natural and human-induced activities.

---

64. The court noted:

> Considerable testimony was presented regarding the effects of regional flooding, reclamation, hydraulic mining in the Sierra Nevada, the gradual rise in sea level and a host of other factors affecting the boundary over this time span. The testimony indicated that the individual effects of these changes were not specifically discernible. Some of these forces were naturally induced; others by the acts of man. *Delta Meadows Decision*, Appendix A, para. C.5.

This chapter will discuss much of that expert testimony in connection with the account of the methodology used to re-create the historic property conditions and the historic property boundary at Delta Meadows.

65. No California case had adopted the "controlling influence" concept of the *Alexander Hamilton* case at the time the *Delta Meadows Case* was tried. And the Supreme Court in *State of Cal.ex rel. State Lands Com. v. Superior Court (Lovelace), supra*, 11 Cal.4th at 79–80, basically rejected this concept. See notes 146 through 152 and accompanying text in Chapter 5.

66. The parties did not question this ruling as there was no appeal. The court apparently based the ruling, although it did not specifically so state, on the implied finding that the accumulation of human-induced alterations of the Delta regime constituted an avulsion or an artificial condition. Thus, according to rules governing boundary movement in California, the court fixed the ordinary high-water mark boundary in location as of 1850, before the onset of such conditions. In *State of Cal.ex rel. State Lands Com. v. Superior Court (Lovelace), supra*, 11 Cal.4th at 79–80, the Supreme Court impliedly rejected this holding. As discussed in notes 146 through 152 and accompanying text in Chapter 5, wholesale adoption of such a rule could eviscerate the rule that accretion belongs to the upland owner and create significant problems of proof.

67. *Delta Meadows Decision*, Appendix A, para. B.5.; see notes 38 through 47 and accompanying text in Chapter 3.

How would it be possible to establish the location of the mean high-water line as of 1850, given such conditions?

According to one of the parties, the answer to this question was simple. The railroad contended that maps of the Delta Meadows area prepared by the U.S. Geological Survey[68] and the United States Coast and Geodetic Survey (USC&GS) established the physical location of the ordinary high-water mark at the cartographic location of the waterline portrayed along the shore of Snodgrass Slough in the Delta Meadows reach. If that mapped waterline were adopted as the physical location of the ordinary high-water mark, the only tidelands that would have existed in the Delta Meadows area would have been located in the bed of Snodgrass Slough. Support for this argument was found in what were termed "settled rules governing the measurement of the boundary of tide waters."[69]

On the other side, it was claimed to be axiomatic that such rules apply only in the measurement of *geographical features*, not in the location of property boundaries.[70] In the *Delta Meadows Case* the court found that such maps did not locate the high-water line in tidal marshes; only the apparent dividing line between the marsh and the water is shown.[71] Because the court did not agree that the ordinary high-water mark was shown on these government maps, just how was the historic ordinary high-water mark property boundary physically located?

6.9 PROOF OF THE PHYSICAL LOCATION OF THE HISTORIC ORDINARY HIGH-WATER MARK—USE OF PHYSICAL MEASUREMENTS

In order to physically locate the historic ordinary high-water mark property boundary, the plaintiff in the *Delta Meadows Case* attempted to situate the

---

68. The U.S. Geological Survey (USGS) is a unit of the Department of the Interior created to assist in the classification of public lands and in the examination of the geological structure, mineral resources, and products of the public domain. 43 U.S.C. § 31(a) and (b); 5 West's Federal Practice Guide (1970) § 5235, p. 21. The USGS has completed a nationwide topographic mapping program. These topographic maps are widely used, not only by surveyors, civil engineers, and lawyers.

69. These "settled rules" were based on a concurring opinion in *Knight v. U. S. Land Association* (1891) 142 U.S. 161, 208 (Field, J., concurring). That opinion claimed that the need to decide the location of the ordinary high-water mark ceases once the banks of the tidal stream are reached. The claim was based on the USCS mapping rule that when a water of a larger dimension intersects a water of a smaller dimension, the line of measurement of the water of the larger dimension crosses the water of the smaller dimension from headland to headland. *Id.* at 207. See also note 124 and accompanying text in Chapter 5.

70. 2 Shalowitz, Shore and Sea Boundaries (Dept. Comm. Pub. 10-1 1962) (herein "2 Shalowitz") § 141, p. 367.

71. *Delta Meadows Decision*, Appendix A, para. B.2. The United States Supreme Court appears to have rejected a similar argument. See notes 124 through 125 and accompanying text in Chapter 5. In addition, the expert civil engineer and the expert oceanographer testified, and the Delta Meadows court found, that the term, "ordinary high-water mark," had no accepted scientific or technical meaning. *Delta Meadows Decision*, Appendix A, para. B.4.

two elements of the boundary at the earliest point in time for which accurate physical measurements were available. Once those elements were established, two questions were asked: First, What would have made the water level different in elevation in 1850 from what has been measured or established in (__)? Second, What would have made the land elevation or form different in 1850 from what has been measured or established in (__)? The first step in this very difficult task was to focus on reconstructing the historic water level.

6.9.1 Establishing Initial Water Levels

With the apparently quixotic goal of locating tide records at the time of statehood,[72] experts made extensive investigations to uncover the existence of tide or water level records for the Delta Meadows reach of Snodgrass Slough[73] or any relevant nearby tidal or water level measurements.[74] These searches resulted in the acquisition of water level records from a tide station[75] located at the northerly edge of Delta Meadows. The records[76] covered a period of $8\frac{1}{2}$ years during the 1960s. Even though a full 19 years of tide records were not available,[77] experts[78] described how it was possible to calculate a 19-year mean high-water datum from this shorter series of observations by comparison with a reference station[79] and how that calculation accounted for the masking

---

72. This effort is not as far-fetched as might be imagined. In many of the nation's estuaries, the USCS or the USC&GS established tide stations and accumulated detailed water level records from almost the moment of the United States' acquisition of the territory encompassing the bay. Many of those stations have been in continuous existence. For example, continuous water level records exist from 1850 to the present for the San Francisco Presidio, virtually at the Golden Gate entrance to San Francisco Bay. See 2 Shalowitz, § 543, n. 36, p. 235.
73. In connection with this discussion of water levels, it will be the convention of this chapter to describe the Delta Meadows reach of Snodgrass Slough by the shorthand formulation "Snodgrass Slough."
74. Among the sources investigated were the U.S. Army Corps of Engineers, the State of California Department of Water Resources, the California State Lands Commission, the NOS, the United States Bureau of Reclamation, and local governmental entities.
75. Tide stations are explained at notes 71 through 73 and accompanying text in Chapter 4.
76. The tide records were collected by a State of California Department of Water Resources tide recorder. The recorder provided a continuous pin trace depicting the elevation of the water surface of Snodgrass Slough over time. These tide records were tabulated and certified by the California Department of Water Resources. They were introduced into evidence through testimony of a civil engineer expert in Delta hydrodynamics, Delta tidal measurements, and calculation of Delta tidal datums.
77. The rationale for obtaining 19 years of tide records is explained in notes 74 through 76 and accompanying text in Chapter 4.
78. Besides the civil engineer familiar with calculation of tidal datums in the Delta, the other expert was a coastal oceanographer. This expert witness was experienced in estuarine and riverine regimes and was proficient in tidal mechanics and water level calculations in such regimes. Particularly significant was the fact that the oceanographer had supervised postgraduate studies about the methods and accuracy of the calculation of tidal datums in the Delta.
79. To calculate a 19-year mean high water datum for Snodgrass Slough, a comparison was made of the Snodgrass Slough water level records and those for a tide station where 19 years of tide records were available. These two stations were chosen for a number of reasons. The Snod-

of the tides by high river flows.[80] This type of calculation is routinely done by the National Ocean Survey (NOS), even in a mixed tide and river environment.[81]

As a result of this study,[82] a mean high water for Snodgrass Slough was determined[83] for 1967, a year in the middle of a National Tidal Datum Epoch.[84] The elevation of mean high water was determined relative to the National Geodetic Vertical Datum.[85]

---

grass Slough tide station was in the Delta Meadows area, and the reference tide station was part of the same waterway system as Snodgrass Slough. Moreover, the reference tide station had 19 years of record encompassing the Snodgrass Slough tide station's period of record.

80. To enable comparison of the two water level records, the Snodgrass Slough tide station and the reference tide station were related to a common vertical reference. To find mean high water at Snodgrass Slough, the high-water tides at the Snodgrass Slough tide station were computer-correlated with those same high tides at the reference tide station. Because of extreme river flows, not all the water level records for Snodgrass Slough were used. The record contained 35 days during which no tide was shown at the tide station. These days were not included in the comparison. By not including the record of the water levels on those days, the calculated mean high water was lower in elevation. Stated another way, if the record of the water level for those 35 days had been included in the calculations, the elevation of mean high water would have been higher. See also note 78 and accompanying text in Chapter 4.

81. NOS tidal bench mark sheets were introduced into evidence without an objection. These tidal bench mark sheets published the result of the NOS calculation of tidal datums in the Delta. According to the oceanographer, tidal datums can be calculated in the Delta although river flows affect the water level. This "super-elevation" is a very minor component of the water level in the Delta and, in any event, is part of the Delta water regime.

82. These calculations established not only tidal datums. They also supported the oceanographer's conclusions that even considering the considerable river flows, the water level in the Delta is controlled by the level of the ocean.

83. This mean high water (MHW) determination was examined for accuracy by the oceanographer. The oceanographer independently calculated MHW at Snodgrass Slough by a different method than the method used by the other expert and obtained the identical result. In addition, the oceanographer tested the MHW determination against certain accuracy elements he had developed in his broad experience. Finally, he confirmed that discarding days when no tide was shown in the water level record was consistent with NOS standard procedure.

84. The National Tidal Datum Epoch is the specific 19-year period adopted by the NOS as the official time segment over which tide observations are taken and reduced to obtain mean values for tidal datums. This standardization is necessary because of periodic and secular trends in sea level; i.e., sea level may be rising over time. Tide and Current Glossary (Dept. of Comm., Nat. Oceanic and Atmos. Admin. 1984), p. 14. The National Tidal Datum Epoch concerned in Delta Meadows was 1960–1978. Relating the MHW determination to the year 1967 was done because this was the mid-epoch year when the long-term mean water level would be predicted to occur.

85. The National Geodetic Vertical Datum (NGVD) used to be known as Sea-Level Datum 1929. NGVD is a fixed reference datum, or plane, adopted as a standard geodetic datum for elevations determined by leveling. Tide and Current Glossary, *supra*, at p. 14. In laymen's terms, NGVD is the United States government's zero (0) point or beginning level of all geodetic elevations. To compare elevations on a relative basis, one must measure the elevations from the same geodetic datum. The Delta Meadows court defined NGVD as "the fixed vertical reference point from which all land elevations are measured." *Delta Meadows Decision*, Appendix A, para. B.10.

6.9.2 Re-creation of Historic Water Levels

Knowing the 1967 elevation of MHW at the Delta Meadows reach of Snodgrass Slough was the first step in physically locating the property boundary in 1850. The next step was to determine what conditions might have influenced mean high-water elevations in the Delta to possibly make the 1967 mean high-water elevation different from a mean high-water elevation of 1850. The experts were required to apply their combined knowledge and long years of experience in the hydraulics of the Delta and other estuarine regions and regimes and in re-creating historic conditions in such regions and regimes.

Based on his experience in and knowledge of the Delta's hydraulics, the civil engineer determined that seven factors could have influenced the 1967 water level to possibly make it different from an 1850 water level.[86] The oceanographer testified that it was reasonable to use this method to re-create historic water levels. In addition, the oceanographer agreed with the civil engineers analysis and the identification of the factors influencing the Delta's water level. He also agreed with the direction and magnitude the civil engineer had attributed to each factor relative to the water level.[87] Having whetted the reader's curiosity, the author does not intend to discuss this testimony in detail. The important point is not what the experts said, but the various technical and scientific disciplines employed to solve the problem. Thus, some examples of how technical expertise was brought to bear on this issue and the corresponding conclusions reached by the experts will be worthwhile.

Recall that the Delta Cross Channel was built during the 1950s to allow high flows from the Sacramento River system to be distributed through the San Joaquin River system.[88] Because Snodgrass Slough is part of the San Joaquin River system, the civil engineer asked: Did the Delta Cross Channel affect the 1967 mean high water in Snodgrass Slough as compared with Snodgrass Slough's mean high water of 1850? The answer to that question required the employment of the seemingly incongruous disciplines of river hydraulics and historical geography.

In exploring the pre-European-settlement regime of the Delta, the historical geographer uncovered historic accounts describing a natural connection be-

---

86. Those factors were federal and state water project reservoirs and pumping of water out of the Delta, hydraulic mining debris changing the configuration of the Delta channels, the rise in mean sea level, Delta reclamation, dredging in the Delta; the Delta Cross Channel, and unnatural seepage from the Sacramento River. *Delta Meadows Decision*, Appendix A, para. B.10.

87. This synergistic approach demonstrates how application of systematic and logically thought out expert advice, appropriately communicated between the experts and the lawyer, can function in proving a conclusion at trial. The court's acceptance of the civil engineer's opinions and the oceanographer's benediction, *Delta Meadows Decision*, Appendix A, para. B.10., was based on the breadth and scope of these experts' knowledge and experience. It would be most welcome to have expert witnesses of such caliber in every case.

88. *United States v. State Water Resources Control Bd.*, *supra*, 182 Cal.App.3d at 99.

tween the Sacramento River and San Joaquin River systems. A slough, similar in function and location to the Delta Cross Channel, had existed in 1850. Advised of this historic fact and based on his experience in the Delta's hydrodynamics, the civil engineer concluded that this natural connection served the same function in 1850 as the Delta Cross Channel did in 1967. That slough (now lying behind a levee within the interior of a Delta island) served to transfer water from the Sacramento River to the San Joaquin River system during flood conditions in the Sacramento system. Thus, the expert civil engineer and the expert oceanographer concluded that the 1967 water level in Snodgrass Slough would not have differed from its water level in 1850 solely as result of the construction and operation of the Delta Cross Channel.

As a further and final example, a refined technical analysis was required in determining the effect of Delta reclamation[89] since 1850 on the comparative water levels in the Delta (and Snodgrass Slough) between 1850 and 1967. The civil engineer and the coastal oceanographer observed that the purpose of reclamation was to exclude the tidal water from the reclaimed areas. Consequently, in 1850, prior to reclamation, a greater area of the Delta was subject to the ebb and flow of the tides than in 1967, after reclamation. Knowing only the 1967 mean high-water level of Snodgrass Slough, how could one account for the effect of this modification of the Delta regime to determine whether the 1850 water level in Snodgrass Slough would have been any different from the 1967 water level, and in what magnitude and direction?

To determine whether Delta reclamation would have made the 1967 water level at Snodgrass Slough different from the Snodgrass Slough water level in 1850, the civil engineer operated a computer model specifically designed to re-create the Delta's hydrodynamics.[90] Relying on advice and consultation with other experts and on his experience in the Delta,[91] the civil engineer concluded that only a certain percentage of the reclaimed Delta islands had been an effective part of the tidal regime before reclamation. The civil engineer varied the computer model to account for this factor. Based on the results from the computer model, the civil engineer concluded that the effect

---

89. The historical geographer testified about the progress and process of reclamation in the Delta. His conclusion was that the Delta was haphazardly reclaimed in fits and starts over a 40–50-year period. Reclamation was essentially complete by about 1910–1920. He also prepared an exhibit that showed the geographic progress of Delta reclamation over time.

90. By way of foundation, the civil engineer described the computer model and the assumptions incorporated into its design. This computer model had been created and used to analyze the effect on Delta water levels of various physical conditions. The model had been found reliable in other litigation.

91. Based on his experience and observations, the civil engineer noted that the prereclamation islands in the Delta along the major rivers and discontinuous natural levees rimmed distributary channels. Water entered these areas through distributary sloughs that breached the levees. In the opinion of the civil engineer, the interiors of the prereclamation Delta islands contained areas of dead storage where water would not be completely exchanged. In addition, the civil engineer also concluded that the interiors of such islands were very inefficient conveyors of water; they were choked with vegetation.

of reclamation was to raise the Delta's water levels over time. Thus, he found that the elevation of mean high water at Snodgrass Slough in 1850 would have been lower than the elevation of mean high water in 1967. According to both the civil engineer and the coastal oceanographer, this rise in relative elevation was a large-scale change, which at any one point could not have been detected by human observation.[92]

Using this process of reasoning back from known facts, the experts were able to reconstruct, with an acceptable degree of precision, the elevation of mean high water in Snodgrass Slough in 1850 relative to NGVD.[93] Each of the factors that could have affected the water level elevation of 1850 as compared with the 1967 water level elevation was accounted for or quantified. As a result, the experts concluded that the elevation of mean high water in Snodgrass Slough in 1850 would have been lower than the elevation of mean high water in Snodgrass Slough in 1967.

Determination of the water level component of the boundary was only part of the battle. It was necessary to physically locate the mean high-water line. That water level had to be established in relation to the topography and relative elevation of the land surface.

6.9.3 Establishing the Location of the Perimeter Watercourse

As a predicate to establishment of the location of the mean high-water line, the perimeter watercourse, Snodgrass Slough, had to be located in its geographic position before it was changed by the works of humans in its Delta Meadows reach.[94] Geographical location of Snodgrass Slough would establish the limits of cartographically determinable tide and submerged lands (the bed of Snodgrass Slough) and their relation to current property boundaries. In addition, establishment of the location of Snodgrass Slough would also establish a link to tidal infiltration through the slough to the interior of Delta Meadows. The surveyor and the civil engineer, both experienced in the reconstruction of historic conditions in the Delta, were requested to locate the last natural channel of Snodgrass Slough.

The task began by researching all available sources to find the earliest authoritative planimetric survey map[95] of the configuration of Snodgrass

---

92. Even construction of reclamation levees near Delta Meadows would not have discernibly raised Snodgrass Slough water levels or caused a ponding effect in the Delta Meadows area. The experts agreed that the water level effect of construction of such levees would have been distributed over the Delta as a whole.
93. *Delta Meadows Decision*, Appendix A, para. B.11.
94. Application of property boundary movement principles established the date in time in which the boundary was to be located. See notes 62 through 66 and accompanying text in this chapter.
95. A planimetric survey or map presents the horizontal position of the physical features represented. 2 Shalowitz, Appendix A, p. 592. Sometimes such maps may show the horizontal position of elevations or contours of equal elevation. In strict definitional terms, however, planimetric surveys or maps omit relief.

Slough in the Delta Meadows reach. Mere cartographic depictions or cadastral surveys[96] were not deemed sufficient. For example, the very earliest surveys discovered were those accomplished in conjunction with the issuance of swamp and overflowed patents by the state. These cadastral surveys purported to meander Snodgrass Slough, but were shown to be unreliable.[97]

As a result of this exhaustive search, the experts ultimately obtained a manuscript map and the survey field books of the U.S. Army Corps of Engineers and the survey field books of the railroad. Both of these sources portrayed the Delta Meadows area and the railroad strip in particular and provided the principal basis for geographically locating the extent and alignment of the last natural channel of Snodgrass Slough just after the railroad roadbed and trestle were constructed.[98] The location was confirmed by field visits to Delta Meadows.[99] The experts also concluded that the Snodgrass Slough channel had been in that location long before the twentieth century.[100]

---

96. Cadastral surveys are government surveys of public land boundaries. Such surveys may also include the horizontal position of physical features.

97. Given the technology at the time, the surveyor prepared an exhibit that plotted these surveys with respect to their individual calls. The exhibit was a clear Mylar overlay referenced to a large aerial photograph (scale 1" = 400'), which served as the base map of the Delta Meadows area. Because of that effort, the surveyor concluded that the surveys claiming to meander Snodgrass Slough did so only in the broadest sense. One of the surveys entirely ignored the existence of the slough. The surveyor also found that the surveys were not internally accurate nor did they connect properly. Comparison of the survey calls as plotted to early and modern cartographic depictions of Snodgrass Slough affirmed that the surveyor's conclusions about the accuracy and veracity of the cadastral surveys' location of Snodgrass Slough were correct. This is an example of how documents disclosed in a chain of title can be used and analyzed in the process of physically settling the property boundary. With the GIS system, assuming there is adequate base information, this task may be accomplished more readily and with somewhat less effort.

98. The civil engineer was furnished with the field notes of both the Corps of Engineers and the railroad surveys. He relied on both sets of field notes to reconstruct the last natural channel of Snodgrass Slough. As a matter of foundation, the civil engineer explained how the railroad and Corps surveyors each accomplished their respective surveys of Snodgrass Slough; that he reviewed the field books of both surveys and compared both sets of notes with the Corps of Engineer's manuscript map to assure himself of that map's accuracy, that he personally plotted the railroad's survey and found it to be a detailed identification of the bank line that carefully measured the sinuosities of Snodgrass Slough. The surveyor testified that the survey was so detailed that it registered the existence of "beaver cuts," breaks or discontinuities, along the bank of the slough and depicted their dimensions in the field notes. Once the civil engineer plotted the railroad survey field notes, he compared that survey plot with contemporary photographs, the Corps of Engineers map, and maps of the USGS for accuracy. The result of that comparison was that the railroad survey of the sinuosities of Snodgrass Slough was very close, in general form, to the sinuosities of Snodgrass Slough depicted in those sources. As a result, the civil engineer used the railroad's own survey to form the last natural channel of Snodgrass Slough as supplemented by the Corps of Engineers map. An exhibit map, an overlay to the base mapping, showed this horizontal position.

99. Such visits assisted the expert by confirming that the slough had well-defined banks, that the beaver cut features described in the field notes still existed, and that the slough was a permanent feature.

100. The civil engineer based this opinion on the fact that the slough's location had not changed since 1907, the date of the railroad survey, as compared with current conditions and a review of earlier general maps. The oceanographer's conclusion was identical.

Establishing the horizontal location of the adjacent watercourse was an important step. Re-creation of the historic mean high-water line in Delta Meadows was a more difficult task. The topography of the land and its relative elevation had to be demonstrated. Tidewater intrusion in and onto that topography had also to be corroborated.[101] These tasks required the combined skills of all the many experts. The first step was re-creation of the historical physical condition and character of Delta Meadows.

6.9.4 Establishing the Historical Physical Conditions of Delta Meadows

The pre-European-settlement physical character and attributes of the area were established by expert testimony. Experts qualified in historical geography, geology, geomorphology, oceanography, plant ecology, and soils science examined, among other resources,[102] state and federal land surveys, topographic mapping by the USGS and the Army Corps of Engineers (then known as the United States Engineering Department), field books of detailed topographic surveys accomplished by railroad surveyors, ground and aerial photographs, and soils and vegetation mapping. These sources were complemented by painstaking (and sometimes painful) field reconnaissance.

All the documents and materials examined by the experts were admitted into evidence, and many were displayed in graphic form. To take just one example, the federal surveyor's field notes of the federal township surveys[103] contained narrative accounts and portrayals describing physical conditions at various geographic locations in or in the vicinity of Delta Meadows. The surveyor, conversant with historical surveying methodology and surveys of both the federal and state governments, prepared an exhibit that was geographically referenced to base maps and aerial photographs of the Delta Meadows area. The surveyor plotted the location of and quoted descriptive accounts of the then-extant physical character of the land contained in the U.S. Deputy Surveyor's field notes.[104] Those field notes established that Delta

---

101. This criterion, that there must be tidal penetration, is a recognition that to be tideland the lands must be tidewater flowed. This criterion also prevents any tideland claim to areas, such as, to pick the most extreme example, Death Valley. These areas are below the elevation of mean high water, but not (presently at least) tidewater flowed.
102. These other resources consisted of historic narrative accounts found in field books of surveyors of nearby special districts. These districts were created to reclaim adjacent lands. Other sources used were field notes of investigations by state and federal agencies concerned with flood control and reclamation and photographs taken by railroad engineers and surveyors in conjunction with the construction of the railroad through Delta Meadows. Obviously, such a wealth of sources of information about historic physical character will not be available in every case. Determined, creative investigation can, more often than not, lead to sources that at first may not have been considered.
103. As a foundation, the surveyor testified in detail on both the purposes of the federal township survey and just how the United States conducted such surveys in the field. The method of surveying townships is described in notes 54 through 65 and accompanying text in Chapter 1.
104. A GIS system would be of considerable assistance (even vital) to accomplishing these tasks today.

Meadows was a tule swamp, covered by water and impenetrable vegetation cover, so thick and overgrown that the surveyor was unable to gain access to complete a portion of the township survey.[105]

Relying on this information, the other experts described the features of Delta Meadows to the court.

6.9.5 Establishing the Topography and Prominent Physical Features and Attributes of Delta Meadows

Geomorphically,[106] the Delta Meadows back swamp was bounded on the west by the natural levee of the Sacramento River and on the east by a sloping alluvial plain. Snodgrass Slough serviced the basin, allowing the ingress and egress of tides and the drainage of seasonal flows. Within those boundaries, the main geomorphic features were identified. One of the identified features of particular significance to the case was the Snodgrass Slough levee. This levee[107] was not a "Chinese wall;"[108] it was irregular in height and discontinuous. The most significant geomorphic feature to the property boundary issue, however, was the marsh flat.

The marsh flat, characterized by saturated soils and aquatic plant communities, constituted most of the Delta Meadows area.[109] It was interlaced with a series of interior channels, all of which connected to Snodgrass Slough through breaches in the discontinuous Snodgrass Slough levee.[110] These in-

---

105. A series of aerial photographs of the Delta Meadows area validated the description in the field notes. These aerial photographs portrayed, at many locations, physical conditions identical to those described in the federal township survey field notes. Field reconnaissance also confirmed the conclusion.
106. This testimony was provided by the coastal oceanographer, who was also a registered geologist and had been trained in geomorphology. This type of testimony has been considered favorably by courts in other cases as well. *Arkansas Land & Cattle Co. v. Anderson-Tully Co.* (Ark. 1970) 452 S.W.2d 632, 637–638.
107. The coastal oceanographer compared the Sacramento River and Snodgrass Slough to point out the differences in the levee building process. To form natural levees, two elements are required: (1) coarse sediments and (2) turbulence to suspend sediments. The Sacramento River has both elements. Snodgrass Slough, on the other hand, has neither a source of coarser sediments nor the velocity to carry them if such sediments had been available. This conclusion was consistent with the historical geographer's knowledge of natural levees in the prereclamation Delta. In the back swamp area, the levees were much less defined than along the distributaries of the Sacramento River. The historical geographer noted that these levees undulated and their continuity was broken by the dendritic pattern of sloughs extending into the interior. The description is consistent with earlier judicial discussions of this physical feature. *Gray, supra,* 174 Cal. at 626–627.
108. The use of this term to describe and denote an impenetrable barrier may be generally accepted, but has been roundly criticized. See, *Peat, Marwick, Mitchell and Co. v. Superior Court* (1988) 200 Cal.App.3d 272, 293 (Low, J., concurring).
109. Assuming sea level continues to rise, the marsh flat will eventually submerge the other geomorphic features. It is the growing part of the Delta.
110. See *Gray, supra,* 174 Cal. at 628.

terior sloughs served to conduct the tidal waters of Snodgrass Slough in and out of the marsh flat on a daily basis and to distribute runoff waters as well. Because the legal character of the marsh flat as either swamp and overflowed land or tideland was the issue, a detailed understanding of its physical character was important.

The coastal oceanographer compared Delta Meadows' marsh flat geomorphology to that of the San Francisco Bay salt marshes.[111] A significant result of this comparison was that the coastal oceanographer concluded that the controlling influence on the geomorphology of the Delta Meadows marsh flat was the tidal regime of San Francisco Bay.

The testimony was confirmed by physical evidence. According to the surveyor, a comprehensive and site-specific description of the character of Delta Meadows was provided by field books and photographs of the Delta Meadows area. The photographs were taken by the railroad immediately before construction of the roadbed across Delta Meadows. The field books contained detailed information on the character of the land and soil and the then-extant vegetation, as well as references to the existence and location of watercourses and the area's then-contemporary cultural use.[112] Photographs,[113] field notes, and manuscript maps of other historic topographic surveys[114] confirmed the

---

111. According to that testimony, both marshes support aquatic vegetation. One difference between the two marshes is that the San Francisco Bay marsh supports salt-tolerant aquatic plants while the Delta Meadows marsh supports tules, an aquatic, but not salt-tolerant, plant. Both marshes, however, sustain only a restricted variety of plants that live within a narrow range of vertical elevation. The soils of the two marshes are highly organic saturated soils. Deposition of sediment and organic debris has built both marshes as sea level has risen. Both are exceedingly flat. Both marshes are right at about the elevation of high tide and will be inundated at least once daily by the daily high waters. Although natural levees are not apparent in the salt marshes, at Delta Meadows there may be subdued natural levees about $\frac{1}{2}$ inch higher than marsh flat terrain. The channel density in both the marshes is very high. In the San Francisco Bay salt marshes, when the neighboring slough is filled with water, sheet flow will occur across the salt marsh. In other marsh environments with subdued levees, such as at Delta Meadows, tidewater enters through the interior channels. The salt marsh interior channels in both marshes are long-lived.

112. The surveyor and the civil engineer analyzed these field books in detail. The surveyor prepared an exhibit that, referenced to base mapping, geographically located the field book descriptions of land character at the site described by the railroad surveyors. According to those field books, portions of the area were blanketed with tule growth, interlaced with sloughs and covered by a lake.

113. The railroad photographs were a telling reminder of the character of the land before and during railroad construction. The surveyor placed and oriented the photographs in the direction from which they were shot with reference to the base mapping. These photographs confirmed the railroad surveyors' field notes' descriptions. The earliest aerial photograph of the property was taken in 1931. This aerial photograph and later aerial photographs were introduced into evidence. The surveyor, the coastal oceanographer, and other experts interpreted the aerials. All these experts were familiar with the interpretation of signatures of features depicted by aerial photographs.

114. The U.S. Army Corps of Engineers and the USGS accomplished these surveys in the first decades of the twentieth century. The surveyor interpreted the then-current cartographic symbolization on the manuscript maps of these surveys.

descriptions contained in the field books. That is not to say that all of the historic sources were perfectly congruent. A short diversion will explain the resolution of apparent inconsistencies in cartographic depiction of the same area by two historic maps created at about the same period of time.

The Corps of Engineers map of 1914 did not show any interior watercourses at Delta Meadows. On the other hand, the 1905–1907 USGS topographic map quite prominently depicted such features. This apparent inconsistency was confronted and explained by the surveyor. That surveyor was knowledgeable in the purpose for and methodology of the preparation of these topographic maps.

The explanation does not involve an inventive USGS surveyor. The expert surveyor testified at trial that this seeming inconsistency arose because the Corps of Engineers depicted watercourses under a different convention than did the USGS. Consequently, the Corps of Engineers and the USGS maps were not inconsistent because the USGS depicted a watercourse feature that did not exist in fact.

The reason for the conflicting portrayals was that the vertical reference datum to which the Corps map and the USGS maps were each referred was not the same. The practice of the Corps was to graphically depict watercourses whose surface was at a relatively higher point than the surface of the same watercourse as depicted by the USGS. Thus, the Corps map would not show the same watercourses as the USGS map. In layman's terms, the Corps of Engineers map would show a "drier" Delta Meadows than the USGS map. The datum disparity of the two maps resulting in an apparently inconsistent cartographic representation of the same area is an object lesson in the inappropriateness of comparing apples to oranges.

Physical evidence confirmed the surveyor's conclusion regarding the datum disparity. The features described by the early USGS surveyors as well as in the railroad field books (but not by the Corps map) persist in more modern depictions of the area.[115] According to the coastal oceanographer, this congruity should be expected because of the persistence of the marsh flat features over time.

6.9.6 Establishing Historic Vegetation and Vegetation Patterns

A more specific study of the character of the vegetation of Delta Meadows was undertaken by an expert plant ecologist.[116] The plant ecologist prepared

---

115. This conclusion was based on a comparison of the signatures depicted on aerial photographs taken over time. For example, features noted by the railroad surveyors, shown on the 1906 USGS topographic map and in a 1931 aerial, were still present in current aerial photography. The existence of these features was also verified on the ground.
116. The plant ecologist was skilled in differentiation of vegetation types from one another and in mapping of vegetation cover types according to dominant species. He was also knowledgeable about the habitat, propagation, and growth of tules in the Delta. See note 119 in this chapter.

a map showing the historic vegetation types in the Delta Meadows area. In order to prepare such a map, the plant ecologist made extensive site visits, taking aerial photographs with him during the visits. These visits allowed him to observe how different vegetation cover types signified themselves in aerial photographs. While in the field he used certain conventional methods[117] of identifying plant cover types and noting their relative abundance and dominant species.

To document the vegetation types as they existed in 1850, the plant ecologist reviewed accounts of persons who lived in that era.[118] In incorporating all these sources, the plant ecologist used his investigations of and experience with the present conditions, content, and configuration of the vegetation to weigh and calibrate the information on past vegetation conditions and configuration. The result was a map re-creating the location and extent of vegetation cover types as of 1850. The map was prepared by using the earliest reliable information to modify or enhance the vegetation pattern shown in the earliest aerial photograph.

The plant ecologist found that tules[119] and cattails were the dominant cover type. Tules and cattails, as opposed to the other cover types present at Delta Meadows, are aquatic plants. This was a notable fact. The plant ecologist testified that tules have a significant demand for water, which must be present during the time the plants are growing. The plant ecologist was of the opinion that the water requirement for tules, their water budget, could be met only by an abundant and reliable year-round source of water. Rainfall and spring and winter high flows would not perpetuate a stand of tules as extensive as the currently existing stand nor the stand that, according to historic accounts, had once been present at Delta Meadows. The ecologist concluded that the source of this water must be Snodgrass Slough, the only perennially available source of water in the area.[120]

---

117. Once one knows the dominant species, it is a relatively straightforward matter to extrapolate from site visits to the general pattern of vegetation observed in aerial photographs. According to the plant ecologist, the dominant species usually shows up conspicuously on aerial photos.

118. These sources included the field notes of the 1859 federal township survey as plotted and located by the surveyor, an 1861 report of the State Board of Swampland Commissioners, field notes of the state engineer prepared in the late 1800s, the railroad survey field books and photographs, and the 1914 mapping by the Corps. The plant ecologist specifically relied on the railroad surveyors' field note descriptions of the landscape features through which these surveyors were passing. The survey notes assisted the plant ecologist in precisely placing the extent of features observed and noted by railroad surveyors.

119. Tules are an emergent aquatic plant, a plant rooted in soil with shoots sticking up above the surface of the water. Courts have taken judicial notice of where tules will grow. *United States v. Otley* (9th Cir. 1942) 127 F.2d 988, 1000.

120. In that connection the plant ecologist observed a lack of cover type on the map by noting the existence of water. He personally observed the waters from Snodgrass Slough enter the interior of Delta Meadows through breaks in the Snodgrass Slough levee. He also observed tidewater channels extending from the breaks in the levee into the interior of Delta Meadows.

Moreover, the plant ecologist analyzed the railroad surveys to see how the descriptive accounts of vegetation in the field notes related to elevations determined by the railroad surveyors. According to the plant ecologist, the range of elevations at which a water surface occurred and at which tules were described by the railroad surveyors were almost congruent.[121]

The plant ecologist's vegetation map did not depict a specific property boundary. It showed the historic configuration of vegetation cover types over time. Because of manuscript references to tules as far back as 1859 and the fact that tules take some time to make their presence known, the plant ecologist believed it reasonable to conclude that the dominant vegetation in the mid-1800s was a tule marsh.[122] Indeed, that witness concluded that there was more extensive aquatic vegetation existing in 1850 than present today at Delta Meadows. The persistent appearance of tules established that the soils were wet virtually year-round and portions of Delta Meadows had been a tule swamp long before California became a state.

6.9.7 Establishing the Soils Types Present

Further attributes of the physical character of Delta Meadows were revealed by analyzing of the types of soils present. A soils scientist[123] was asked to classify and map the soils at Delta Meadows. The soils scientist conducted extensive investigations preliminary to the preparation of the soils map. These investigations included site visits[124] and review of manuscript sources (topo-

---

121. There was, however, an interesting anomaly. Pasture, a cover type usually appearing higher in relative elevation than tules, sometimes shared the same elevation as tules, as reported by the railroad surveyors. The plant ecologist explained this seeming irregularity by noting that horses or cattle had been grazed in the tules. The historical geographer confirmed this pattern of cultural use. According to that expert's testimony, before reclamation occurred, settlers allowed livestock to feed in the tules. Thus, reference to pasture in the surveyors' field notes could well have referred to tules used for pasturage.

122. This opinion concerning the long-term persistence of tules was borne out by soils borings taken by another expert, a soils scientist. These borings established that tules had been present at Delta Meadows for hundreds, if not thousands, of years. See note 127 in this chapter.

123. The soils scientist had experience in the soils of California and their origins, particularly the soils in the Delta. He was skilled in conducting soils investigations, in mapping soils types and conditions, and in reconstructing soils types in disturbed areas. In addition, he had conducted extensive research in the Delta and had a particular familiarity with the aging of the peat soils that have formed as the Delta has grown. Peat is described as the partially carbonized vegetable tissue that forms as plants decompose in water and are deposited and compacted. The deposition of the Delta peat soils began about 10,000 years ago as marsh plants died, decayed, and accumulated under oxygen-deficient conditions.

124. On such site visits the soils scientist sampled soils at more than 50 different sites. The soils samples were taken by hand auger to a depth of 5 feet. These borings disclosed the makeup of the soils, the hydrology of the area, and the patterns of deposition and drainage. Individual boring sites were chosen to relate to different soils types that might be shown by the presence of different vegetation or signatures on aerial photographs. To help in the preparation of a soils map, the boring sites were also geographically located by field survey accomplished by the civil engineer.

graphical maps, aerial and ground photographs, and the railroad survey).[125] This is an accepted method of investigation to determine soils types.[126]

The soils map depicted the soil types present in 1850, but did not show a particular property boundary. On the map, the soils scientist denominated the soil types present by choosing terms commonly used to refer to soils of the types found at Delta Meadows. In his testimony, the soils scientist also discussed the drainage characteristics of each soil type and the type of vegetation it supported.

For example, the soils scientist noted that a particular soil type, Burns soil, occupied the major part of Delta Meadows. This soil type is identified by its highly organic composition. In Burns soil, the water table is at or near the surface. The Burns soil supported an aquatic plant community of tules and confirmed their long-term existence.[127] The soils scientist also represented on his map the existence of water bodies in the interior of Delta Meadows.[128]

Although the testimony of these experts established the pre-European-settlement physical character of Delta Meadows, such testimony did not in and of itself end the court's inquiry. The fact that the area had been, in large part, a tule marsh for hundreds of years was of considerable significance in providing support for the court's boundary determination. Yet that fact alone did not settle the physical location of the ordinary high-water mark property boundary, the mean high-water line, within that marsh.[129] The establishment of the historic mean high-water line by the placement of the elevation of mean high water on the historic topography would resolve its location. It is to re-creation of that topography and location of the historic mean high-water line that this chapter turns next.

---

125. The soils scientist also used the railroad's soils investigation to reconstruct the soil conditions in the railroad strip area disturbed by the construction. For example, the soils scientist concluded that the railroad built a trestle at a particular location because of the depth of peat soil in other locations disclosed by the railroad's soils investigation. That expert opined that peats of the depth shown at other locations in Delta Meadows could not support the massive railroad work.
126. The soils scientist did not rely on government soils maps. These maps were not as detailed as the soils scientist's map. Nor were they prepared using aerial photographs.
127. This Burns soil had, on the average, an organic thickness between 2 and 3 feet. The soils scientist had studied how long it takes to establish such organic layers. Such layers required a rising water level to avoid peat oxidation, which occurs when peats are exposed to air. According to the scientist, it takes 150 to 200 years to develop 1 foot of peat. It was his opinion that the age of the peats in the Delta Meadows area was about 800 to 1000 years. Thus, he concluded that tules were not a recent phenomenon at Delta Meadows.
128. The soils scientist closely examined the interior channels both by boat and by personal depth sounding. He fell into a channel 3 feet deep and 6 to 8 feet wide. He believed these interior channels had been there for hundreds of years. It was the soils scientist's opinion that the saturation of the Burns soil related to the height of the water in Snodgrass Slough.
129. Some would end the inquiry at this point and draw the conclusion, from the evidence of the physical condition and attributes, that Delta Meadows is swamp and overflowed in legal character rather than tidelands. On the other hand, as previously noted, the physical character of the land is claimed to be equally consistent with land of the legal character of tidelands. See notes 23 through 39 and accompanying text in Chapter 5.

6.9.8 Re-creation of the Historic Topography of Delta Meadows

To the untrained observer, because of its overgrown condition, Delta Meadows may appear wild and primeval. In fact, the basic geomorphic and physical features of Delta Meadows as described by the experts remained essentially unchanged. Nevertheless, like the rest of the Delta, Delta Meadows had been altered by the intrusion of humans; its current topography was not the same as the topography that had existed in 1850.[130] Areas within or adjacent to Delta Meadows had been leveed, drained, and reclaimed. A massive railroad roadbed had been built on the railroad strip, over and through Delta Meadows.

On a more reduced scale, two low and subdued levees had been installed in another portion of Delta Meadows. Dredger cuts had been made to obtain the materials necessary to build certain of those levees as well as the railroad roadbed. These dredger cuts provided additional pathways for intrusion of tidal waters into the Delta Meadows interior. Moreover, watercourses that were distributaries of Snodgrass Slough and Snodgrass Slough itself had been dredged and the configuration of their beds altered.[131]

In addition to these certain pronounced and readily identifiable topographic changes, there were some very subtle, not easily recognized, but significant regional topographic changes. These regional changes had been taking place over years in the Delta and could have affected Delta Meadows. In order to reliably re-create the Delta Meadows topography of 1850, it was necessary to acknowledge and account for those regional topographic changes. The skills of expert witnesses were required to discover the factors to be considered and to account for them. The first step in this process was to establish the topography to which expert-determined factors would be applied.

The 1850 topography of Delta Meadows was developed through the same methodology used to derive the 1850 water level: Find the earliest reliable topographic mapping and ask how that topography would have differed from a topography of 1850. The civil engineer prepared the topographic map. He based his depiction of the topography of Delta Meadows on two principal sources: the 1914 Corps of Engineers survey and the 1907 railroad survey of the railroad strip.[132] These sources were supplemented by the 1905–1907

---

130. The same sources of information discussed in note 102 and accompanying text in this chapter supported expert testimony about specific alterations humans had made to the immediate area of Delta Meadows.
131. The oceanographer concluded that the levees and dredged channel had virtually no impact on the geomorphology of the Delta Meadows area. Certainly, some alteration in the interior channels would have occurred, but water would have flowed into interior marsh areas through marsh channels that preexisted the dredger cuts just as readily as via the dredger cuts.
132. As a matter of foundation, the civil engineer recounted how the Corps and railroad surveyors accomplished such topographic surveys. The objective of these surveys was to learn the average elevation of the ground in the area surveyed. Where possible, the civil engineer verified the vertical (elevation shown) and horizontal accuracy (geographic location of the vertical elevation) of the topographic information contained in these surveys.

6.9 USE OF PHYSICAL MEASUREMENTS   **225**

USGS topographic map, a 1965 photogrammetric survey[133] of Delta Meadows, aerial photos, and the field notes of the civil engineer's own field survey.

The civil engineer testified in great detail how that topography was reconstructed and how different survey information was used in the preparation of the topographic map. For example, topographic elevations from the Corps of Engineers' survey were used and had to be related to information from the railroad's survey and vice versa. Unfortunately, relating the topographic elevations from the two surveys to each other was not straightforward. The elevations shown on the two surveys were not measured from the same vertical reference plane or datum.[134]

Datum differences are a potential problem any time one wishes to compare or relate two surveys of the same area by different surveyors.[135] In order for the elevations from the two surveys to be used together and, in turn, related to a water level elevation, such as a tidal datum (which is referred to another entirely different datum), all the sources of vertical elevations must be related to the same vertical reference datum. In Delta Meadows the datum chosen was NGVD, the datum to which water level elevations had been referenced.[136]

To reconcile the elevations shown in the surveys and the water level elevations, it was necessary to obtain a conversion factor between the datums used in the topographic surveys and the datum used in the water level determination. Use of the conversion factor allowed the civil engineer to relate the elevations in the two topographic surveys to each other and to re-create the historic topographic conditions depicted in the two surveys, all with reference to NGVD.

With respect to the topography, as far as the area now lying under the railroad roadbed in the railroad strip, the civil engineer had the most detailed topographic information. In areas outside the railroad survey, the Corps survey was used as the basis for the topography. In areas where neither the Corps nor the railroad survey provided topographic information, the topography was derived from the USGS topographic map or the 1965 Department of Water Resources photogrammetric survey (for contour shape and feature character).[137]

---

133. A photogrammetric survey is a survey of a land area utilizing aerial photographs. The survey is reduced to map form by use of stereoscopic or other similar equipment. 2 Shalowitz, Appendix A, p. 591.
134. The Corps' survey was referred to what was known as USED (United States Engineering Department) datum. The railroad had its own vertical reference datum, differing from USED datum.
135. Recall the discussion of the datum disparity between the Corps of Engineers topographic map and the railroad's topographic map (comparison of apples and oranges). See the discussion in Section 6.9.5, above.
136. Any vertical reference datum could have been chosen if all the elevations could be related to such a datum.
137. To confirm the geographic extent, contour shape, and relative elevation of features not shown in detail in the other basic topographic information, aerial photography and field inspection were incorporated.

As a result of these efforts, the topography of Delta Meadows was recreated for the 1910 era, the era of the last reliable, detailed topographic surveying. Investigation by the experts did not discover any specific, detailed topographic depiction of any portion of Delta Meadows that had been created in the mid or even late 1800s.[138] The question of how the 1910-era Delta Meadows topography reconstructed by the civil engineer would have differed from the 1850 Delta Meadows topography was left to be answered by the other expert witnesses.

In response to that inquiry, both the geologist and the soils scientist noted that the topographic features at Delta Meadows were persistent. That is, the basic geomorphic character of Delta Meadows had remained essentially unchanged since 1850, if not before.[139] This conclusion did not eliminate the possibility that more subtle changes had occurred over the years. Such changes, although not affecting the basic topographic features of Delta Meadows, probably affected its relative elevation. The experts concluded that there were two such possible subtle changes: subsidence and deposition.

From the modern perspective, it is a well-known fact that parts of the Delta have been considerably lowered in elevation. Some consider this to be "subsidence." As a consequence, the soils scientist examined whether, what has been popularly referred to as "Delta subsidence"[140] had occurred at Delta Meadows. Because Delta Meadows had not been reclaimed, the soils scientist concluded that there had been no "Delta subsidence" in Delta Meadows. This conclusion did not eliminate the possibility that since 1850, subsidence, as geologist's define the term,[141] had occurred at Delta Meadows.

To determine if regional subsidence had occurred the surveyor and the geologist compared elevations of bench marks in the Delta and in the Delta

---

138. For example, around 1887, the state engineer produced a topographic map of the Central Valley that included Delta Meadows. This sounds like a much more valuable resource than it actually was. The value of the map for topographic information in the litigation was considerably diminished because its contour interval was 20 feet and its scale was so small that the entire Central Valley of California was shown on a manuscript map less than 6 feet long and 2 feet wide.

139. This conclusion was based, in part, on the cartographic depictions and written descriptions of the physical character of the Delta Meadows area by early surveyors and the persistence of those same features in more modern depictions of the area. The surveyor interpreted the symbolization contained in those early survey maps. According to the geologist, one should expect persistence of such features. The geologist specifically noted persistence of the marsh flat features over time. His opinion was supported by a comparison of aerial photographs taken 46 years apart. The features shown in the earliest aerial photograph persisted in more recent photographs taken almost a half-century later.

140. According to the soils scientist, the term "Delta subsidence" referred to loss of soil surface caused by oxidation and decomposition of underlying peat soils. As peat soils become exposed to air as a result of reclamation and drainage, they begin to decompose and desiccate. Moreover, farming these soils exposes them to wind erosion. These occurrences lower the surface; the soil simply disappears.

141. The geologist defined subsidence as the lowering of the relative elevation of the entire region. He contrasted regional or tectonic subsidence with Delta subsidence caused by loss of surface soil.

6.9 USE OF PHYSICAL MEASUREMENTS   **227**

Meadows area to stable bench marks in the surrounding foothills: The purpose of this comparison was to determine whether Delta Meadows was subsiding, and if so, its rate of subsidence.[142] By this method, the geologist and the surveyor determined the regional subsidence rate for Delta Meadows as a whole. The geologist testified that this rate had been uniform and that regional subsidence would not have been caused nor its rate influenced by the works of humans in the Delta Meadows area.

Besides this natural phenomenon of subsidence, the other subtle topographic change occurring at Delta Meadows was deposit of a veneer of sediment over low-lying areas. Some of this sediment assuredly came from hydraulic mining.[143] It was not possible, however, to determine the exact source of the deposition.[144] Relying on opinions of other experts,[145] his long experience in and familiarity with the Delta, and his field survey, the civil engineer quantified the rate at which such deposition occurred in Delta Meadows.[146] The civil engineer's determination concerning deposition was borne out by the testimony of other experts.[147]

---

142. A brief explanation of this complex investigation may be of assistance to others. The surveyor compiled and examined long-term leveling records for certain bench marks in the Sierra foothills. The purpose of so doing was to find relative vertical elevation changes. Based on such studies, the geologist and the surveyor determined which of such bench marks were stable. By comparison with the stable foothill bench marks, the geologist and the surveyor determined that Delta bench marks were subsiding relative to the stable bench marks. One of those subsiding bench marks was close by Delta Meadows. Knowing the amount of subsidence over a number of years, the geologist was able to calculate a yearly rate of tectonic subsidence at Delta Meadows.

143. *Gray, supra*, 174 Cal. at 629 ("The immense quantities of detritus brought down by the streams raised enormously the bed of the river, necessarily lessened its depth, aggravated flood conditions, [and] caused the precipitation of large quantities of 'slickens' upon the adjacent farming lands to their great injury, [etc.]").

144. Recall the discussion in *State of Cal.ex rel. State Lands Com. v. Superior Court (Lovelace), supra*, 11 Cal.4th at 77–78, that specifically noted the practical impossibility of tracing deposition to its source. See notes 146 and 152 and accompanying text in Chapter 5 and note 66 in this chapter.

145. Based on his boring logs and field investigation, the soils scientist concluded that deposition had occurred at and on Delta Meadows since 1850. He also concluded that the rate of deposition would have been greater between 1850 and 1914 than between 1914 and today.

146. The civil engineer's field survey also had horizontally positioned the elevations the Corps surveyed in the 1914 field survey. The civil engineer then compared his surveyed elevations with the elevations determined by the Corps of Engineers. Knowing the rate of subsidence provided by the geologist, the civil engineer could determine the rate of deposition since 1914. To find the rate of deposition between 1850 and 1914, the civil engineer consulted with the soils scientist and reviewed Gilbert's work on the effects of hydraulic mining. Gilbert found that substantially more debris was present than under natural conditions. The soils scientist and the geologist also agreed that there would be an increased rate of deposition during that era. Courts have acknowledged this fact as well. *Gray, supra*, 174 Cal. at 629.

147. The historical geographer recounted that under natural conditions, Delta Meadows received seasonal runoff and overbank flow from two river drainages. Both the historical geographer and the geologist testified that a backwater effect would take place during high river flow conditions when there was a great deal of suspended sediment being carried. When the river waters lost their velocity, the suspended sediments would be deposited. The Delta Meadows area would acquire a thin veneer of sediment during each flood period. Floods during the period of hydraulic mining

Relying on these conclusions, the civil engineer prepared an 1850 topographic map of Delta Meadows based on the same sources of topographic information as the 1910-era topographic map. In re-creating the topography as it existed in 1850, he accounted for the long-term, steady, but not readily discernable, processes of deposition and regional subsidence.[148] The civil engineer quantified the rate of deposition and applied it against the subsidence rate determined by the surveyor and the geologist. The result was an opinion that the 1850 ground surface was very slightly higher with respect to the reference datum, NGVD, than the 1910-era ground surface of Delta Meadows. The opinion was reflected in the exhibit map that depicted the topography of 1850 as slightly higher in relative elevation than the topography of the 1910 era.

6.9.9 Re-creation of the Location of the Historic Mean High-Water Line

The afore-mentioned topographic map provided the basis for applying the elevation of the historic mean high-water line as a contour line of equal elevation.[149] To establish a mean high-water line, the civil engineer required that there be a hydraulic connection[150] of Snodgrass Slough to the interior. To establish this hydraulic connection, the testimony and conclusions of all the experts were required.

First, the topographic maps prepared by the civil engineer depicted the existence of a prominent channel entering Snodgrass Slough from the interior of Delta Meadows.[151] Second, the geologist concurred with the existence and location of the channel from a geomorphological standpoint, recounting how

---

were more frequent than after flood control project levees and other works were constructed. Thus, these experts opined that it was reasonable to assume that more sediment would have been deposited during rather than before periods of hydraulic mining. See *Gray, supra*, 174 Cal. at 629.

148. In reconstructing the topography, the civil engineer accounted for the subsidence and deposition uniformly. That is, no particular area of Delta Meadows was treated differently from any other area. The decision on how to treat the deposition and subsidence factors was based on the opinion of other experts, the civil engineer's long experience in the Delta, and his field investigations.

149. Elevation measurements were not obtained for every inch of Delta Meadows. Therefore, the contour line was based on the topographic features and contours shown in aerial photography, on the 1965 photogrammetric survey, and on field observation of such topographic features by the civil engineer.

150. What the civil engineer meant by hydraulic connection was that if the water level of Snodgrass Slough rose to a certain elevation, it would penetrate into the interior of Delta Meadows without any intervening higher ground preventing such infiltration. See note 101 in this chapter.

151. The civil engineer testified in great detail about the support for this channel. This support included the railroad survey information from 1906, a 1906 cartographic depiction by the USGS, the channel's persistent signature in aerial photography, and its present existence confirmed by field reconnaissance. According to the civil engineer, it was inconceivable that such a well-defined channel with vertical banks did not also exist in 1850.

the railroad surveyors noted its existence in separate surveys and how this channel could be readily traced in aerial photos.[152] Third, as to the channel's existence in 1850, the soils scientist identified channels on his soils map and opined that such channels had persisted for hundreds of years. Fourth, the geologist, in discussing marsh channels in general, specifically noted their long-term persistence. Thus, it was shown that portions of Delta Meadows were below the mean high-water line and were hydraulically connected to a source of tidewater, and that lands below the mean high-water line were subject to the ebb and flow of the tides on a daily basis.[153]

6.9.10 Physical Justification of Location of the Historic Mean High-Water Line

The plaintiff did not rest on its achievements to this point. Statistical analysis and re-created historic topography and water levels, even though accomplished with relative scientific precision and technical accuracy, should be tested, if possible, in an actual physical context. Verification was especially necessary in the re-creation of a marsh area where topographic features may be indistinct. In examining that re-creation, the geologist and the coastal oceanographer analyzed how Delta Meadows' geomorphic features and their associated plants and soils related to the 1850 mean high-water line re-created by the civil engineer.

First, the geologist agreed that the topography reconstructed by the civil engineer reflected existing physical features of Delta Meadows.[154] That the topography accurately depicted such features was affirmed when the soils and plant information was compared with the topography. Soils and plants associated with geomorphic features closely meshed, further corroborating the accuracy of the civil engineer's topographic depiction as compared with physical fact.[155] Thus, actual physical conditions discovered through investigation

---

152. Attacking this problem another way, the oceanographer analyzed the Snodgrass Slough levee elevations shown by the Corps of Engineers' survey and determined the mean and range of elevation of that levee. Applying the daily high waters to this levee, he concluded that 40 percent of the daily high waters exceed the lowest elevation of the levee and therefore overtopped that portion of the levee on a daily basis. Since the levee was irregular and had channels that penetrated it, it was the oceanographer's conclusion that the slough levee was not a barrier to tidal penetration of the interior of the Delta Meadows back swamp. Indeed, the existence of marsh channels over the more-than-40-year time period of the series of aerial photos, coupled with the oceanographer's knowledge of the formation and persistence of marsh channels, all supported the conclusion that tides at mean high water penetrated the marsh flat area.
153. *Delta Meadows Decision*, Appendix A, para. B.13–16.
154. One method used to support this conclusion was the geologist's preparation of an area elevation curve. An area elevation curve relates the elevation of the topography as a percentage of the total area. The geologist could distinguish the geomorphic units as reflected in the topography through this area elevation curve.
155. For example, the marsh flat geomorphic unit comprised the largest geomorphic unit at Delta Meadows. Soils and plant mapping established that the Burns soil and the tule and cattail marsh also comprised the great majority of the area of soils and vegetation types present at Delta

by independent experts in different disciplines all substantiated the topographic depiction.

The oceanographer also compared the marsh flat elevation with the elevation of daily high waters[156] that would have occurred in Snodgrass Slough at the reach of Delta Meadows had the more-than-5000 daily high waters of 1958 through 1966 occurred in historic times.[157] His conclusions were telling: nine of the ten daily high waters fell within the vertical range of the marsh flat. This had considerable geomorphological significance, as it showed that the year-round water level for the entire range of the marsh flat was very stable.[158] As a consequence, the higher portions of the marsh flat would be inundated with the frequency required for the existence of Burns soils and an aquatic plant community.[159] The result of application of these various scientific disciplines and extensive investigations was the finding by the court that there was an area within Delta Meadows below the ordinary high-water mark in 1850 and thus constituted lands that were tidelands in legal character.[160] As a consequence, the State of California's public trust ownership interest had to be accounted for in the valuation of the property being condemned.

6.10 CONCLUDING THOUGHTS AND SUGGESTIONS

Presentation of evidence to the court, including the evidence presented by the railroad, took almost eight weeks of trial time. Putting aside the expense of such a lengthy proceeding, maintaining the interest of the trier of fact had always to be considered. One must thoroughly and carefully prepare exhibits that graphically depict expert testimony. Perhaps more important, one must thoroughly and carefully prepare experts to succinctly, understandably, and simply describe their conclusions and the methods used to reach those conclusions.

Meadows. These soils and vegetation types are associated with saturated soil conditions found in the marsh flat.

156. The elevation of mean high water fell in the lower portion of the marsh flat. The relief was so slight that relatively small increases in elevation of the water surface would submerge a disproportionally larger area of the marsh flat.

157. The oceanographer translated those high waters back to historic times, using the same method as had the civil engineer.

158. Mean high water represents only the mean of all high waters. That is, half of all daily high waters were higher than the mean elevation. Such high waters would submerge a much greater area of the Delta Meadows marsh flat than was submerged at mean high water.

159. The soils scientist testified that such soils require saturation by both surface and subsurface flows. The plant ecologist testified that tules require an abundant and reliable water source. The oceanographer testified that the daily high waters will inundate more of Delta Meadows than is shown as below the mean high-water level. In fact, 50 percent of the daily high waters would inundate a greater area.

160. *Delta Meadows Decision*, Appendix A, para. A.6–16.

6.10 CONCLUDING THOUGHTS AND SUGGESTIONS

Even considering the graphics and courtroom theatrics, it is the substance of the testimony and the weight of the evidence that decide cases. The *Delta Meadows Case* endorsed one approach: In establishing the historic physical location of the ordinary high-water mark in tidal marshes, it was necessary to bring to bear on the issue applied physical and scientific disciplines and to have those disciplines exercised by persons intimately familiar with the area.[161] The case also bears witness to the complexity and difficulty of re-creating historic conditions.

For both the lawyer describing the work to be performed to the expert and the expert being informed as to what is expected, a very careful and measured approach is recommended. Thus, experts should obtain or require data be obtained on which to base their opinions. When there is no such evidence or data reasonably available, experts should be requested, in appropriate cases, to rely on their knowledge and experience concerning similar facts or circumstances in reaching their conclusions. Unsupported speculation should be avoided, as it raises questions regarding the credibility of the expert and may impact the trier of facts' view of the case of the party sponsoring such evidence.

In presenting expert testimony, simplicity and clarity are the watchwords. The goal is to accurately and fully present to the trier of fact, whether a judge or a jury, not only the bare conclusions of the experts but the essential foundations on which those conclusions are based. If adopted, these suggested approaches will, it is submitted, assist the judge or jury in arriving at as fully informed a decision as possible. Using these guidelines, it is feasible, as in the *Delta Meadows Case*, to bring to bear applied scientific disciplines and to re-create historic conditions in as understandable and interesting a manner as necessary to satisfy a court of law that the burdens of proof and persuasion have been satisfied.[162]

As we move out of the geographic area where the tides are the greatest influence to entirely freshwater regimes, the problems, both legal and factual, become somewhat different but just as interesting and obdurate.

---

161. *Delta Meadows Decision*, Appendix A, para. B. This testimony need not be scientifically indisputable or absolutely free from all error. See notes 203 through 204 and accompanying text in Chapter 4. In a highly complex and technical case such as the *Delta Meadows Case*, some mistakes were bound to be and were made. If and when such mistakes are discovered, they should be immediately brought to the attention of the court; explained, and their consequences accounted. See *Alexander Hamilton, supra*, 757 F.2d at 543.

162. In the *Delta Meadows Case*, the railroad's record title was not disputed. This left the burden of proof to the plaintiff to establish a prima facie case that part of the Delta Meadows area was tidelands. Once plaintiff established such a prima facie case, the burden of producing contrary evidence shifted to the railroad. In addition, at that point, the plaintiff state was aided by a presumption in favor of sovereign ownership. *Delta Meadows Decision*, Appendix A, para. B.16.

7

PROPERTY BOUNDARY DETERMINATION ALONG AND IN NAVIGABLE, NONTIDAL RIVERS AND STREAMS

7.1 INTRODUCTION

Perhaps no other type of water body has fostered as much title and boundary litigation as the nation's navigable rivers and streams. This has been especially true along the most traveled and storied rivers—the Mississippi and the Missouri Rivers.

In the formative years of our country, exploration and migration, trade and commerce were transported along (or against) the current of these great watercourses. For this reason, and much like the ubiquitous suburban shopping centers that seem to spring up overnight adjacent to the beltways and interstate highways on which we travel today, clusters of civilization sprouted at propitious intervals along the riverbanks. These communities served the needs and desires of the explorers, immigrants, travelers, merchants, gamblers, and others borne by the river. Villages grew into towns and eventually into prominent and thriving cities. Even a partial listing of these sites evokes the westward rush of this nation's settlement. New Orleans and St. Louis on the Mississippi River, Kansas City and St. Joseph on the Missouri River, all were jumping-off points for countless immigrants who turned these locations into centers of merchant and manufacturing wealth.

Along virtually every river, this bumptious progress and development increased geometrically both the desirability and the value of river-fronted land. The ever growing pressure of progress and development stimulated competition for such land. This competition led to frequent disputes over the limits of riparian ownerships and highlighted the need for definite and stable property boundaries within which progress and development could be fostered and enhanced.

7.1 INTRODUCTION

These property boundary disputes were occasioned by the very nature of rivers. In their uncontrolled state, the rivers were unimaginably powerful and entirely unpredictable;[1] their levels and margins changed periodically, sometimes in time with the seasons, but often erratically.[2] These immutable physical facts are reflected in the many reported river boundary dispute cases.[3] In fact, the Missouri River alone has given rise to numerous cases important enough to have been accepted by and decided in the United States Supreme Court.[4] Although many of these cases did not concern property boundary location per se,[5] courts deciding property boundary cases often relied on the property boundary movement principles decided in those non-property boundary cases.[6]

---

1. The account of one traveler crossing the Missouri River in 1846 noted:

   . . . [T]he weather soon became clear, and showed distinctly the broad and turbid river, with its eddies, its sand-bars, its ragged islands, and forest covered shores. The Missouri is constantly changing its course—wearing away its banks on one side, while it forms new ones on the other. Its channel is continually shifting. Islands are formed, and then washed away; and while the old forests on one side are undermined and swept off, a young growth springs up from the new soil on the other. Parkman, F., *The Oregon Trail* (Feltskog, ed.) (Univ. of Wisconsin Press, 1969) pp. 2–3.

2. The courts have seconded these observations. E.g., *Kansas v. Missouri* (1944) 322 U.S. 213, 216; *Jefferis v. East Omaha Land Co.* (1890) 134 U.S. 178, 189; *St. Louis v. Rutz* (1891) 138 U.S. 226, 231. The seasonal nature of rivers is noted in many cases. E.g., *Howard v. Ingersoll* (1851) 54 U.S. (13 How.) 409, 447; *United States v. Gerlach Live Stock Co.* (1950) 339 U.S. 725, 730; *United States v. Fallbrook Public Utility Dist.* (S.D.Cal. 1952) 109 F.Supp. 28, 37; *Rank v. Krug* (S.D. Cal. 1950) 90 F.Supp. 773, 784 (herein "*Rank I*") (judicial notice of seasonal nature); *State v. Cain* (Vt. 1967) 236 A.2d 501, 504. At least one court has taken judicial notice of the fact that river waters vary in their stages. *Joyce-Watkins Co. v. Industrial Commission* (Ill. 1927) 156 N.E. 346, 348.

3. Among the early volumes of the United States Reports is this sampling of river cases that reached the Supreme Court in the nineteenth century alone: e.g., *New Orleans v. United States* (1836) 35 U.S. (10 Pet.) 662 (New Orleans, La.); *Jones v. Soulard* (1860) 65 U.S. (24 How.) 41 (St. Louis, Mo.); *Railroad Company v. Schurmeir* (1868) 74 U.S. (7 Wall.) 272 (St. Paul, Minn.); *Missouri v. Kentucky* (1870) 78 U.S. (11 Wall.) 395 (interstate boundary); *Jefferis, supra,* 134 U.S. 178 (Pottawattamie County, Iowa); *St. Louis v. Rutz, supra,* 138 U.S. 226 (St Louis, Mo.); *Nebraska v. Iowa* (1892) 143 U.S. 359 (interstate boundary dispute). Of course, there are many other cases as well. These citations do not do justice to the many cases decided in courts in the states later formed in new territories. Both editions of Clark, Robillard, W., and Bouman, L., Clark on Surveying and Boundaries (7th Ed., 1997), Chapter 24 (hereinafter "Clark 7th"), and Grimes, *Clark on Surveying and Boundaries* (4th ed., 1976) (herein "Clark 4th"), Chapter 24, as well as Skelton, *The Legal Elements of Boundaries and Adjacent Properties* (1930)(herein "Skelton"), Chapter V, contain citations to many of the numerous state cases.

4. *Jefferis, supra,* 134 U.S. 178; *Nebraska v. Iowa, supra,* 143 U.S. 359; *Missouri v. Nebraska* (1904) 196 U.S. 23; *Missouri v. Kansas* (1909) 213 U.S. 78; *Kansas v. Missouri, supra,* 322 U.S. 213; *Wilson v. Omaha Indian Tribe* (1979) 442 U.S. 653.

5. Non-property-boundary cases arise out of interstate boundary disputes and concern the geographic extent to which a state may exercise its political authority. E.g., *Nebraska v. Iowa, supra,* 143 U.S. 359; *Kansas v. Missouri, supra,* 322 U.S. 213. See Chapter 2, Section 2.2.

6. Remember that property boundary problems along rivers are different from those in other physical regimes, such as the open ocean shoreline. Cases considering property boundary location

This chapter will concern itself with property boundary determination for those properties adjacent to or underlying navigable, nontidal rivers.[7] While private persons are frequently contending parties, the principal boundary litigants are most often the state or federal governments or their successors in title. Such litigants' riverbed ownership is in competition with riparian upland title.[8] The crucial preliminary question in these disputes is whether the boundary watercourse is navigable. Thus, the legal definition of navigability, the court interpretation of that definition, and the types of evidence that have established navigability of particular watercourses will first be discussed.

After considering the many facets of the question of navigability, this chapter analyzes, in some detail, principles of property boundary movement in relation to rivers. Many of the rules determining the property boundary effect of change in geographic location of the boundary watercourse with which we are already familiar have been generated by the ever changing course of rivers and will not be repeated. What this chapter discusses are some reasons for the still considerable uncertainty concerning the application of these principles to particular species of changes in the geographic location of a river property boundary. The causes of the uncertainty will be discussed and suggestions made on how to avoid them.

With those suggestions in mind, this chapter turns to a most vexing question: In the context of river boundaries, what differentiates an accretive river process from one that is avulsive? This chapter examines in detail three leading cases that have considered this issue and the contradictory results reached by the United States Supreme Court. Close inspection of facts is a necessary evil where legal consequences of the same, or remarkably similar, facts can differ quite strikingly. The mystery of how to distinguish an accretive process

---

questions concerning one type of physical regime are not necessarily directly applicable to property boundary location questions arising in another physical regime. *Smith Tug & Barge v. Columbia-Pacific Tow. Corp.* (Wash. 1971) 482 P.2d 769, 773–774, *cert. den.* (1971) 404 U.S. 829; *Fogerty v. State of California* (Cal. 1986) 187 Cal.App.3d 224, 240, n. 12, *cert. den.* (1987) 484 U.S. 821 ("*Fogerty II*"). Notwithstanding this principle, it is not unusual for river property boundary cases to be decided based on principles derived from another physical regime, and vice versa. E.g, *California ex rel. State Lands Comm'n v. U.S.* (1982) 457 U.S. 273, 278 (U.S. Supreme Court relies on river cases to decide a case concerning the open coast littoral boundary). As will be discussed in the next chapter, this problem has become acute in the case of lakes. See *Provo City v. Jacobsen* (Utah 1947) 176 P.2d 130, 132. In this chapter, the common convention will be to direct the reader to river cases, where possible. If river cases are not in point, this chapter will refer the reader to analogous lake cases because of the rough similarity of their respective regimes. Occasionally, where the principle expressed is relevant, cases concerning tidal water bodies will also be specified.

7. For purposes of this chapter, the term "rivers" will also include navigable streams. The term "navigable" will be defined later in this chapter. See text accompanying notes 9 through 59, in this chapter.

8. This chapter does not intend to discuss title and boundary disputes between contending riparian property owners along nontidal, nonnavigable rivers. Also see note 14.

from one that is avulsive and the quagmire generated in determining what is gradual and imperceptible furnish the framework in which the burden of proof is the crucial component.

The final topics in this area concern two interesting legal questions, the answers to which are critical to the type of proof that must be produced: Is there any reason to be concerned about what caused the change in the physical location of the river boundary, and what is the effect on the property boundary of the reemergence of once submerged land?

The analysis of boundary movement principles in the context of rivers lays the foundation for a discussion of exactly what is the boundary that will be located by application of those principles. Again we will chase down the elusive ordinary high-water and ordinary low-water marks and attempt to translate these legal words of art, without any accepted scientific or physical meaning, into a precise and definite physical location. The kinds of proof that have been used in boundary cases to define a particular boundary movement or to supply the physical indicia for the location of the ordinary high- or ordinary low-water mark river property boundary constitutes the concluding subject of this chapter.

7.2 THE MATTER OF NAVIGABILITY

As in most endeavors, there is a starting point. In the case of disputes over the titles to and boundaries of land adjacent to or in the bed of a nontidal river, the question of navigability of the river is that starting point. The reason to begin with this question is that its answer must be affirmative for the inquiry to proceed; it is *the* bedrock issue.

7.2.1 Significance of Navigability in Title and Boundary Disputes Along Nontidal Water Bodies

At the risk of repeating what has been related in a slightly different context in earlier chapters, navigability of the river is an essential prerequisite to sovereign ownership of the riverbed.[9] By virtue of the Equal Footing Doctrine,

---

9. Some have stated that the concept of navigability was created for demarcating that reach of the watercourse that could be privately owned from that reach of the watercourse belonging to the public and held in trust by the sovereign. E.g., MacGrady, "The Navigability Concept in the Civil and Common Law" (1975) 3 Fla.St.U.L.Rev. 511, 511–512; *State of Alaska v. U.S.* (D. Alaska 1987) 662 F.Supp. 455, 459; *Lamprey v. Metcalf* (Minn. 1893) 53 N.W. 1139, 1143; *Flisrand v. Madson* (N.D. 1915) 152 N.W. 796, 799. Another purpose for this concept, at least in the western United States, was to differentiate lands owned by the federal government from those held by the states. *State of Alaska, supra,* 662 F.Supp. at 460. For an excellent discussion of the historical antecedents of the navigability doctrine in the United States, see *Ibid.*; Althaus, *Public Trust Rights* (U.S. Fish & Wildlife Service 1978), pp. 120–121.

the states, as a matter of their sovereign being, possess title to the lands underlying navigable watercourses.[10] Congress confirmed and recognized this sovereign ownership.[11] Moreover, the sovereign title to lands beneath navigable river waters is held and is subject to the public trust to the same extent as lands beneath the ocean or navigable lakes.[12] Thus, in the case of nontidal rivers,[13] before determining the location of the boundary of the river, the initial question for title purposes will always be whether the river is navigable. If the river is not navigable, the boundary questions may be somewhat different and are only tangentially discussed in this work.[14]

7.2.2 Questions About the Concept of Navigability

What does the term "navigable" mean? As with virtually every aspect of the title and boundary questions discussed in this book, the meanings of terms adopted and used by courts to describe a particular physical condition or effect in those less sophisticated (and less litigious) times have been continually refined or, as some may say, redefined. That refinement of meaning is not only the product of developing scientific knowledge and sophistication, but

---

10. See notes 111 through 121 and accompanying text in Chapter 1. The question of whether, for purposes of sovereign ownership, the watercourse must be navigable even if it is unquestionably tidal was discussed in notes 82 through 96 and accompanying text in Chapter 5.

11. 43 U.S.C. § 1311(a); *Furlong Ent. v. Sun Exploration & Prod.* (N.D. 1988) 423 N.W.2d 130, 132.

12. Many cases have recognized the existence of the public trust over the beds of rivers. E.g., *New Orleans, supra*, 35 U.S. at 480; *Barney v. Keokuk* (1876) 94 U.S. 324, 338; *Bonelli Cattle Co. v. Arizona* (1973) 414 U.S. 313, 321–324, *ovrld on other grnds* (1977) 429 U.S. 363; *State Land Board v. Corvallis Sand & Gravel Co.* (1977) 429 U.S. 363, 370; *Montana v. United States* (1981) 450 U.S. 544, 551; *People v. Gold Run D. & M. Co.* (1884) 66 Cal. 138, 146–147, 151–152; see *National Audubon Society v. Superior Court* (1983) 33 Cal.3d 419, 435, *cert. den., sub. nom. Los Angeles Dept. of Water & Power v. National Audubon* (1984) 464 U.S. 977. Congress recognized this public trust beginning with the Northwest Ordinance of 1787. 1 Stat 50 (1789) (". . . navigable waters . . . shall be common highways, and forever free . . .") and in each Act of Admission, e. g. (1850) 9 Stat. 452 (". . . navigable waters within the said State shall be common highways, and forever free . . ."). Recognition of public rights in rivers is not limited to this country. For example, in *The Brothers Karamazov* Dostoevsky made special note of the public right to fish in rivers in tsarist Russia. Dostoevsky, F., *Brothers Karamazov* (Penguin 1958), p. 7.

13. The question of navigability will also arise in the case of lakes. As a consequence, this discussion of navigability is equally applicable to lakes.

14. For example, the boundary of lands adjacent to a nontidal, nonnavigable river may not be the same as the boundary of lands adjacent to a nontidal, navigable river. In general, lands adjacent to nontidal, nonnavigable rivers or streams are bounded by the center line or middle of the boundary watercourse. E.g., Clark 7th, *supra*, § 25.03; Clark 4th, *supra*, at § 584; Skelton, *supra*, at § 287; Cal. Civ. Code § 830. For a further explanation of the boundaries of such nonnavigable watercourses, the author commends the two editions of Clark and Skelton and the authorities cited therein as a good starting point.

also a reflection of the changing views and values of society. The definition of the legal term "navigable" has evolved through such a process.

It is easy to acknowledge that the great oceans are "navigable." Personal observation and film or video clips of great ocean liners, container ships, or aircraft carriers plying the waters indelibly confirm that knowledge. But what about rivers? Must the aircraft carrier *Nimitz* nuclear-power itself from the Gulf of Mexico to St. Louis to establish that the Mississippi River is navigable? What if various reaches of the river are broken up by riffles or rapids? Do federal courts or do each state's courts determine what waters are navigable and thus which lands the state owns by virtue of the Equal Footing Doctrine? This last question should be answered first so that at least the reader will be aware of what set of casebooks should be examined to answer the prior questions.

7.2.3 Federal Courts Establish Test for Navigability

The test for navigability for sovereign title purposes is established in and by federal law.[15] In other words, in order to determine whether a particular waterway came to a state by virtue of the Equal Footing Doctrine, federal cases deciding the question of navigability of the waterway are authoritative.

The federal test for navigability was first set out in an 1870 case concerning the extent of jurisdiction of an admiralty[16] court:

> Those rivers must be regarded as public navigable rivers in law which are navigable in fact. And they are navigable in fact when they are used, or are susceptible of being used, in their ordinary condition, as highways for commerce over which trade and travel are or may be conducted in the customary modes of trade and travel on water. . . .[17]

---

15. *United States v. Utah* (1931) 283 U.S. 64, 75; *United States v. Oregon* (1935) 295 U.S. 1, 14; *United States v. Champlin Refining Co.* (10th Cir. 1946) 156 F.2d 769, 773; *Utah v. United States* (1971) 403 U.S. 9, 10; *State ex rel. Brunquist v. Bollenbach* (Minn. 1954) 63 N.W.2d 278, 287; *Luscher v. Reynolds* (Ore. 1936) 56 P.2d 1158, 1161.
16. In very general terms, admiralty or maritime courts hear cases arising on the sea, the ocean, or navigable water bodies. *The Steamboat Thomas Jefferson* (1825) 23 U.S. (10 Wheat.) 428; *The Propeller Genesee Chief et al. v. Fitzhugh et al.* (1851) 53 U.S. (12 How.) 443, 452–454.
17. *The Daniel Ball* (1870) 77 U.S. (10 Wall.) 557, 663. Although *The Daniel Ball* was an admiralty case, the test for navigability of rivers adopted in that case has been used, with some modification, by subsequent courts in boundary and title cases where navigability of the watercourse was in issue. E.g., *Utah v. United States, supra*, 403 U.S. at 10; *United States v. Utah, supra*, 283 U.S. at 76; *United States v. Holt State Bank* (1925) 270 U.S. 49, 56; *Oregon v. Riverfront Protective Ass'n* (9th Cir. 1982) 672 F.2d 792, 794; *Alaska v. United States* (9th Cir. 1985) 754 F.2d 851, 853, *cert. den.* (1986) 474 U.S. 968; *Phillips Petroleum Co. v. Mississippi* (1988) 484 U.S. 469, 488–489, *reh. den.* (1988) 486 U.S. 1018 (O'Connor, J., dissenting).

In fact, much case authority defining navigability has arisen in nonproperty boundary dispute contexts in which the question of navigability of a watercourse was a crucial element to be established.[18]

These cases provide valuable guidance in deciding the issue of navigability for sovereign title purposes. The reason is that standards to determine navigability in non-sovereign-title litigation are quite similar to the tests for navigability for sovereign title purposes, and thus, for this purpose, the cases are claimed to be analogous.[19]

7.2.4 Relevance of State Tests of Navigability

State courts, however, have also decided questions of navigability. These state rules of navigability are not directly relevant for deciding sovereign title questions. These tests are valuable for other important reasons.

States promulgated rules of navigability to determine the extent of non-property-based public rights and interests in watercourses that, under the federal test of navigability, are perhaps *not* navigable.[20] In California, to pick just one state, courts have considered the utility of such waterways for other vital purposes, such as recreation, and provided for public rights of boating, swimming, and water recreation in such waters.[21] These rights are not rights

---

18. Other than in admiralty litigation, e.g., *The Montello* (1874) 87 U.S. (20 Wall.) 430, 438, or sovereign title litigation, e.g., *Martin v. Waddell* (1842) 41 U.S. (16 Pet.) 367, 408, the other instance in which such cases arise is in defining the parameters of the commerce clause powers of the United States, e.g., *United States v. Appalachian Electric Power Co.* (1940) 311 U.S. 377, 407–408; *Puget Sound Power and Light Co. v. Federal Energy Regulatory Commission* (9th Cir. 1981) 644 F.2d 785, 787, *cert. den.* (1981) 454 U.S. 1053. The test is also set out in federal regulations defining the extent of U.S. Army Corps of Engineers' jurisdiction by virtue of the Clean Water Act. 33 C.F.R. § 329.
19. The Supreme Court has even found that the "title" and "commerce clause? tests of navigability are the same with but two minor differences. First, under the commerce clause test, it is not necessary that the water way be navigable in its "natural and ordinary" condition; it is sufficient if the waterway can be made navigable in fact by "reasonable improvements." Second, unlike the title test, commerce clause navigability does not require the watercourse be susceptible to navigation at time of statehood; the watercourse may become navigable at some later time. *United States v. Appalachian Electric Power Co., supra,* 311 U.S. at 407–408. By accounting for these differences, one may argue that judicial decisions about navigability for commerce clause purposes are applicable to the question of navigability for sovereign title purposes.
20. E.g., Frank, "Forever Free, Navigability, Inland Waterways, and the Expanding Public Interest" (1983) 16 U.C. Davis L. Rev. 579, 589–590; *Hitchings v. Del Rio Woods* (Cal. 1976) 55 Cal.App.3d 560, 567; *Southern Idaho Fish & Game Ass'n v. Picabo Livestock Co.* (Idaho 1974) 528 P.2d 1295, 1298.
21. E.g., *People ex rel. Baker v. Mack* (1971) 19 Cal.App.3d 1040, 1045–1046; *National Audubon Society, supra,* 33 Cal.3d at 435, n.17. On the other hand, other states do not endorse such rights. E.g., *State of Kansas ex rel. Meek v. Hays* (Kan. 1990) 785 P.2d 1356 (use of nonnavigable stream by public for recreation up to legislature); *People v. Emmert* (Colo. 1979) 597 P.2d 1025 (upheld criminal trespass for rafting down nonnavigable stream without owner's permission).

in property. The underlying bed ownership is not determined. The public has, however, the right to use the waters. As stated by one California court:

> The streams of California are a vital recreational resource of the state. . . . [M]embers of the public have the right to navigate and to exercise the incidents of navigation in a lawful manner at any point below high water mark on waters of this state which are capable of being navigated by oar or motor propelled small craft.[22]

Because property rights are not concerned, such state rules are said to be more expansive than the federal test.[23] In promulgating these broad rules of navigability, some state courts have relied on the recognized benefits of public ownership of watercourses.[24]

State navigability rules have sometimes been broadly characterized as a "recreational boating test."[25] As we shall see, however, there is some support in federal courts for this more expansive view of navigability.[26] And some state courts considering sovereign title have adopted this test.[27] On the other hand, at least one state court refused to adopt a recreational boating test in a sovereign title case,[28] notwithstanding a vigorous dissent.[29]

7.2.5 Basic Parameters of Navigability

Now that we know on what kind of case authority to rely (and what kind of decisions should be regarded with some misgivings), that authority discloses

---

22. *People ex rel. Baker v. Mack*, *supra*, 19 Cal.App.3d at 1049. Other cases affirm the right of the public to passage, in a lawful manner (without trespassing), over water usable only for recreational boating. See *Pacific Gas & Electric Co. v. Superior Court* (1983) 145 Cal.App.3d 253, 258.
23. Frank, *supra*, at 590; *Lamprey*, *supra*, 53 N.W. at 1144; *Southern Idaho Fish & Game Ass'n v. Picabo Livestock*, *supra*, 528 P.2d at 1297–1298; *People ex rel. Baker v. Mack*, *supra*, 19 Cal.App.3d at 1045–1046. As stated by one California appellate court:

    > However, . . . the federal test of navigability, involving as it does property title questions, has always been much more restrictive than state tests dealing with navigability for purposes of the right of public passage. . . . The federal test of navigation does not preclude a more liberal state test establishing a right of public passage whenever a stream is physically navigable by small craft. *People ex rel. Baker v. Mack*, *supra*, 19 Cal.App.3d at 1049.

24. E.g., *Hillebrand v. Knapp* (S.D. 1937) 274 N.W. 821, 822 (citing *Flisrand*, *supra*, 152 N.W. at 798, and *Lamprey*, *supra*, 53 N.W. at 1143).
25. *People ex rel. Baker v. Mack*, *supra*, 19 Cal.App.3d at 1045; *Hitchings*, *supra*, 55 Cal.App.3d. at 568, *Southern Idaho Fish & Game Ass'n v. Picabo Livestock*, *supra*, 528 P.2d at 1297–1298.
26. See notes 47 through 57 and accompanying text, in this chapter.
27. *Lamprey*, *supra*, 53 N.W. at 1143; *Flisrand*, *supra*, 152 N.W. at 799–800; *Hillebrand*, *supra*, 274 N.W. at 822.
28. See *Bott v. Com'n of Natural Resources, etc.* (Mich 1982) 327 N.W.2d 838, 843–853.
29. *Id.* at 853–875 (Williams, J., and Ryan, J., dissenting).

several key elements of the test for navigability.[30] Only the principal constituents will be discussed in detail in this chapter. The first element relates to the manner in which the question of navigability is treated by the courts.

7.2.5.1 *Navigability—a Question of Fact.* Even from a brief review of the test for navigability, it can be readily seen that the federal test of navigability does not provide a magic formula to infallibly determine whether any particular water body is navigable for purposes of sovereign title. In other words, with the conspicuous exceptions of, for example, major rivers, one cannot simply apply the test and, as a consequence, expect to have the navigability of a particular reach of a waterway irrefutably determined. The test provides a basic framework only. Navigability is a question of fact that the court must, based on evidence presented to it, determine in connection with each watercourse.[31]

There are some actions taken by the state and federal governments that provide assistance in this task. For example, some state legislatures declared that certain rivers are navigable.[32] While such a legislative declaration supplies some evidence of the reputation of the stream for its ability to support navigation, it is not determinative.[33] On the other hand, failure of the legislature to include a watercourse or any portion of that watercourse in a list of navigable waterways does not, by omission, necessarily grant a carte blanche recognition of private ownership of the watercourse or constitute a cession of sovereign title in such a waterway to private ownership.[34]

With respect to the federal government, recall that government surveyors, in conducting surveys to determine the extent of the public lands, were instructed to meander navigable streams.[35] The fact that a United States Deputy

---

30. These elements were set out in an excellent law review article, Frank, "Forever Free, Navigability, Inland Waterways, and the Expanding Public Interest" (1983) 16 U.C. Davis L. Rev. 579, cited frequently in this chapter.
31. *The Daniel Ball, supra,* 77 U.S. at 563. And courts have not made this determination inch by inch. Thus, it is said that navigable bodies of water include nonnavigable areas at their boundaries. E.g., *Phillips Petroleum, supra,* 484 U.S. at 490 (O'Connor, dissenting). In other words, if a court finds a particular reach of a watercourse to be navigable, the finding does not exclue those ancillary portions of the watercourse over which it is not possible to navigate, such as its shallows, minor distributary channels, or banks.
32. I.e., Cal. Harbors & Nav. Code §§ 101–106.
33. *United States v. Utah, supra,* 283 U.S. at 75; *Newcomb v. Newport Beach* (1936) 7 Cal.2d 393, 399; *People ex rel. Baker v. Mack, supra,* 19 Cal.App.3d at 1048–1049; *State v. Bunkowski* (Nev. 1972) 503 P.2d 1231, 1238.
34. *People ex rel. Baker v. Mack, supra,* 19 Cal.App.3d at 1049.
35. Instructions for Deputy Surveyors (1815), p.4, reprinted in Minnick, *A Collection of Original Instructions to Surveyors of the Public Lands 1815–1881* (Landmark) (courses of all navigable rivers must be accurately surveyed); Instructions to the Surveyors of General of Public Lands (1855), *Id.* at 365–367 (both banks of navigable rivers to be meandered); Manual of Surveying Instructions (U.S. Dept. of Int., Bur. of Land Mgmt 1973) (Landmark Reprint), p. 93 (navigable bodies of water are segregated from public lands at mean high water elevation).

Surveyor meandered a riverbank and did not extend township and section lines across the river is not a conclusive determination of the navigability of the river; such officers had no power to settle questions of navigability.[36] Although not determinative, meandering of the riverbank should not be ignored. The fact that a stream has been meandered may provide evidence of navigability or may shift the burden of proof to the party claiming the stream is not navigable.[37] Township surveys may be especially valuable because they were often closely contemporary to the time of statehood, a crucial date for navigability.

7.2.5.2 Date to Establish. Navigability is established by reference to the time of statehood.[38] Notwithstanding this rule, to the extent the watercourse has remained in a natural or near natural condition, more recent events and history are also relevant to the navigability inquiry.[39]

7.2.5.3 Natural and Ordinary Condition. To be considered navigable, watercourse must be navigable in its natural and ordinary condition.[40] For sovereign title purposes, courts ignore artificial changes in rivers such as damming, leveeing, and filling of the waterway that may have made a nonnavigable watercourse navigable.[41] In any event, that a watercourse is wholly in one state or is geographically isolated is of no consequence to a determination of its navigability.[42]

7.2.5.4 Effect of Obstructions or Seasonal Flows. Watercourses are still navigable for title purposes even if there exist occasional obstructions to navigation or even if the watercourse is not navigable during certain seasons of

---

36. *Oklahoma v. Texas* (1922) 258 U.S. 574, 585.
37. See e.g., *United States v. Utah* (1975) 420 U.S. 304; Report of Special Master in *United States v. Utah*, reproduced in 1976 Utah L.Rev. 245, 306–307; *Harrison v. Fite* (8th Cir. 1906) 148 F. 781, 784; *State v. McIlroy* (Ark. 1980) 595 S.W.2d 659, 663, *cert. den.* (1980) 449 U.S. 843; *Mood v. Bancero* (Wash. 1966) 410 P.2d 776, 779; *Provo City, supra,* 176 P.2d at 131–132. Navigability is not the only context in which courts have recognized these surveys as having evidentiary value. In one case, the fact that a government agency had surveyed an area as an island in the Missouri River and issued a patent for that island was given substantial, but not conclusive, weight in considering the issue of whether there had ever been an island at that location in the river. *Conran v. Girvin* (Neb. 1960) 341 S.W.2d 75, 79.
38. *Utah v. United States, supra,* 403 U.S. at 9; *State of Alaska v. Babbitt* (9th Cir. 2000) 201 F.3d 1154, 1156.
39. *Utah v. United States, supra,* 403 U.S. at 9; *United States v. Utah, supra,* 283 U.S. at 82. In one case, the United States argued that navigability must be established with reference to the nature of commerce and the customary modes of trade and travel in existence *at the time of statehood! State of Alaska v. U.S., supra,* 662 F.Supp at 459. This argument was rejected. *Id.* at 461–463. In *State of Alaska v. Babbitt, supra,* 201 F.3d at 1156–1158, the Court accepted evidence of Athabascan trading in the first half of this century, long before Alaska became a state.
40. *United States v. Holt State Bank, supra,* 270 U.S. at 56.
41. *Ibid., United States v. Utah, supra,* 283 U.S. at 75–79.
42. *The Daniel Ball, supra,* 77 U.S. at 565.

the year.[43] In one case, a river was held to be navigable even in light of the facts that the river froze during the winter and was subject to very low flow during the summer.[44]

Given these basic parameters, what exactly are the types of activities that will establish susceptibility for navigation?

7.2.6 Actual Use Is Not Required, Only Susceptibility for Navigation Is Essential

While obviously relevant and helpful to the determination of whether a watercourse is navigable, evidence of actual use is not indispensable. It is the watercourse's susceptibility for navigation, rather than actual use, that is essential.[45] The basis for this rule is founded in the manner in which the western portion of this nation was settled.

Many areas in the western United States were remote and essentially unpopulated until well into the twentieth century. Depending on time-of-statehood evidence of historic use of a watercourse located in these regions would support a claim of navigability on the thin reed of the relative desirability and accessibility of an area to settlement. The actual physical characteristics of the watercourse would essentially be ignored. Indeed, the Supreme Court rejected the argument that evidence of past use was required to establish navigability.[46]

---

43. *The Montello, supra,* 87 U.S. at 441, 442; *United States v. Utah, supra,* 283 U.S. at 86–87; *Economy Light & Power Co. v. United States* (1921) 256 U.S. 113, 122; *State of North Dakota v. Andrus* (8th Cir. 1982) 671 F.2d 271, 277, *rev'd on other grounds, sub. nom. Block v. North Dakota ex rel. Bd. of Univ. and State Lands* (1983) 461 U.S. 840. *Economy Light & Power* is important because of its rule of "indelible navigability." In that case there was evidence that long ago the river had been used in a vast fur trade. Due to changes that had occurred in the physical condition of the river, the river was no longer capable of being used for commerce. Because of the river's *past history* of navigability, the Supreme Court held that the river was navigable although it was not navigable in its then-present condition. *Ibid.*
44. *State of North Dakota v. Andrus, supra,* 671 F.2d at 278. One California case noted that there was no authority provided to support the argument that a river must be designated as "nonnavigable" because it may be navigated only seasonally. *Bess v. County of Humboldt* (1992) 3 Cal.App.4th 1544, 1549 n. 2.
45. *The Daniel Ball, supra,* 77 U.S. at 563; *The Montello, supra,* 87 U.S. at 441; *United States v. Utah, supra,* 283 U.S. at 83.
46. The Supreme Court stated:

> . . . [A]s the title of a State depends on the issue, the possibilities for growth and future profitable use are not be ignored. [A State], with its equality of rights as a State of the Union, is not to be denied title to the beds of such of its rivers as were navigable in fact at the time of admission of the State either because the location of the rivers and the circumstances of exploration and settlement of the country through which they flowed had made recourse to navigation a late adventure or because commercial utilization on a large scale awaits future demand. The question remains one of fact as to the capacity of the

With this understanding, we turn to the standard for and the uses that have satisfied the federal navigability standard.

7.2.7 Use for Commerce or Transportation Will Establish Susceptibility

To establish navagibility of a watercourse, the watercourse must be susceptible to navigation as a highway for commerce or, arguably, for transportation.[47] There is another view that holds navigability conferring sovereign title does not turn on merely one of the range of uses that can be made of a particular watercourse; the watercourse need merely be useful as a highway for transportation, which may include recreation.

In a federal district court it was argued that commerce is not limited to freight hauling activities, but could include such activities as fishing, camping, hunting, sightseeing, trapping, and governmental activities such as surveying and enforcing game laws.[48] Responding to that argument, the district court stated: "When a water body is susceptible of being used as a route for transporting people or goods, it generally will also be susceptible to use as a 'highway for commerce.' . . . [I]n the ordinary case there is no material difference between susceptibility to use as a route of transportation and susceptibility to use as a 'highway for commerce.' "[49] This view has been adopted by one federal circuit court case as well.[50] In addition, actual private or recreational use can be used to show the susceptibility of a watercourse for navigation for commerce.[51]

Arguments expanding the scope of uses supporting claims of navigability have gained a favorable reception in the courts. Such arguments, when

---

rivers in their ordinary condition to meet the needs of commerce as these may arise in connection with the growth of the population, the multiplication of activities and the development of natural resources. And its capacity may be shown by physical characteristics and experimentation as well as by the uses to which the streams have been put. *United States v. Utah, supra*, 283 U.S. at 83.

47. For example, *The Daniel Ball, supra*, 77 U.S. at 663, holds that a waterway must be susceptible to navigation for *commerce* to be navigable-in-fact. In more recent decisions, such as *Utah v. United States, supra*, 403 U.S. at 10 and *State of Alaska v. Babbitt, supra*, 201 F.3d at 1157–1159; *State of Alaska v. U.S., supra*, 662 F.Supp. at 463–465, courts focused on the watercourse's ability to serve as a highway for transportation, not limited to commercial transportation.

48. *State of Alaska v. U.S., supra*, 662 F.Supp. at 465.
49. *Ibid.*
50. See, *State of Alaska v. Ahtna, Inc.* (9th Cir. 1989) 891 F.2d 1401, 1405 (recreational river rafting); *State of Alaska v. Babbitt, supra*, 201 F.3d at 1158 (used for trapping even when frozen).
51. *Ibid*; *United States v. Utah, supra*, 283 U.S. at 75; *State of North Dakota v. Andrus, supra*, 671 F.2d at 277–278; *State of Alaska v. U.S., supra*, 662 F.Supp. at 467–468.

adopted, maintain the public nature of waterways and do not dispossess states of lands allegedly held in trust for public benefit.[52]

7.2.8 Evidence of Navigability

While the definition of navigability seems to have remained the same, it can be seen that the qualifying uses that establish "susceptibility" for "commerce" or "transportation" appear to have been refined considerably and to have been expanded. This proof of navigability varies from case to case and is limited only by the imagination and creativity of the lawyer presenting the case and the expert consultants assisting that lawyer.[53]

A watercourse may be deemed navigable based exclusively on its susceptibility for use by small watercraft such as canoes.[54] Navigability does not depend on the evidence of use by large watercraft.[55] Indeed, courts have held that navigation may be by any customary mode of trade or travel.[56] This

---

52. Some may argue that the United States Supreme Court agreed with this approach. *Montana v. United States, supra,* 450 U.S. at 552. In that vein, one state court has noted:

> ... If, under present conditions of society, bodies of water are used for public uses other than mere commercial navigation, in its ordinary sense, we fail to see why they ought not to be held to be public waters, or navigable waters, if the old nomenclature is preferred. Certainly, we do not see why boating or sailing for leisure should not be considered navigation, as well as boating for mere pecuniary profits. Many, if not most, of the meandered lakes of this state, are not adapted to, and probably never will be used to any great extent for, commercial navigation; but they are used ... by the people for sailing, rowing, fishing, fowling, bathing, ... and other public purposes which cannot now be enumerated or even anticipated. To hand over all these lakes to private ownership, under any old or narrow test of navigability, would be a great wrong upon the public for all time, the extent of which cannot, perhaps, be now even anticipated. ... *Lamprey, supra,* 53 N.W. at 1143; *accord, Bohn v. Albertson* (1951) 107 Cal.App.2d 738, 744.

53. Many cases about the issue of navigability depend on conclusions to be drawn from historical evidence that is subject to differing interpretations. In such cases, which party has the burden of proof will be critical. There are concepts that may assist in determining who has the burden of proof. For example, in considering the navigability question, the fact that public land surveys meandered the watercourse in question may be claimed to shift the burden of proof to those who contend the watercourse was not navigable. See notes 35 through 37 in this chapter.

54. *United States v. Holt State Bank, supra,* 270 U.S. at 56–57; *Utah v. United States, supra,* 403 U.S. at 11; *United States v. Utah, supra,* 283 U.S. at 79, *State of North Dakota v. Andrus, supra,* 671 F.2d at 278; *State of Alaska v. Ahtna, supra,* 891 F.2d at 1405 (recreational rafting); *State of Alaska v. Babbitt, supra,* 201 F.3d at 1158 (prestatehood fur trading).

55. "It would be a narrow rule to hold that in this country, unless a [watercourse] was capable of being navigated by steam or sail vessels, it could not be treated as a public highway." *The Montello, supra,* 87 U.S. at 441. A more recent case stated: "To deny that this use of the River [guided fishing and sightseeing trips] is commercial because it relates to the recreation industry is to employ too narrow a view of commercial activity." *State of Alaska v. Ahtna, Inc., supra,* 891 F.2d at 1405.

56. *United States v. Appalachian Electric Power Co., supra,* 311 U.S. at 405; *United States v. Holt State Bank, supra,* 270 U.S. at 56.

evidence may vary from usage by large ships to the rafting of logs or recreational rafting.[57] Thus, it does not have to be shown that the *Nimitz* can sail from the Gulf of New Mexico to St. Louis for the Mississippi River to be deemed navigable. Huck Finn's raft would probably be enough.

Some of the evidence of navigation that has been used with success to support claims of navigability has been found in accounts contained in personal journals, in the writings of travelers, in books, and in newspaper articles.[58] Surveyors' field notes often contain descriptive accounts of the watercourses being surveyed.[59] Less obvious, but just as potentially useful, is the opinion testimony of persons with specialized training and skill in navigation such as river pilots and ship's masters or mates. Such expert consultants, based on their experience, review of available evidence, or inspection of the watercourse (if it is still relatively undisturbed), can provide opinion testimony on the susceptibility of a watercourse for navigation in its ordinary condition as a highway for commerce over which trade or travel was, or could have been, conducted in the customary modes of trade or travel on water.

Considering the tendency to favor public ownership of navigable waters, riparian or littoral landowners along rivers or lakes would be well advised to think very carefully before deciding to question the navigability of a particular watercourse. On the other hand, as can be seen by the profusion of cases, the question is not at all free from doubt.

---

57. *Puget Sound Power and Light Co., supra,* 644 F.2d 785; *Oregon v. River Front Protective Association, supra,* 672 F.2d 792. In the *Puget Sound Power* case, a river was found to be navigable based on evidence of use by Indian canoes and periodic and difficult flotation of quartered sections of logs. *Puget Sound Power, supra,* 644 F.2d at 789. In the *River Front Protective* case, evidence of seasonal log drives established navigability, even though the river was used neither during high water periods because of unsafe conditions nor during low water periods because of insufficient water levels. *Oregon v. River Front Protective Association, supra,* 672 F.2d at 795. More recent cases continued the same trend. *State of Alaska v. U.S., supra,* 662 F.Supp. at 461; *City of Centralia, Wash. v. F.E.R.C.* (9th Cir. 1988) 851 F.2d 278, 281–282; *State of Alaska v. Ahtna, Inc., supra,* 891 F.2d at 1405 (recreational rafting evidenced navigability); *State of Alaska v. Babbitt, supra,* 201 F.3d at 1158 (prestatehood fur trading). However, other opinions are not quite as expansive. *Alaska v. United States, supra,* 754 F.2d at 855 (use of Slopbucket Lake by floatplanes does not establish its navigability).

58. For example, Theodore Roosevelt reported on his navigation of the Little Missouri River. Morris, *The Rise of Theodore Roosevelt* (1979) pp. 322–326, as had Lewis and Clark, *State of North Dakota v. Andrus, supra.* 677 F.2d at 277–278. both accounts were used to support a claim of navigability.

59. Both the surveyor and the lawyer should be aware that evidence introduced to support claims of navigability must be admissible. It would be to little avail if, because of improper or sloppy preparation, a crucial item of evidence was not admitted because it could not be authenticated. Both surveyors and lawyers should keep this constantly in mind during the course of their investigations and pretrial investigation and preparation.

7.3 PROPERTY BOUNDARY CONSEQUENCES OF RIVER MOVEMENT—IN GENERAL

The question of navigability is a model of certainty, however, when compared with the apparently contradictory results in applying the rules created to account for the property boundary effect on riparian ownerships of the geographic movement of a river. Seemingly identically described physical situations appear to have remarkably different legal consequences. Thus, in all but the most straightforward case, one cannot always reliably predict the ultimate riparian property boundary consequence of any particular physical river movement.

It is not that boundary movement rules for rivers are any different or more complex than for other physical regimes. Generally, the riparian property boundary follows the physical location of the river as long as that physical movement is gradual and imperceptible; only when there is a sudden and perceptible change in physical location of the river does the riparian boundary become fixed in its former geographic location.[60] Indeed, many of the principles determining the effect of the physical movement of the boundary watercourse on the property boundary of littoral or riparian land were developed in the context of rivers.[61] And most of these oft-cited principles are quite venerable.[62] What, then, is the cause of the uncertainty?

7.4 RULES TO AVOID UNCERTAINTY IN DECIDING THE PROPERTY BOUNDARY CONSEQUENCES OF RIVER MOVEMENT

The uncertainty that has plagued courts, lawyers, surveyors, and public and private property owners in considering the property boundary effect of specific changes in the geographic location of the boundary watercourse has, in the author's estimation, resulted from two main causes: mechanical and conceptual. Two cardinal rules are proposed that, it is hoped, should prevent confusion in this area.

7.4.1 Understand the Physical Process; Retain Expert Consultants

If the mechanism, the physical process, that caused a particular change in the geographic location of a river is not understood, then the property boundary

---

60. See notes 94 through 127 and accompanying text in Chapter 3.
61. *Ibid.*
62. For example, one of the principal cases establishing the property boundary effect of a particular physical process, *New Orleans v. United States, supra,* 35 U.S. at 486, was argued 15 years before California became a state by one of the most accomplished advocates and politicians in our early history, Daniel Webster. *New Orleans,* by the author's personal count of its subsequent history, has been cited for this principle in at least 200 cases.

effect of that change in location would also be difficult to determine. But simply understanding (and then persuading the trier of fact to arrive at the same understanding) just how, when, and what physical process happened to have caused a river to change its geographic configuration or move from one geographic location to another is hardly a simple matter or one entirely free from doubt.[63] To offer just one example, it may be that no person was either present to observe and recount or could have observed the change or changes in the geographic location of the river as such change or changes occurred. As a consequence, application of many of the disciplines of the physical and social sciences is required in order to analyze, reconstruct, and explain the past and present physical situations.

Of course, retention of expert consultants[64] will not necessarily end the confusion. In an adversarial situation, the expertise of a retained consultant is unlikely to go unchallenged.[65] And despite application of similar skills or experience to the same physical situation, a conclusion drawn by one expert consultant may be diametrically opposed or made much less certain as the result of a conclusion drawn by another expert consultant.[66] Still, retention of well-qualified expert consultants who can educate the lawyer, the surveyor, the riparian landowner, or the court in understanding the physical processes that cause or evidence a certain type of river movement will provide the solid basis upon which to apply the legal principles of river boundary movement.[67]

---

63. Witness the case of *Omaha Indian Tribe, Treaty of 1854, etc. v. Wilson* (8th Cir. 1978) 575 F.2d 620, 639–650, *vacated and remanded* (1979) 442 U.S. 653. The Eighth Circuit spent a considerable effort in a detailed analysis of the complex technical evidence regarding the mechanism by which the geographic location of the Missouri River moved almost 2 miles over a long period.

64. In some cases experts' fees are recoverable as costs. E.g., Cal. Code of Civ. Proc. § 998; *State v. Placid Oil Company* (La. 1973) 300 So.2d 154, 165, *cert.den.*, (1975) 419 U.S. 1110, *rehg den.* (1975) 420 U.S. 956. Just because fees are paid to the experts by a party does not necessarily mean the full, billed amount will be recovered as costs. *Placid Oil, supra*, 300 So.2d at 165 (reduction of statutorily authorized fees because of "excessive amount of time expended in preparation").

65. E.g., *Arkansas Land & Cattle Co. v. Anderson-Tully Co.* (Ark. 1970) 452 S.W.2d 632, 634 (unsuccessfully argued that the following person was not a suitable expert: a civil engineering graduate of the University of Arkansas who, among other matters, had been long employed by the Mississippi River Commission in connection with navigation and construction problems on that river and its tributaries, was a member of many professional societies (American Society of Civil Engineers, etc.), had been engaged for 30 years in studying land formations along the River for the purpose of determining the "genesis" of such lands, and had been retained in three court cases and in numerous nonlitigation boundary disputes concerning accretion issues).

66. *Arkansas Land & Cattle Co., supra*, 452 S.W.2d at 637–639; *Conran, supra*, 341 S.W.2d at 83; *Carpenter v. Ohio River Sand & Gravel Corp.* (W.Va. 1950) 60 S.E.2d 212, 214.

67. At least at this stage, little time will and can be devoted to an explication of specialized disciplines that one can bring to bear to understand the physical process that might have caused the river boundary movement. River hydrodynamics, fluvial geomorphology, hydraulics, and civil engineering are only some of the specialized disciplines that can be applied. A detailed discussion of such disciplines, and others not mentioned, is beyond the scope of this book. Later in this chapter, the discussion on proof of the location of the boundary will include some facts and

Having a clear understanding of the physical process, the mechanics, if you will, of the river movement will also help in avoiding the second element of the uncertainty surrounding river boundary movement determination.

7.4.2 Understand the Property Boundary Movement Issue

The second cause of uncertainty surrounding the property boundary consequences of the physical movement of a river boundary is the difficulty for surveyors, lawyers, and, consequently, courts in correctly conceptualizing and framing exactly what *is* the boundary movement issue. Faulty conception of the boundary movement issue has led many courts to beg the question or even to answer the wrong question. As a consequence, these courts have improperly decided cases, have avoided deciding the real issue, or have not allowed counsel to fully and fairly develop the facts so that the court might apply established legal principles to decide the legal consequences arising from such facts. At the risk of exemplifying the adage "Fools go where wise men fear to tread," this chapter turns to the seemingly intractable conceptual problem.

The conceptual problem seems to be intractable because, among other reasons, one can never be certain of the form in which the boundary question may arise or exactly how to place the factual circumstances creating the question in the proper context. The ways of nature and the minds of surveyors and lawyers are too unpredictable and inventive. Better to review what has occurred in the past to derive a lesson for the future.

Nowhere, perhaps, is the conceptual problem better exemplified than in a case concerning a portion of the Omaha Indian Tribe's vast reservation which, at one time at least, formed the Missouri River's Blackbird Bend, shown in Figure 7.1. In the Blackbird Bend area, the tribe's reservation in Nebraska was bounded on the east[68] by the thalweg[69] of the Missouri River. East of Blackbird Bend, farmers in Iowa had, by virtue of a chain of conveyances stemming from federal public land patents, come to own land whose western boundary was also the Missouri River.[70] The dispute arose because, over a

---

deductions that may be developed by these experts and the application of these conclusions in actual cases.

68. Actually, the Missouri River in the Blackbird Bend reach flowed in an easterly, southerly, and then westerly direction. The direction "east" is adopted here solely for ease of reference.
69. The center or main channel of the river. See note 86 in Chapter 3.
70. Besides the Iowa farmers, the State of Iowa also claimed an ownership interest—sovereign riverbed ownership—in the Blackbird Bend area. *Wilson v. Omaha Indian Tribe, supra,* 442 U.S. at 657, n.2. The State of Nebraska was not a party, as Nebraska did not have any bed ownership. This was because the Tribe and the United States had entered into the treaty creating the boundaries of the Omaha Tribe's reservation before Nebraska became a state. *Id.* at 658. Before statehood, the United States had authority to cede such sovereign riverbed lands, e.g., *Shively v. Bowlby* (1894) 152 U.S. 1, 48, but not after a state was admitted, *Montana v. United States, supra,* 450 U.S. at 553–554.

7.4 PROPERTY BOUNDARY CONSEQUENCES OF RIVER MOVEMENT

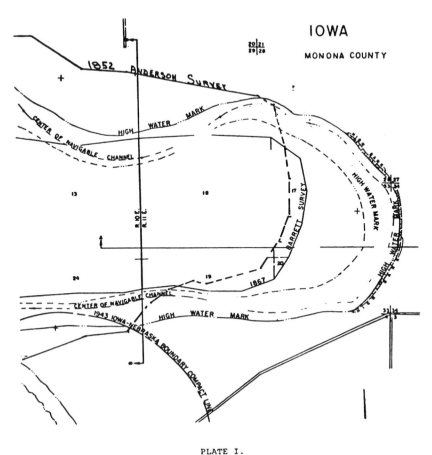

PLATE I.
Sketch of river's position in approximately 1875
showing 1852 Anderson and 1867 Barrett Survey meander lines.

Figure 7.1 U.S. v. Wilson Plate I at 575 F.2d 625

period of some 120 years, the Missouri River had significantly changed its geographic location.

From its geographic location in 1854, the river had moved more than 2 miles to the west to the geographic location shown in Figure 7.2 shown above. In effect, the river migrated over and obliterated the Blackbird Bend portion of the reservation.

As the river changed its location, the Iowa farmers farmed land that appeared along the Missouri River margin of their land. On the other side of the river, the area of the tribe's reservation was being reduced in size, although such a reduction was not, at any one time, immediately apparent. It was only when the tribe, assisted by the United States Bureau of Indian Affairs, "dispossessed" the Iowa farmers that it was evident that there were some in high

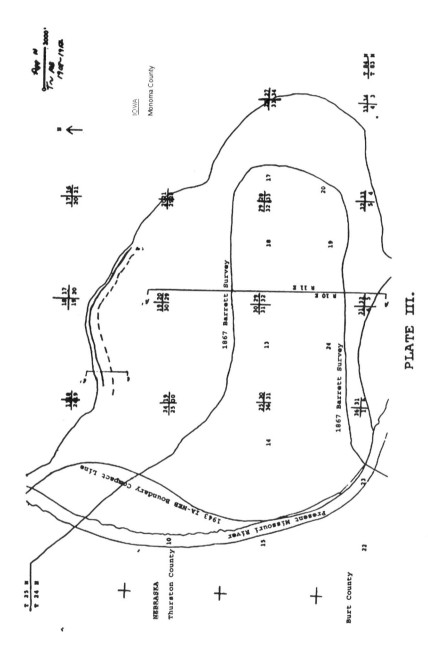

Figure 7.2 [See *U.S. v. Wilson*, Plate III at 575 F.2d 627]

7.4 PROPERTY BOUNDARY CONSEQUENCES OF RIVER MOVEMENT 251

places who viewed the property boundary of the tribe's reservation as unchanged, regardless of the change in geographic location of the physical indicia of that boundary, the Missouri River. This dispute alone generated eight separate court opinions,[71] including one in the United States Supreme Court.[72] It is, however, the United States Supreme Court's decision in this matter that exemplifies the misconception of the boundary issue and provides a lesson from which we can learn.

In this dispute, the Supreme Court considered[73] that the United States had never yielded its title or otherwise terminated its interest in the Blackbird Bend geographic area that had originally been part of the tribe's reservation in Nebraska but had long been farmed by the Iowa farmers.[74] This finding begs the question of whether the geographic area existing in 1975 was the *same* land in which the United States first held title in trust for the Indians as part of the Blackbird Bend area of the tribe's reservation in 1854.

The Court's rigid focus on the original geographic location of the Indian reservation without respect to the river's geographic location did not take into account the fact that the Missouri River boundary of the reservation had moved 2 miles to the west. After all, in 1975, land in that longitudinal and latitudinal location of Blackbird Bend lay some 2 miles *east* of the Missouri River, the eastern boundary of the Omaha Indian Tribe's reservation.

To reach the desired result, the Supreme Court had to assume that the Missouri River property boundary of the reservation had remained fixed in location. This assumption prejudged the property boundary effect of the change in geographic location of the Missouri River.[75] Such misconceptions can readily be avoided.

---

71. *United States v. Wilson* (N.D. Iowa 1977) 433 F.Supp. 57; *United States v. Wilson* (N.D. Iowa 1977) 433 F.Supp. 67; *Omaha Indian Tribe, Treaty of 1854, etc., supra,* 575 F.2d 620; *Wilson v. Omaha Indian Tribe* (1979) 442 U.S. 653; *Omaha Indian Tribe v. Wilson* (8th Cir 1980) 614 F.2d 1153, *rehg den., cert. den.* (1980) 449 U.S. 825; *United States v. Wilson* (N.D. Iowa 1981) 523 F.Supp. 874, *rev'd* (8th Cir. 1982) 707 F.2d 304; *United States v. Wilson* (8th Cir. 1983) 707 F.2d 304, *rehg den., cert. den.* (1984) 465 U.S. 1101; *United States v. Wilson* (N.D. Iowa 1984) 578 F.Supp. 1191, *cert. den.* (1984) 465 U.S. 1025.
72. One of the main issues in the United States Supreme Court opinion was to decide what was the proper choice-of-law. The *Wilson* case and the Supreme Court's resolution of the choice-of-law dispute are discussed in notes 61 through 79 and accompanying text in Chapter 2.
73. This finding was made as part of the Supreme Court's consideration of the choice-of-law question and was central to the Court's ultimate conclusion.
74. *Wilson v. Omaha Indian Tribe, supra,* 442 U.S. at 670.
75. Only if the Court treated the boundary movement as avulsive in character would land in the geographic location of the Blackbird Bend portion of the Omaha Indian reservation have been land in which the United States' interest would never have been terminated. On the other hand, if the Missouri River movement had been found to have occurred through the gradual and imperceptible process of accretion, the reservation's river property boundary would have continued to migrate to its 1975 geographic location. Then the United States would not have had any interest in the land in the geographic location of Blackbird Bend as such land lay east of the new geographic location of the Omaha Indian reservation's Missouri River property boundary.

The source of the misconception is confusion between the legal property boundary and the physical boundary. To avoid this misconception, one must recognize or conceive that the change in geographic location of the boundary watercourse is a discrete physical phenomenon, to account for which the law has developed rules and principles.

For example, rather than characterize the change in geographic location as "accretive" or "avulsive," or, as the United States Supreme Court did in the Blackbird Bend case, prejudge the issue, the client, the surveyor, the lawyer, and the judge should ensure that the expert consultants determine, reconstruct, and then sensibly describe the physical process or conditions instrumental in the change in geographic location. The contestants' lawyers will then seek to have the court apply to such physical conclusions the appropriate rules determining the legal effect on the property boundary of such physical process or conditions.[76] In other words, the issue is at least twofold: What was the physical nature of the boundary movement process, and what is, given an answer to the first question, the legal effect, if any, on the geographic location of the property boundary?[77]

With these admonitions in mind, we now turn to application of the rules concerning the property boundary consequences of river movement.

7.5 PROPERTY BOUNDARY CONSEQUENCES OF RIVER MOVEMENT—ACCRETION AS COMPARED WITH AVULSION

The majority of boundary cases arising in the context of rivers seem to concern one central question: Was the boundary movement gradual and imperceptible (accretive), or was it sudden and perceptible (avulsive)? This is because one of the parties to the dispute has commonly contended that the riparian property boundary should have been fixed in a different (and presum-

---

76. *Kitteridge v. Ritter* (Iowa 1915) 151 N.W. 1097, exemplifies a case in which the court correctly understood the issue:

> The land in dispute . . . is concededly on the [Iowa] side of the Missouri River. At the time of the original government survey in the [18]50s much of this land, and perhaps all of it, was not existent in Iowa. Its location under the sky—that is to say, its points of latitude and longitude—were within the boundaries of Nebraska. The question . . . is whether this land was thrown up as accretion on the [Iowa] side of the river by the gradual recession of the river towards the west, or whether there was a sudden avulsion of the river whereby the body of land upon this location was cut off from the state of Nebraska in such a manner that it can be identified as a body of land existing in Nebraska before such avulsion.

77. Thus, in the *Wilson* case the Court should have stated the question: What law governs to decide the effect on the mutual property boundary of an Indian reservation held in trust by the United States in one state and farmland in another state when the physical location of their common boundary watercourse has changed? On the framing of questions for court decision, see note 108 in this chapter.

ably more advantageous) geographic location and at an earlier date than its present geographic location.

In answering this question, judges, like most people, do not want to disappoint seemingly deserving individuals—in this instance, the litigants. Thus, courts have not slavishly followed and rigidly applied rules developed for or in a particular situation. Rather, judges in this area have often refashioned the rules or applied existing rules in a manner that resulted in an outcome tailored to specific physical conditions or certain reaches and regimes. The consequence of this method of judicial river boundary determination is that there seem to be as many "rules" as there are boundary cases.

A review of a few leading cases will establish that this practical approach has led to rulings in what appear to be similar situations that may seem quite hard to reconcile with one another. Before analyzing these cases, however, it will help to keep in mind the purpose behind the rules of accretion and avulsion.

7.5.1 Purposes of Accretion and Avulsion Rules

Long ago the United States Supreme Court theorized the rationale of the accretion rule: Those who own riparian property may suffer because of changes in the boundary watercourse. Consequently, those owners should be allowed to retain any benefit resulting from the geographical proximity of their property to the watercourse in order to maintain the valuable riparianess of their property.[78] This theory has become known as the compensation theory.[79]

On the other hand, the rule of avulsion was created for a somewhat different purpose. The basis for the principle of avulsion is the assumed need to mitigate hardship to riparian or littoral property owners caused by a sudden change in location of the boundary watercourse whereby one property owner appears to lose some land without fault.[80] Thus, the accretion rule focuses on the property owner's expectation of a continued relationship to the water-

---

78. *New Orleans, supra,* 35 U.S. at 717; *County of St. Clair v. Lovingston* (1874) 90 U.S. (23 Wall.) 46, 64 (". . . The owner takes the chances of injury and of benefit arising from the situation of the property. If there be a gradual loss, he must bear it; if a gradual gain, it is his"); *Bonelli Cattle Co., supra,* 414 U.S. at 326.

79. Other rationales support the accretion doctrine. Those theories are (1) the expectation of riparian owners, (2) the great value of continuing to be riparian, (3) a vested right to future deposition, and (4) the interest of the community that all lands have an owner and the consequent rule of convenience that the riparian owner has title to such additions. E.g., *Jefferis, supra,* 134 U.S. at 191; *Bonelli, supra,* 414 U.S. at 326; *State v. Sause* (Ore. 1959) 342 P.2d 803, 826. As a practical matter, these other theories are subsumed into the compensation theory.

80. *Bonelli, supra,* 414 U.S. at 327; *Nebraska v. Iowa, supra,* 143 U.S. at 362. The rationale of these cases appears to be founded on rules established for international political boundaries between nations. *Nebraska v. Iowa, supra,* 143 U.S. at 362–365. It may be that the international political basis distinguishes this theory. After all, what nation can one recall that has voluntarily, without any protestation, given up any part of its national territory?

course; the avulsion doctrine focuses on the necessity for the stability of property boundaries and the unfairness to property owners if one should gain and another should lose land innocently and suddenly.

So far as concerns the purpose of the rule of avulsion, one other point should be mentioned in this context. As long ago as Roman times it was considered immutable that in the case of an avulsion, public ownership of the riverbed followed the flowing water.[81] This rule has been somewhat modified over the centuries.[82] Some recent cases appear to hold that in the case of an avulsion, the public retains ownership of the former bed under the doctrine of avulsion and acquires ownership of the new bed by prescription.[83] The reason for this rule is the claim that no individual person should have control of a navigable watercourse.[84]

With these purposes in mind, how similarly described physical changes in the location of riparian property boundaries in different river regimes have been treated by some leading cases will show the uncanny ability of courts (and the litigants' lawyers) to manipulate and massage the facts to reach a desired result. This practice not only satisfies particular litigants, but appears to provide considerable leeway to other property owners and their retained consultants and lawyers searching for creative solutions to especially nasty riparian boundary problems.

7.5.2 Three Conflicting Cases

There is a body of United States Supreme Court cases[85] concerning lands along the Mississippi and Missouri Rivers in which application of the accretion and avulsion doctrines as they concerned rivers was defined (but hardly settled). More than 100 years ago alluvion belonging to the owner of Mississippi River riparian lands was described by the United States Supreme Court as ". . . an addition to riparian land, gradually and imperceptibly made by the water to which the land is contiguous. . . ."[86] In three cases decided just

---

81. MacGrady, *supra*, 3 Fla.St.U.L.Rev. at 575–576.
82. *Arkansas v. Tennessee* (1918) 246 U.S. 158, 174–175; see *Furlong Ent., supra,* 423 N.W.2d at 140, n.30 (by virtue of statute, state probably gains title to bed of new channel).
83. See., e.g., *State of California v. Superior Court (Fogerty)* (1981) 29 Cal.3d 240, 248–249; cert. den. sub nom. *Lyon v. State of California* (1981) 454 U.S. 865, *rehg den.* (1981) 454 U.S. 1094 ("*Fogerty I*"); *Fogerty II, supra,* 187 Cal.App.3d at 238.
84. Cf. *Bohn, supra,* 107 Cal.App.2d at 744; *Lamprey, supra,* 53 N.W. at 1143; *Flisrand, supra,* 152 N.W. at 799–800.
85. In this area, the choice-of-law question is often more important than the rule of boundary movement. This is because, in certain instances, the boundary rule may be different or the burden of proof changed, depending on which state's law applies or whether there is a federal rule of decision. Maloney, "The Ordinary High Water Mark: Attempts at Settling an Unsettled Boundary Line" (1978), 13 Land & Water L.Rev. 466, 476. For a particularly graphic example, see *Omaha Indian Tribe Treaty of 1854, etc., supra,* 575 F.2d at 632, n.22. Except where necessary, this chapter does not intend to deal with such nuances in any detail. For an overview of the choice-of-law issue, refer to Chapter 2.
86. *County of St. Clair v. Lovingston, supra,* 90 U.S. at 68.

three years apart, this test, sometimes referred to as the *Lovingston* test, was applied by the Supreme Court to similarly described Mississippi and Missouri river processes with decidedly different results.

7.5.2.1 The Jefferis Case—Application of the Accretion Rule. Because of the nature of the Missouri River (and the inventiveness of one property claimant's lawyer), the basic accretion rule described earlier was required to be further refined.[87] The facts were these: A public land lot was bounded on the north by the Missouri River.[88] For about 20 years new land continued to appear along that lot's river boundary until about 40 additional acres had been "formed by accretion."[89] Then the river suddenly cut through its bank. The new geographic location of the river left a part of the lot and the additions to the lot, which had once all been south of the riverbank, now north of that bank. The plaintiff, successor to the original owner of the lot, claimed all of the land within the original lot boundaries and all land that had attached itself to those boundaries.[90]

The defendants, on the other hand, had occupied the land formed by accretion without any authority and, considering the flimsiness of their right of occupancy, audaciously threatened to interfere with plaintiff's plans to improve the land in dispute. Taking a particularly contentious approach, the defendants insisted that the "well-settled" rule of accretion did not apply to

---

87. In this area it is particularly important for the facts of cases to be clearly understood. Rules of boundary movement may be clear in the abstract. In real life, application of such rules is hardly so cut and dried. Consequently, knowledge and understanding of the factual context in which such rules have been developed and applied are critical to ensure their correct application or in determining which rules will or should be asserted in a particular situation. Indeed, once the physical process has been defined and then described by one legal term or the other, application of the boundary movement rule should be routinely determined.

88. *Jefferis, supra*, 134 U.S. at 180.

89. *Id.* at 181. The opinion stated:

The said land was so formed by natural causes and imperceptible degrees, that is to say, by the operation of the current and waters of the river, washing and depositing earth, sand, and other material against and upon the north [boundary]; and the waters . . . receded therefrom, so that the new land so formed became high and dry above the usual high water mark and the river made for itself its main course far north of the original meander line. Such process went on so slowly [over a 17 year period] that it could not be observed in its progress; but, at intervals of not less than three or four months, it could be discerned by the eye that additions greater or less had been made to shore. *Id.* at 181–182.

90. *Id.* at 182. It is important to remember that unless the base title conveyed is contiguous to the watercourse, there is no foundation on which to claim these depositions. E.g., Clark 7th, *supra*, § 23.05; Clark 4th, *supra*, § 556; *Saulet v. Shepherd* (1866) 71 U.S. (4 Wall.) 502, 508; *Slauson v. Goodrich Transp. Co.* (Wisc. 1897) 69 N.W. 990, 992. There is a conflict of authorities over whether land separated from the watercourse by a public road, street, or highway is considered riparian. E.g., *City of Missoula v. Bakke* (Mont. 1948) 198 P.2d 769, 771–772 (setting forth majority and minority rules and holding with the minority that an alley between the property and the watercourse (a nonnavigable river) will not prevent the property from being considered riparian).

land adjacent to the Missouri River because of circumstances peculiar to that region.[91] The Court did not accept this novel argument; it held that the rule of accretion also applied to the Missouri River[92] even considering the comparatively great and rapid changes occurring in that river regime.[93]

The Court also provided further refinement to what was meant by the term "imperceptible." It concluded that land that had formed as additions to the lot over the almost 20-year period was formed imperceptibly, even though every three months or so one could see that additions to the upland had occurred.[94] One of the elements that appeared to be important to the Court in concluding that the land was formed by the imperceptible deposition of alluvion was that such land could not be identified as having been previously the land of a particular person.[95]

In the *Jefferis* case it was easy to follow the "well-settled" rule. One claimant was a mere trespasser. The hard case would come when application of those "well-settled" rules would "dispossess" a truly innocent landowner of riparian land that in all equity should be his or hers. Such a case arose just one year later with predictable results.

---

91. The opinion noted:

> ... [T]he course of the river [is] torturous, the current rapid, and the soil a soft, sandy loam, not protected from the action of the water either by rocks or the roots of trees; the effect being that the river cuts away its banks, sometimes in a large body, and makes for itself a new course, while the earth thus removed is almost simultaneously deposited elsewhere, and new land is formed almost as rapidly as the former bank was carried away. *Jefferis, supra*, 134 U.S. at 189.

92. These cases have not been limited in their applicability only to property boundary issues along the Mississippi and Missouri Rivers. Missouri and Mississippi River property and interstate boundary cases have been constantly cited by later courts as the source of the accretion rule or, when appropriate, the avulsion rule in the case of other rivers, lakes, and even the open coast. E.g., *Bonelli, supra*, 414 U.S. at 325–327 (Colorado River); *Cal. ex rel. State Lands Com'n v. United States* (9th Cir. 1986) 805 F.2d 857, 865, *cert. den.* (1987) 108 S.Ct. 70 (lake); *California ex rel. State Lands Comm'n v. U.S., supra*, 457 U.S. at 278 (open coast).

93. *Jefferis, supra*, 134 U.S. at 190. In fact, it was noted that the need for application of the rule of accretion was even more imperative because of these great changes. *Id.* at 191.

94. *Id.* at 192–194. The Supreme Court relied on Roman and English law. According to the opinion, English law did not require that, for gradual and imperceptible deposition, formation of the land must be indiscernible by comparison at two distinct points in time. *Id.* at 193. The Court also relied on the definition found in the *Lovingston* case. *Ibid.* See note 135 in this chapter.

95. "There is no suggestion ... that the land made by the accretion can be identified as having been previously the land of any particular person. There can be no identification unless there is a sudden change, and that is the very opposite of an imperceptible accretion." *Id.* at 194. Much has been made of this statement. Some have argued that to be determined an avulsion, it is required that the land cut off must be capable of being identified as land that a particular owner owned before the cutoff. E.g., *Nebraska v. Iowa, supra*, 143 U.S. at 368–369; *Kitteridge, supra*, 151 N.W. at 1097. On the other hand, one can also argue that the Court did not mean that identification of the cutoff land was necessary for an avulsion, only that the ability to identify such land was *evidence* of suddenness of the change. *Omaha Indian Tribe, Treaty of 1854, etc., supra*, 575 F.2d at 637–638. Some courts have not required that there be any identifiable land in order to find that an avulsion has occurred. *Dartmouth College v. Rose* (Iowa 1965) 133 N.W.2d 687, 689.

7.5.2.2 The Rutz Case—Application of the Avulsion Rule. In the next case, *St. Louis v. Rutz*, plaintiff's title to lands near St. Louis, Missouri, was derived from public land patents for certain lots in Illinois, the surveys of which extended to the Mississippi River.[96] From the time plaintiff acquired the lots, "accretions" had formed on their river frontage and had been possessed by the plaintiff.[97] Subsequently, over about a 20-year period, the riverfront of plaintiff's lots was washed away and caved in[98] until the riverfront of the lots had retreated east, beyond its originally surveyed location.[99] Dikes constructed by defendant City of St. Louis had caused this washing away.[100] Interestingly, the Court specifically found that the washing away and caving in were rapid and perceptible, not slow and imperceptible.[101]

---

96. *St. Louis v. Rutz, supra,* 138 U.S. at 227–228. Plaintiff's title claim extended not just to the bank, but to the center thread of the Mississippi River. This claim provides a graphic example of the importance of considering the choice-of-law question. Illinois, like some other states, had, as the Supreme Court put it, "established and steadily maintained as a rule of property" that title of a riparian owner along the Mississippi River extended to the middle line of the main channel of that river. *Id.* at 242. Some states that also extended riparian ownership waterward of the ordinary high-water mark are listed in *Shively, supra,* 152 U.S. at 18–25; Clark 7th, *supra,* § 23.02; Clark 4th, *supra,* § 553, p. 741, § 566, p. 772; and Skelton, *supra,* § 287, p. 321. Some variants of riparian ownership are recounted in *Conran, supra,* 341 S.W.2d at 80, and *United States Gypsum v. Uhlhorn* (E.D. Ark. 1964) 232 F.Supp. 994, 1001, n.4, *aff'd* (8th Cir. 1966) 366 F.2d 211, *cert. den.* (1967) 385 U.S. 1026. This question, whether riparian ownership based on public land title extends to the middle thread of the stream or to the high- or low-water mark, is governed by state, not federal, law. *St. Louis v. Rutz, supra,* 138 U.S. at 242; *Barney, supra,* 94 U.S. at 337; *Packer v. Bird* (1891) 137 U.S. 661, 669.
97. *St. Louis v. Rutz, supra,* 138 U.S. at 229.
98. The Supreme Court appeared to choose carefully the words describing the physical process by which plaintiff's Mississippi River property boundary changed its geographic location. The Court did not use terms such as "erosion" or "avulsion" in describing this process. Instead the Court adopted the expression ?washing away.? This appears to be an implicit recognition that certain terms used to describe a physical process may also have taken on unintended legal connotations. Thus, to some, if there is "erosion" of the bank of a river, title is lost. E.g., Skelton, *supra,* § 567, p. 773. In this case, the Supreme Court avoided that pitfall and thoughtfully chose terms without legal consequence.
99. *St. Louis v. Rutz, supra,* 138 U.S. at 230.
100. *Id.* at 231–232.
101. *Id.* at 231. This conclusion was based on the following statement:

> ... [T]he count further finds that such washing away of said river bank occurred principally at the spring rises or floods of high water ... ; that during each flood there was usually carried away a strip of land from off said river bank from two hundred and fifty to three hundred feet in width, which loss of land could be seen and perceived in its progress; that as much as a city block would be cut off and washed away in a day or two; that blocks or masses of earth from ten to fifteen feet in width frequently caved off and fell into the river and were carried away at one time ..... *Ibid.*

At this juncture in the opinion the Court drew no legal conclusion as to the property boundary effect of this rapid and perceptible washing away. The reader should compare this description with the description of the physical process in *Jefferis* at note 91 in this chapter. Thus, the reader will be able to assess the different property boundary effects of quite similarly described physical processes.

The competing claimant, the City of St Louis, had a more complicated situation. The City of St. Louis owned an island in the bed of the river a mile or so upstream from plaintiff's lots; the island lay in Missouri on the western side of the main channel of the river.[102] Before plaintiff's lands were washed away, a large portion of the city's island was washed away by the river[103] and a bar formed in the river downstream, but joined to the island.[104] Critically, the navigable channel of the Mississippi was on the west side of the bar, and thus the bar, unlike the island, was located in Illinois, not in Missouri.[105]

As a result of the actions of a third party, the river bar ultimately became dry land in front of plaintiff's lots.[106] Figure 7.3 on the following page depicts the situation in Rutz. The question, as stated by the Court, was whether this land was part of the plaintiff's lots on the Illinois side of the Mississippi River or was owned by the City of St. Louis as accretion to the city's island, which lay more than a mile upstream from plaintiff's lots and on the Missouri side of the Mississippi River.[107] From the way the question was posed, the answer is obvious.[108]

---

102. *Id.* at 232.
103. *Ibid.*
104. *Id.* at 234. The opinion noted that this bar was not formed by accumulation of soil on the lower end of the island, but rather by deposition of sediment on the bed of the river downstream from the island. *Ibid.* This is a critical distinction in such cases. If the bar were found to be deposition to the island, title would have been quieted in the owner of the island. Skelton, *supra*, § 294; Clark 7th, *supra*, § 25.07; Clark 4th, *supra*, § 588; *Benne v. Miller* (1899) 50 S.W. 824, 827; *Peterson v. Harpst* (Mo. 1952) 247 S.W.2d 663, 668; *People v. Ward Redwood Co.* (Calif. 1964) 225 Cal.App.2d 385, 395. On the other hand, if the bar were found to be deposition to the bed of the river, the bar or island would be owned by the riverbed owner. *Kansas v. Missouri*, *supra*, 322 U.S. at 229; *Dartmouth College*, *supra*, 133 N.W.2d at 690; *Anderson-Tully Company v. Walls* (N.D. Miss. 1967) 266 F.Supp. 804, 813–814; *Conran*, *supra*, 341 S.W.2d at 81; *People v. Ward Redwood*, *supra*, 225 Cal.App.2d at 388; Clark 7th, *supra*, § 25.19; Clark 4th, *supra*, § 600. Islands that were in existence in the bed of a river (or other navigable water body) prior to statehood are public lands of the United States and were not transferred to the states under the Equal Footing Doctrine upon admission. E.g., Manual of Surveying Instructions, 1973 (Landmark) § 3-122, p. 97; *Scott v. Lattig* (1913) 227 U.S. 229, 244; *Moss v. Ramey* (1916) 239 U.S. 538, 546. Riparian owners claiming by virtue of United States' public land title or through United States' surveys that extend to the river margin, do not take title to such islands by virtue of such patents. *Packer*, *supra*, 137 U.S. at 672–673; *Callahan v. Price* (Idaho 1915) 146 P. 732, 734. Occasionally, such islands have passed to the states. *United States v. Chandler-Dunbar* (1908) 209 U.S. 447, 451–452; *Whitaker v. McBride* (1905) 197 U.S. 510, 513; *Flisrand*, *supra*, 152 N.W. at 801.
105. *St. Louis v. Rutz*, *supra*, 138 U.S. at 234.
106. *Id.* at 234–235.
107. *Id.* at 241.
108. Non-attorney readers who have never had the opportunity (or the penance) of reviewing a court brief composed by a lawyer should note that one of the great skills of which lawyers claim to be endowed is the ability to state the question in a case so that the court deciding the issue can give only the answer desired by the lawyer who wrote the question. For example, one could have stated the question here, not as the Supreme Court formulated it, but rather: Who is the

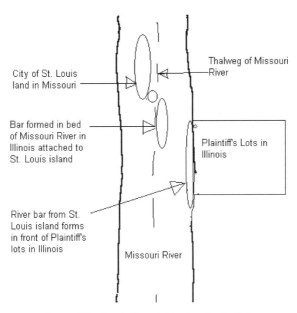

Figure 7.3 Depiction of the situation in Rutz

The Court held[109] that the sudden and perceptible loss of plaintiff's land, "which was visible in its progress," did not work a change in the boundaries of plaintiff's lots as the lots existed at the time the washing away commenced.[110] That is, an avulsion fixed the geographic location of the Missis-

---

owner of the land formed as accretions to an island in the bed of a navigable river, but through no action on the part of the owner of the island become attached to land on the opposite bank of the river? The answer to that question is also obvious—the owner of the island. E.g., *People v. Ward Redwood, supra,* 225 Cal.App.2d at 395. Thus, in order to be able to pose or to respond to such questions in a manner benefiting your client, as emphasized by this chapter, it is necessary to obtain a complete understanding of the physical process, the boundary issue, and the application of riparian boundary movement principles.

109. Another important question in the case concerned the interpretation of plaintiff's deed. The deed description called to the "low water mark of the Mississippi River" and "thence down to the extended line between [the lots]." The defendant contended that the deed did not convey the bed of the river beyond the low water mark. *St. Louis v. Rutz, supra,* 138 U.S. at 243. The Court found that this description made the river the boundary and, in addition, the description that included ". . . all rights as riparian owner to the accretion or sand bar . . ." showed an intent by the grantor not to retain any interest in the bed of the river beyond the low water mark. *Id.* at 243–244. This discussion provides further emphasis to the earlier injunction that an understanding of the instruments in the chain of title is a critical prerequisite to a successful boundary determination. See notes 222 through 225 and accompanying text in Chapter 1. The decision also established that plaintiff's title was contiguous to (or in) the river with all the benefits of riparianess. See note 90 in this chapter.

110. *Id.* at 245. The defendant had argued that when the plaintiff's lots were swept away by the river and ceased to exist, plaintiff's ownership of that property also ceased. Without using the word "avulsion," the opinion responded:

sippi River boundary of plaintiff's lots. Plaintiff was entitled to land that arose in the bed of the river within that boundary.

The Court was good enough to provide reasons for this result. Following the rationale of both the accretion and the avulsion rules[111] and concentrating on the rapidity of the "washing away," the Court reasoned that unless plaintiff's title to the land was sustained, plaintiff, without being in any way at fault in the loss of the land area, would lose his riparian rights[112] and access to the navigable part of the river.[113]

While this case seems to make perfect sense on its own, it appears to be at odds with both earlier[114] and, as we shall see, later rulings of the United States Supreme Court. This case provides further evidence that there are no cut-and-dried results when it comes to answering the accretion/avulsion question.[115]

... It is laid down, however, by all the authorities, that, if the bed of the stream changes imperceptibly by the gradual washing away of the banks, the line of the land bordering upon it changes with it; but that, if the change is by reason of a freshet, and occurs suddenly, the line remains as it was originally. *Ibid.*

111. See notes 78 through 80 and accompanying text in this chapter.
112. The court listed some of those riparian rights: "Among his rights as a riparian owner are access to the navigable part of the river from the front of his land, and the right to make a landing, wharf, or pier, for his own use or the use of the public. [Citations omitted.]" *St. Louis v. Rutz, supra,* 138 U.S. at 246.
113. There may, however, have been an additional factor in this case. The opinion specifically acknowledged plaintiff had not done anything to cause the loss of his land or to transfer title to that land to the defendant. *Ibid.* On the other hand, the defendant City of St. Louis, by building dikes across the river, had caused the washing away of plaintiff's lots. The Court implied that the city should not be allowed to gain title to land that later came to be formed in the geographic location where plaintiff's lots existed before they had been washed away because of the city's own actions. *Ibid.*
114. *St. Louis v. Rutz* deviated from the Court's earlier decisions' result in three notable fashions. First, there appeared to be no requirement in this case that to be treated as an avulsion there be any identifiable land cut off or deposited as the result of the sudden and perceptible loss of land. See note 95 in this chapter. Second, in *Rutz* the Court treated the sudden and perceptible physical process described in the opinion as an avulsion. In *Jefferis*, however, the Court treated a virtually identically described process as rapid accretion. See notes 91 and 101 in this chapter. The third deviation was that in comparison with cases in which an avulsion had been found, e.g., *Missouri v. Nebraska, supra,* 196 U.S. at 35–36 (within 24 hours during very high water cut through neck of land), and *Arkansas v. Tennessee, supra,* 246 U.S. at 162 (sudden and violent cutting through neck of land in 30 hours), concluding that an avulsion occurred where the process took place in stages over 20 years is what one may call "smoking gun" evidence that the Court was willing to refashion the rules to reach a desired result. On the other hand, it may also be that the Court wrongly decided *Jefferis.*
115. The Court treated the rapid washing away, not as an erosion as it had previously done and would do in the future, but as an avulsion. The Supreme Court could have reached the same result in another fashion and avoided the uncertainty that appeared to result when one compares *St. Louis v. Rutz* with the Court's earlier and later decisions about the property boundary effect of

7.5.2.3 Nebraska v. Iowa—Accretion Found Once Again.
The final example concerns an interstate boundary dispute along the Missouri River. The boundary line between Nebraska and Iowa was[116] the middle of the main channel of the Missouri River.[117] Between 1851 and 1877 there were marked changes in the course of the river channel; its geographic location in 1877 was much different from its 1851 geographic location.[118] A political jurisdiction dispute arose between the two states.

The Court's description of the peculiarities of the Missouri River processes was remarkably similar to the Court's previous description of the Mississippi River process in *St. Louis v. Rutz*.[119] Its conclusion, however, was not the same. In *Nebraska v. Iowa*, decided barely one year after *St Louis v. Rutz*, the Court held that the rule of accretion *did* apply to the rapid and visible physical deterioration of the riverbank.[120] In reaching this conclusion the

---

similarly described physical processes. The Court could have found that erosion had occurred and plaintiff's river boundary had retreated. Then, based on the doctrine of reemergence, see notes 164 through 166 and accompanying text in this chapter, the Court could have found plaintiff to be the owner of the land that later appeared at the geographic location of the previously eroded land.

116. This boundary has now been fixed by interstate boundary compact. *Nebraska v. Iowa* (1972) 409 U.S. 285. Other states with mutual river boundaries have also established such boundaries' location by compact. E.g., Interstate Compact Defining the Boundary Between the States of Arizona and California, 80 Stat. 340. Boundary compacts determine only the extent of political jurisdiction, *not* questions of ownership of real property. E.g., *Nebraska v. Iowa* (1972) 406 U.S. 117, 120–121; *California v. Arizona* (1979) 440 U.S. 59, 61.

117. *Nebraska v. Iowa, supra*, 143 U.S. at 359–360.

118. *Id*. at 360.

119. Compare the United States Supreme Court's description of the physical process of the Mississippi River in *St. Louis v. Rutz*, see note 101 in this chapter, with the following description of the Missouri River in *Nebraska v. Iowa*:

> . . . The large volume of water pouring down . . . with the rapidity of its current has great and rapid action upon the loose soil of its banks. Whenever it impinges with direct attack upon the bank at a bend of the stream, and that bank is of the loose sand obtaining in the valley . . . , it is not strange that the abrasion and washing away is great and rapid. Frequently, where above the loose substratum of sand there is a deposit of comparatively solid soil, the washing out of the underlying sand causes an instantaneous fall of quite a length and breadth of the superstratum of soil into the river; so that it may, in one sense of the term, be said the diminution of the banks is not gradual and imperceptible, but sudden and visible. Notwithstanding this, two things must be borne in mind . . . : that, while there may be an instantaneous and obvious dropping into the river of quite a portion of its banks, such portion is not carried down the stream as solid and compact mass, but disintegrates and separates into particles of earth borne onward by the flowing water . . . ; and, also, that while the disappearance . . . of a mass of bank may be sudden and obvious, there is no transfer of such a solid body of earth to the opposite shore, or anything like an instantaneous and visible creation of a bank on that shore. . . . *Id*. at 368–369.

120. *Id*. at 369–370.

Court carefully noted that whatever the fact was as to "diminution," the deposition was imperceptible even though it was rapid.[121] What meaning, if any, can be drawn from these contradictory results?

7.5.3 The Flexible Approach to the Accretion Versus Avulsion Issue

These three cases, decided over a relatively short period of time, form the basis of many property boundary decisions reached by courts in modern times. But it is in these three cases that the United States Supreme Court provided a most concrete example of the flexible approach to accretion/avulsion questions. The theoretical basis and the results of those decisions shift forth and back much like the rivers the decisions discuss. The pragmatism shown in these decisions strongly suggests that the courts will decide each case on its own peculiar facts and equities. Past precedent will be given due regard, so long as the precedent and the specific physical process concerned in the property boundary dispute can be massaged to fit one another to attain the desired result.

This flexible approach to the accretion/avulsion issue leaves property owners, their lawyers, and surveyors who are trying to predict the property boundary consequences of the inevitable changes in the physical location of a river boundary, in what could be seen as an enviable position. Because there are no cookie-cutter formulas, creative solutions may be proposed. By careful analysis of the title documents and painstaking investigation of the physical circumstances leading to or causing the change in the geographic location of

---

121. The Court stated:

There is no heaping up at an instant, and while the eye rests upon the stream, of acres or rods on the forming side of the river. No engineering skill is sufficient to say where the earth in the bank washed away and disintegrating into the river finds its rest and abiding place. . . . There is, no matter how rapid the process of subtraction or addition, no detachment of earth from the one side and deposit of the same upon the other. The only thing which distinguishes this river from other streams, in the matter of accretion, is in the rapidity of the change caused by the velocity of the current; and this in itself, in the very nature of things, works no change in the principle underlying the rule of law in respect thereto. *Id.* at 369.

In *St. Louis v. Rutz*, the Supreme Court had focused on the rapid washing away of the land; no mention was made of any deposition. See notes 87 through 92 in this chapter. The different focus of the *Rutz* court may be one reason for the different result than in *Nebraska v. Iowa*. Also in *Nebraska v. Iowa* the Supreme Court once again appeared to require there be some identification of the land detached and later deposited as evidence of an avulsive change.

the riparian land's property boundary, the surveyor, the retained consultants, and the lawyer should be able to discover, refashion, or develop legal theories by which the principles of property boundary movement can be applied. There are some other doctrines that may well assist them in that effort.

7.6 IMPACT OF THE BURDEN OF PROOF ON THE PROPERTY BOUNDARY CONSEQUENCES OF RIVER MOVEMENT

Perhaps the most important factor in these river accretion/avulsion cases, a factor even more important than the substantive rule applied, is just who has the burden of proof[122] to establish that the change in the physical location of the boundary should be treated as either accretive or avulsive in legal effect.[123] The general rule is that there is a presumption[124] in favor of accretion, and the party claiming that an avulsion has occurred has the burden of proving the facts supporting such a claim.[125] This presumption is based on long experience and observation of rivers[126] and the conclusion from experience and observation that river movement occurs gradually rather than suddenly.[127] This is not to say that the presumption can never be overcome,[128] but it is extremely

---

122. See notes 132 through 135 and accompanying text in Chapter 3.
123. E.g., *Pannell v. Earls* (Ark. 1972) 483 S.W.2d 440, 442 ("considerable burden" to show avulsion); *Kansas v. Missouri, supra,* 322 U.S. at 230 (Kansas does not carry burden to show avulsion after accretion).
124. See note 132 in Chapter 3.
125. *Ibid.*; Skelton, *supra,* § 298; *Shapleigh v. United Farms Co.* (5th Cir. 1938) 100 F.2d 287, 288 (citing 9 C.J. 271); *Mitchell v. Smale* (1891) 140 U.S. 406, 414 (assume change in level of lake not sudden); *Mississippi v. Arkansas* (1974) 415 U.S. 289, 295 (Douglas, J., dissenting) (quoting from Report of Special Master approved by the majority); *Dartmouth College, supra,* 133 N.W. 2d at 689; *Pannell, supra,* 483 S.W.2d at 442; *Bone v. May* (Iowa 1929) 225 N.W. 367, 368; *Plummer v. Marshall* (Tex. 1910) 126 S.W. 1162, 1163.
126. Clark 7th, *supra,* § 24.02; Clark 4th, *supra,* at § 567, p. 775, sets forth, in simplified terms, a basic explanation of one river process. A more graphic example is set forth at note 119 in this chapter.
127. *Pannell, supra,* 483 S.W.2d at 442; *United States Gypsum Co. v. Reynolds* (Miss. 1944) 18 So.2d 448, 449; Skelton, *supra,* § 298. Other reasons supporting the presumption are that a person asserting a material change has the burden to show it, and it is assumed that if an avulsion had occurred the change would have been so noticeable that there would be witnesses or written accounts of the event. E.g., *Oklahoma v. Texas* (1923) 260 U.S. 606, 638; *Mississippi v. Arkansas, supra,* 415 U.S. at 293; see note 142 in this chapter.
128. In *Arkansas Land & Cattle Co. v. Anderson-Tully Co., supra,* 452 S.W.2d at 635–636, it was noted that the presumption in favor of accretion operates only in the absence of countervailing evidence. Even then, it is only when "the lay of the land, the length of the elapsed time as related to the distances of the movement, and the general correspondence of the location and directions of the river at the later period as compared with that of the earlier . . . ," that permitting the presumption to operate is appropriate. *United States Gypsum Co. v. Reynolds, supra,* 18 So.2d at 449. In another case, where the evidence showed the river "literally cut a new channel . . . over a 50 year period . . . ," the presumption of accretion was overcome. *Nesbitt v. Wolfkiel* (Idaho 1979) 598 P.2d 1046, 1049. And the presumption has been held not to apply when there is evidence of an avulsion. *McCafferty v. Young* (Mont. 1964) 397 P.2d 96, 100.

difficult to do so.[129] And in certain special circumstances the presumption is not even applicable.[130]

Assuming that there is no presumption in favor of accretion, what is meant by the phrase "gradual and imperceptible" so that one may decide whether a given physical process should be termed an accretive process in the eyes of the law?

7.7 THE "TRUE" MEANING OF "GRADUAL AND IMPERCEPTIBLE"

The landowner contending that a gradual and imperceptible physical process has increased the absolute size of his or her landholding is aided not only by the presumption in favor of accretion, but also by certain court decisions that provide guidance as to what is meant by gradual or imperceptible. The Supreme Court has said, at least in cases involving land along the Mississippi and Missouri Rivers, that accretion and erosion can be quite rapid.[131] And other cases have applied this principle to rivers other than the Missouri and

---

129. To emphasize the point, one could examine the litigation over the private ownership of what was known as the Luna Bar in the Mississippi River between Arkansas and Mississippi. *Arkansas Land & Cattle Co. v. Anderson-Tully Co.*, *supra*, 452 S.W.2d 632. The question was whether Arkansas' courts had jurisdiction to decide the case. *Id.* at 634. As might be expected, Arkansas' jurisdiction depended on whether the land was in Arkansas. It was claimed that the land was in Arkansas because the land had been separated from Arkansas by an avulsion. *Id.* at 633. The Arkansas Supreme Court determined that the presumption in favor of accretion did not apply because of countervailing evidence and upheld the trial court decision that the Mississippi landowner had not shown that accretion to the Mississippi shore had formed the land in question. *Id.* at 634–636. Therefore, the Arkansas court had jurisdiction to try the case. *Id.* at 640.

Because of that litigation, the State of Mississippi commenced an original action in the United States Supreme Court against the State of Arkansas to decide the political boundary line between the two states in the Luna Bar area of the Mississippi River. *Mississippi v. Arkansas*, *supra*, 415 U.S. at 289–290. Further proceedings in the *Arkansas Land and Cattle* case were stayed pending the Supreme Court decision. *Id.* at 293–294, n.4.

The Supreme Court upheld a Report of a Special Master finding for Mississippi. The presumption that the change in the location of the thalweg of the river was the product of accretion aided Mississippi. *Id.* at 295–296 (as quoted by Douglas, J., dissenting). This placed a "considerable burden" on Arkansas to show that the change in location of the thalweg was other than by the process of accretion, a burden the state could not carry. *Id.* at 294. For other cases in which the burden of proof was significant, see *Omaha Indian Tribe, Treaty of 1854, etc.*, *supra*, 575 F.2d at 650; *Dartmouth College*, *supra*, 133 N.W.2d at 691; *Kitteridge*, *supra*, 151 N.W. at 1098.

130. E.g., *Gaskill v. Cook* (Mo. 1958) 315 S.W.2d 747, 755 (burden of proof to show accretion as part of proof of claim of title by adverse possession); *Gaines v. Dillard* (Tex. 1976) 545 S.W.2d 845, 851 (had burden to prove accretion because of peculiarities of purchase and sale transaction).

131. E.g., *Jefferis*, *supra*, 134 U.S. at 189–191. One case has also noted that, as a technical matter, a reliction ". . . refers to the sudden baring of land resulting from a sudden change in the course of a waterbody." *Furlong Enterprises*, *supra*, 423 N.W.2d at 133, n.4. That court also noted that the term "reliction" is used to refer to gradual receding of water resulting in gradual exposure of formerly submerged land. *Ibid.*

Mississippi Rivers.[132] As a result, some astoundingly rapid building up of land over a relatively short period of time has been treated as accretion.[133]

The key appears to be that the rapidity of the washing away or deposition is not critical as long as the deposition is imperceptible.[134] Trying to define or describe "imperceptibility," an illusive concept, is, as can be imagined, a difficult task. How can one describe something that cannot be perceived? Yet surveyors, lawyers, and jurists have endeavored to do so.

The most enduring description of what is imperceptible is found in the *Lovingston* case: "The test as to what is gradual and imperceptible in the sense of the rule is, that though the witnesses may see from time to time that progress has been made, they could not perceive it while the process was going on."[135] On its face, this statement merely begs the question as to what is perceivable. How long is from "time to time"? Is it from year to year, day to day, or from one "instant" to another?

There are some cases, such as *County of St. Clair v. Lovingston*,[136] in which the determination seems relatively easy.[137] Others are less clear.[138] To explain their rulings and to aid others in making the determination, some courts have attempted to describe the process by contrast between the dissolution of the

---

132. See note 92 in this chapter; *Washington v. Oregon* (1909) 214 U.S. 205, 215 (Columbia River); *Philadelphia Co. v. Stimson* (1911) 223 U.S. 605, 624 (Allegheny, Monongahela, and Ohio Rivers); *Oklahoma v. Texas, supra*, 260 U.S. at 636–637 (Red River); *United States v. Aranson* (9th Cir. 1983) 696 F.2d 654, 659-660, *cert. den., sub. nom., Colorado River Indian Tribes v. Aranson* (1983) 464 U.S. 982 (Colorado River); *Hancock v. Moore* (Tex. 1941) 146 S.W.2d 369, 370 (Red River).
133. *Oklahoma v. Texas* (1925) 268 U.S. 252, 256; *Hancock, supra*, 146 S.W.2d at 370; *Dartmouth College, supra*, 133 N.W.2d at 689–690.
134. *Nebraska v. Iowa, supra*, 143 U.S. at 369; *Sieck v. Godsey* (Iowa 1962) 118 N.W.2d 555, 557 (". . . the length of time during formation is not material. An increment which either slowly or rapidly results from floods is an accretion if it is beyond the power of identification.") The *St. Louis v. Rutz* decision would seem to call this conclusion into some question. See notes 101 and 109 and accompanying text in this chapter.
135. *County of St. Clair v. Lovingston, supra*, 90 U.S. at 64. The factual basis for this conclusion is found in the Supreme Court's statement of the case: "The fact that additions were a making was perceptible at certain intervals though the additions were too gradual to strike the eye as they were in the actual process of formation." *Id.* at 50.
136. In *Lovingston*, the land area had about doubled in a 55-year period (surveyed in 1815, Congress passed a law in 1870 confirming title to additions, claim made that surveys could not embrace an addition as large as or larger than the survey itself). *Id.* at 49–51.
137. For example, in *Jefferis*, the deposition took place over a 20-year period and could only be observed as occurring every three to four months; it was treated as accretion. See notes 88 through 95 and accompanying text in this chapter. For some other cases see e.g., *McBain v. Johnson* (Mo. 1900) 55 S.W. 1031, 1033 (98 acres formed over 83-year period); *Beaver v. United States* (9th Cir. 1965) 350 F.2d 4, 10, *cert. den.* (1966) 383 U.S. 937 (11.8 acres over 34-year period). On the other hand, when a river cuts through a neck of land in a 24-hour period, there is little difficulty in finding this that is a sudden and, presumably, perceptible change. *Missouri v. Nebraska, supra*, 196 U.S. at 34-35.
138. E.g., *Omaha Indian Tribe, Treaty of 1854, etc., supra*, 575 F.2d 639–650; *Omaha Indian Tribe, etc., supra*, 614 F.2d at 1160–1161.

bank into individual particles (imperceptible) and the tearing away of an area of riverbank and the deposit of that land on the opposite shore (perceptible).[139] Other courts describe the process as imperceptible if it is not possible to fix a period when the deposition began and no one can tell the exact margin of the channel at any time.[140] On the other hand, with respect to erosion, there appears to be a rule that the process can be quite rapid and even "startling to view" and still not be considered an avulsion.[141]

No firm and fast rule can be drawn. There is, however, a practical solution: Unless one has witnessed or has witnesses[142] to the avulsive event or can otherwise establish the sudden and perceptible character of the change in physical location,[143] the physical movement of the boundary will usually be

---

139. *Nebraska v. Iowa, supra*, 143 U.S. at 369 ("The accretion, . . . is always gradual and by the imperceptible deposit of floating particles of earth. . . . There is no heaping up at an instant, and while the eye rests upon the stream. . . . No engineering skill is sufficient to say where the earth in the bank washed away and disintegrating into the river finds its rest and abiding place. . . . There is, no matter how rapid the process of subtraction or addition, no detachment of earth from the one side and deposit of the same upon the other.") Modern cases have followed the lead of the Supreme Court in *Jefferis* and *Nebraska v. Iowa*. Thus, in *Peterson v. Morton* (C.D. Nev. 1979) 465 F.Supp. 986, 998, *vacated, sub nom. Peterson v. Watt* (9th Cir.) 1982) 666 F.2d 361, it did not seem to matter how rapid and visible the deposition as long as the land being deposited was dissolved into small particles. Other cases have described imperceptibility by reference to analogies in nature. Some judges have compared the deposition process to the growing of a plant. These jurists concluded that if one cannot see the development of the plant as it takes place, the process is imperceptible. If the same is true with deposition, that process is also imperceptible. *Esso Standard Oil Co. v. Jones* (La. 1957) 98 So.2d 236, 242, *aff'd on rehg* (1957) 98 So.2d 244.

140. *State v. Sause, supra*, 342 P.2d at 819 (citing *Adams v. Frothingham* (Mass. 1807) 3 Mass. 352, 362).

141. *Nebraska v. Iowa, supra*, 143 U.S. at 368–369; *Ussery v. Anderson-Tully Co.* (E.D. Ark. 1954) 122 F.Supp. 115, 119; Clark 7th, *supra*, § 24.07; Clark 4th, *supra*, § 572. One case noted:

> . . . [I]t is not literally correct to say that the rule of accretion does not apply if any part of the process is perceptible. For instance, the change of the bed of the river takes place by attrition, causing the caving of the banks on one side and by accretion, the act of deposit, on the other side, . . . Therefore, the fact that the caving of the banks on one side is observable at any given time does not prevent the formation on the other side of the river from being accretion which gives title to the land to the riparian owner. *Yutterman v. Grier* (Ark. 1914) 166 S.W. 749, 751.

St. Louis v. Rutz appears to come to the opposite conclusion. See notes 98 through 109 and accompanying text in this chapter.

142. For example, in *Mississippi v. Arkansas, supra*, 415 U.S. at 292–293, the Special Master found that no avulsion had occurred based, in part at least, on ". . . the total absence of any known historical reference to an avulsion in this area that changed the course of the river by the necessary half mile." Even a witness is sometimes not sufficient to carry the burden. In *Bone, supra*, 225 N.W. at 368, the testimony of the sole witness to an alleged avulsion was termed improbable when no other witness corroborated him. As in *Mississippi v. Arkansas*, the court reasoned that such a sudden change in the course of a river would have attracted considerable attention and would be known to others in the vicinity had such a change occurred. *Ibid.*; accord, *Dartmouth College, supra*, 133 N.W.2d at 691.

143. *Missouri v. Nebraska, supra*, 196 U.S. at 35–36 (cut through neck of land in 24-hour period); *Arkansas v. Tennessee, supra*, 246 U.S. at 162 (sudden and violent cutting through of

treated as imperceptible and, consequently, accretive in legal effect. In addition, if one is not able to identify land that may have been quite suddenly and rapidly washed away or separated, it would appear to be evidence supporting a holding that such circumstances, despite the rapidity and violence of the changes in the course of the channel, constituted an accretive rather than an avulsive process.[144] There are, however, cases in which the inability to identify land was not critical.[145]

What is important is evidence that at a particular time there was a sudden, violent, and visible change. Such evidence would support a deviation from the rules of accretion and erosion.[146] This is a reflection of the rule that the party who asserts that the course of a river has changed by avulsion has the burden of proving it.[147] Consequently, while no absolute, firm and fast rules can (or should be) provided, the tendency is to favor the rule of accretion preserving the riparianess of the landowner.

Even assuming the physical process is determined to be gradual and imperceptible, there is another question that may legitimately be asked: Is it of any legal significance to the location of the property boundary that the gradual and imperceptible change in boundary location was the result of nonnatural, human-caused events?

---

neck of land over 30 hours); *State Land Board v. Corvallis Sand & Gravel Co.*, supra, 429 U.S. at 367 (major flood suddenly and with great force and violence); *Heckman v. Swett* (Cal. 1893) 99 Cal. 303, 305 (sudden unusual and excessive rise or freshet that continued for several days and cut, washed, and carried away land; "... loss could be plainly seen and perceived; that blocks or masses of earth caved off ... and fell into the river and were washed and carried away ..."). On the other hand, some courts have considered an event that has occurred "in such a manner as to be readily perceived in a departure not requiring the passage of much time" much like an avulsion. *Dickson v. Sandefur* (La. 1971) 250 So.2d 708, 719 (under a unique statute a 5-month flood was apparently treated like an avulsion); see *Furlong Ent., supra*, 423 N.W.2d at 139.

144. Witness the United States Supreme Court's treatment of rapid changes of the Missouri River, *Nebraska v. Iowa*, supra, 143 U.S. at 369–370, and the implication that because the eroding bank was dissolved into particles and not identifiably transmitted to another location, that the process, no matter how rapid, would be treated as accretive in legal effect. See notes 118 through 121 and 139 and accompanying text in this chapter. But also see notes 98 through 109 and accompanying text in this chapter, appearing to question the basis for that theory.

145. *Philadelphia Co. v. Stimson*, supra, 223 U.S. at 625; see note 95 in this chapter.

146. "The determining words are that the land was 'washed away from time to time by heavy floods and freshets,' and the reference is to what occurred in many years. This is far from a statement that any particular time there was such a sudden, violent, and visible change as to justify a departure from the ordinary rule which governs accretion and diminution albeit the stream suffer wide fluctuations in volume, the current be swift, and the banks afford slight resistance to encroachment." *Philadelphia Co. v. Stimson*, supra, 223 U.S. at 625.

147. *Oklahoma v. Texas*, supra, 260 U.S. at 638 (party contending that material changes have taken place has burden of proving them). A somewhat unusual twist on this theme occurred in *Kansas v. Missouri*, supra, 322 U.S. at 228, in which Kansas was asserting accretion followed by an avulsion; Kansas was held to have the burden of proof. See also *New Jersey v. New York* (1998) 523 U.S. 767, 140 L.Ed.2d 993, 1016.

7.8 SIGNIFICANCE OF THE CAUSE OF THE RIVER'S CHANGE IN GEOGRAPHIC LOCATION ON THE PROPERTY BOUNDARY

Recall that in the case concerning the St. Louis, Missouri, waterfront lots (*St. Louis v. Rutz*), there was some implication that the landowner who caused the washing away of another owner's land should not be able to gain title to land that later formed and would prevent the innocent owner's access to the river.[148] In some states a rule has grown up in property boundary controversies between private parties and the state as riverbed owner that examines the cause of the change in the watercourse regime. In those states, where the cause of the change in physical location was artificially induced[149] (that is, by works of humans), the rule is that the property boundary is fixed in its geographic location prior to introduction of artificial conditions.[150]

The basis for the "artificial accretion" doctrine is founded on at least two claims: Sovereign lands cannot be alienated without a legislative authoriza-

---

148. See note 113 in this chapter. *Lovingston*, where the city was not a party, reached the exact opposite conclusion based on the same exact cause—the construction of dikes by the City of St. Louis. Compare *County of St. Clair v. Lovingston*, supra, 90 U.S. at 50, with *St. Louis v. Rutz*, supra, 138 U.S. at 231–232. The inclination of the Court in *Rutz* has become the rule in later cases. For example, one opinion stated: "It is undoubtedly the general rule that additions to land fronting on bodies of water, caused by artificial means by the owner of the adjoining shore land, do not come under the laws as to the ownership of accretions caused by purely natural causes." *Marine Ry. Co. v. United States* (1921) 257 U.S. 47, 65; *United States v. 222.0 Acres of Land* (D. Md. 1969) 306 F.Supp. 138, 151; *Brundage v. Knox* (Ill. 1917) 117 N.E. 123, 127; *State v. Sause*, supra, 342 P.2d at 827; *People ex rel. Blakslee v. Commissioners of the Land Office* (N.Y. 1892) 32 N.E. 139, 140; *State v. George C. Stafford & Sons, Inc.* (N.H. 1954) 105 A.2d 569, 574; *Forgeus v. County of Santa Cruz* (Cal. 1914) 24 Cal.App. 193, 199. In an aberrant case, when the upland owner was the United States, the United States Supreme Court awarded accretion caused by the upland owner to the United States. *California ex rel. State Lands Comm'n v. U.S.*, supra, 457 U.S. at 288.

149. In California, the rule is that the accretion must be "directly caused by human activities . . . that occurred in the immediate vicinity of the accreted land." *State of Cal. ex rel. State Lands Com. (Lovelace)*, supra, 11 Cal.4th at 79–80. See Section 5.8 in this chapter 5. One case quite narrowly defined a natural, as opposed to an artificial, process as follows: ". . . [W]hile some of the alluvium was undoubtedly comprised of dredge spoil material, nevertheless, this material was deposited as a result of the natural effects of running water rather than as a result of having been pumped into or placed directly in the . . . area." *State v. Placid Oil Co.*, supra, 300 S.2d 154, 163. Other courts have accepted some quite indirect effects as artificial, *United States v. Aranson*, supra, 696 F.2d at 660 (accretion caused by dam miles away); still other courts have not (*Brainard v. State of Texas* (1999) 12 S.W.3d 6, 22 (accretion downstream from a dam not treated as artificial).

150. *State of Cal. ex rel. State Lands Com. (Lovelace)*, supra, 11 Cal.4th at 56; *Furlong Ent.*, supra, 423 N.W.2d at 133 (NDCC 47-06-05)(river); *United States v. Aranson*, supra, 696 F.2d at 660 (river); Cal. Civ. Code § 1014; *Lakefront Trust, Inc. v. City of Port Arthur* (Tex. 1974) 505 S.W.2d 606, 608–609 (Sabine Lake); *Curry v. Port Lavaca Channel & Dock Co.* (Tex. 1930) 25 S.W.2d 987, 988; see *Carpenter v. City of Santa Monica* (1944) 63 Cal.App.2d 772, 789 (open coast); *City of Los Angeles v. Anderson* (Cal. 1929) 206 Cal. 662, 667 (bay); *Garrett v. State* (N.J. 1972) 289 A.2d 542, 546 (tidal creek); *Lorino v. Crawford Packing Co.* (Tex. 1943) 175 S.W.2d 410, 414. Some cases have even held that a landowner whose once submerged lands have

7.8 SIGNIFICANCE OF THE CAUSE OF THE RIVER'S CHANGE IN GEOGRAPHIC LOCATION

tion, and, between the public and an individual landowner, the public should not suffer from the result of public works accomplished to improve a waterway or as the result of off-site works constructed by a third party.[151] This is not the rule in most states, nor do federal courts follow it.[152] And despite the holding of many early cases that the rules of accretion, and so on, are matters of state law,[153] there is now some question as to whether, at least in the public land states, the artificial accretion doctrine has any continued viability as a matter of state law.[154] Moreover, California has narrowed the scope of the artificial accretion rule.[155]

---

become exposed has the right to have the lands maintained in the exposed condition. *Natural Soda Prod. Co. v. City of L.A.* (Cal. 1943) 23 Cal.2d 193, 197–199, *cert.den.* (1944) 321 U.S. 793, *rehg den.* (1944) 322 U.S. 768.

151. *State of Cal. ex rel. State Lands Com.* (*Lovelace*), *supra*, 11 Cal.4th at 66–73; see *Dana v. Jackson Street Wharf Co.* (1866) 31 Cal. 118, 120 and notes 143 through 157 and accompanying text in Chapter 5.

152. *County of St. Clair v. Lovingston*, *supra*, 90 U.S. at 63; *Kansas v. Merriwether* (8th Cir. 1910) 182 F. 457, 464; *Beaver, supra,* 350 F.2d at 11; *State v. Sause, supra,* 342 P.2d at 819; *Esso, supra,* 98 So.2d at 241; *Brainard, supra,* 12 S.W.3d at 23; *State Engineer of Nevada v. Cowles* (Nev. 1970) 478 P.2d 159, 161; *Burket v. Krimlofski* (Neb. 1958) 91 N.W.2d 57, 61; *City of Missoula v. Bakke, supra,* 198 P.2d at 772 (construing statute substantially identical to Cal. Civ. Code § 1014); *Whyte v. City of St. Louis* (Mo. 1899) 54 S.W. 478, 480; *Mather v. State* (Iowa 1972) 200 N.W.2d 498, 500; *Sieck, supra,* 118 N.W.2d at 557; *Solomon v. City of Sioux City* (Iowa 1952) 51 N.W.2d 472, 476; *Brundage, supra,* 117 N.E. at 128; *Tatum v. City of St. Louis* (Mo. 1894) 28 S.W. 1002, 1003; *Frank v. Smith* (Neb. 1940) 293 N.W. 329, 333; *United States v. Claridge* (9th Cir. 1969) 416 F.2d 933, 935, *cert. den.* (1970) 397 U.S. 961. Recall, however, that a riparian landowner cannot gain the benefit of additions he or she causes. See note 148 in this chapter.

153. To recount briefly, there is a long line of still viable Supreme Court cases that have left to state law questions about ownership of additions to riparian land whose source of title is a federal patent. E.g., *Barney, supra,* 94 U.S. at 337; *St. Louis v. Rutz, supra,* 138 U.S. at 250; *Hardin v. Shedd* (1903) 190 U.S. 508, 519; *Joy v. St. Louis* (1906) 201 U.S. 332, 342. And a Ninth Circuit case reaffirmed that a federal patent to land along a river will be construed according to the law of the state in which the land lies. *United States v. Pappas* (9th Cir. 1987) 814 F.2d 1342, 1345, n. 8. For a much more detailed discussion of this contention and the opposing cases, see Chapter 2.

154. E.g., Chapter 2. *Hughes v. Washington* (1967) 389 U.S. 290, 291; *California ex rel. State Lands Commission v. United States, supra,* 805 F.2d at 865; *California ex rel. State Lands Comm'n v. U.S., supra,* 457 U.S. at 278.

155. See notes 143 through 157 and accompanying text in Chapter 5. The difficulties with application of the artificial accretion rule were explained by the trial court in *Lovelace*. The issue in that case arose because of the claimed effects of hydraulic mining (explained in notes 28 through 32 and accompanying text in Chapter 6. Was the change in geographic location of riparian property caused by the effects of hydraulic or from "natural causes?" The court said: "Given the great passage of time since [hydraulic] mining dislodged these materials and the distance from the source of the materials, and the time the accretions have existed, it would be unreasonable to interpret [Cal.Civ. Code] § 1014 to require tracing these accretions to their sources. *This would place an inappropriate burden on the parties and the courts.*" *Lovelace v. State of California* (1990 Sacramento County Superior Court) No. 506331 (emphasis supplied). The California Supreme Court agreed with this decision. The Supreme Court held that the artificial accretion rule

Taking a somewhat different tack, other courts have considered such circumstances to be an "artificial avulsion."[156] In what may become a most prominent example of the application of the "artificial avulsion" rule, the United States Supreme Court appears to have retreated from what had been thought to be the settled federal rule.[157] Although the Supreme Court acknowledged that an avulsive action "ordinarily calls to mind something somewhat sudden or, at least, of short duration, whereas accretion has as its essence the gradual deposit of material over a period by action of water flow no matter what its cause, ..."[158] the Supreme Court did not apply these principles in *Georgia v. South Carolina*.[159]

The facts were these: The Corps of Engineers undertook navigation improvement work in the bed of the Savannah River forming the boundary between Georgia and South Carolina. This improvement work, including deposit of dredged materials, caused land to grow up in the bed of the river. Over a 40-year period, the land became connected to South Carolina.[160] Contrary to the general rules set out earlier in this chapter, the lands were awarded to Georgia.[161] The basis of the ruling was the Court's determination that the change in the river's bed was caused by the Corps and its conclusion that the rapidity of some aspects of the dredging and other processes would be treated as avulsive in character.[162] This holding calls into question some recent decisions of the Supreme Court concerning property boundary effects of humans' works.[163]

---

does not apply to accretion caused by "human activities . . . far from the site of the accretion." *State of Cal. ex rel. State Lands Com. (Lovelace), supra*, 11 Cal.4th at 77.

156. E.g., *Puyallup Tribe of Indians v. Port of Tacoma* (9th Cir. 1983) 717 F.2d 1251, 1263, *cert. den., sub. nom. Trans-Canada Enterprises Ltd. v. Muckleshoot Indian Tribe* (1984) 465 U.S. 1049, *reh. den.* (1984) 466 U.S. 954; *Witter v. County of St. Charles* (Mo. 1975) 528 S.W.2d 160, 161–162; *Furlong Ent., supra*, 423 N.W.2d at 134, n. 9; see *Riverland Co., Inc. v. McAlexander* (Ark. 1983) 661 S.W.2d 451, 452. A Texas Supreme Court case refused to find that a dam caused an artificial avulsion. *Brainard, supra*, 12 S.W.3d at 24 (?We also reject the State's related contention that changed conditions in a river brought about or influenced by a public works project are inherently avulsive . . .").

157. *Georgia v. South Carolina* (1990) 497 U.S. 376, 404 (when the bed of the river is changed by the natural, gradual process of accretion and erosion, the boundary follows the varying course of the stream; if the stream leaves its bed and forms a new one by avulsion, there is no boundary change; citing *Arkansas v. Tennessee* (1918) 246 U.S. 158).

158. *Ibid.*

159. *Georgia v. South Carolina* was an interstate political boundary case. Courts often use principles adopted in such cases in deciding property boundary questions along navigable waters. For example, *Nebraska v. Iowa, supra*, 143 U.S. 359, an interstate boundary case, is cited in this work alone more that ten times for various principles relating to property boundary disputes, not necessarily arising from interstate boundary contests.

160. *Id.* at 334.

161. *Id.* at 335.

162. *Ibid.*

163. See note 154 and accompanying text in this chapter.

7.9 REEMERGENCE

Although a bit obscure, one last matter[164] requires mention. Suppose once-riparian land adjacent to nonriparian property becomes submerged by a gradual movement of the river. Sometime later, land in the same geographic location of the once-submerged land reappears as a result of a gradual recession of the river. To whom does this newly emerged property belong? Does it belong to the former nonriparian owner or to the landowner whose property has again become riparian as a result of the river's gradual movement?

Some courts have held that when land reemerges, such land belongs to its former record owner and not to the recently (and now formerly) riparian owner.[165] Other courts, perhaps the majority, have not followed this rule.[166] That is, the former riparian owner would gain the newly formed land. Each state's rule must be checked before you advise your client.[167]

Having explored the legal effect of a change in the physical location of the boundary, the time is ripe to examine exactly what is the boundary whose movement we have been examining. Until now we have been discussing river property boundary movement in purely geographic terms: a river's change in physical location as measured in latitude and longitude. Now we will refine our focus to examine how the law and litigants describe and define the phys-

---

164. Certainly, there is more than "one last matter" that could be discussed. Just as an example, a rule that is not addressed in any great detail in this work, but has its own quite substantial body of case law, is the "island rule." The island rule is an exception to the accretion rule. It holds that despite movement of the thalweg of a river from one side of an island to the other, the original channel remains the boundary even though the migration of the thalweg has been gradual and imperceptible. E.g., Clark 7th, *supra*, § 24.08; Clark 4th, *supra*, § 573; *United States Gypsum v. Uhlhorn* (8th Cir. 1966) 366 F.2d 211, 218, *cert. den.* (1967) 385 U.S. 1026. This basic work cannot possibly completely discuss all rules and variants of the rules concerning riparian property boundaries.
165. Clark 7th, *supra*, § 24.14; Clark 4th, *supra*, at § 579; *Furlong Ent.*, *supra*, 423 N.W.2d at 133; *Mikel v. Kerr* (10th Cir. 1974) 499 F.2d 1178, 1181 (river); *Coastal Industrial Water Author. v. York* (Tex. 1976) 532 S.W.2d 949, 954 (land subsidence along nontidal ship channel); *Arkansas v. Tennessee*, *supra*, 246 U.S. at 174.
166. E.g., *Wilcox v. Penney* (Iowa 1959) 98 N.W.2d 720, 726; *Wemmer v. Young* (Neb. 1958) 93 N.W.2d 837, 848; *Payne v. Hall* (Iowa 1921) 185 N.W. 912, 915; *Bone*, *supra*, 225 N.W. at 368–369. *Furlong Ent.*, *supra*, 423 N.W.2d at 133, n.6, lists some further authorities.
167. Somewhat related is the contention that a landowner has the right to reclaim property lost as the result of an avulsion. See, e.g., Cal. Civ. Code § 1015 (appears to permit reclamation of an identifiable portion of a riverbank within one year); *Bohn*, *supra*, 107 Cal.App.2d at 749; *Arkansas v. Tennessee*, *supra*, 246 U.S. at 174. The *Bohn* case held that since the waters that occupied the flooded land were navigable, the public had the right to use the water until the owner reclaimed the area. *Bohn*, *supra*, 107 Cal.App.2d at 749. At some point, the long occupation of the area by navigable waters may create a "new" natural situation or prescriptive occupancy that one may argue would prevent the former owner from reclaiming long submerged land. *State v. Superior Court (Fogerty)*, *supra*, 29 Cal.3d at 248–249; *Natural Soda Prod. Co.*, *supra*, 23 Cal.2d at 197–199.

ical feature demarking the mutual legal property boundary of the river and riparian uplands, the ordinary high-water mark.

7.10 PHYSICAL INDICIA OF THE ORDINARY HIGH-WATER MARK IN A RIVERINE ENVIRONMENT

The waterward boundary of public lands adjacent to navigable rivers and the landward boundary of the bed of the river owned by the state is the ordinary high-water mark.[168] Of course, each state may choose to convey lands waterward of the ordinary high-water mark.[169] It has long been recognized that this legal term[170] and its equivalent, "the bank" of the river,[171] are not scientific terms capable of ready physical description.[172] Much surveying, legal, and judicial effort has gone into providing some scientifically supportable physical indicia of this legal property boundary.[173]

---

168. E.g., *Barney, supra*, 94 U.S. at 836; *Jefferis, supra*, 134 U.S. at 181; *Packer, supra*, 137 U.S. at 672; *Hardin v. Jordan* (1891) 140 U.S. 371, 382; *Heckman, supra*, 99 Cal. at 307; *Callahan, supra*, 146 P. at 733; 43 U.S.C. §§ 1301, 1311.

169. See note 96 in this chapter. For example, the California legislature extinguished the state's proprietary title, the *jus privatum*, to low water. Cal. Civ. Code § 830; *State of California v. Superior Court (Lyon)* (1981) 29 Cal.3d 210, 225, *cert. den.* (1981) 454 U.S. 865. A more detailed discussion of the effect of such an extinguishment is found in the text accompanying notes 29 through 36 in Chapter 8. For now it is sufficient to say that as a matter of law, the legislature did not alienate the State of California's public trust easement in such lands between the ordinary high-water mark and the ordinary low water mark and that lands between the ordinary high-water mark and ordinary low-water mark remained subject to the public trust easement. *Ibid.*

170. Equally unhelpful in defining the physical location of the ordinary high-water mark was one United States Supreme Court opinion that termed the boundary the "usual high-water mark." *Jefferis, supra*, 134 U.S. at 181.

171. Cases have held that "the bank" of the river is the landward extent of the bed of the river and that, as far as defining the extent of the river owned by the state, the terms "bank" and "ordinary high-water mark" describe the identical location. E.g., *Morrison v. First Nat. Bank of Skowhegan* (Me. 1895) 33 A. 782, 784; *Richards v. Page Inv. Co.* (Ore. 1924) 228 P. 937, 942.

172. As early in our judicial history as 1851, one Supreme Court Justice stated:

> The term high water, when applied to the sea or to a river where the tide ebbs and flows, has a definite meaning. But in respect to fresh-water rivers, the term is altogether indefinite, and the line marked uncertain. It has no fixed meaning . . . and should never be adopted as a boundary in the case of fresh-water rivers . . . . It may mean any stage of the water above its ordinary height, and the line will fluctuate with every varying freshet or flood that may happen. *Howard supra*, 54 U.S. at 454 (Nelson, J., concurring).

The majority opinion in *Howard* agreed with this statement, as do much later cases. *Id.* at 448; *Borough of Ford City v. United States* (3rd Cir. 1965) 345 F.2d 645, 649, 650, *cert. den.* (1965) 382 U.S. 902; *United States v. Washington* (9th Cir. 1961) 294 F.2d 830, 834, *cert.den.* (1962) 369 U.S. 817.

173. See, e.g., Maloney, *supra*, 13 Land & Water L. Rev. 465; *United States v. Cameron* (M.D. Fla. 1978) 466 F.Supp. 1099.

7.10.1 Physical Location of Ordinary High-Water Mark Property Boundary of Federal Lands Probably Determined by Federal Law

Before explaining how to physically define the ordinary high-water mark boundary along nontidal navigable rivers, a brief digression into the choice-of-law question is necessary. The question is, What law will supply the physical definition of the ordinary high-water mark property boundary—state law or federal law? Even without a lengthy review of the choice-of-law question,[174] the short answer to this question is that one does not know for certain. Although the answer is not free from substantial doubt,[175] federal law will probably supply the description of the physical location of the ordinary high-water mark property boundary between federal public lands and state sovereign lands.[176] The rationale is that what is being determined is the limit of the federal upland grant; this is a federal question.[177] Regardless (and maybe even heedless) of potential federal preemption, states have propounded their own definitions of the term "ordinary high-water mark."[178] Both sources will be used in the following discussion.

7.10.2 The *Howard v. Ingersoll* Test—Geomorphic Features

Federal law begins with the 1851 case of *Howard v. Ingersoll*. In that case, two private litigants disputed the physical location of the Chattahoochee River boundary between the States of Alabama and Georgia.[179] Georgia had generously ceded to the United States its territory "west of line beginning on the western bank of the Chattahoochee River."[180] The question was what was intended by this cession language.[181] Three separate opinions resulted, each of which attempted to define what was meant by the language of the grant.

The most notable description of the boundary is found in the second concurring opinion. Because of its importance to later opinions that also attempt

---

174. For such a review, refer to Chapter 2. Also see notes 153 and 154 in this chapter.
175. See note 153 in this chapter.
176. See *Borax, Ltd. v. Los Angeles* (1936) 296 U.S. 10, 22; *California ex rel. State Lands Comm'n v. U.S.*, supra, 457 U.S. at 278; *California ex rel. State Lands Comm'n v. United States*, supra, 805 F.2d at 865.
177. *Borax, supra*, 296 U.S. at 22.
178. A listing of such cases is found in note 183 in this chapter.
179. *Howard, supra*, 54 U.S. at 443. At least one justice was not at all pleased that the Court did not have the benefit of the participation of the two states concerned. *Id.* at 451 (Nelson, J., concurring). In today's legal environment, it is highly unlikely the Court would decide an issue of such magnitude without complete participation by all affected parties. See note 129 in this chapter. But *Howard* was decided in the days of a more roughhewn jurisprudence.
180. *Id.* at 443. The *Howard* opinion contains a lengthy history of the genesis and progress of the cessions of the western territories of the thirteen original states to the United States. *Id.* at 429–441.
181. *Id.* at 451–452 (Nelson, J., concurring).

to describe the physical location of the ordinary high-water mark, that portion of the opinion is set forth at some length:

> That the banks of a river are those elevations of land which confine the waters when they rise out of the bed; and the bed is that soil so usually covered by water as to be distinguishable from the banks, by the character of the soil, or vegetation, or both, produced by the common presence and action of flowing water. But neither the line of ordinary high-water mark, nor of ordinary low-water mark, nor of a middle stage of water, can be assumed as the line dividing the bed from the banks. This line is to be found by examining the bed and banks, and ascertaining where the presence and action of water are so common and usual, and so long continued in all ordinary years, as to mark upon the soil of the bed a character distinct from that of the banks, in respect to vegetation, as well as in respect to the nature of the soil itself. Whether this line between the bed and the banks will be found above or below, or at a middle stage of water, must depend on the character of the stream. . . . But in all cases the bed of a river is a natural object, and is to be sought for, not merely by the application of any abstract rules, but as other natural objects are sought for and found, by the distinctive appearances they present; the banks being fast land, on which vegetation, appropriate to such land in the particular locality, grows wherever the bank is not too steep to permit such growth, and the bed being soil of different character and having no vegetation, or only such as exists when commonly submerged in water.[182]

Just from reading this description, it can readily be observed that there could be as many locations of the ordinary high-water mark as there are observers. It seems apparent the *Howard* concurring opinion assumes that each river, throughout the length of that particular river, will have substantially similar physical characteristics and that those characteristics are exemplified by the paradigmatic Chattahoochee River near Columbus, Georgia. Nature is not so neat and tidy. The struggles of courts to describe the physical location of the ordinary high-water mark property boundary along other rivers could well be traced, at least in part, to a somewhat slavish adherence to the *Howard* test.[183]

---

182. *Id.* at 458 (Curtis, J., concurring). On the other hand, some state courts have decided the boundary is the highest level the river ever obtained in its known history. *State v. Edwards* (Wash. 1936) 62 P.2d 1094, 1096; *Pierce v. Central National Bank* (Ill. 1906) 79 N.E. 296, 299; *State v. Longfellow* (Mo. 1902) 69 S.W. 374, 377.

183. Some federal cases follow, accept, or use the *Howard* test as a basis for physically locating the ordinary high-water mark: *Oklahoma v. Texas, supra*, 260 U.S. at 632; *Borough of Ford City, supra*, 345 F.2d at 648; *Harrison, supra*, 148 F. at 783. State cases include *Seaman v. Smith* (Ill. 1860) 24 Ill. 521, 525 (the line where the water usually stands when unaffected by any disturbing cause); *Wilcox, supra*, 98 N.W.2d at 723 (high-water mark coordinates with bed, and bed is that "which the water occupies sufficiently long and continuously to wrest it from vegetation, and destroy its value for agricultural purposes"); *Houghton v. RR Co.* (Iowa 1877) 47 Iowa 370, 374 (riverbed is land that a river occupies long enough to be unvegetated and to be valueless for agriculture); *Sun Dial Ranch v. May Land Co.* (1912) 119 P. 758, 761 (determine by ascertaining

7.10.3 Variants of *Howard v. Ingersoll* Test Including Use of Stage Data as Indicia of Ordinary High-Water Mark

There are other theories or variants of the *Howard v. Ingersoll* test that have been offered with some success. Cases emphasizing the geomorphic criteria of *Howard*[184] have adapted that standard to fit the physical regime being considered.[185] And, in certain cases, some courts have held that the ordinary high-water mark is evidenced by government meander line surveys,[186] or an observable change in the character of the soil,[187] or that particular flows or water stage elevations provided indicia of its physical location.[188]

---

where presence and action of water are so common and usual as to mark on the bed a character distinct from that of the banks in respect to vegetation and soil); *Carpenter v. Board of Commissioners* (Minn. 1894) 58 N.W. 295, 297 (coordinate with bed that the water occupies so long and continuously as to "wrest it from vegetation, and destroy its value for agricultural purposes"); *Dow v. Electric Co.* (N.H. 1899) 45 A. 350, 357 (same test); see *Carpenter v. Ohio River Sand & Gravel Corp., supra,* 60 S.E.2d at 216. One commentator has concluded that state and federal definitions are very similar. Maloney, *supra,* 13 Land and Water L.Rev. at 468, 475.

184. ". . . But in all cases the bed of the river is natural object, and is to be sought for, . . . as natural objects are sought for and found, by the distinctive appearance they present . . . ." *Howard, supra,* 54 U.S. at 458 (Curtis, J., concurring).

185. For example, the Supreme Court in *Oklahoma v. Texas, supra,* 260 U.S. at 631–632, determined, in the case of the Red River, that the bank is:

> . . . the water-washed and relatively permanent elevation or acclivity at the outline of the river bed which separates the bed from the adjacent upland . . . and serves to confine the waters within the bed and to preserve the course of river, and that the boundary intended is on and along the bank at the average or mean level attained by the waters in the periods when they reach and wash the bank without overflowing it. When we speak of the bed we include all of the area which is kept practically bare of vegetation by the wash of the waters of the river from year to year in their onward course, although parts of it are left dry for months at a time . . .

Further, in this context, the word "bank" does not describe a boundary location distinct from the word "river." *Oklahoma v. Texas, supra,* 268 U.S. at 255–256 ("up the river" interpreted as equivalent to "bank" of the river). *Brainard, supra,* 12 S.W.3d at 16, describes what it calls the gradient boundary methodology: This method requires determining two factors, (1) the location of the bank and (2) the gradient ("rate of fall") of the water. "The boundary line is defined as 'a gradient of the flowing water in the river,' located halfway between the lowest level where the flowing water first touches the bank the highest point where the water reaches the top of the bank without overflowing it."

186. See, e.g., *Utah v. United States supra,* 425 U.S. 948; Report of Special Master, *supra,* 1976 Utah L.Rev. at 306–307 (lake); *People ex rel. Dept. Pub. Wks. v. Shasta Pipe etc. Co.* (1968) 264 Cal.App.2d 520, 531 (survey of Mexican land grant along a river); *United States v. Cameron, supra,* 466 F.Supp. at 1114; *Mood, supra,* 410 P.2d at 779 (lake); *Trustees of Internal Improvement Fund v. Wetstone* (Fla. 1969) 222 S.2d 10, 14–19 (coastal); *Provo City, supra,* 176 P.2d at 141 (Larson, J., dissenting) (lake).

187. *Willis v. United States* (S.D. W.Va. 1943) 50 F.Supp. 99, 101.

188. *Alabama v. Georgia* (1859) 64 U.S.(23 How.) 505, 515; *Oklahoma v. Texas, supra,* 260 U.S. at 633; *United States v. Cameron, supra,* 466 F.Supp. at 1112 ("[F]or a body of water whose

River flows or river stage data appear to have been used in California as the indicia of the ordinary high-water mark of a river. Two cases used flow or river stage evidence to define an elevation that was described as the ordinary high-water mark.[189] And even *Howard v. Ingersoll* refers to river stage evidence.[190] Further, as to what types of flows or stages should be excluded or included in that calculation, the California Supreme Court noted long ago that spring flood runoff flows constitute part of the usual and ordinary flows of the river.[191]

Consequently, as noted by one case: "In the few cases in which the lower courts have been called upon to apply the [ordinary high-water mark] concept, one finds little in the way of comprehensive definition. Instead the courts have been satisfied to tailor their definitions to meet the precise set of facts before

---

levels fluctuate considerably with changes in climate, accurate water stage and elevation data may provide the most suitable method for determining the ordinary high water mark."); *United States v. Fallbrook Public Utilities Dist., supra,* 109 F.Supp. at 38 (for construing a contract term "usual high-water line"); *Payette Lakes Protective Ass'n v. Lake Reservoir Co.* (Idaho 1948) 189 P.2d 1009, 1017, 1019. Advocates must show the methodology used to be scientifically sound. *United States v. Cameron, supra,* 466 F.Supp. at 1114; *Conran, supra,* 341 S.W.2d at 81. Early federal cases disapproved use of such a statistical technique because it was claimed that results would depend on an arbitrarily selected time period and frequency of occurrence. *Willis, supra,* 50 F.Supp. at 101.

189. *People v. Ward Redwood Co., supra,* 225 Cal.App.2d at 390 ("In California, ordinary high water mark is normal high water. [Citation omitted.] It is the " 'average level of the water attained by such river in its annual seasonal flow . . .' " [citation omitted]"); *Mammoth Gold Dredging Co. v. Forbes* (Cal. 1940) 39 Cal.App.2d 739, 752 ("The ordinary maximum flow of water in a river during the wet season of each year may not be deemed to be mere " 'flood' " or " 'storm waters.' " The average level of the water attained by such a river in its annual seasonal flow establishes its high water mark as defined by the authorities.") But another case used a variant of the *Howard* test. "The high-water mark does not vary with the season, but is defined as the place where the river bed ends and the riverbank begins. This method involves examining the riverbank to find the highest point where the water's flows have prevented the growth of vegetation. [citation omitted]." *Bess, supra,* 3 Cal.App.4th at 1549 n. 2.

190. *Howard, supra,* 54 U.S. at 458 (Curtis, J., concurring).

191. In a famous (some say infamous) opinion, that court observed:

These annually occurring accretions in the amount and flow of the waters of said river are natural and regular, and occur in their usual, expected, and accustomed seasons and result in an increased amount and flow of the waters of said river . . . , lasting through several months in the annual change of seasons of every year. The conclusion is inevitable that the waters of the . . . [r]iver annually flowing therein before and during and after those regularly occurring accretions in the volume thereof constitute the usual and ordinary flow of said river and are in no sense " 'storm' " or " 'flood' " or " 'vagrant' " or " 'enemy' " waters. . . . *Herminghaus v. Southern California Edison Co.* (Cal. 1926) 200 Cal. 81, 88, *cert. dismd.* (1927) 275 U.S. 486; *accord, United States v. Gerlach Live Stock, supra,* 339 U.S. at 730; *Joslin v. Marin Mun. Water Dist.* (1967) 67 Cal.2d 132, 136.

This statement was made in a water rights case. Even so, it may be claimed that high flows or stages during spring runoff should be included in calculating the stage or elevation of the river in determining the physical indicia of the ordinary high-water mark along California nontidal rivers.

them. One may thus find varied definitions in the decisions."[192] Consequently, to separate the river and its bed from privately held uplands, it is left to the inventive surveyor, lawyer, or property owner to propose a methodology and to establish that this methodology is based on rational and provable scientific criteria.[193]

7.11 PHYSICAL INDICIA OF THE ORDINARY LOW-WATER MARK IN A RIVERINE ENVIRONMENT

In those states that treat riparian ownership as extending to the ordinary low-water mark, the same problem exists. How is this legal term to be translated to a locatable physical position?

Some states have held that the low-water mark is the lowest elevation to which the waters have receded at any time.[194] Other courts held that the term meant the lowest elevation that the water reaches during a season not affected by droughts or floods.[195] And some courts have even permitted, over objection, the use of a statistical averaging technique to calculate the ordinary low-water mark.[196]

Thus, from this scattering of opinions, it can be seen that the resourceful surveyor or lawyer proposing a scientifically sensible test, tailored to the facts and circumstances of the case at hand, will have a reasonable chance of convincing the judge to adopt that test as the method for determining the physical indicia of the legal term "ordinary low-water mark."[197]

---

192. *United States v. Cameron, supra,* 466 F.Supp. at 1111.

193. Subject to the admissibility of the testimony. See note 187 in Chapter 4.

194. *Paine Lumber Co. v. United States* (W.D. Wis. 1893) 55 F. 854, 864; *Carpenter v. Ohio River Sand & Gravel Corp., supra,* 60 S.E.2d at 215; *Union Sand & Gravel Co. v. Northcott* (W.Va. 1926) 135 S.E. 589, 593.

195. *Vermont v. New Hampshire* (1934) 290 U.S. 579, 580; *Hillebrand, supra,* 274 N.W. at 822; *Flisrand, supra,* 152 N.W. at 801; *Slauson, supra,* 69 N.W. at 992; *Appeal of York Haven Water & Power Co.* (Pa. 1905) 62 A. 97, 98, 100; *Stover v. Jack* (Pa. 1869) 60 Pa. 339, 343 (". . . to bound title by a mark which is set by an extraordinary flood, or extreme drought, would do injustice and contravene the common understanding of the people"); 43 Ops.Cal.Atty.Gen. 291, 296 (1964).

196. *Conran, supra,* 341 S.W.2d at 80–81; *State v. Placid Oil, supra,* 300 S.2d at 164–165; *State v. Cain, supra,* 236 A.2d at 503. In *Placid Oil,* an engineer obtained the gauge records for a 29-year period from a nearby stage gauge. He tabulated the daily highs and lows from that record. From this tabulation a graph was prepared, showing the water surface from year to year and each year's high- and low-water season. The engineer then averaged the daily lows during the low-water season. This method was objected to on grounds that there was no formula or objective method in determining the low-water season and that the method adopted was completely subjective. The court, however, found this method of averaging was the "fairest and most reasonable." *State v. Placid Oil, supra,* 300 S.2d at 165.

197. See, e.g., *United States v. Cameron, supra,* 466 F.Supp. at 1111–1112; *Provo City, supra,* 176 P.2d at 139 (Larson, J., dissenting).

Whether considering the determination of the physical location of the ordinary high- or ordinary low-water mark, a common factor should be that, wherever possible, physical fact should support proof. It is the type of evidence submitted in riverine boundary cases, the provable physical facts, that will be the last subject of this chapter. The location of this topic at the end of the chapter does not reflect its relative importance.

7.12 PROOF IN A RIVERINE ENVIRONMENT—CASE STUDIES

The nature and kinds of proof can and may vary with the specific physical regime being considered, the issue in the case, the availability of evidence and expert consultants, (or the competence, experience, or knowledge of the lawyer or surveyor representing the litigants). Consequently, there can be no set formula or pattern universally applicable to each and every river boundary case. Some examples of how others, property owners, lawyers, and surveyors, once confronted with these complicated questions, responded to them should prove informative and will reinforce the axiom of this work: Proof in title and boundary litigation is limited only by the resourcefulness, creativity, and imagination of the surveyor, the lawyer, and the retained expert consultants.

Oklahoma and Texas are bitter football rivals. But even football rivalry pales when questions of oil and money are involved between the two states. Thus, in an interstate boundary dispute concerning the Red River boundary between Oklahoma and Texas where location of an oil field was in question, it should come as no surprise that "the parties . . . exhausted the avenues of research and speculation in presenting testimony. . . ."[198]

Oklahoma was claiming that a particular oil-rich area had originally been located in Oklahoma, north of the Red River, but the river had suddenly abandoned that channel.[199] Because there were no surveys or human witness[200] at the relevant time,[201] the parties and the Court were forced to rely

---

198. *Oklahoma v. Texas, supra,* 260 U.S. 638.
199. *Ibid.*
200. There are some considerable risks in relying on human witnesses in such cases. Unless properly prepared and instructed on how correctly to describe their reference to portions of particular exhibits, such as maps or photographs, their testimony may be worthless. *Conran, supra,* 341 S.W.2d at 98 (". . . much of the examination, both questions and answers, was conducted by pointing, which means nothing to the reader of the written transcript although it doubtless had some significance to those who observed it"). In *Mississippi v. Arkansas, supra,* 415 U.S. at 295 (Douglas, J., dissenting), lay witnesses, who were quite familiar with the area in question, unsuccessfully opposed "highly qualified experts."
201. This is not an unusual circumstance. *Mississippi v. Arkansas, supra,* 415 U.S. at 295 (Douglas, J., dissenting).

on physical evidence.[202] Based in part on the age of trees,[203] the Court found that Oklahoma's claim was not sustained.[204]

In many other cases, physical evidence has also been used by courts to support findings on the cause of movement in the geographic location of a property boundary. Thus, testimony of an expert forester about the age of trees was determined to have negated the possibility of a slow migration of the river. Had such a slow migration occurred, it would have eroded the soils supporting the roots and washed away the trees.[205] Geological and potamological[206] experts also provide significant evidence.[207] If there is conflicting physical evidence and interpretations thereof, some courts have not considered such evidence and testimony.[208]

In some cases quite elaborate proof has been used. In a case about substantial change in the location of the Missouri River, the existence of identifiable land in place as evidence of an avulsion was in question.[209] The parties offered, among other matters, ancient surveys and letters, lay testimony from persons who lived in the area, expert testimony concerning the geomorphic formations typical of avulsive river movement, the dendrochronology[210] of trees, predictability of erosive changes in the river by use of certain hydrological principles, analysis of soils as an indicator of accretive movement and topographic cross sections showing the land surface gradient.[211] Even on the basis of such evidence, the Court determined that it would be speculative to determine when or how the river had changed its location.[212]

Cases in which the courts have ignored conflicting physical evidence or found expert testimony to be speculative pose a considerable problem to those

---

202. "[T]here are inanimate witnesses, such as old trees, which tell a good deal." *Ibid.* Other cases, even very early cases in what was supposedly a less sophisticated era, also used such evidence, *Missouri v. Kentucky, supra,* 78 U.S. at 408, as do more modern courts, *Provo City, supra,* 176 P.2d at 139 (Larson, J., dissenting); *United States v. Cameron, supra,* 466 F.Supp. at 1106–1110.

203. The Court also relied on the fact that Texas had been exercising jurisdiction over the area and asserting proprietorship of soil for more than 50 years. *Oklahoma v. Texas, supra,* 260 U.S. at 639. Long acquiescence in the location of a boundary is an important evidentiary point to consider in these cases. E.g., *Missouri v. Kentucky, supra,* 78 U.S. at 402–403; *Indiana v. Kentucky* (1890) 136 U.S. 479, 510; *New Jersey v. New York, supra,* 140 L.Ed.2d at 1016. See notes 189 through 192 and accompanying text in Chapter 1.

204. In some considerable part at least, the reason for the finding was that because the Supreme Court placed the burden of proof on Oklahoma as the party who was asserting material changes. *Oklahoma v. Texas, supra,* 260 U.S. at 638. See notes 122 through 130 and accompanying text in this chapter.

205. *McCafferty, supra,* 397 P.2d at 99.

206. An expert in rivers.

207. *Arkansas Land & Cattle Co., supra,* 452 S.W.2d at 637–639.

208. *Id.* at 640.

209. *Omaha Indian Tribe, Treaty of 1854, etc., supra,* 575 F.2d at 639.

210. The study of time span within trees. *Id.* at 648.

211. *Id.* at 648–649.

212. Thus, defendants were held to have failed to carry their burden of proof. *Id.* at 650.

trying to advise property owners concerning the risks and probabilities of success in property boundary disputes. It is an unusual case in which the disputing parties' interpretations of evidence are in agreement. Indeed, it is even more uncommon for contending property owners' experts to reach identical conclusions based on the same set of physical circumstances. And the expense of the cases, with the necessary investigations by expert consultants, can be startling.[213] How, then, is it possible to prove a case in a fashion convincing to a trier of fact and yet affordable by most litigants?

If one had "the answer" to that question, one would be rich beyond dreams of avarice and as sought after as a Hollywood superstar. This is not to say that there is no answer. *An* answer to this difficulty is provided by the practical approach adopted by the Supreme Court in another case. The issue was the geographic location of an interstate boundary just prior to the occurrence of an avulsion.[214]

One of the parties contended that the interstate boundary could not be located with reasonable certainty even though a U.S. Army Corps of Engineers survey[215] had been made of the area just prior to the avulsion.[216] The Court reviewed the purpose for that survey and the methodology used, which did not include any actual measurements.[217] It was asserted that to be of any value for the purpose of locating a particular object shown on the map, the map must have been based on a survey accurately made by ". . . measurement and instrumental observations, and must be a faithful and correct delineation of a faithful survey, in every detail."[218] Because no field survey had been done and the Corps of Engineers' map was not tied to a monument, it was claimed that the Corps' map was worthless.

In words to be carefully considered by those concerned with the types and accuracy of evidence and proof, the Supreme Court noted:

> . . . [T]he standard demanded is not applicable. The thing to be done must be regarded. It is to locate the boundary along that portion of the bed of the river that was left dry as a result of the avulsion . . . Absolute accuracy is not attainable. *A degree of certainty that is reasonable as a practical matter, having regard for the circumstances, is all that is required* . . . . The determination of its

---

213. In one case with which the author is familiar, almost $300,000 was spent by the parties in consulting fees alone.
214. *Arkansas v. Tennessee* (1925) 269 U.S. 152, 153. The fact that an avulsion had occurred was based on the stipulations of the two states. *Arkansas v. Tennessee, supra,* 246 U.S. 161–162.
215. Courts have treated the efforts of the Corps of Engineers quite differently. Compare *Missouri v. Kentucky, supra,* 78 U.S. at 411 (report, authorized by the government and "prepared with great learning and industry" given determinative weight) with *Carpenter v. Ohio Sand & Gravel Corp., supra,* 60 S.E.2d at 215 (court refused to give government charts controlling effect).
216. *Arkansas v. Tennessee, supra,* 246 U.S. at 154–155.
217. *Arkansas v. Tennessee, supra,* 269 U.S. at 156.
218. *Ibid.*

location involved more than mere measurement and required the exercise of judgment based on experience.[219]

On this note of the proper perspective to be taken and weight to be given to the uncertainties of proof in riverine boundary and title litigation,[220] we turn to the final subject of this work. That subject, littoral property boundaries and titles along navigable lakes, is one also fraught with uncertainty and the consequent room for creativity and resourcefulness.

---

219. *Id.* at 157 (emphasis supplied). Other cases agree. See *Alexander Hamilton Life Ins. Co. v. Govt. of V.I.* (3rd Cir. 1985) 757 F.2d 534, 543; *Matthews v. McGee* (8th Cir. 1966) 358 F.2d 516, 518 (review evidence to see if there are true conflicts in the evidence not one where opinion evidence ignores or is inconsistent with proven physical facts).
220. This type of finding is not untypical, e.g., *Conran, supra,* 341 S.W.2d at 85; *Slauson, supra,* 69 N.W. at 993, and underscores the predicament of these complex cases.

8

PROPERTY BOUNDARY DETERMINATION ALONG NAVIGABLE LAKES

8.1 INTRODUCTION

The lakes speckling the United States have a contradictory nature. At times they appear serene and tranquil, seemingly unchanging in their scope and beauty. Yet Lake Michigan's wind-spawned, oceanlike waves have jeopardized property along Chicago's Lakeside Drive, some of America's most valuable real estate.[1] And a seemingly relentless increase in the level of the Great Salt Lake in Utah has threatened to submerge two major east-west interstate transportation arteries (and, in the fertile imagination of some limnologists, reestablish prehistoric Lake Bonneville).[2] The perplexing disorderliness of lakes is also exhibited in lake-related boundary and title disputes.

Lakes have not sparked the frequent or protracted littoral boundary or title disputes we associate with rivers. A lakebed title case,[3] however, is the lodestar case about the nature and scope of the public trust doctrine.[4] And cases

---

1. Cobb, "The Great Lakes Troubled Waters," National Geographic (July, 1987) v. 172, n. 1, pp. 2, 6.
2. Gore, "The Rising Great Salt Lake: No Way to Run a Desert," National Geographic (June, 1985), v. 167, n. 6., p. 694.
3. *Illinois Central Railroad v. Illinois* (1892) 146 U.S. 387.
4. *City of Berkeley v. Superior Court* (1980) 26 Cal.3d 515, 521, cert. den., sub. nom. *Santa Fe Land Improv. Co. v. Berkeley* (1980) 449 U.S. 840; *Wade v. Kramer* (Ill. 1984) 459 N.E.2d 1025, 1027.

considering the location and character of lake littoral boundaries, while relatively rare, are not all that infrequent.[5] Many of these disputes arise in the context of state lakebed ownership competing with private upland ownership[6] or between the United States and a state or private party, each claiming lakebed ownership.[7]

Title and boundary contests, however, are not the only contexts in which questions concerning lake boundaries arise. By way of example only, other litigation has concerned the effect of lake water level regulation by dams,[8] the effect of deliberate diversions from lake tributaries on the value of once-littoral lands surrounding a dwindling or rising lake,[9] or on the type of zoning of lakeside lands.[10] Finally, although many lake littoral title or boundary disputes are of relatively recent origin, some of the important cases were decided nearly 100 or more years ago.[11] To a large extent this disjointed polyglot of cases is explained by the physical character of lakes and by geography and demographics.

The serenity of lakes is not only part of their timeless beauty, it is their very nature.[12] Unlike the open ocean and tidal estuaries and rivers, most lakes are not subject to the ofttimes visible or regular fluctuations of these other physical regimes. Although some lakes are subject to large seasonal vacillations (much the same as nontidal rivers), in the absence of human-induced effects, shoreline changes taking place around the majority of lakes are understood to be on a more geologic than historic time schedule.

Moreover, with the notable exception of the Great Lakes, most lakes have not served as transportation mediums to anywhere near the same extent as

---

5. E.g., *Utah Div. of State Lands v. U.S.* (1987) 482 U.S. 193; *Cal. ex rel. State Lands Com'n v. United States* (9th Cir. 1986) 805 F.2d 857, *cert. den.* (1987) 484 U.S. 816; *State of California v. Superior Court (Fogerty)* (1981) 29 Cal.3d 240, *cert. den.* (1981) 454 U.S. 865, *reh. den.* (1981) 454 U.S. 1094 ("*State v. Superior Court (Fogerty I)*").
6. *State v. Superior Court (Fogerty I)*, *supra*, 29 Cal.3d at 243. One unusual case concerned the competition between state ownership of the bed of Lake Michigan and submerged wrecks that had become imbedded in the lakebed on the one hand, and federal admiralty and maritime jurisdiction on the other. *People v. Massey* (Mich. 1984) 358 N.W.2d 615. The matter of the ownership of imbedded shipwrecks appears to have been resolved. Abandoned Shipwreck Act of 1987 (1988) P.L. 100-298, 102 Stat. 432. *California v. Deep Sea Research* (1998) ___ U.S. ___, ___ S.Ct. ___, 140 L.Ed.2d 626 discusses this statute and its application.
7. *Utah Div. of State Lands*, *supra*, 482 U.S. at 195; *Calif. ex rel. State Lands Comm.*, *supra*, 805 F.2d at 859–860; *United States v. Otley* (9th Cir. 1942) 127 F.2d 988, 992.
8. *Payette Lakes Protective Ass'n v. Lake Reservoir Co.* (Idaho 1948) 189 P.2d 1009, 1014.
9. *City of Los Angeles v. Aitken* (1935) 10 Cal.App.2d 460, 462; *Carpenter v. Board of Commissioners* (Minn. 1894) 58 N.W. 295, 296.
10. *State of Wisconsin v. Trudeau* (Wisc. 1987) 408 N.W.2d 337, 340, *cert. den.* (1988) 98 L.Ed.2d 652.
11. *Jones et al. v. Johnston* (1855) 59 U.S. (18 How.) 150; *Banks v. Ogden* (1864) 69 U.S. (2 Wall.) 57; *Hardin v. Jordan* (1891) 140 U.S. 371; *Live Stock Co. v. Springer* (1902) 185 U.S. 47.
12. John Muir called Mono Lake a "... burnished metallic disc...." *City of Los Angeles v. Aitken*, *supra*, 10 Cal.App.2d at 463. This description implies an immutable tranquility not usually associated with rivers or the open ocean coast.

rivers. This is due to the fact that, particularly in the western United States, lakes are geographically isolated or in mountainous or desertic terrain, lands that were inhospitable to early or large and sustained settlement.[13] Other lakes were just too small in size to support the establishment of significant long-term communities that would require or be the initial point or terminal destination for transportation. As a consequence, considerable settlement that would give rise to title and boundary disputes did not occur along many lakes until well into the twentieth century.

It was not until after the Second World War that the country's economic growth, improved and more affordable means of transportation, and the changing social environment led to the increasing population of the lands adjacent to our lakes. Vacation homes and supporting communities sprang up, and so did title and boundary disputes. This final chapter will consider some of these lake title and boundary disputes and many of the somewhat peculiar rules adapted to resolving such disputes.

First, the chapter will consider the basis of public ownership of the beds of lakes from a policy standpoint. This discussion will be an important foundation to understanding lake boundary principles and a vivid reminder of the potent effect of the public trust doctrine. Because lakebed ownership turns on navigability, questions about lake navigability will be briefly discussed. The public policy reasons that the courts have given in support of lake navigability decisions will be the main topic. The chapter will then move on to a very basic matter: What physical attributes makes a water body a lake, not just a wide spot in a river? After a short description of the different kinds of lakes, there is a longer discussion of the ever slippery concept of the ordinary high- and ordinary low-water mark littoral property boundaries and the methodology for their physical location.

Movement of the geographic location of the littoral shoreline and the effect of such movement on the property boundary will be discussed in the unique case of lakes. Particular attention will be devoted to the relevance of the cause of movement in geographic location of the lake shoreline and the property boundary effect of lake recession and lake encroachment. Finally, the types of proof that have been used in lake boundary cases to physically locate the ordinary high-water mark property boundary will be analyzed.

But before this chapter gets that far, it is important to understand the derivation of title to lakes. For the nature of lakebed title supplies a significant rationale, almost a presumption according to some, appearing to favor public, as opposed to private, rights and interests in the lakebed and shorezone.[14]

---

13. *State v. Bollenbach* (Minn. 1954) 63 N.W.2d 278, 290. Exemplary of this situation was the fate of the emigrant party that tried to winter at what is now known as Donner Lake, near Lake Tahoe, California.

14. In *State v. Superior Court (Fogerty I)*, *supra*, 29 Cal.3d at 245, the California Supreme Court defined the "shorezone" as the lands between high and low-water. The court noted the importance of the shorezone:

8.2 CHARACTER OF TITLE TO THE BEDS OF NAVIGABLE LAKES

Ownership of the beds of navigable lakes resides in each of the states[15] and is confirmed by the United States Congress in the Submerged Lands Act.[16] Sovereign lakebed ownership is said to be a function of the basic sovereign character of the states. This is the same foundation of ownership as in the cases of the beds of navigable rivers, tidal estuaries or open ocean coast, tidelands or the seabed.[17] Federal and state court decisions hold that state title in such lands, such as the beds of navigable lakes, is unique.

One prominent case concerned a grant by the Illinois state legislature of the City of Chicago's entire lakefront to the Illinois Central Railroad Company for commercial development by the railroad.[18] Subsequently, the Illinois legislature had second thoughts about this conveyance and belatedly repealed what had been thought of as an irrevocable legislative grant.[19] The resulting lawsuit against the state by the surprised and disgruntled railroad ended up in the United States Supreme Court.

The opinion, one of the United States Supreme Court's longest,[20] decided whether the Illinois legislature had authority to revoke the earlier grant to the railroad. In holding that the railroad's seemingly ironclad legislative grant

---

The shorezone is a fragile and complex resource. It provides the environment necessary for the survival of numerous types of fish . . . , birds . . . , and many other species of wildlife and plants. These areas are ideally suited for scientific study, since they provide a gene pool for the preservation of biological diversity. In addition, the shorezone in its natural condition is essential to the maintenance of good water quality, and the vegetation acts as a buffer against floods and erosion. *Ibid.*

---

15. E.g., *Illinois Central, supra*, 146 U.S. at 435–437; *Utah Div. of State Lands, supra*, 482 U.S. at 195–196; *State of California v. Superior Court (Lyon)* (Cal. 1981) 29 Cal.3d 210, 217–222, *cert. den.* (1981) 454 U.S. 865, *reh. den.* (1981) 454 U.S. 1094; *Matter of Ownership of Bed of Devils Lake* (N.D. 1988) 423 N.W.2d 141, 142–143; *People v. Massey, supra*, 358 N.W.2d at 618; *Utah State Road Commissioners v. Hardy Salt Company* (Utah 1971) 486 P.2d 391, 392; *Breese v. Wagner* (Wisc. 1925) 203 N.W. 764, 766; *State of Wisconsin v. Trudeau, supra*, 408 N.W.2d at 341; *Hazen v. Perkins* (Vt. 1918) 105 A. 249, 251; *State v. Korrer* (Minn. 1914) 148 N.W. 617, 621; *Lakeside Boating and Bathing Inc. v. State* (Iowa 1984) 344 N.W.2d 217, 220; *Idaho For. Indus. v. Hayden Lk. Watershed Imp.* (Idaho 1987) 733 P.2d 733, 737; *State v. Thomas* (Iowa 1916) 155 N.W. 859; *Martin v. Busch* (Fla. 1927) 112 So. 274, 283; Skelton, *The Legal Elements of Boundaries and Adjacent Properties* (1930)(herein "Skelton") § 286; Robillard, W., and Bouman, L., *Clark on Surveying and Boundaries* (7th Ed., 1997) § 25.07 (hereinafter "Clark 7th"); Grimes, *Clark on Surveying and Boundaries* (4th Ed., 1976) (herein "Clark 4th") § 588, p. 849.
16. 43 U.S.C. §§ 1301(a), 1311(a); see *Furlong Ent. v. Sun Exploration and Prod.* (N.D. 1988) 423 N.W.2d 130, 132.
17. *Illinois Central, supra*, 146 U.S. at 435–437; *Utah Div. of State Lands, supra*, 482 U.S. at 195–196; *State v. Superior Court (Lyon), supra*, 29 Cal.3d at 219–220.
18. *Illinois Central, supra*, 146 U.S. at 454–455.
19. *Id.* at 389.
20. The reported opinion, with all the arguments and statement of the case, is 89 pages long.

was terminable at the will of a later state legislature, the Supreme Court laid out the doctrinal foundation of the public trust doctrine. That foundation has been at the core of many, including some quite recent, title and boundary cases decided in various courts in states extending from coast to coast and north to south.[21] Succinctly stated, the Supreme Court held that the state cannot abdicate its sovereign rights in lands underlying navigable waters to private interests, nor can such rights ever be completely and irrevocably alienated, except in some extremely limited circumstances.[22]

According to the Supreme Court, the basic reason for this rule is that the lands underlying navigable and tidal waters are public in character and are held subject to a public trust for the common use. The state cannot give away this trust, nor can one state administration bind or sell the discretion of its successor administrations with respect to such lands; each administration must have authority to deal with the public trust interest as needs and circumstances require.[23] In the case of lakes, the principles expressed in *Illinois Central*

---

21. E.g., *People v. California Fish Co.* (Cal. 1913) 166 Cal. 576, 584; *City of Berkeley, supra*, 26 Cal.3d at 521; *State v. Superior Court (Lyon), supra*, 29 Cal. 3d at 227–228; *Kootenai Environ. Alliance v. Panhandle Yacht* (Idaho 1983) 671 P.2d 1085, 1088–1089; *Thomas v. Sanders* (Ohio 1979) 413 N.E.2d 1224, 1228; *State v. Southern Sand & Material Co.* (Ark. 1914) 167 S.W. 854, 856; *New York, N.H.H.R. Co. v. Armstrong* (Conn. 1922) 102 A. 791, 794; *Wade, supra*, 459 N.E. 2d at 1027; *Cinque Bambini Partnership v. State* (Miss. 1986) 491 So.2d 508, 512–513, *aff'd, sub nom. Phillips Petroleum Co. v. Mississippi* (1988) 484 U.S. 469, *reh. den.* (1988) 486 U.S. 1018; *Lake Mich. Fed. v. U.S. Army Corps of Engineers* (N.D. Ill. 1990) 742 F.Supp 441, 444–446.

22. *Illinois Central, supra*, 146 U.S. at 453; *City of Berkeley, supra*, 26 Cal.3d at 524; *People v. California Fish Co., supra*, 166 Cal. at 597. The United States Supreme Court observed:

> A grant of all the lands under the navigable waters of a State has never been adjudged to be within the legislative power; and any attempted grant of the kind would be held, if not absolutely void on its face, as subject to revocation. The State can no more abdicate its trust over property in which the whole people are interested, like navigable waters and soils under them, so as to leave them entirely under the use and control of private parties, except in the instance of parcels mentioned for the improvement of the navigation and use of the waters, or when parcels can be disposed of without impairment of the public interest in what remains, than it can abdicate its police powers in the administration of government and the preservation of peace.

Illinois Central, supra, 146 U.S. at 453.

23. *Illinois Central, supra*, 146 U.S. at 460. This rationale also accounts for a rule of strict construction of statutes that purport to grant away such sovereign rights. The rule of statutory construction is said to be that courts will carefully scrutinize statutes attempting to grant away sovereign rights or interests, to decide whether that was in fact the legislative intent and whether that intent was clearly set forth in the statute or is necessarily implied from its terms. No such inference will be made if any other inference is reasonably possible that would not involve a destruction of public trust rights and interests. E.g., *City of Berkeley, supra*, 26 Cal. 3d at 525; *People v. California Fish, supra*, 166 Cal. at 597; see note 162 in Chapter 1. One case held that even a legislative finding that determined that a grant did not violate the public trust was not entitled to deference. *Lake Mich. Fed. v. U.S. Army Corps of Engineers, supra*, 742 F.Supp at

have been affirmed by both state and federal courts. If there was any question in California of whether the state owned lakebed lands underlying navigable, but not tidal, lake waters to the ordinary high-water mark in trust, that question has been answered.

8.3 RESOLUTION OF THE QUALITY AND CHARACTER OF STATE TITLE TO THE BEDS OF NAVIGABLE LAKES

Showing what may be considered admirable inventiveness, in one California case littoral landowners argued that when California adopted English common law, California also accepted the English common law rule of private ownership of nontidal navigable waters such as rivers and lakes.[24] The California Supreme Court, citing both federal and state authorities, noted that courts had never autonomously adhered to the tidal test[25] of navigability to determine sovereign ownership.[26] Thus, according to the California Supreme Court,

---

446. This was because the purpose of the public trust doctrine was "to police the legislature's disposition of the public lands." *Ibid.*
24.   *State v. Superior Court (Lyon)*, *supra*, 29 Cal.3d at 217–218; Clark 7th, *supra*, § 25.10; Clark 4th, *supra*, at §591, p. 851. As can be seen from the *State v. Superior Court (Lyon)* case and from others as well, e.g., *Furlong Ent.*, *supra*, 423 N.W.2d at 134–136, the history of development of a state's legal system may provide valuable information in construing and interpreting statutes concerning title to or property boundaries of lands along navigable water bodies.
25.   Using a type of "heads, I win; tails, you lose" argument, littoral landowners had argued in another case that tidal waters must be navigable, no matter their tidality, for the underlying lands to be owned by the state. *Phillips Petroleum Co. v. Mississippi* (1988) 484 U.S. 469, 473, *reh. den.* (1988) 486 U.S. 1018. Had the arguments made in the *State v. Superior Court (Lyon)* case and in the *Phillips Petroleum* case been adopted by the courts, some have claimed that sovereign titles and public trust rights would be narrowed in geographic extent to the main navigation channels of the ocean and harbors. See note 26 in this chapter.
26.   *State v. Superior Court (Lyon)*, *supra*, 29 Cal.3d at 218–222 citing *The Propeller Geneses Chief et al. v. Fitzhugh et al.* (1851) 53 U.S. (12 How.) 443, 454–458 and *American Water Co. v. Amsden* (1856) 6 Cal. 443, 446. In addition, the lake waters did not have to be entirely navigable-in-fact to be included as part of the lakebed over which sovereign ownership extended. The Supreme Court of Wisconsin held that the state owned a particular area underlying Lake Superior even though the waters of the lake overlying the particular site were not navigable in fact; the court treated the area as part of the lakebed as it was waterward of the ordinary high-water mark. *State of Wisconsin v. Trudeau*, *supra*, 408 N.W.2d at 342 ("An area need not be navigable to be lakebed"). *The United States Supreme Court decision in Phillips Petroleum has reinforced this view.* The Supreme Court recognized that there are areas along the borders of navigable or tidal waters that "by no means could be considered navigable" but that "[i]t [was] obvious that these waters are part of the sea, and the lands beneath them are State property." *Phillips Petroleum*, *supra*, 484 U.S. at 480. The Florida Supreme Court also held that ". . . navigable waters include lakes . . . and all waters capable of practical navigation for useful purposes, . . . whether the water is navigable or not in all its parts towards the outside lines or elsewhere, or whether the waters are navigable during the entire year or not." *Martin*, *supra*, 112 So. at 283.

adoption of English common law did not restrict California's sovereign ownership only to lands underlying tidal waters; California also owned the beds of navigable lakes to the ordinary high-water mark.[27] The United States Supreme Court reaffirmed this principle of sovereign lakebed ownership in a case concerning a navigable lake in Utah, far from the ebb and flow of the tide.[28]

The California Supreme Court did not end its opinion on that point. The court went on to note the principle had long been recognized that each state had absolute authority to grant lands waterward of the ordinary high-water mark as it might or might not choose.[29] Accepting this rule as the gospel (or assuming it was the case, anyway), in 1872 the California legislature enacted a statute (California Civil Code section 830). That statute set the boundary of lands bordering on navigable lakes or streams at the ordinary *low*-water mark.[30] It took more than 100 years for a situation to arise in which the effect of this statute could be finally decided by the courts.

Upland owners of littoral lands bounding California lakes, such as Lake Tahoe, claimed this statute was a grant in fee simple absolute by the California legislature of the state's entire and complete property interest in the lands between the ordinary high- and ordinary low-water marks.[31] The state claimed there had been no grant.[32]

In a Solomon-like decision, the California Supreme Court resolved the conflict by giving both parties something. The California court upheld a portion of the littoral landowners' claim in deciding that the legislature had granted *fee* title to such lands.[33] The court also held that the state's grant of

---

27. *State v. Superior Court (Lyon), supra*, 29 Cal.3d at 219.
28. *Utah Div. of State Lands, supra*, 482 U.S. at 195–196.
29. *State v. Superior Court (Lyon), supra*, 29 Cal.3d at 219–220; *Hardin, supra*, 140 U.S. at 382.
30. Cal. Civ. Code § 830; *State v. Superior Court (Lyon), supra*, 29 Cal.3d at 222. Many other states are also low-water states. E.g., *Matter of Ownership of Bed of Devils Lake, supra*, 423 N.W.2d at 142; *S.D. Wildlife Federation v. Water Mgmt. Bd.* (S.D. 1986) 382 N.W.2d 26, 30. Some of the states that have also extended riparian ownership waterward of the ordinary high-water mark are listed in *Shively v. Bowlby* (1894) 152 U.S. 1, 18–25; *Clark 7th, supra*, § 23.02; Clark 4th, *supra*, § 553, p. 741, § 566, p. 772; and Skelton, *supra*, § 287, p. 321. Some variants of riparian ownership are recounted in *Conran v. Girvin* (Neb. 1960) 341 S.W.2d 75, at 80, and *United States Gypsum v. Uhlhorn* (E.D. Ark. 1964) 232 F.Supp. 994, 1001, n.4, *aff'd* (8th Cir. 1966) 366 F.2d 211, *cert. den.* (1967) 385 U.S. 1026.
31. *State v. Superior Court (Lyon), supra*, 29 Cal.3d at 223. This case arose because upland owners wanted to develop part of their property and were not granted a permit to do so because of a claim of state ownership waterward of the high-water mark. *Id.* at 215.
32. The state argued that California Civil Code § 830 was merely a rule of statutory construction. It was urged that the statute contained no granting language and to construe the statute as a grant would violate the rule that grants to private persons by the state are to be construed in favor of the public. *Ibid.*; see note 23 in this chapter.
33. *Id.* at 226. The court based this holding, in large part, on evidence of an opinion by the state attorney general and "hundreds of letters" by the state stating, either expressly or by implication, that the state's ownership extended waterward of the ordinary low-water mark. *Id.* at

fee title was not absolute; instead, the littoral owners' fee title to the lands between the high- and low-water marks was held subject to a reserved public trust easement[34] retained by the state.[35] This was because of the rule that statutes in derogation of public rights will not be interpreted to abandon those rights unless no other interpretation is reasonably possible.[36]

Finally, the private landowners also argued that their long, allegedly undisputed ownership of the lands between the ordinary high- and ordinary low-water marks had ripened into a "rule of property."[37] But, as we recall from the results of similar claims detailed earlier,[38] the court decided that failure

---

225. This "administrative construction" of California Civil Code § 830 was determinative to the court. *Id.* at 224–225. The court would not ". . . ignore [the] long-continued and frequently expressed views to the effect that [the section] constitutes a grant to private persons of title to the beds of navigable, non-tidal bodies to low-water mark." *Id.* at 225. The United States Supreme Court has used the administrative construction doctrine. *United States v. Gerlach Live Stock Co.* (1950) 339 U.S. 725, 739–742. Another California case concerning a lake also used "administrative construction" to establish the location of the ordinary low-water mark of the lake. *County of Lake v. Smith* (1991) 238 Cal.App.3d 214, 236-237. Do not confuse the "administrative construction" rationale with the doctrine of equitable estoppel. See notes 194 through 210 and accompanying text in Chapter 1. States cannot be prevented from asserting sovereign trust title by the rule of equitable estoppel (*Ibid.*; *State v. Superior court (Fogerty I), supra,* 29 Cal.3d at 244), except in limited circumstances. See notes 211 through 220 and accompanying text in Chapter 1.

34. Recall the significance of the public trust doctrine. See, e.g., notes 117 through 121 and accompanying text in Chapter 1, notes 16 through 17 and accompanying text in Chapter 3, notes 6 through 10 and accompanying text in Chapter 4, and notes 19 through 20 and accompanying text in Chapter 5; notes 15–23 and accompanying text in this chapter.

35. *State v. Superior Court (Lyon), supra,* 29 Cal.3d at 226–231. The littoral landowners argued that courts had never held the public trust applicable to navigable but nontidal waters. *Id.* at 226. Citing *Illinois Central*, the California Supreme Court held to the contrary. *Id.* at 227–228. The court stated that the alleged distinction between tidal and nontidal waters for public trust purposes had been "thoroughly discredited in this country.? *Id.* at 230–231. Responding to the argument that *Illinois Central* was a special case because it concerned Lake Michigan, a lake of great size and importance to commerce, the court said: "The application of the trust doctrine to tidal waters is not confined to those bodies which are huge in size and important for purposes of commerce; we can see no reason why such a test should not be applied to non-tidal waters." *Id.* at 228. In two relatively recent cases, the Supreme Court appeared to reaffirm this principle. *Utah Div. of State Lands, supra,* 482 U.S. at 195–196 (inland lake); *Phillips Petroleum, supra,* 484 U.S. at 476. Other state courts have also accepted this principle. E.g., *S.D. Wildlife Federation, supra,* 382 N.W.2d at 30. Public rights in much, much smaller bodies of water have long been recognized. Leighty, L., Public Rights in Navigable State Waters—Some Statutory Approaches (1971) 6 Land & Water Law Rev. 459, 471 (great ponds); *Gratt v. Palangi* (Me. 1958) 147 A.2d 455; *Conant v. Jordan* (Me. 1910) 77 A. 938. There are some limits, however. *Golden Feather Community Association v. Thermalito Irrigation District* (1989) 209 Cal.App.3d 1276, 1284–1286 (public trust not applied to artificial reservoir).

36. *State v. Superior Court (Lyon), supra,* 29 Cal.3d at 231. See note 23 in this chapter. Other states had already recognized a similar rule prior to *State v. Superior Court (Lyon).* E.g., *State v. Korrer, supra,* 148 N.W. at 623.

37. *State v. Superior Court (Lyon), supra,* 29 Cal.3d at 231. The term "rule of property" means a settled rule or principle based on precedents or decisions regulating the ownership of real property. *Abbott v. City of Los Angeles* (1958) 50 Cal.2d 438, 456.

38. See notes 139 through 149, 151 through 172, and accompanying text in Chapter 1.

of the state to assert its sovereign trust rights in such lands did not constitute a "rule of property."[39]

One final example of the application of the public trust doctrine to navigable lakes will confirm the unique character of sovereign title interests. Based on water rights permits and licenses that appeared to have been regularly issued by the State of California, the City of Los Angeles had long diverted all (or nearly all) of the water flow of the *nonnavigable* tributaries of Mono Lake, a navigable lake in remote central eastern California.[40] This diversion severely disrupted the natural regime of the lake. The level of the lake declined dramatically over the course of years, and its surface area dwindled noticeably. As just one part of the effect of the changed lake regime, the former lakebed became exposed, allowing predators to attack a seagull rookery on what previously had been an island in the bed of the lake. Other alleged public trust interests were also foreseeably and adversely affected.[41]

Forty years after the issuance of the permits and license allowing the diversions and after the city had constructed diversion and waterworks worth many hundreds of millions of dollars, these governmental authorizations were challenged.[42] It was argued that the water rights permits and license should be reconsidered in light of the effect of the tributary stream diversions on the public trust interests in Mono Lake, matters that had specifically not been considered in the original permit and license proceedings.[43] In what many considered to be the end of the organized water rights system in California, the California Supreme Court agreed with that contention. The California court held that because of the importance of the public trust interests in the navigable lake, the permit and license to divert water from the lake's non-navigable tributaries could be reconsidered in light of those public trust interests.[44]

These cases are only exemplary of the impact of the public trust doctrine in lake boundary and title disputes. And such examples of the application of the public trust doctrine are not limited to California.[45]

---

39. *State v. Superior Court (Lyon), supra,* 29 Cal.3d at 231.
40. *National Audubon Society v. Superior Court* (1983) 33 Cal.3d 419, 424, *cert. den., sub nom. Los Angeles Dept. of Water & Power v. National Audubon Soc.* (1984) 464 U.S. 977. Mono Lake, like many western lakes, is a terminal lake; it has no outlet. *Cal ex rel. State Lands Com'n, supra,* 805 F.2d at 859.
41. *National Audubon Society, supra,* 33 Cal.3d at 429–431.
42. *Id.* at 426–431.
43. *Id.* at 447.
44. *Id.* at 445–447.
45. Across the country in nominally "conservative" New England, the state and private landowners were disputing placement of some fill in Lake Champlain, a navigable lake. The case was tried on the theory that location of the boundary between private and public ownership would define the geographic extent of public rights in the bed and waters of the lake. *State v. Cain* (1967) 236 A.2d 501, 502–503. On appeal, the state changed its theory. It sought to have the case remanded because the trial court had not considered the issue of the public's claim of ownership of the waters in which the fill had been placed. The state had not made this argument

8.4 NAVIGABILITY OF LAKES

The court's desire to protect and preserve public rights in lakes has had other consequences as well, especially concerning the question of the navigability of lakes.[46] In one South Dakota case[47] the question was whether the small and, in places quite shallow, lake, Lake Albert,[48] was a "navigable lake" within the meaning of a South Dakota statute.[49]

Using its innate knowledge that South Dakota is far from the crash and roar of the ocean surf, the court quickly dispensed with what it called "the common law test of navigability"—whether the lands were "tide ebbed and flowed"—as inapplicable to the geophysical circumstances in South Dakota.[50] In the direct and plainspoken language one expects from the Midwest, the court went right to the heart of the matter. Although there was no specific

---

at trial. *Id.* at 505. Defendants argued that the case had been tried, correctly or incorrectly, on a certain theory; the trial court and counsel had acquiesced in and adopted the theory and thus it became the "law of the case." *Ibid.* "Law of the case" is a rule of practice and policy holding that once a legal issue is resolved at one stage of the litigation, a party is bound to that outcome at later stages and cannot again and again try the same question. *Ibid.* Even considering this rule, because of the asserted vital public interest of both the recreating public and littoral owners in the determination of the right to use the waters of Lake Champlain, the Vermont Supreme Court remanded the case to allow the state to assert its new theory. *Id.* at 507. A similar result occurred in *State of Wisconsin v. Trudeau, supra*, 406 N.W. at 345 (Wisconsin Supreme Court overturned a variance granted to a condominium project where the public trust had not been originally considered). Other cases are *Flisrand v. Madsen* (S.D. 1915) 152 N.W. 796, 799–800; *Lamprey v. Metcalf* (Minn. 1896) 53 N.W. 1139, 1145; *United Plainsmen Assoc. v. North Dakota State Water Conservation Comm.* (N.D. 1976) 247 N.W.2d 457, 461–463. The future impact of these cases may be weakened by the Supreme Court's ruling in *New Hampshire v. Maine.* That case held that, as a matter of judicial policy, a state could be estopped from asserting a position contrary to a position it had taken in the same case. *New Hampshire v. Maine* (2001) ___ U.S. ___, ___ S. Ct. ___, ___ LE.2d ___, 2001 U.S. Lexis 3981.
46. The test for navigability for title purposes is discussed at considerable length and in great detail in notes 9 through 59 and accompanying text in Chapter 7. The following discussion will consider some refinements of the concept of navigability specifically related to lakes, but not specifically to lakebed title.
47. *Flisrand, supra*, 152 N.W. 796.
48. *Id.* at 798. Lake Albert was 5 miles long; its width was barely $2\frac{1}{2}$ miles. The amount of water in the lake varied according to the climate, but at times of high-water the lake's depth varied from 1 to 10 feet. *Ibid.*
49. *Id.* at 799. That statute was virtually identical in content to California Civil Code § 830 construed in the *State v. Superior Court* (*Lyon*) case. Compare Cal. Civ. Code § 830 with So. Dak. Civ. Code § 289. Note that the question of sovereign lakebed ownership under the federal test of navigability was not in issue in the *Flisrand* case.
50. *Flisrand, supra*, 152 N.W. at 799. The United States Supreme Court has also yet to give any weight to this argument. In a recent case concerning the ownership of the bed of an inland navigable lake, neither the majority nor the dissenting Supreme Court justices felt it necessary to note in their opinions whether the waters of the lake were tidal or not. Yet the Court still had no trouble deciding that the lakebed came to the state under the Equal Footing Doctrine as lands underlying navigable waters (unless they had been otherwise reserved). *Utah Div. of State Lands, supra*, 482 U.S. at 195–196, 209 (White, J., dissenting). The lakebed includes all of the lands

statutory test for what constituted navigability, the purpose of the test of navigability in South Dakota was to determine ". . . whether waters were public or private."[51] Applying the test of whether or not such waters were more reasonably adapted to public than to private use, the South Dakota court found little Lake Albert to be navigable.[52]

Courts in other states have embraced this public/private test,[53] although not necessarily in the context of the determination of one of the elements of lakebed ownership.[54] Adoption of the public/private test by some courts does not necessarily mean that this test is uniformly accepted as the standard to determine whether a lake is navigable for purposes of determining title to the lakebed. In cases about sovereign title to lakebeds federal courts and some state courts, for example, have a much narrower view of navigability.[55] The restricted view of the federal courts may be based, in part at least, on the fact that in the western states, where many of these cases arise, the federal government, not a private person, is the littoral owner.[56]

Opinions from such courts may leave the impression that strict compliance with each element of the federal test of navigability must be met before a

---

covered by the navigable waters, including that part of the lakebed toward the shore or the outside lines of the lake that are not in fact navigable. See notes 26 and 35 in this chapter.

51. In this country many courts have held that whether or not certain waters are navigable depends on the natural availability of such waters for public purposes, taking into consideration the natural character and surroundings of such waters." *Flisrand, supra*, 152 N.W. at 799. The court relied on a line of cases beginning with the Minnesota case of *Lamprey v. Metcalf, supra*, 53 N.W. 1139. In the *Lamprey* opinion, the Minnesota Supreme Court broadly defined navigability to include boating or sailing for pleasure in order to preserve and protect public rights and interests in lakes:

> . . . To hand over all these lakes to private ownership, under any old or narrow test of navigability, would be a great wrong upon the public for all time, the extent of which cannot, perhaps, be now even anticipated. *Lamprey, supra*, 53 N.W. at 1145; see note 52 and accompanying text in Chapter 7; accord, *Hillebrand v. Knapp* (S.Dak. 1937) 274 N.W. 821, 822.

52. *Flisrand, supra*, 152 N.W. at 800.
53. *Idaho For. Indus. Inc., supra*, 733 P.2d at 739; *Provo City v. Jacobsen* (Utah 1947) 176 P.2d 130, 135 (Larson, C.J., dissenting); *Montana Coalition for Stream Access, Inc. v. Curran* (Mont. 1984) 682 P.2d 163, 169; *State v. McIlroy* (Ark. 1980) 595 S.W.2d 659, 663, *cert. den.* (1980) 449 U.S. 843; *Kelley, ex rel. MacMulan v. Halledin* (Mich. 1974) 214 N.W. 2d 856, 862; but see *Bott v. Com'n of Natural Resources, Etc.* (Mich. 1982) 327 N.W.2d 838 for some retrenchment by Michigan. In *State v. Bollenbach, supra*, 63 N.W.2d 287–288, the Minnesota Supreme Court refused to follow *Lamprey*, instead using a federal test of navigability.
54. *State v. McIlroy, supra*, 595 S.W.2d at 663; see *People ex rel Baker v. Mack* (1971) 19 Cal.App.3d 1040, 1049; *Hitchings v. Del Rio Woods* (1976) 55 Cal.App.3d 560, 571.
55. See note 23 and accompanying text in Chapter 7.
56. *Cal. ex rel. State Lands Com'n, supra*, 805 F.2d at 859; *Utah Div. of State Lands, supra*, 482 U.S. at 205, n. *; *Provo City, supra*, 176 P.2d at 141 (Larsen, C.J., dissenting). For a case in which the United States was claiming bed ownership against private landowners and made some seemingly desperate arguments to preserve its claim of ownership, see *United States v. Otley, supra*, 127 F.2d at 998.

decision can be made that a particular waterbody is navigable. Notwithstanding these opinions, there are some cases in which, because of the physical configuration and attributes of the lake, the lake's navigability is sufficiently self-evident to be judicially noticeable, even when federal lands are adjacent.[57] And a wide variety of evidence of navigability has been accepted by the courts to support and sustain ownership claims even under the narrower federal test of navigability.[58] One should recognize that the concept of navigability is dynamic and evolving and that the last case on the subject has still not been decided, nor has the last word been written.

8.5 WHAT IS A LAKE?

As we know by now, at least to a certain extent, the question of navigability is related to the nature and regime of the water body. But some courts posed even more basic questions about the nature of the water body with which they were dealing. These jurists asked: What set of physical attributes makes a lake a "lake" as opposed to a river or even a swamp?[59] As with most matters discussed in this work, there is no standard assortment of physical endowments that courts have adopted that would universally determine when a particular water body should be considered a lake. Surveyors and lawyers, however, are not left entirely to their own devices in advising their clients on whether a particular water body is or is not a lake.

---

57. In *City of Los Angeles v. Aitken, supra*, 10 Cal.App.2d at 466, the court took judicial notice of the fact that Mono Lake, which was ". . . ten miles wide and fifteen miles in length with an average depth which will readily float large vessels and is susceptible of use as a public highway for transporting persons and property for commercial purposes . . . is a navigable lake." Although Mono Lake lay in the high desert in remote eastern California, far off the commercial track, a later federal case also held the lake to be navigable. *Cal. ex rel. State Lands Com'n, supra*, 805 F.2d at 859. This holding was based on a stipulation by the United States.
58. For example, in *State v. Longyear Holding Co.* (Minn. 1947) 29 N.W.2d 657, 663, a court affirmed the State of Minnesota's ownership of the bed of a navigable lake. The evidence of navigability in that case established that Indians and others used the relatively small lake as part of a trading route for travel, for transportation of commerce (furs, fish, and supplies) by canoe, and by timber companies for floating logs. *Id.* at 663. A wide variety of evidence supported this conclusion. The source of much of that evidence was excerpts from certain United States House of Representative documents containing ancient letters and affidavits of surveyors and trappers describing the trade route of which the lake was a part. In addition, ancient maps and an expert cartographer's interpretation of such maps, coupled with firsthand eyewitness testimony and on-the-ground verification of the existence of portage sites, confirmed the trade route that included the lake. *Id.* at 664–665. For further discussion, see notes 53 through 57 and accompanying text in Chapter 7.
59. The dictionary defines a lake as an inland body of water, usually fresh water, formed by glaciers, river drainage, etc., larger than a pool or pond. Webster's New World Dictionary (College Ed.). One writer described a lake as "an aneurysm in a river." McPhee, John, *Basin and Range* (Farrar, Straus, Giroux 1980), p.38.

Some of the physical indicia considered by the courts and commentators are the surface area covered by the water body, the depth of the water, the permanence of coverage and existence of the water body, and the source of water supply to the water body.[60] In one case in which it was important to determine whether a water body was a river or a lake,[61] a Louisiana court noted several additional ingredients that distinguished lakes from rivers or streams: Lakes were more or less stagnant and supplied by drainage. On the other hand, rivers contained flowing water in a permanent bed or channel between well-defined banks with a current of sufficient capacity or velocity to form accretions.[62] That court first held that the water body in issue was a stream, based on the existence of an accretion-forming current.[63] In a subsequent opinion, however, the court changed its mind and found, based in part on the water body's designation as a lake on official maps, that the water body was a lake, not a stream.[64]

---

60. Clark 7th, *supra*, § 25.10; Clark 4th, *supra*, § 591. One case considered whether a 40-acre slough was part of a lake. Doing so was important because of competing contentions about public use of the slough. Using maps and aerial photographs, it was shown before a state water management board that the slough was separated from the lake. A trial court, however, reversed the board's decision. The board had found that the ordinary high-water mark of the lake was at a certain elevation. According to the court, at that elevation the lake and the slough were "one body of water" and consequently the slough was necessarily part of the lake. *In the Matter of the Determination of the Ordinary High-water Mark and Outlet Elevation for Beaver Lake* (So. Dak. Sixth Judicial Circuit, 1990) Civ. No. 89-94. The regulations of the U.S. Army Corps of Engineers contain a definition of a lake. 33 C.F.R. § 323.2(b). In part, the regulations describe a lake as a standing body of open water occurring in a natural depression fed by a stream and from which a stream may flow, that occurs due to widening or blockage of a river or stream or is located in an isolated natural depression. *Ibid.*

61. If the water body in question was determined to be a river, the state would own to the ordinary low-water mark; if the water body was determined to be a lake, the state would own to the ordinary high-water mark. *State v. Placid Oil Company* (La. 1973) 300 So.2d 154, 157, *cert.den.* 419 U.S. 1110, *reh. den.* (1975) 420 U.S. 956. In addition, in Louisiana, the doctrine of accretion was applicable to rivers, but not to lakes. This distinction was important, as the upland owner would not own the uncovered bed if the body of water was characterized as a lake. *Id.* at 158. On the answer to these questions depended many millions of dollars in oil revenues. *Id.* at 172.

62. *Id.* at 160–162.

63. *Id.* at 162, 172–173.

64. *Id.* at 175. The court considered the water body's size, its width as compared with the streams that entered it, its depth, its banks, its channel, its current as compared with the current of the streams that entered it, and its characterization in historical documents. *Ibid.* In *Placid Oil*, the water body was, in addition to its longtime designation as a lake on official maps, about 30 miles long and 3 to 10 miles wide; it had a channel depth of about 8 feet; it was about 20 times wider than the river that entered it, and the current in it was substantially reduced. *Ibid.* Another court stated, "A lake is differentiated from a water course only in that it is simply an enlarged water course wherein . . . the waters are quiescent." *Roberts v. Taylor* (N.D. 1921) 181 N.W. 622, 625.

Not in every case will the lawyer or the surveyor have to use this knowledge. Indeed, in most cases, the character of the water body as a lake is usually self-evident.[65] This is not to say that all lakes are the same. Nature is not that neat and tidy.

The major difference between lakes is the existence or nonexistence of an outlet from the lake and the type of surrounding terrain. Some lakes in the West, Mono Lake and the Great Salt Lake, for example, are terminal lakes; these lakes have no outlet. All water that flows into the lake remains within the confines of the lakebed until it evaporates or is lost through underground seepage. In addition, the terrain and vegetation surrounding these lakes is desertic.

Do such physical differences in the regime of and terrain surrounding terminal lakes make a difference in the physical indicia of the legal property boundary, as compared with lakes that have an outlet and are surrounded by forests or grass plains? Are such distinctions significant in determining the physical location of the property boundary of littoral properties surrounding the lake when there are shoreline changes reflecting a change in the lake's regime? These questions will be discussed in order.

8.6 LOCATION OF THE ORDINARY HIGH-WATER MARK PROPERTY BOUNDARY OF LITTORAL LANDS

8.6.1 In General

The property boundary of lands adjacent to navigable lakes, the beds of which are owned by the states, is, in most cases,[66] the ordinary high-water mark.[67]

---

65. There are some exceptional cases in which appearances were deceiving. In *Carr v. Moore* (Iowa 1903) 93 N.W. 52, 53, deciding whether there was a lake present was necessary to determine whether the principles of accretion or reliction applied. Although federal surveyors had meandered a body of water, this fact alone was not determinative. *Id.* at 54. In finding that there was not a lake present (*Ibid.*), the court relied on testimony that, among other matters, showed the following: the area where a "lake' was supposedly located was swampy in character; there was no definite shoreline; the depth of water (except in one or two locations) did not exceed 5 to 6 feet; the water body was shallower in the center than on the sides; and there was no subterranean source of water and no definite inlet. *Id.* at 53. In another case, a particular area that had been surveyed as land, sectionalized, and subdivided, and was covered with trees and undergrowth was not held to be part of a lake. *State v. Parker* (Ark. 1918) 200 S.W. 1014, 1018, *cert. den.* (1918) 247 U.S. 512.
66. See note 30 in this chapter.
67. E.g., *Illinois Central, supra*, 146 U.S. at 435–437; *Determining Natural Ordinary Highwater Level* (Minn. 1986) 384 N.W.2d 510, 516 (boundary of public waters); *State of Wisconsin v. Trudeau, supra*, 408 N.W.2d at 341; *Martin, supra*, 112 So. at 283; *Miami Corporation v. State* (La. 1937) 173 So. 315, 325, *cert. den.* (1937) 302 U.S. 700; *State v. McFarren* (1974) 215 N.W.2d 459, 463; *Provo City, supra*, 176 P.2d at 136 (Larsen, C.J., dissenting); *Callahan v. Price* (Idaho 1915) 146 P. 732, 734; *Flisrand, supra*, 152 N.W. at 800–801; see *Cal. ex rel. State Land Com'n, supra*, 805 F.2d at 859.

In the case of lakes, defining the physical location of this property boundary appears, at least at first impression, to be relatively simple. Considering for example a lake surrounded by evergreen trees and summer cottages, one could readily see where the evergreen trees stopped and the rocky and sandy shore began. Thus, there appears little need to spend a great deal of time in discussing the geographic location of the ordinary high-water mark. The physical location of that boundary should be apparent.

Although, after reaching this point in the book, the reader should be familiar with the litany, it bears one last incantation. The ordinary high-water mark is a legal boundary describing no particular geographic or physical location.[68] In stark contrast to the example above envision the difficulty in locating the ordinary high-water mark of a lake in the arid regions of the western United States. There are no evergreen trees or lawns that so conveniently cease at a rocky or sandy beach, only intermittent sagebrush and endless sand or rock.

And even in the case of the evergreen-surrounded lake exactly where is the ordinary high-water mark? Is it at the line where the evergreens stop growing or along the lower portion of beach at the edge of the water? What happens when the lake level fluctuates seasonally? Should that line be determined in the fall or in the spring? And where is the physical location of the boundary in the many marshy areas that lie adjacent to lakes?

8.6.2 Value of Meander Lines

At the outset, one would do well to put aside the potential source of confusion caused by the existence of federal public land surveys. Those surveys delineate and give apparent official government endorsement to what appears to the untutored observer to be an exactly defined "boundary" of the lake and the littoral lands surrounding the lake. Don't be misled! While useful, that seemingly precisely portrayed line is not the property boundary line of the littoral lands or the lakebed.[69]

Recall that in determining the extent of public lands, government surveys, until recently, meandered the lake margin or shore.[70] Now United States sur-

---

68. See *Borough of Ford City v. United States* (3rd Cir. 1965) 345 F.2d 645, 649, 650, *cert.den.* (1965) 382 U.S. 902.
69. In one case, the meander line meant even less. In *Carr v. Moore, supra*, 93 N.W. at 54, although federal surveyors had meandered a body of water, the court found this meander line was not determinative of whether there was a lake present as a matter of physical fact. Another court has observed: "[W]e note that a water line, rather than a meander line, ordinarily forms the boundary of a tract of land abutting a navigable body of water." *Matter of Ownership of Bed of Devils Lake, supra*, 423 N.W.2d at 143; see notes 44 through 65 and accompanying text in Chapter 3.
70. Circular from General Land Office to Surveyor General (1831) as reprinted in Minnick, *A Collection of Original Instructions to the Surveyors of the Public Lands, 1815-1881* (Landmark) p. 11; General Instructions (1834) to Deputy Surveyors in Illinois and Missouri as reprinted in

veyors are instructed to meander the lakeshore at "mean high-water elevation."[71] Exactly what was meant by the terms "shore,"[72] "bank,"[73] or "mean high-water elevation"[74] has been left to the fertile imagination of the government surveyor. As a result, the location of meander lines in relation to the actual location of the lake boundary has been somewhat random.

The geographical location of the meander line could depend, at least in part, on the time of year the United States surveyor happened to be at a particular site or, for example, just how intrepid that surveyor was when faced with locating a meander corner in a quagmire. In fact, at least one court has found a lake meander line to be an "extremely" poor and inaccurate approximation of the ordinary high-water mark.[75] The uncertainty and randomness of location of meander lines have given rise to the time-honored rule that a meander line is not a property boundary; it is only an approximation of the sinuosities of the shore for the purposes of estimating the amount of public

---

Id. at p. 83; General Instructions to Deputy Surveyors from Office of the Surveyor General, Territory of Florida (1842) as reprinted in *Id.* at 129 (lakes and ponds of sufficient magnitude are to be meandered); General Instructions of 1846 (Wisconsin and Iowa) as reprinted in *Id.* at 156 (meander lakes of 40 acres or more in size that cannot be drained or are not likely to fill up); General Instructions to His Deputies by the Surveyor General of the United States for the States of Ohio, Indiana and Michigan (1850) as reprinted in *Id.* at 174 (meander lakes of 40 acres or more in size); Instructions to Surveyor of Public Lands in Oregon (1851) reprinted in *Id.* at 251 (meander lakes of 25 acres or more in size); Instructions to the Surveyors General of Public Lands of the United States (1855) reprinted in *Id.* at 366 (meander lakes of 25 acres or more in size). There is an excellent narrative description of how a lake meander line was constructed by public land surveyors in *Utah v. United States*, Special Master's Report II, as reprinted in 1976 Utah L.Rev. 245, 262–263.

71. Manual of Instructions for the Survey of the Public Lands of the United States, 1973 (Landmark reprint), §3-115, p. 93 (meander lakes of 50 acres or more in size); *Utah State Road Com'n v. Hardy Salt, supra*, 486 P.2d at 392.

72. In the case of lakes, "shore" has been defined as that portion of the bank that touches the margin of the lake at low- or high-water (depending on whether one is in a high- or low-water state). See *City of Peoria v. Central Nat. Bank* (Ill. 1906) 79 N.E. 296, 298–300; *Freeman v. Bellegarde* (Cal. 1895) 108 Cal. 179, 187; *Noyes v. Collins* (Iowa 1894) 61 N.W. 250 (natural shore).

73. Bank" has been defined as the equivalent of the ordinary high-water mark. *Morrison v. First Nat. Bank of Skowhegan* (Me. 1895) 33 A. 782, 783. Another case more broadly construed the term "bank" to mean an elevation of land confining the waters in their natural channel when the waters rise to their highest level but do not overflow the banks. *State v. Faudre* (W.Va. 1903) 46 S.E. 269, 270.

74. The 1973 Manual of Instructions provides slim guidance as to what is meant by the phrase "mean high-water elevation." The Manual notes that all inland bodies of water will go through annual cyclical changes, ". . . between the extremes of which will be found mean high-water." Manual of Instructions for the Survey of the Public Lands of the United States, 1973 (Landmark reprint), §3-116. The many unanswered questions include the following: Whether a visual "mean" of the lake levels was intended and, if not, what is the length of time (weeks, months, years) over which the mean should be determined? Should hourly, daily, weekly, or monthly gauge readings comprise the record? What is the limnological or climatological basis for such a procedure?

75. *Matter of Ownership of Bed of Devils Lake, supra*, 423 N.W.2d at 142.

land embraced in a particular survey; the lake itself forms the property boundary.[76]

There are, however, some significant deviations from this rule. In one case a federal court adopted the meander line as the location of the ordinary high-water mark.[77] The court held that when the United States approves a meander line public land survey, the United States assumes a heavy burden of proof to show that the survey was fraudulently procured or grossly erroneously approved.[78]

In another case a high-water mark could not be ascertained because of physical circumstances peculiar to a lake.[79] As a consequence, it was specifically found that some other basis had to be devised to determine the property boundary. In that case, the meander line was adopted as the geographic location of the boundary.[80]

It has also been successfully argued that an official meander line, surveyed for the purpose of ascertaining, locating, and establishing the line between lands underlying a navigable lake and adjacent lands, should be regarded as the true location of ordinary high-water mark, unless impeached by fraud.[81]

---

76. *Hardin, supra,* 140 U.S. at 380; *Niles v. Cedar Point Club* (1899) 175 U.S. 300, 308. In litigation concerning the location of the boundary of the Great Salt Lake, the United States stipulated: "The boundary of a navigable body of water is represented on the public land surveys by a surveyed meander line. In surveying, a meander line represents an approximation of the ordinary high-water mark. The legal measure of the boundary of a body of water is the high water line, which is located at the ordinary high-water mark." *Utah v. United States, supra,* 1976 Utah L.Rev. at 255; *Matter of Ownership of Bed of Devils Lake, supra,* 423 N.W.2d at 143.
77. *United States v. Otley, supra,* 127 F.2d at 999–1000.
78. *United States v. Otley, supra,* 127 F.2d at 995–996. The contending party did not show that there was an unusual amount of land between the location of the meander line and the shore of the lake. *Id.* at 998. In addition, that party did not establish that, in a particular marshy area, the area between the meander line and the shore was disproportionate to the acreages of the patents bounded by the meander line in that shoreline reach. *Id.* at 1000. But compare *Otley* with *Live Stock, supra,* 185 U.S. 47. See notes 66 through 79 and accompanying text in Chapter 3.
79. The high-water mark ordinarily is indicated by a line of vegetation or wave-action erosion line formed when the water level reaches its ordinary high cycle for the year. Neither a vegetation nor an erosion line can be identified on the shores of the Great Salt Lake. The exceedingly high salt content of the Lake accounts for absence of a vegetation line, and various other factors, such as the flat shorelands, account for the lack of erosion line." *Utah v. United States, supra,* 1976 Utah L.Rev. at 255-256; *Utah State Road Com'n v. Hardy Salt, supra,* 486 P.2d at 392.
80. *Utah v. United States, supra,* 1976 Utah L.Rev. at 295–305. In summary, the Special Master found (and the United States Supreme Court agreed) that, among other matters, the surveyed meander line approximated the ordinary high-water mark; the United States did not seek to change that approximation; Utah acted as if it had complete sovereignty landward of the meander line; the United States recognized this action by Utah; the meander line was recognized as the boundary of public lands on official U.S. township plats, and the meander line was considered as the boundary between the state and the nation by Congress.
81. *Martin, supra,* 112 So. 283–284. In describing the line surveyed, the surveyor testified:

> ... that "there are marks and monuments still in existence by which you could testify to the ordinary water mark of the lake prior to the drainage operations," and that "the old rim of the lake, or the line at which the water of the lake usually stood before affected by the drainage operations of the state, were then and are still indicated by lines of trees

And there are also some cases in which meander lines have been treated as boundaries.[82] At the very least, it can be argued that meander lines should be considered as evidence of the location of the ordinary high-water mark.[83]

Notwithstanding such cases and although it has been forcefully argued and sometimes held otherwise, government meander lines normally will not be treated as the physical location of the property boundary of littoral lands.[84] Thus, government surveys of the public domain do not provide a definitive answer. Perhaps judicial opinions in littoral property boundary cases containing discussions of the physical location of the ordinary high-water mark property boundary will be instructive.

8.6.3 Physical Indicia of the Ordinary High-Water Mark as Described by the Courts—In General

Courts have provided no clear and unmistakable instructions or guidelines by which one can decide exactly what physical indicia comprise the ordinary high-water mark property boundary or where exactly along the littoral shoreline the ordinary high-water mark property boundary should be physically

---

growing about high-water mark, . . . also by the more pronounced slope at the margin of the lake, where the level lands . . . sloped through the shore to the bottom of the lake, forming the marginal line between the flats of the lake bottom and the level lands . . . , and also by the change of character of soil, which formed the land . . . above the water of the lake, which soil was usually muck, and the soil forming the shore and bottom of the lake, which soil was usually sand"; that the state survey of the water line of the lake "was established for the purpose of marking, as nearly as practicable, the line of the lake at ordinary water level or for showing the original shore of the lake." *Id.* at 282.

82. When a meander line has been located through fraud or error at a location where a body of water does not exist, courts will treat the meander line as a boundary, but no riparian or littoral rights will be created. *Niles, supra,* 175 U.S. at 307; *Live Stock, supra,* 185 U.S. at 53–54. For an unsuccessful attempt by the United States to show that there was an erroneous meander survey, see *United States v. Otley, supra,* 127 F.2d at 996–1000. See notes 72 through 79 and accompanying text in Chapter 3.

83. *Martin, supra,* 112 So. 283–284; *Provo City, supra,* 176 P.2d at 139 (Larsen, C.J., dissenting).

84. *Provo City, supra,* 176 P.2d at 132; *Matter of Ownership of Bed of Devils Lake, supra,* 423 N.W.2d at 142. There is an interesting twist to this rule provided by a Washington State court. In *Mood v. Banchero* (Wash. 1966) 410 P.2d 776, 777, party A's property boundary was along a government section line that did not mention the existence of a water body. In the survey of the adjoining northern section, the surveyor noted existence of a lake that extended south almost to the section line. *Ibid.* Eventually the lake rose and intruded across the section line; Party A used the lake as if a littoral owner. *Ibid.* After a number of years the lake level was lowered and the shoreline receded. *Ibid.* Party A claimed the property boundary followed the lake recession and that the government meander line survey was evidence of the actual high-water line. *Id.* at 779. The court found that the meander line survey ". . . may be considered as evidence of the actual high-water lines, as they existed at that time, [but] they are not conclusive as to the actual waterline as it existed then or later." *Ibid.* More important, the court noted that this rule concerned the establishment of the boundary of one's own land. Thus, the rule had no application to Party A, as the nonmeander line character of Party A's section line boundary land was undisputed. *Ibid.*

located. For example, in a case decided in the latter part of the nineteenth century defining the limits of state lakebed ownership in California, the California Supreme Court grandly pronounced that state sovereign title extended to the entire lakebed.[85] For those who deal solely in abstractions, that pronouncement would be perfectly fine. The more interested investigator or the littoral property owner may insist on knowing just what, exactly, is the true physical extent of the lakebed. Where, exactly, does the "lakebed" end and the "upland" begin?

Anticipating this question, the opinion only added to the problem by enigmatically describing the lakebed as the land covered and uncovered by the ordinary rise and fall of the tide, stream, or lake.[86] It may be unfair to dwell on such an example, but only a little. This example accentuates the fact that attempts by courts and lawyers (aided and abetted by surveyors) to define the ordinary high-water mark property boundary of lakes in terms that are intended to describe a particular and consistently locatable physical feature have not entirely been successful.

In some respects, this lack of success is a reflection of the physical nature of lakes. Lakes are unlike tidal water bodies where there is an accepted,[87] relatively precise indicator that can be predictably calculated through statistical methods to determine the water level of the adjacent water body. Unlike the tides, the water level fluctuations of lakes can be predicated only in the grossest of terms.

For example, all that can be said of lakes not controlled by the works of humans is that in summer and fall, on the average, these lakes will be lower in elevation than they are in winter and spring.[88] This seasonal fluctuation varies with, among other matters, the precipitation in the lake's water-supplying drainage basin. Thus, predicting long-term lake fluctuations will

---

85. *Churchill Co. v. Kingsbury* (1918) 178 Cal. 554, 558. A later case shows that judicial treatment has not greatly improved over time. The bed of Clear Lake in California was described as the lake bottom under the high-water mark. *Lyon v. Western Title Ins. Co.* (1986) 178 Cal.App.3d 119, 120.
86. The court further increased the confusion by issuing a later "clarification" in the opinion. That court stated the lake consisted of a body of water confined within its banks as they existed at the stage of ordinary high-water. *Churchill, supra,* 178 Cal. at 559. The "clarification" raises many questions, such as: What physical feature comprises the "banks" of a lake? What is the "stage" of ordinary high-water? Does use of the word "stage" imply that use of water stage gauge information is appropriate to determine what is "ordinary high-water"? If so, does this imply that the values from stage gauges should be mathematically meaned to determine an "ordinary high" stage? On what scientific basis?
87. E.g., *Borax Ltd. v. City of Los Angeles* (1935) 296 U.S. 10, 22–26.
88. E.g., *United States v. Otley, supra,* 127 F.2d at 999 ("It is a matter of judicial knowledge that the rainfall in the mountain area varies greatly from year to year. In a single year a hot thaw may supply snow water and raise the lake level and widen its boundaries, and a later freeze in the mountains thins their streams so that the lake boundaries recede, to widen again with succeeding warmer weather"); *Carpenter v. Board of Commissioners, supra,* 58 N.W. at 295; *Utah State Road Com'n v. Hardy Salt, supra,* 486 P.2d at 393.

probably be just as successful and accurate as predicting long-term weather patterns.[89] And in the case of nonterminal lakes with outlets, there is a limit on how extreme high-water can become in a particular year; the water can only reach the level of the lake's outlet before it is released. As a consequence, it should be understood that the highs and lows of these nonterminal lakes do not parallel weather patterns in the same manner or extent as do the highs and lows of terminal lakes.

8.7 VEGETATION/EROSION LINE TEST

Despite this somewhat grim picture, one should not be discouraged. Courts have supplied some guidance. This is true even though, except in the case of federal public lands bounded by tidal waters, the United States Supreme Court has not given the term "ordinary high-water mark" any precise meaning.[90]

It has been urged that federal cases concerning non-tidal rivers[91] lend considerable weight to a contention that the ordinary high-water mark property boundary of a lake should be physically located with reference to what are known as the complementary vegetation line or erosion line tests. Those tests are usually expressed in the following terms: The ordinary high-water mark is the margin of the land over which the waters have visibly asserted their dominion or the mark impressed on the soil by the effect of water covering it for a sufficient period to deprive the land of vegetation and to destroy its value for agriculture purposes.[92] Some state courts have adopted and combined these tests and have defined the ordinary high-water mark as the ". . . outer line or limit of the lake bed, and the lake bed is that body of land which the water occupies or covers sufficiently long or continuously to denude of ordinary vegetation and makes unfit for agricultural purposes."[93] Some states

---

89. In one case, a court rejected the use of rainfall tables to predict water levels of a lake as without foundation and not proper expert testimony. *State v. Bollenbach, supra,* 63 N.W.2d at 285.
90. See e.g., *Borough of Ford City, supra,* 345 F.2d at 649, 650; *United States v. Cameron* (M.D. Fla. 1978) 466 F.Supp. 1099, 1011. In the limited case of tidelands, the Supreme Court has accepted a particular scientifically determinable and calculable physical location of the ordinary high-water mark, the mean high-water line. *Borax, Ltd. v. Los Angeles, supra,* 296 U.S. at 22–26. No similar expression is found in any case concerning navigable but nontidal rivers or lakes. See notes 170 through 172 and accompanying text in Chapter 7.
91. E.g., *Howard v. Ingersoll* (1851) 54 U.S. (13 How.) 380, 415; *Oklahoma v. Texas* (1922) 260 U.S. 606, 632; *Borough of Ford City, supra,* 345 F.2d at 648.
92. *Ibid.; Harrison v. Fite* (8th Cir. 1906) 148 F. 781, 783. On the complementary nature of these tests, see *Borough of Ford City, supra,* 345 F.2d at 648.
93. *State v. Thomas, supra,* 155 N.W. at 861; *Diana Shooting Club v. Husting* (Wisc. 1914) 145 N.W. 816, 819. By way of further example, the Wisconsin Supreme Court combined both tests in stating ". . . the ordinary high-water mark is . . . the point on the bank or shore up to which the presence and action of the water is so continuous as to leave a distinct mark either by erosion, the destruction of terrestrial vegetation or other easily recognized characteristic." *Id.* at 820. Of

have even adopted statutes defining the ordinary high-water mark that also combine the two tests.[94] The U.S. Army Corps of Engineers incorporates this test in its regulations regarding the extent of its jurisdiction in dredging and filling in waters of the United States under section 404 of the Clean Water Act.[95]

Particular locations along the lakeshore where it is not possible to locate the ordinary high-water mark by reference to the vegetation/erosion line test may also be accounted for. Thus, in marshy areas or areas that have been disturbed, courts have permitted extrapolation of the elevation of the described physical feature from one area to another.[96]

While of some value in situations concerning particular types of lakes and surrounding lake terrain, these complementary tests have some major failings.

---

course, what is a "characteristic" "easily recognized" by one person may not be the same "characteristic" "easily recognized" by another perhaps less (or more) observant person. Other cases are *Rutten v. State* (N.D. 1958) 93 N.W.2d 796, 799; *Matter of Ownership of Bed of Devils Lake, supra*, 423 N.W.2d at 144–145; *State v. McFarren, supra*, 215 N.W.2d at 463; *Carpenter v. Board of Commissioners, supra*, 58 N.W. at 297; *Provo City, supra*, 176 P.2d at 132.

94. In Minnesota the ordinary high-water level is

 . . . the boundary of public waters and wetlands, and shall be an elevation delineating the highest water level which has been maintained for a sufficient period of time to leave evidence upon the landscape, commonly that point where the natural vegetation changes from predominantly aquatic to predominantly terrestrial. *Determining Natural Ordinary High-water Level, supra*, 384 N.W.2d at 516 (quoting Minn. Stat. § 105.37, subd. 16 (1984)).

In South Dakota the ordinary high-water mark is defined as

 . . . the high level reached by the waters of a lake under ordinary and continuous conditions, unaffected by periods of extreme and periodic freshets. The ordinary high-water mark is indicated by the continuous presence and action of water which leaves a distinct mark either by erosion, destruction of terrestrial vegetation, or some other easily recognized characteristic . . . " *S.D. Wildlife Federation, supra*, 382 N.W.2d at 27 (quoting SDCL 43-17-20(2)).

Compare note 93 in this chapter.

95. 33 C.F.R. § 328.3(e) provides as follows: "The term ordinary high-water mark means that line on the shore established by the fluctuations of water and indicated by physical characteristics such as clear, natural line impressed on the bank, shelving, changes in the character of soil, destruction of terrestrial vegetation, the presence of litter and debris, or other appropriate means that consider the characteristics of the surrounding areas."

96. In *Diana Shooting Club*, the Wisconsin Supreme Court stated that mark would be defined by erosion, destruction of terrestrial vegetation, or "other easily recognized characteristic." *Diana Shooting Club, supra*, 145 N.W. at 820. ". . . [W]here the bank or shore at any particular place is of such character that it is impossible or difficult to ascertain where the point of ordinary high-water mark is, recourse may be had to other places on the bank or shore of the same stream or lake to determine whether a given stage of water is above or below ordinary high-water mark." *Ibid.*; *State of Wisconsin v. Trudeau, supra*, 408 N.W.2d at 342; *Borough of Ford City, supra*, 345 F.2d at 348.

8.7.1 Problems with Vegetation/Erosion Test in the Case of Lakes

First, although the use of the term "vegetation or erosion line test" lends itself to the implication there is only one all-purpose set of physical attributes or features defining the shores of lakes, that is not the case. Even widespread judicial sanction of a line of vegetation or some particular topographic feature, such as an erosion line, as the sine qua non of the physical location of the ordinary high-water mark does not necessarily mean that each court choosing to adopt that test will consistently and predictably chose the same physical features or attributes as did earlier courts.[97] Basically, these different expressions or variants of the vegetation/erosion line tests are merely ill-disguised versions of the "I know it when I see it" rule.[98]

Inconsistency and lack of predictability are not the only failings of the vegetation/erosion line rule. States that adopted these tests for locating the property boundary of navigable lakes seem to have done so either as a result of the conditions peculiar to the geography and geomorphology of a particular region and its surroundings or without giving close consideration to the particular or peculiar geography and geomorphology of the individual lake and its surrounding terrain.[99] Thus, many cases using the vegetation/erosion line rule are found in the midwestern Lake States such as Minnesota and Wisconsin. In these states, lakes are, more often than not, bounded by forests or grass plains and have fairly steep shorelines.[100] It is also apparent that the assumption on which many of these midwestern cases were decided, either

---

97. A comparison of the expressions of the vegetation or topographical tests by the courts will be instructive. *Borough of Ford City, supra*, 345 F.2d at 648, identifies the ordinary high-water mark as a clear, natural line impressed on the bank, as shown by erosion, shelving, soil characteristics and litter, as complemented by the vegetation test, described as the line below which the waters have so visibly asserted their dominion that terrestrial plant life ceases to grow and the value of the land for agricultural purposes is destroyed. On the other hand, another federal court cryptically identified the ordinary high-water mark as a natural physical characteristic placed on the lands by the action of the river in its ordinary flow. *United States v. Claridge* (D. Ariz. 1966) 279 F.Supp. 87, 91, *aff'd* (9th Cir. 1967) 416 F.2d 933, *cert. den.* (1970) 397 U.S. 961. A Florida case concerning a lake defined the ordinary high-water mark as marks on the ground or on local objects that are more or less permanent. *Martin, supra*, 112 So. at 283.
98. See note 22 and accompanying text in Chapter 3.
99. See note 119 in this chapter.
100. *State of Wisconsin v. Trudeau, supra*, 408 N.W.2d at 344. Current instructions for the survey of public lands make the same assumption—there will be a definite feature to locate. The Manual of Instructions for the Survey of the Public Lands of the United States, 1973 (Landmark reprint), §3-116 notes that in forested areas all timber growth normally ceases at the margin of permanent water. The instructions also note that where the shoreline bordering the lake is relatively flat, the most reliable indication of mean high-water elevation is a mark made on the soil by the water. In timbered localities, a very certain indication is found in the belting of native species that are progressively more (or less) overflow tolerant. In one case that adopted these tests, the seasonal variance of the lake was a mere 6 feet. *Carpenter v. Board of Commissioners, supra*, 58 N.W. at 296. No mention was made of any significant land exposure or submergence as the lake fluctuated

expressly or implicitly, was that the water body to which the test was applied was relatively stable and permanent in size and water level.[101] The contrast with some lakes in the western United States is dramatic.[102]

Certainly, using a vegetation/erosion line test as an indicator of the ordinary lake level presumes that the water level is sufficiently stable to "impress" itself on the ground or to "render" an area valueless for agriculture. The basic assumption of this test is that although the level of the lake will fluctuate somewhat over the course of the year as the result of various factors, the low and high levels for one year will be about the same as they were for any other year.[103] This assumption accounts for the phrase "ordinary high-water" in the term "ordinary high-water mark."

By the same token, it has also been accepted by courts employing the vegetation/erosion line tests that a particular topographical feature will evidence the lake's constant return to a particular level after recession. This assumption is based on the theory that a stable water level forms the topographical feature.[104] This is the "mark" in the term "ordinary high-water mark." It seems all so neat and tidy. This standard, however, does not travel well. Nature refuses to be confined within the straitjacket of these lawyer-made and court-approved assumptions.

Although cited by some western state courts to support the physical location of the ordinary high-water mark, the vegetation/erosion line test is not well adapted to determination of lake boundaries in the more arid western states.[105] In those areas there are many lakes in which there are substantial,

---

in level in that range. Other Midwestern cases are equally silent on the shoreline exposure effect of lake level fluctuations. *Determining Natural Ordinary High-water Level, supra,* 384 N.W.2d at 512; *S.D. Wildlife Federation, supra,* 382 N.W.2d at 27, 29.

101. *Determining Natural High-water Level, supra,* 384 N.W.2d at 517; *S.D. Wildlife Federation, supra,* 382 N.W.2d at 31 (". . . the ordinary high-water mark will be set at the level where there is a distinct mark which evidences erosion, *and* changes in the character of the soil, *and* destruction of terrestrial vegetation, which have occurred under the *ordinary* and *continuous* conditions of the lake . . .") (emphasis in original).

102. The shoreline of one western lake was so flat that a 1-foot rise or fall in elevation would submerge or expose 11,000 acres of land. *United States v. Otley, supra,* 127 F.2d at 999. At Mono Lake, in California, a 37-foot vertical recession exposed 12,000 acres. *Cal. ex rel. State Lands Com'n, supra,* 805 F.2d at 861. At the Great Salt Lake, a recession of 2 vertical feet exposed more than 140,000 acres of land! *Utah State Road Com'n v. Hardy Salt, supra,* 486 P.2d at 393.

103. E.g., *Determining Natural High-water Level, supra,* 384 N.W.2d at 517; *S.D. Wildlife Federation, supra,* 382 N.W.2d at 31; *Carpenter v. Board of Commissioners, supra,* 58 N.W. at 816; see *United States v. Chicago B&Q R. Co.* (7th Cir.) 90 F.2d 161, 170, *cert. den.* (1937) 302 U.S. 714 (action of water is "constant"); see *United States v. Claridge, supra,* 279 F.Supp. at 91 (action of river in its ordinary flow); *Harrison, supra,* 148 F. at 783 (soil is "usually covered"). See also *Oklahoma v. Texas, supra,* 260 U.S. at 632.

104. *Ibid.*

105. *Provo City, supra,* 176 P.2d at 138 (Larsen, C.J., dissenting); *Utah v. United States, supra,* 1976 Utah L.Rev. at 255–256.

even wild, fluctuations in lake level as a consequence of, among other matters, weather patterns.[106] Hence, one year's lake level is no indicator of what the past level of the lake may have been or what the future lake level may be. And when these lakes are terminal lakes, with no outlet, there is no reasonable likelihood that the high-water level will return to the same elevation from one year to the next, making the term "ordinary high"-water mark virtually meaningless.[107]

Further, many of these lakes lie in relatively dry regions, areas where there may not be any vegetation, only sand and sagebrush. Not only is there no "ordinary high-water" as that term has been given content by various court decisions, but there is no "mark." The vegetative and topographical features are just not available or reliable. Vegetation does not grow or is reflective of only one year's moisture. And topographic features or indicators may have relevance only in a geological time frame.

Thus, the cases using the vegetation/erosion line standard provide only general guidance in locating the ordinary high-water mark on lakes in arid regions that are subject to striking variations in water level in both the short and long term.[108] Are there any other tests that have been recognized by the courts to assist lawyers, surveyors, and others in attempting to determine the location of the ordinary high-water mark property boundary in this type of geophysical situation?

---

106. *Idaho For. Indus., supra,* 733 P.2d at 740 (case sent back for determination of ordinary high-water mark where there was a difference of 23 feet between observed highest and lowest levels of lake over a 30-year period); *Utah State Road Com'n v. Hardy Salt, supra,* 486 P.2d at 393 (lakes without outlet more susceptible to effect of climatic changes); see note 102 in this chapter.

107. See *Utah State Road Com'n v. Hardy Salt, supra,* 486 P.2d at 393.

108. One of the best explanations of the defects in applying the vegetation/topographical tests to ever varying water bodies is the one tendered by the United States Department of the Interior in an opinion regarding the boundary of the Great Salt Lake.

> ". . . [C]ustomary methods of determining the high-water mark . . . are not capable of application to the Great Salt Lake. The principle embodied in the [Manual of Instructions for the Survey of the Public Lands of the United States]. . . is that the annual flux and reflux of a lake carves upon its shores guidelines to the location of the mean high-water mark. This is based on the assumption that each year's cycle is repeated with the same range; that the low and high-water levels for one year will be about the same low and high-water levels for any other year, and that marks on the ground will result from, and reflect, the lake's constantly receding from, and returning to, the same levels. But this assumption, as we have seen, is not valid for Great Salt Lake. Similarly, the use of vegetation as a guide to determining the mean high-water mark is applicable only in situations where the chief deterrent to the growth of vegetation is the presence and action of water against a shore, whereas the absence of vegetation on hundreds of square miles of land adjacent to Great Salt Lake is due to other reasons, not connected with the location of the shoreline in historic times. *State of Utah* (1963) 70 Int. Dec. 27, 62.

8.7.2 Other Court-Approved Indicia of the Ordinary High-Water Mark in the Case of Lakes

There are some cases that have considered the location of the ordinary high-water mark property boundary in the case of such lakes and have not used the vegetation/erosion line tests. These cases have rejected the traditional tests because of the physical conditions and character of the lakes concerned.[109] Instead, these courts have adopted a variety of approaches to resolve the geographic location of the upland/lakebed ordinary high-water mark property boundary.

One court adopted the meander line as the location of the ordinary high-water mark.[110] In that case it was asserted that the meander line was faulty because it did not accurately depict the ordinary high-water mark at the time the meander line survey was accomplished.[111] The court determined, in light of the fluctuating nature of the lake level, that the surveyor had accurately meandered the lake.[112] In so holding, the court defined the ordinary high-water mark in terms of the mean or average of the lake's level.[113] This decision gives support to those courts that, as in some river cases,[114] also used stage or gauge reading averages to determine the elevation of the "ordinary high-water" of lakes.[115]

Nonetheless, use of stage or gauge readings to determine an average or mean lake level raises certain questions. It has been said that the statistical averaging of lake stage data, unlike statistical averaging of long-term tidal records, lacks a scientifically verifiable relationship with the physical regime of the water body.[116] For example, how many years should be averaged?

---

109. *Rutten, supra*, 93 N.W.2d at 797, 799 (lake had no outlet); *United States v. Otley, supra*, 127 F.2d at 999–1000; *Utah v. United States, supra*, 1976 Utah L.Rev. at 10–11.
110. *United States v. Otley, supra*, 127 F.2d at 999–1000.
111. *Id*. at 999.
112. According to the court, the surveyed meander line represented an average of the yearly fluctuating lake levels and was found to be an accurate representation of the ordinary high-water mark. *Id*. at 999–1000.
113. Customary water mark of any lake fluctuates with the waters of its supplying rivers and streams and with its evaporation. "The 'ordinary high-water mark' is a mean or average of these fluctuating [l]ake's extreme rises." *Id*. at 1000. A 1-foot fluctuation in the lake level would expose or cover 11,000 acres of land. *Id*. at 999. Thus, averaging the fluctuations may have been the most equitable way of apportioning the huge gain or loss due to such relatively insignificant fluctuations.
114. E.g., *Conran v. Girvin* (Mo. 1960) 341 S.W.2d 75, 81; *Mammoth Gold Dredging Co. v. Forbes* (1940) 39 Cal.App.2d 739, 752 (average level of the water attained by the river in its annual seasonal flow); see notes 188 through 191 and accompanying text in Chapter 7.
115. *Payette Lakes Protective Ass'n, supra*, 189 P.2d at 1019; *Flisrand, supra*, 152 N.W. at 800–801 ("Neither high nor low-water mark means the highest or lowest point reached by the waters of a lake during periods of extreme and continued freshets, or periods of extreme and continued drought, but does mean the high and low points of variation of such waters under ordinary conditions, unaffected by either extreme").
116. See *Determining Natural Ordinary High-water Level, supra*, 384 N.W.2d at 517 (quoting from memorandum of Minnesota Commissioner of Natural Resources).

Should only certain years when the lake is above a certain level be included? At what time of year should the gauge readings be taken, or should readings be taken hourly for the whole year and then meaned? Indeed, what is a "year," and when does it begin or end? These questions, and many others yet unasked, have not been completely answered by the courts in these property boundary disputes.[117]

As a consequence, while stage or gauge information is extremely useful, courts have not seen fit to rely on it as the sole indicia of the location of the ordinary high-water mark. Some courts have used such information to support the adoption of the meander line as the location of the ordinary high-water mark.[118] While some other courts have rigidly adhered to the vegetation/erosion line test, even when its application was questionable, given the physical character of the lake and the surrounding shore,[119] other courts have not. These courts recognized that in reaching a decision on the location of the ordinary high-water mark, all sources of information should be explored as fully as possible.[120] It is the latter approach that should be recommended.

8.8 PHYSICAL INDICIA OF ORDINARY LOW-WATER MARK PROPERTY BOUNDARY OF LITTORAL LANDS

An even more difficult area than the determination of the physical location of the ordinary high-water mark is the determination of the physical location of the ordinary low-water mark. Until recently this topic could only briefly be discussed because of the paucity of authority and the very character of what was being determined.

---

117. In the case of dam-controlled lakes, courts have provided some answers. See notes 186 through 202 in this chapter.

118. *United States v. Otley*, *supra*, 127 F.2d at 999–1000.

119. The majority in *Provo City*, *supra*, 176 P.2d at 132, used the traditional vegetation/erosion line test. The majority opinion noted that the high-water mark is not determined by the average over a period of years of the highest levels that the lake's water reached each year. Nevertheless, the dissent noted that the traditional test was particularly inappropriate, given the physical character of the lake and the surrounding shorelands: "It is evident that the application of this old established vegetation rule is of only relative value in solving this case for, in territory as flat as this is in the vicinity of the lands involved in this action, we have no distinct or deciding mark. . . ." *Id.* at 138.

120. In commenting on the use of water level readings, one opinion said:

> "During any particular year, it is easy to make precise water level measurements, but without a reliable historical basis, those measurements are simply one-time observations and have a very limited utility for accurately predicting a lake's water level range over a period of years. The only way to avoid being fooled by nature is by collecting the greatest number of reliable measurements over the longest period of time possible before coming to a conclusion.'" *Determining Natural Ordinary High-water Level*, *supra*, 384 N.W.2d at 517 (quoting from memorandum of Minnesota Commissioner of Natural Resources).

Defining, describing, or locating the physical indicia of the ordinary low-water mark is much more difficult because of the very nature of that property boundary. By definition, the ordinary low-water mark must be covered with water most of the time. How else could the action of water on the land form the ordinary high-water mark? Further, if the water surface receded to a level so as to mark on the ground a physical feature and permit the establishment of vegetation, could it not be argued that this mark has now become the ordinary high-water mark? Definitional problems such as these are not solved by a review of court decisions that discuss, mainly in the context of rivers, the ordinary low-water mark.

In many states there are no cases that define the ordinary low-water mark.[121] Those courts that have considered the question in the context of nontidal water bodies have variously defined the term "ordinary low-water mark." For example, some courts have decided that the ordinary low-water mark is the lowest elevation to which water has receded at any time.[122] But other courts have held that the ordinary low-water mark is the lowest elevation the water reaches during a season that is not affected by droughts or floods.[123] One California case considered the issue in the context of the California statutory scheme.[124] That case, which decided a dispute about the ordinary low-water mark of Clear Lake, described the line as ". . . a 'mark' created during the time of 'low-water' which occurs with regularity, so that the land beyond it is 'below the water' in normal years . . . ," a mark set without reference to unusual conditions, such as droughts or floods.[125] As the California court noted:

> . . . persuasive authority and common sense support the conclusion that the "low-water mark," which forms the boundary here with the sovereign ownership of

---

121. Indeed, until *County of Lake v. Smith* (1991) 238 Cal.App.3d 214 decided the location of the ordinary low-water mark of Clear Lake there, were no such cases in California. *Id.* at 221; 43 Ops.Cal.Atty.Gen. 291, 295 (1964). There are, however, some cases that define the ordinary low-water mark of tidal waters. See notes 113 through 135 and accompanying text in Chapter 4.
122. *Paine Lumber Co. v. United States* (E.D. Wisc. 1893) 55 F. 854, 864 (charge to jury); *Joyce-Watkins Co. v. Industrial Comm'n* (Ill. 1927) 156 N.E. 346, 348; *Carpenter v. Ohio River Sand & Gravel Corp.* (W.Va. 1950) 60 S.E.2d 212, 215; *Union Sand & Gravel Co. v. Northcott* (W.Va. 1926) 135 S.E. 589, 593.
123. *State v. McFarren, supra*, 215 N.W.2d at 462–463; *Vermont v. New Hampshire* (1933) 289 U.S. 593, 619–620 (actions of Congress and state legislature); *Vermont v. New Hampshire* (1934) 290 U.S. 579, 580; *Kentucky Lumber Co. v. King* (Ky. 1901) 65 S.W. 156, 157; *Appeal of York Haven Water & Power Co.* (Pa. 1905) 62 A. 97, 98; *Flisrand, supra*, 152 N.W. at 801; *Slauson v. Goodrich Transp. Co.* (Wisc. 1897) 69 N.W. 990, 992. Some commentators agree. Maloney, "The Ordinary High-water Mark: Attempts at Settling an Unsettled Boundary Line" (1978) 13 Land and Water Law Rev. 465, 466, n.8.
124. In *County of Lake*, the court interpreted Civil Code sections 830 and 670 as being part of a complete ". . . specification of property rights in shorezone areas . . ." *County of Lake, supra*, 238 Cal.App.3d at 225.
125. *Id*, at 225–226.

"the land below the water," is not some extraordinary "lowest" level reached during a severe drought, but is a regular or ordinary level which property owners could use for practical purposes in setting a realistic boundary."[126]

Even when the definition of the ordinary low-water mark has been decided, the issue is not settled. Where (or how), as a physical matter, is this boundary to be located? Courts have conceived of various solutions. Some may claim that, in an analogy to the approach used with rivers,[127] where courts should permit the use of stage data to define the location of the ordinary low-water mark.[128] Thus, it can be argued that the use of lake level data to determine the ordinary low-water mark is appropriate.[129] For example, in a New England lake case one of the parties challenged the court's determination of the physical location of the ordinary low-water mark by use of the average of only the lowest levels reached by the lake during each of some 37 years (not including droughts).[130] It was held that this procedure was incorrect and that the trial court should have used a low-water level representing the arithmetic mean or average of *all* the daily water level readings below the mean lake level.[131]

As seems to be the case concerning the use of stage data to determine the ordinary high-water mark of navigable lakes, however, there does not appear to be any climatological or limnological reason for any particular set of lake stage heights to be averaged for any particular period of time to determine the ordinary low-water mark. Notwithstanding this scientific skepticism, one comes to realize that there is very little else that can be relied upon in deter-

---

126. *Id.* at 227–228.
127. In one midwestern state, the court examined the rationale for locating the ordinary low-water mark of a river in light of the riparian owner's ownership of upland to the "waters edge" and held that the boundary must be physically located in a manner to always ensure the right of the riparian owner to have access to the watercourse. *Conran, supra,* 341 S.W.2d at 80–81. The reasoning of this case can be argued to be equally applicable to lakeshore owners. *State v. Slotness* (Minn. 1971) 185 N.W.2d 530, 532 (main right of lakeshore owner is right of access to water in front of his or her land); *State v. Korrer, supra,* 148 N.W. at 621.
128. In *Conran, supra,* 341 S.W.2d at 81, the court adopted physical definition of the ordinary low-water mark water level as "mean low-water" on a certain river staging gauge. The court saw no material difference between the statistical mean low-water elevation derived from river stage data and the term "ordinary stages of low-water or the lowest ordinary stage of the water." *Ibid.* Some commentators support the use of such data. Maloney, *supra,* 13 Land & Water L.Rev. at 495.
129. See, *United States v. Otley, supra,* 127 F.2d at 1000. A California case agrees. *County of Lake, supra,* 238 Cal.App.3d at 232–233.
130. *State v. Cain, supra,* 236 A.2d at 503.
131. *Id.* at 503–504. In so holding the Vermont Supreme Court relied on a Massachusetts case that discussed determination of the ordinary low-water mark on a tidal water body. The court did not discuss the physical differences between a tidal regime and a lake regime. *Ibid.* This fact may blunt somewhat the authority of this case for use of the statistical approach to determination of the ordinary low-water mark. The one California court that considered the issue agreed with the Vermont Counts approach. *County of Lake, supra,* 238 Cal.App.3d at 233.

mining the ordinary low-water mark other than data concerning the height of the water in the lake. At least such an approach does provide some certainty to both public and private landowners. And, as previously noted, there is case authority concerning both rivers and lakes that seems to support such a technique.[132] Although the one California case that considered the issue did not use stage data in establishing the physical location of the ordinary low-water mark because of circumstances peculiar to the case,[133] such use was favorably considered.[134]

The court suggested the mean[135] or average level of low-water over the period for which records had regularly been kept, should be used.[136] The opinion, however, did not provide any guidance on how the lake stage data was statistically manipulated, the appropriate number of years of data to be used in such analysis, or the scientific basis for using any particular period of years.[137]

To reiterate, the rule that seems best adapted to the peculiar situation of western arid lakes is not to restrict the kinds of proof that may be offered to establish the physical location of the ordinary high- or low-water marks. Until there is a scientific basis for a court to adopt a particular physical feature, all available relevant indicia should be considered. Thus, the vegetation pattern, the type of vegetation, the water level measurement, or any combination thereof, should be considered in the determination of the physical location of the ordinary high- or ordinary low-water mark property boundary. Indeed, there is no case authority that would permit a judge to place greater reliance on or accord greater weight to a particular type of physical feature, pattern, or water level in determining the physical location of the littoral property boundary. All types of pertinent and competent evidence should be permitted

---

132. See notes 127 through 131 and accompanying text in this chapter.
133. *County of Lake, supra*, 238 Cal.App.3d at 235–237 (an "... agglomeration of longstanding administrative acquiescence and extensive private reliance on a prior unappealed trial court decree, which established the normal minimum elevation widely accepted as the lakeward boundary of littoral parcels on Clear Lake at Zero Rumsey").
134. *Id.* at 232–233.
135. The basis for meaning the data was: "Fixing the 'low-water mark' with reference to the mean or average low-water level comports with the proper interpretation of the word 'mark' as signifying a level to which the water tends to return, on average; it also avoids giving undue weight to droughts or floods, which will affect an average of many years only slightly." *County of Lake, supra*, 238 Cal.App.3d at 233. The opinion also noted that averaging the data would make the parties equally unhappy; it had "Solomonic rectitude." *Ibid.*
136. *County of Lake, supra*, 238 Cal.App.3d at 232. Records had been kept for more than 100 years.
137. The court, in discussing *Fogerty II*, noted that there are problems in determining how far back to go in gathering values for purpose of averaging. *Id.* at 233. In the Clear Lake case there were records that dated back to the time when Civil Code § 830 was adopted and did not "... reveal any marked bias in the level of the lake. . . ." *Ibid.* The court asserted argued that use of an average would "... avoid the possibility of excessive divergence between the fixed value for purposes of a fee, and the actual value in any specific year." *Ibid.* The opinion also observed that a mean value is not some abnormal or aberrational figure. *Ibid.*

to establish the physical location of the ordinary high- or ordinary low-water mark littoral property boundaries.[138]

8.9 PROPERTY BOUNDARY CONSEQUENCES OF LITTORAL SHORELINE CHANGES—IN GENERAL

The need to establish a physical location of either the ordinary high-water mark or the ordinary low-water mark of a lake arises when the geographic location of the property boundary of uplands adjacent to a lake must be determined. Such a decision is typically made necessary because of some change in the regime of the lake impacting the geographic location of its shoreline. In examining the basis for this determination, recall that littoral property boundaries are ambulatory; they change their geographic location as a lake naturally recedes or encroaches onto the upland, gradually and imperceptibly.[139] There are instances, however, when the property boundary of littoral lands has been argued to be fixed in geographic location.

When land is uncovered as a lake recedes, what happens to the geographic extent of the upland owner's title? Does that title increase in geographic extent as the lake recedes or, as a matter of law, is there no increase in the geographic extent of upland property ownership caused by lake recession? Does it matter what the cause of or how rapid the recession is in determining the property boundary effect of change in geographic location of the shoreline? At first blush, one case appears to provide some answers to those questions. We shall see, however, in light of other authority, that the case may raise many more questions than it answers.

8.10 THE MONO LAKE RECESSION CASE—APPLICATION OF THE SO-CALLED FEDERAL COMMON LAW RULE

A case concerning Mono Lake, the site of much significant California litigation,[140] considered the effect of diversion of Mono Lake's tributary streams

---

138. See note 120 in this chapter. Some types of evidence suggested are maps, historical data, testimony of knowledgeable residents or observers, expert testimony concerning the physical regime of the lake and the adjacent lands, and lake stage data. 43 Ops.Cal.Atty.Gen. at 299.
139. E.g., notes 106 through 123 and accompanying text in Chapter 3; *Cal. ex rel. State Lands Com'n, supra,* 805 F.2d at 864; *Wilbour v. Gallagher* (Wash. 1969) 462 P.2d 232, 238, *cert. den.* (1970) 400 U.S. 878; *Lakeside Boating & Bathing Inc., supra,* 344 N.W.2d at 220; Cal. Civ. Code § 1014; Clark 7th, *supra,* § 24.02; Clark 4th, *supra,* at § 567, p. 773. As one court observed: "The concept of an ambulatory ordinary high-water mark forming the boundary of land along a . . . navigable body of water is not contrary to [certain] cited decisions of the United States Supreme Court . . . [and] is not novel in this State [North Dakota]." *Matter of Ownership of Bed of Devils Lake, supra,* 423 N.W.2d at 143–144.
140. *City of Los Angeles v. Aitken, supra,* 10 Cal.App.2d 460; *National Audubon Society, supra,* 33 Cal.3d 419.

and its consequent recession on the property boundary between uplands owned by the United States and the lakebed owned by California.[141] California argued that state law should apply to determine this question.[142] The choice-of-law decision was important in the outcome of the case.

According to the arguments of the parties, there could have been a substantially different result, depending on whether state or federal law was applied in deciding the issue.[143] The federal circuit court decided that federal law controlled the decision and, as a consequence, the federal government, not the State of California, gained title to lands exposed as Mono Lake receded.[144]

Close examination of the court's decision is important. Such an examination will prove instructive in gauging how different courts have treated, and may in the future treat, the property boundary effect of lake recession. More to point, this analysis will disclose serious flaws in the opinion and the lack of authoritative support for its conclusion, which should call into question the opinion's precedential value and the weight it should be accorded in the future.

8.10.1 The So-Called Federal Law of Reliction

According to the circuit court, there was a "well-settled" body of federal law awarding land exposed by the natural and gradual recession of a lake's

---

141. *Cal. ex rel. State Lands Com'n, supra,* 805 F.2d at 859–860.
142. *Id.* at 861.
143. *Id.* at 860. Under state law, California claimed it would have obtained title to the lands exposed as Mono Lake receded. E.g., Cal. Pub. Res. Code § 7601; *Churchill, supra,* 178 Cal. at 559–560; *People v. Los Angeles* (Cal. 1950) 34 Cal.2d 695, 696; *Natural Soda Prod. Co. v. City of L.A.* (Cal. 1943) 23 Cal.2d 193, 196, cert.den. (1944) 321 U.S. 793; see *Carpenter v. City of Santa Monica* (Cal. 1944) 63 Cal.App.2d 772, 787. On the other hand, under the so-called federal common law of reliction, the United States argued that the exposed land would belong to the upland owner, the United States. See *California ex rel. State Lands Comm'n v. U.S.* (1982) 457 U.S. 273, 284 (relying on, among other cases, *County of St. Clair v. Lovingston* (1874) 90 U.S. (23 Wall.) 46, 68). California contended, however, even under federal common law, the reliction doctrine was inapplicable. See notes 149 through 172 and accompanying text in this chapter.
144. *Cal. ex rel. State Lands Com'n, supra,* 805 F.2d at 864. A more complete discussion of the choice-of-law decision is found in notes 99 through 106 and accompanying text in Chapter 2. Cases decided after *Cal. ex rel. State Lands Com'n* reach a contrary conclusion on the choice-of-law question. One state court observed that after a state enters the Union, state law governs title to land underlying its navigable waters. *Matter of Ownership of Bed of Devils Lake, supra,* 423 N.W.2d at 143, citing *Montana v. United States* (1981) 450 U.S. 544, 551. More important, the United States Supreme Court itself deferred to state law in a quiet title action between the state and federal governments that concerned title to a riverbed. The Supreme Court made specific note that state law would govern a claim of adverse possession against the state by the federal government. *Block v. N. D. ex rel. Bd. of Univ. and Sch. Lands* (1983) 461 U.S. 273, 292 n.28.

waters—termed a reliction[145]—to the littoral owner.[146] That law made no exceptions for recessions that were caused by artificial (human-instigated or induced) means.[147] It may be asked, with some considerable justification, Is there a "well-settled" body of federal common law rule of reliction?[148] Many of the cases cited by the court in its opinion do not deal with lake recessions,[149] and those that do are subject to some question.[150] And, as the reader shall see, there is a considerable body of law, including cases approved by the United States Supreme Court, that appears to reach a quite different con-

---

145. Recall that the legal term "reliction" describes a physical process—the exposure of land by the imperceptible recession of water. E.g., *Cal. ex rel. State Lands Com'n, supra,* 805 F.2d at 861 n. 1. The terms "accretion" and "reliction" are sometimes used interchangeably; the law relating to the property boundary effect of accretion applies to reliction. *Ibid.* "The law of reliction is generally the same as that of accretion as it concerns contiguity, imperceptibility and naturalness of the process of the water's recession and the exposure of additional land." *Utah State Road Com'n v. Hardy Salt, supra,* 486 P.2d at 392. One opinion, however, observed that " 'dereliction' is the process by which land is bared by the gradual recession of water; that the baring of new land by a sudden change in the location of water is " 'reliction,' " as contrasted with " 'dereliction' " above and that " '[m]any courts and writers today refer to this swift change process in all of its aspects as avulsion and use reliction in the situation described over where they should be using dereliction.' " *Matter of Ownership of Bed of Devils Lake, supra,* 423 N.W.2d at 143, n.1 (quoting a commentator); see note 131 and accompanying text in Chapter 7. As we will see, the elements of reliction and accretion are not entirely coincidental and may be distinguishable in both composition and application. See notes 160 through 175 and accompanying text in this chapter.
146. *Cal. ex rel. State Lands Com'n, supra,* 805 F.2d at 863; *Matter of Ownership of Bed of Devils Lake, supra,* 423 N.W.2d at 143; see *United States v. Ruby Co.* (9th Cir. 1978) 588 F.2d 697, 701, n.4, *cert. den.* (1979) 442 U.S. 917; *Bear v. United States* (D.Neb. 1985) 611 F .Supp. 589, 593 n.2, *aff'd* (8th Cir. 1987) 810 F.2d 153.
147. *Cal. ex rel. State Lands Com'n, supra,* 805 F.2d at 864 (citing *California ex rel. State Lands Comm'n, supra,* 457 U.S. at 285); *County of St. Clair v. Lovingston, supra,* 90 U.S. at 66; *United States v. Claridge* (9th Cir. 1967) 416 F.2d 933, 935, *cert. den.* (1970) 397 U.S. 961.
148. It may also be asked, what authority was there for the federal court to create such a law in the first place? See notes 107 through 109 and accompanying text in Chapter 2.
149. *County of St Clair v. Lovingston* concerned a river; *State Lands Commission* concerned an open ocean coastal shoreline. Other courts have noted that there should be a distinction between rules of property boundary law developed for different physical regimes. *Lamprey, supra,* 53 N.W. at 1141 ("What the relative rights of the state and of riparian owners in the waters and beds of these lakes are, largely depends upon the question of whether the rules of law as to the rights of grantees of lands bordering on running streams are applicable to grants of land bordering on lakes.") *Lamprey* adopted such rules without explanation. See also note 6 in Chapter 7.
150. *Banks, supra,* 69 U.S. at 58 (ground formed by accretion); *Johnston v. Jones* (1862) 66 U.S. 209, 210 (land formed by accretion); *Jones v. Johnston, supra,* 59 U.S. at 151 (tract of alluvial ground). None of these cases was decided as a matter of federal common law; rather, they appear to have been decided as a matter of state or common law. Moreover, state lakebed ownership claims were not concerned; the cases were between competing private owners. None of these cases concerned a recession, especially one of the vastness described in the Mono Lake case. Finally, the cause of deposition of alluvium in such cases was either not discussed or completely tangential to the issue.

clusion about the property boundary effect of lake recession caused by human activities.[151]

That body of law concerns particular aspects of the reliction doctrine. First, the reliction rule does not appear to apply to intentional drainage that lowers the lake's level or to the filling of a lake. Second, there is some question about how the perceptibility standard is applied in the case of lake recession. Third, it also seems that the recession of water must be permanent for the doctrine to apply. Fourth and finally, a decision approved by the United States Supreme Court concerning a lake recession of similar magnitude to that of Mono Lake came to an entirely different conclusion based on federal common law.

8.10.2 The Reliction Rule Possibly Inapplicable to Intentional Filling or Drainage

Filling of the bed of a navigable lake has never operated to transfer title to the upland owner at the expense of the public.[152] Perhaps more important, where the bed of a navigable lake is drained or reclaimed, state courts have held[153] that the state, not the littoral owners, retained ownership of the exposed lakebed.[154] And the United States has even recognized that artificial lowering

---

151. See notes 152 through 175 and accompanying text in this chapter.
152. *County of Lake, supra*, 238 Cal.App.3d at 228 (". . . the dry land appellant . . . created with fill was always in state ownership . . . ; it did not later pass into private ownership as a result of filling, or because . . . the adjoining lake fell to a low level . . ."); *New York v. Wilson & Co.* (N.Y.App. 1938) 15 N.E.2d 408, 413, *reh. den.* (N.Y. App. 1938) 16 N.E.2d 850; *Sage v. Mayor of City of New York* (N.Y. 1897) 47 N.E. 1096, 1103; *Winberger v. Passaic* (N.J. 1913) 86 A. 59, 60; *Ray v. State* (Tex. Civ. App. 1941) 153 S.W.2d 660, 662–663; *Brundage v. Knox* (Ill. 1917) 117 N.E. 123, 128; *Lakeside Boating & Bathing Inc., supra*, 344 N.W.2d at 220 (case sent back to determine whether fill had reasonable relationship to a navigational or other paramount purpose).
153. Where a lake recedes naturally, the geographic area of the littoral owner's property increases to maintain access to the lake's waters. *Lamprey, supra*, 53 N.W. at 1143.
154. In *Martin, supra*, 112 So. at 287, the Florida Supreme Court held that a state conveyance of swamp and overflowed lands adjacent to a lake did not include the lakebed uncovered because of drainage operations. "If to serve a public purpose, the state, with the consent of the federal authority, lowers the level of navigable waters so as to make the water recede and uncover lands below the original high-water mark, the lands so uncovered below such high-water mark, continue to belong to the state." *Ibid.* Many Florida cases follow *Martin. Conoley v. Naetzker* (Fla.App. 1962) 137 So.2d 6, 7; *Padgett v. Central and Southern Florida Control Dist.* (Fla.App. 1965) 178 So.2d 900, 904; *State v. Contemporary Land Sales* (Fla.Ap. 1981) 400 So.2d 488, 492. Other states have the same rule. *Garrett v. State* (N.J. 1972) 289 A.2d 542, 544–546 (artificially caused exposure of the bed of a tidal creek did not cause the state to lose ownership of the exposed land); *State v. District Court of Kandyiohi County* (Minn. 1912) 137 N.W. 298, 299-300; *State v. Placid Oil Co., supra*, 300 So.2d at 175. And in still other states, elements of the rule have appeared. *Noyes supra*, 61 N.W. 250; *State v. Longyear Holding Co., supra*, 29 N.W.2d at 667. Further, in California a statute expressly provides that lands exposed by recession or drainage remain property of the state and are available for sale. Cal. Pub. Res. Code § 7601. English common law also recognizes that the reliction doctrine does not apply where the shoreline changes

of a water body is not a reliction.[155] One of the reasons for this result is that application of the reliction rule in such circumstances would cause the state to forfeit the right to use its sovereign land in any manner it chooses. As a consequence, the state would no longer be able to provide for reclamation operations without fear of losing the lands reclaimed as a consequence. Such a result does not appear to be the law.[156]

The reasoning behind the intentional drainage rule is important to understand. Recall that in order to be a reliction, a recession must be gradual and imperceptible.[157] Imperceptibility is a large part of the reason for the reliction rule—that is, the recession was not significant enough to notice at any one time.[158] In both human-induced drainage and human-initiated diversion of lake tributaries, however, the lowering of the lake level and the consequent exposure of land between the prediversion or predrainage boundary has been well known, fully understood, and accurately forecasted.[159] Because the effect of such drainage or diversion was well known—"perceived," if you will—application of the doctrine of reliction to such a lake recession may be argued not to be proper.

In addition, when one examines other aspects of the reliction doctrine, it will be found that there appear to be different standards for perceptibility. There also appears to be a requirement that the recession be permanent before it can be considered a reliction. The federal court has ignored both of these aspects of the reliction rule. Consequently, one may claim that the foundation of the so-called well-settled federal common law appears quite flimsy.

8.10.3 The Imperceptibility Necessary for the Application of the Reliction Doctrine

Recall the earlier discussions of the perceptibility element of the doctrine of reliction.[160] The test for what is or what is not perceptible is generally ac-

---

its geographic location as the result of deliberate artificial reclamation. Halsbury's Laws of England, ¶ 297, at 162 (4th Ed. 1984).
155. *State of Utah, supra,* 70 Int. Dec. at 43–44 (". . . lands involved . . . were uncovered artificially as result of drainage operations, [it was] correctly held that 'such operations do not result in the transfer of title to the now dry land from the State to the United States . . .'").
156. *Hardin, supra,* 140 U.S. at 381-382 (state has authority to reclaim or otherwise use lands below ordinary high-water mark).
157. E.g., *Matter of Ownership of Bed of Devils Lake, supra,* 423 N.W.2d at 143, n.1; *Utah State Road Com'n v. Hardy Salt, supra,* 486 P.2d at 392–393; *Flisrand, supra,* 152 N.W. at 798; see *Bonelli Cattle Co., supra,* 414 U.S. at 325–326.
158. In legal parlance this is known as *de minimis non curat lex*—the law shall not take notice of little things. *State v. Sause* (Ore. 1959) 342 P.2d 803, 826.
159. In the case of Mono Lake, littoral landowners even received prediversion compensation for the effect of the lowering of the lake on their "riparian" rights. *City of Los Angeles v. Aitken, supra,* 10 Cal.App.2d at 465–466, 474–475 .
160. See note 130 in Chapter 7.

cepted to have been set forth in *County of St. Clair v. Lovingston*: ". . . though witnesses may see from time to time that progress has been made, they could not perceive it while the process was going on."[161] Most cases have applied this test to situations where the changes were not observable over a short time interval.[162] The application of the test of perceptibility to the physical consequences of the tributary stream diversion such as occurred at Mono Lake was not at all clear.

For example, contrary to the decision in the Mono Lake case, there is other authority holding that short-term changes that may have occurred over the period of a day or within a week, and possibly had exposed a lakebed, were "sudden" enough to require more evidence on what was the process of the "reliction."[163] Other cases, concerning both lakes and other regimes, have found that some relatively long-term changes were *not* gradual and imperceptible and were neither relictions nor would they give rise to the application of the reliction doctrine.[164] And some cases have even treated such changes as in the nature of an avulsion.[165]

---

161. *County of St. Clair v. Lovingston, supra*, 90 U.S. at 50.
162. See notes 135 through 147 and accompanying text in Chapter 7.
163. *Boorman v. Sunnuchs* (1877) 42 Wisc. 233, 245 ("It is very manifest that the water might have receded altogether too suddenly to have vested the title in the uncovered land in the plaintiff, and yet a person watching the pond be quite unable to see the waters recede. We take it that had the whole fall in the water occurred within a week, perhaps within a day, at a uniform rate, the human eye is not sufficiently acute to have detected the process. Yet if a portion of the bed was laid bare by a process so sudden, no one will contend that the portion thus uncovered became thereby the property of the owner of the land adjoining").
164. See *People v. Wm. Kent Estate Co.* (Cal. 1966) 242 Cal.App.2d 156, 160 (80-foot seasonal increases in ocean shoreline were not gradual and imperceptible); *McCafferty v. Young* (Mont. 1964) 397 P.2d 96, 100 (court inclined to label river migration of a mile in less than 100 years as perceptible); *St. Louis v. Rutz* (1891) 138 U.S. 226, 251 (movement of mass of land in bed or river over course of several years "not imperceptible"); *Nesbitt v. Wolfkiel* (Ida. 1979) 598 P.2d 1046, 1049 (evidence that channel moved several hundred feet over a 50-year period was enough to overcome the presumption of accretion); *Carr, supra*, 93 N.W. at 54 (drying up of lake over indeterminate time over 54 years not a reliction); *Utah State Road Com'n. v. Hardy Salt, supra*, 486 P.2d at 393 (substantial movement or lake recession over three- or four-month period does not give rise to application of reliction doctrine).
165. See *Puyallup Indian Tribe v. Port of Tacoma* (9th Cir. 1983) 717 F.2d 1251, 1263, *cert. den., sub nom. Trans-Canada Enterprises Ltd. v. Muckleshoot Indian Tribe* (1984) 465 U.S. 1049, *reh. den.* (1984) 466 U.S. 954 (under Washington law, river movement of half its width over two-year period treated as an avulsion); *Witter v. County of St. Charles* (Mo. 1975) 528 S.W.2d 160, 161–162 (dike caused abandonment of a channel over relatively short time, analogous to avulsion); *McCafferty, supra*, 397 P.2d at 100 (later migration of a river perceptible over one generation). Some cases even support the argument that any shoreline change that is not found to be an accretion can be labeled an avulsion as a matter of convenience, if nothing else. See e.g., *James v. State* (Ga.Ct.App. 1911) 72 S.E. 600, 602 (change in course of river by humans analogous to avulsive change); *Durfee v. Keiffer* (Neb. 1959) 95 N.W.2d 618, 624; *Garrett, supra*, 289 A.2d at 546–547; *Barakis v. American Cynamid Co.* (N.D. Tex. 1958) 161 F.Supp. 25, 29; *State v. Bonelli Cattle Co.* (Ariz. 1971) 489 P.2d 699, 702–703. *rev'd* (1973) 414 U.S. 313, *ovrld.* (1977) 429 U.S. 363; see *Georgia v. South Carolina* (1990) 497 U.S. 376, 404.

Moreover, the Supreme Court of the United States adopted a Special Master's Report,[166] based on so-called federal common law.[167] This Special Master's Report came to two quite significant conclusions concerning a lake recession, conclusions that are in stark conflict with the Ninth Circuit's decision concerning the property boundary effect of the recession of Mono Lake.

8.10.4 The Great Salt Lake Case—The Requirement That Reliction Be Permanent

The Great Salt Lake Special Master's Report concerned the resolution of competing ownership claims of the State of Utah and the United States to that portion of the bed of the Great Salt Lake that had become uncovered waterward of the United States' meander line. First, the Special Master found that shoreline movements of the Great Salt Lake were not imperceptible within the parameters of the reliction doctrine.[168] This finding was made even

---

166. This Special Master's Report ("Great Salt Lake Special Master's Report I") contains a comprehensive discussion of the law concerning reliction. The Great Salt Lake Special Master's Report I was adopted by a decree consistent with it by the United States Supreme Court. *Utah v. United States* (1971) 420 U.S. 304. The opinion confirming the Great Salt Lake Special Master's Report I has been cited by the Supreme Court as substantive authority, see *Ohio v. Kentucky* (1980) 444 U.S. 335, 340, and by the Ninth Circuit and Supreme Court of North Dakota in boundary cases. *United States v. Ruby, supra,* 588 F.2d 701, n.4; *Matter of Ownership of Bed of Devils Lake, supra,* 423 N.W.2d at 145. The Great Salt Lake Special Master's Report I has never been published. Because of the interest in this question and the applicability of the Special Master's discussion of the reliction question to the discussion in this chapter, the Great Salt Lake Special Master's Report I is set out in full in Appendix B of this book. All citations to the report will reference the page numbers in the original report.
167. Appendix B, Great Salt Lake Special Master's Report I at pp. 5–6.
168. *Id.* at 20. After noting that an avulsion was not involved, the Special Master discussed the reliction rule citing *County of St. Clair v. Lovingston,* 90 U.S. 46, *Jefferis v. East Omaha Land Co.* (1890) 134 U.S. 178, *Philadelphia Co. v. Stimson* (1912) 223 U.S. 605, and other cases frequently cited as containing formulations of the accretion/reliction rule. *Id.* at 10–14. Then the Special Master discussed the evidence in the case:

> The movements of the water during any year, and the resulting exposure or inundation of the shorelands, are not mirrored in annual averages. . . . [T]he yearly average of these movements is not their visual movement. Changes in elevation may be gradual and imperceptible but only a part of a process, the whole of which includes the effect upon the shorelands of such changes. *If the effect on the shorelands is not imperceptible then the whole of the process is not imperceptible within the meaning of the doctrine of reliction.* *Id.* at 16–17 (emphasis supplied). . . .

> To place the matter . . . in a light most favorable to the United States is to consider what may be referred to as the net result of the movements of the Lake over a substantial period of time. . . . At the end of the period there was a very perceptible area which had become exposed during the period, but perhaps there had been at no particular moment a separate perceptible component of this total movement of exposure. To accept this . . . as meeting the imperceptible element of reliction would . . . be an adaptation of that element of the doctrine to the unique character of the Great Salt Lake rather than to apply the imperceptible element as it has been developed in the law. While there are indeed features in the

though these shoreline movements, even at their most extreme, were less than 15 inches per hour. Second, the Special Master also found as an element of the reliction rule that the recession (or in the case of accretion, the deposition of soil) must be permanent in character.[169] Some courts have refused to apply the reliction doctrine where the recession has not been sufficiently permanent.[170] And the test for determining what is sufficiently permanent is extremely stringent.[171] Because recession of the Great Salt Lake was not permanent, the Special Master did not apply the reliction doctrine to the lake's recession.[172]

In contrast with that Special Master's Report, the Ninth Circuit's expression of the federal common law of reliction accepted, without question, the test of perceptibility as stated in *County of St. Clair v. Lovingston* and the finding of the district court, based on the *Lovingston* test, that an average person would not have observed movement of the shoreline as it was occurring.[173] The Ninth Circuit held that the district court was correct in averaging measurable changes in the shoreline of Mono Lake over its entire lakeshore and

---

over-all process and its result . . . which arguably favor the Court's acceptance of the imperceptible element of the doctrine, nevertheless the continued rise and fall of the elevation of the Lake on an hourly, daily and weekly basis, is unprecedented in the historical development and previous application of this element of the doctrine. *Id.* at 19.

169. *Id.* at pp. 22–23. The Special Master inferred this element of permanence from the nature of the process. ("The omission of the case law to refer uniformly to permanence is understandably due to this quality being implied as a result of the gradual and imperceptible process.") For some cases that specifically require permanence of the recession as an element of the reliction rule, refer to note 170 in this chapter.

170. *Salp v. Frazier* (La. 1899) 26 So. 378, 380; *State v. Longyear Holding Co., supra,* 29 N.W.2d at 667; *Flisrand, supra,* 152 N.W. at 798; *Herschman v. State Department of Natural Resources* (Minn. 1975) 225 N.W.2d 841, 843; *In re Judicial Ditch Proceeding No. 15* (Minn. 1918) 167 N.W. 1042, 1044; *Hillebrand, supra,* 274 N.W. at 823; *State v. Thompson* (Iowa 1907) 111 N.W. 328, 329. By way of example, in *Rutten, supra,* 93 N.W.2d 798–799, the court had a stipulation before it that set forth the elevation of the lake's waters over a period of years. This stipulation showed that the level of the lake fluctuated. Thus, the court found that there had been no permanent recession.

171. One court stated that for the recession to be of a permanent nature, it must be without the possibility of the water again filling in or covering the exposed area. *State v. Longyear Holding Co., supra,* 29 N.W.2d at 667. Although a lake may be completely dry for a considerable period, if the lake were refilled at any time, it has been held to be conclusive that the recession was not permanent. *Herschman, supra,* 225 N.W. 2d at 843. Even evidence of lake levels over a 90-year period that showed the lake had only exceeded a particular level *once* was held to be insufficient to find that the lake had permanently receded. *Rutten, supra,* 93 N.W.2d at 799.

172. Appendix B, Great Salt Lake Special Master's Report I at pp. 25–26. ". . . [T]he reinundation . . . of almost the entire statehood bed of the Great Salt Lake confirms the Special Master in his recommendation adverse to application of the reliction doctrine, because the public benefit of Utah entitled to protection has moved along with the reinundating waters of the lake. . . ." *Id.* at p. 26.

173. *Cal. ex rel. State Lands Com'n, supra,* 805 F.2d at 865.

in determining as a result that shoreline migration over the lake as a whole had "clearly not been sudden and perceptible."[174] In reaching its decision, the Ninth Circuit also failed to discuss the impermanence of the recession. Thus, the court concluded that the human-caused recession, which exposed more than 11,000 acres of former lakebed, was gradual and imperceptible and the property boundary of the lands of the United States followed the lake level as it receded.[175] This result conflicts with the result approved by the United States Supreme Court in the Great Salt Lake case. In addition, the Ninth Circuit's decision appears not to have correctly applied the so-called well-settled federal rule of reliction.

8.10.5 So-Called Federal Law of Reliction Misapplied in the Mono Lake Case

Assuming that the Mono Lake recession case properly described the content of "federal common law," that rule is that the riparian or littoral owner has the right to any "natural and gradual accretion formed along the shore."[176] Since the word "natural" appears in parallel with the word "gradual," it can be argued that "natural" as used in this formulation of the rule refers to the rate or manner in which the land has been created or has become exposed.

Stated another way, under the so-called federal common law rule, an accretion or a reliction must occur at a rate that, regardless of cause, is "natural." Thus, the rule excludes abnormal or unusual increases from being considered as an accretion or a reliction. In the case of Mono Lake, the recessions were claimed to be unusual or abnormal in nature. The court, however, treated

---

174. *Ibid*. This methodology ignored evidence that at particular locations the shoreline migrated more than 2 feet in an hour; that there were 1400 days during which there was a shoreline migration of 10 feet or greater; and more than 150 days during which the shoreline migrated more than 40 feet. Although these locations were chosen where the slope was the flattest (which would necessarily emphasize the shoreline movement), it was argued that considering areas where changes were most prominent was appropriate because such areas would confirm the existence of a dramatic recession of the lake. On the other hand, use of average rates of migration was argued to be misleading. The land surrounding the lake was a mix of extremely flat and extremely steep slopes. If one averages migration from steep and flat slopes, the result will be distorted. As noted by the Special Master in the Great Salt Lake Case:

> [T]he yearly average of these movements is not their visual movement. Changes in elevation may be gradual and imperceptible but only a part of a process, the whole of which includes the effect upon the shorelands of such changes. If the effect on the shorelands is not imperceptible then the whole of the process is not imperceptible within the meaning of the doctrine of reliction. Appendix B, Great Salt Lake Special Master's Report I at pp. 16–17.

175. *Cal. ex rel. State Lands Com'n, supra*, 805 F.2d 864–865.
176. See *Hughes v. Washington* (1967) 389 U.S. 290, 293; *Bonelli Cattle Co. v. Arizona* (1973) 414 U.S. 313, 325–326, *ovrld* (1977) 429 U.S. 363.

the recessions as if they were natural and ordinary occurrences in the lake's regime.

In sum, there are arguments calling into question the result of the Mono Lake recession case. Consequently, there appear to be good reasons to assert that the Mono Lake case provides little assistance, if any, to those seeking guidance on the property boundary effect of human-caused lake recessions.

8.10.6 Property Boundary Consequences of Littoral Shoreline Changes—Submergence

The Mono Lake recession case deals with the possible property boundary effect of lake recession. But what happens to the ordinary high- or low-water mark property boundary when there is a submergence of the shorelands because of a rise in the level of the lake on account of artificial controls? Some answers to this question are provided by two California cases[177] that concerned California's (and Nevada's) Lake Tahoe. Because it is likely that other states will have lakes in a similar state of regulation due to demands of recreational users or due to demands for water storage, conservation or delivery, or flood control, this chapter explores these cases.

In the late 1800s, partly in prescient recognition of the recreational value of Lake Tahoe and the need to have the level of the lake close to that of the blackjack tables and slot machines so gamblers could have a sense of the great outdoors while losing their weekly paychecks, a dam was built at Lake Tahoe's outlet at the Truckee River.[178] This dam raised the level of the lake.[179] It was not until the 1970s that California courts were forced to grapple with the property boundary effect of Lake Tahoe's regulation by the dam. The problem was taken to the courts in two stages. First, the property boundary effect of lake level regulation was decided. Then, given that decision, the physical location of the property boundary was determined.

Turning first to the property boundary effect of lake level regulation, the California Supreme Court recognized the validity of the argument that it would be extremely difficult, if not impossible, to reconstruct the natural level of the lake before the dam was built.[180] Therefore, the court decided that the

---

177. *State v. Superior Court* (*Fogerty I*), *supra*, 29 Cal.3d 240; *Fogerty v. State of California* (1986) 187 Cal.App.3d 224, *cert. den.* (1987) 484 U.S. 821 ("Fogerty II").
178. There is a rumor that the dam was built in the 1870s as part of a planned project to cut a tunnel through the Sierra Nevada and bring Lake Tahoe water to San Francisco. Instead, San Francisco built the Hetch-Hetchy project that brought San Francisco its drinking water by submerging a wilderness valley near Yosemite Valley. The fight over the Hetch-Hetchy project led to the rise of the Sierra Club. For the Sierra Club's connection with Hetch-Hetchy, see the following Web site: http://www.sierraclub.org/chapters/ca/hetchhetchy/history.html.
179. *State v. Superior Court* (*Fogerty I*), *supra*, 29 Cal.3d at 240; *Fogerty II*, *supra*, 187 Cal.App.3d at 230. Some argue that the dam was built in the river below the natural sill of lake. If this claim could be proven, the lake would still exist in its natural condition.
180. *State v. Superior Court* (*Fogerty I*), *supra*, 29 Cal.3d at 248. The court noted that "[t]he monumental evidentiary problem which would be created by measuring the boundary line between public and private ownership in accordance with the water level which existed prior to the construction of these [many] dams. . . ." See also note 66 in Chapter 6 and note 149 in Chapter 7.

current level of the lake would be the standard for the location of the high-water mark.[181] In reaching that conclusion, the court relied on a venerable legal doctrine.

The court's legal basis for locating the littoral owners' property boundary by reference to current lake levels was provided by the doctrine of prescription.[182] The California Supreme Court held that littoral owners lost ownership of land submerged by waters that had risen as the result of construction of the dam and had remained at that level for the period required for the acquisition of prescriptive rights.[183]

The court also appeared to follow the reasoning of two earlier California cases. Those cases held that when an artificial condition, such as a diversion, had been long maintained, those acting in reliance on that condition have the right to have the condition maintained.[184] But this decision did not end the dispute over the effect of submergence on littoral property boundaries. How was the physical location of the new, natural, current ordinary high-water mark of the lake to be determined? The California court's treatment of this question provided valuable insight into the types of evidence available to prove the location of the littoral boundary on both regulated and nonregulated lakes.

8.11 PROOF IN LAKE BOUNDARY AND TITLE CASES[185]

Based on the widespread acceptance of an earlier water rights determination that adopted a particular lake level, it was argued that a specific lake level

---

181. *Ibid.*
182. Some argued that the opinion reflects the fact that the court adopted the current elevation of the lake as the physical location of the boundary in order to avoid the difficult process of taking evidence to determine the physical location of the ordinary high-water mark of the lake before the construction of the dam.
183. *Ibid.* The court relied on two non-California cases, *State v. Parker, supra,* 200 S.W. at 1016; *State v. Sorenson* (Iowa 1937) 271 N.W. 234, 238–239. *Sorenson* stated that under such circumstances "the artificial condition is . . . stamped with the character of a natural condition, and the title to lands covered by the waters of the lake is deemed to have passed from private ownership to the same trust as that of lands covered by the waters of natural navigable lakes." *Ibid.*
184. *Natural Soda Prod., supra,* 23 Cal.2d at 197; *Chowchilla Farms, Inc. v. Martin* (Cal. 1918) 219 Cal. 1, 18. In the *Natural Soda Products* case, Los Angeles had diverted the water-supplying river to a once-navigable lake for a public water supply. As a result, the lake dried up and its bed became exposed. The exposed bed contained valuable minerals. Natural Soda Products leased the mineral rights in the exposed bed from the state and spent a considerable amount of money on equipment to extract the minerals. Those extraction operations depended on continuance of the desiccated state of the lakebed. Continuance of that state in turn depended on the absence of any substantial flow of water from the river onto the dry lake bed. Los Angeles erroneously allowed a large amount of water onto the lake bed, leading to the litigation. *Natural Soda Prod. Co., supra,* 23 Cal.2d at 196–197.
185. There are examples of other kinds of evidence concerning other issues, such as the proof of navigability and the proof of the speed and perceptibility of a lake recession cited at notes 58 and 174 in this chapter.

had been acquiesced in by the littoral landowners or had become an agreed boundary.[186] That argument was not accepted.[187] Rigorously adhering to the doctrine of prescription,[188] the court determined that only actual incursion of water would establish the level of the ordinary high-water mark.[189] But what water level is appropriate where the lake level fluctuated from year to year and from season to season based on weather and drainage conditions and the water needs of downstream water users?

The court found that in order for prescriptive rights to attach, it was not necessary for the water level created by the dam to remain in place continuously for the five-year period.[190] All that need be established was that the dam or reservoir operator returned waters of the lake to the highest point each water year and maintained the water level at that elevation for the duration of the reservoir's needs. This showing would establish that use of the reservoir is "continuous" up to and including that highest water level.[191] Thus, the court set the ordinary high-water mark at the highest elevation actually reached by the lake in five sequential years.[192]

In contrast, another California court refused to apply the doctrine of prescription to establishment of the location of the ordinary low-water mark of

---

186. *Fogerty II, supra*, 187 Cal.App.3d at 236. The doctrine of agreed boundaries is discussed in Chapter 1. See notes 189 through 192 and accompanying text in Chapter 1. In short, to establish an agreed boundary there must be uncertainty as to the true boundary line, an express or implied agreement between the adjoining landowners to accept the line as the boundary, and acceptance or acquiescence in that line for a period equal to the statute of limitations. *Fogerty II, supra*, 187 Cal.App.3d at 236.

187. *Id*. at 236–237. The agreed boundary doctrine was held inapplicable because when the lake level rose there was no boundary controversy and the littoral owners had no reason to believe their property boundary was being adjusted. *Ibid*. "In the circumstances it would be manifestly unfair, if not disingenuous, to justify the imposition of public trust rights upon private owners' land on the basis that the owners, by their ignorance and inaction, somehow 'agreed' to imposition of the public trust boundary at [the elevation contended by the state.]" *Ibid*. In another lake case, a California court also refused to apply the doctrine of agreed boundaries where there was no evidence that the private landowners ever agreed to make a certain lake level the lakeward edge of their parcels. *County of Lake, supra*, 238 Cal.App.3d at 235 (court was ". . . not willing to expand the doctrine of agreed boundaries in a shorezone context so as to allow the state or landowners to establish retrospectively some purely fictional new boundary nowhere marked out or made manifest to the parties").

188. According to the court, the level set by the water rights determination was set "only on paper" and was unaccompanied by any assertion that it would affect property rights. Consequently, the level set by the water rights determination did not fulfill the notice requirement established by the rules for prescription. *Fogerty II, supra*, 187 Cal.App.3d at 238.

189. *Id*. at 238–239.

190. *Id*. at 239.

191. *Ibid*. [Citations omitted.]

192. *Ibid*. The court took care to make sure this average was ordinary in the sense of not representing the level reached by the lake in unusual floods. "The undisputed data indicate the figure . . . is not markedly different from other elevations of the lake reached . . . and does not represent an inappropriate unusual condition of the lake." *Id*. at 240. Somewhat the same approach was followed in the Delta Meadows case, described in detail in Chapter 6.

Clear Lake.[193] The court distinguished[194] cases dealing with the location of the ordinary high-water mark. According to the court, those cases were motivated by important public interests that are absent in the case of the ordinary low-water mark.[195] In such a case, the court "is simply determining a low-water boundary for purposes of rent, within the area subject to state control either by fee or by public trust."[196] And the court found no basis for prescription, as there was nothing hostile or notorious about natural cyclical variations in the water level.[197]

The use of lake stage levels by some courts, rather than more traditional methods of determining the ordinary high-water mark, is worthy of mention.[198] Turning first to the technical side, these courts used different measures in making the stage level determination. One court adopted a five-year period of record, as opposed to a longer period, because of the peculiar circumstances of that case.[199] Where such peculiar circumstances were not present, another court, without discussing any scientific foundation, advocated the use of long-term stage data.[200] On the other hand, it has been recognized that there is no way of rationally determining the number of years to average.[201]

What is most worthy of note, however, is that the vegetation/erosion line method of determining the ordinary high-water mark was specifically rejected by these courts.[202] One opinion noted that the vegetation/erosion line test should be resorted to in determining historic water levels "where more accurate measurements are unavailable."[203] But the opinion noted that in the

---

193. *County of Lake, supra*, 238 Cal.App.3d at 231–232.
194. The court also distinguished *Fogerty II*, where the level of Lake Tahoe had been permanently raised, from the case of Clear Lake where the lake rose and fell within its natural range. *Id.* at 231.
195. "The considerations which motivated [*Fogerty II*] . . .—the fragility of shorezone areas and their importance for public interests [citation omitted]—are essentially absent here, where we need not set a high-water mark at a sufficient elevation to ensure the maintenance of public control over destructive private exploitation . . . ." *Id.* at 231–232.
196. *Id.* at 232.
197. *Id.* at 232.
198. The court swiftly disregarded the traditional method of determining the ordinary high-water mark in tidal waters by averaging tides over an 18.6-year period. Not only did the court note that this method was inconsistent with the prescriptive rights rationale, but nontidal waters impounded in Lake Tahoe know of no tidal rhythmic regularity. "Waters stored in reservoirs, like that in uncontrolled lakes and streams, fluctuates with the weather, but is also under the direction of man." *Fogerty II, supra*, 187 Cal.App.3d at 240, n. 12.
199. *Ibid.* Use of five years of data was consistent with the prescriptive rights rationale of the opinion.
200. *County of Lake, supra*, 238 Cal.App.3d at 233.
201. *Fogerty II, supra*, 187 Cal.App.3d at 240, n. 12; see notes 116 through 117 and accompanying text in this chapter. In addition, the court felt there was a danger in that a landowner would have to endure excess encroachment in years of above-average lake levels. *Ibid.*
202. Among the reasons that the vegetation/erosion line test was unacceptable was that the water level was under control of humans and could fluctuate. Thus, a visible vegetation/erosion line may reflect only recent events in the reservoir. *Id.* at 240, n. 12.
203. *Ibid.*

case of Lake Tahoe, there was data "accurate to two decimal points" and there was no need to rely on physical inspection of vegetation to determine how high the water had risen over time.[204]

Some cases, however, have used a vegetation/erosion line test. A brief examination of those cases will depict what sorts of proof was used to evidence the vegetation/erosion line.[205] In one recent case, the area that was the subject of dispute had been substantially altered by the construction of condominiums and a parking lot.[206] To establish the ordinary high-water mark at the site, expert testimony[207] was introduced. This testimony established an erosion line ordinary high-water mark at a location a half mile from the site.[208] A determination of the elevation[209] of that mark was made and transferred to the site.[210] Based on that determination, the analysis of several aerial photographs of the site as it existed prior to its alteration, government survey maps, and stereo photographs offering a three-dimensional view of the site, the expert concluded that portions of the site were originally part of Lake Superior.[211] Because there was no other evidence, the court determined that portions of the site were waterward of the ordinary high-water mark (OHWM).[212]

Other cases have used different approaches to proving the elements of the vegetation/erosion line test. In one case the age of trees above a particular elevation was particularly persuasive of the location of the ordinary high-water mark.[213] Other litigants have presented not only the evidence of the age

---

204. *Ibid.*
205. See also note 81 in this chapter, detailing the evidence on which a surveyor relied to determine the location of the ordinary high-water mark of a lake that had been drained. Soils and vegetation types and topographic character were all of assistance.
206. *State of Wisconsin v. Trudeau, supra,* 408 N.W.2d at 340. There was still some water on the site, the source of which was uncertain. It could have come from adjacent Lake Superior or from golf course drainage.
207. The expert was a "water management specialist." *Id.* at 344. The expert's qualifications were not set forth in the opinion. Types of experts who may be useful in lake cases are hydraulic engineers, *State v. Bollenbach, supra,* 63 N.W.2d at 283, dendrochronologists, *Matter of Ownership of Bed of Devils Lake, supra,* 423 N.W.2d at 145, hydrographic surveyors, *Lindberg v. Department of Natural Resources* (Minn. 1986) 381 N.W.2d 494, 495, and limnologists.
208. *State of Wisconsin v. Trudeau, supra,* 408 N.W.2d at 344.
209. The elevation was determined with respect to IGLD, International Great Lakes Datum, the reference system used for expressing elevations in the Great Lakes area. *Id.* at 344, n.7.
210. The ordinary high-water mark transfer rule was developed to promote certainty and ascertain property rights in riparian lands." *Id.* at 344, n.8; see note 96 and accompanying text in this chapter.
211. *State of Wisconsin v. Trudeau, supra,* 408 N.W.2d at 344, n.8.
212. "The positive and uncontradicted testimony of [the expert] . . . that the OHWM of Lake Superior is [a certain elevation] . . . that the project site was and is hydraulically connected to and is in fact part of Lake Superior is not discredited nor against reasonable probability. . . . The state therefore properly determined the lake's OHWM at 'other places on the . . . shore of the same . . . lake' and transferred that finding to the project site [citations omitted]." *Id.* at 345.
213. *Matter of Ownership of Bed of Devils Lake, supra,* 423 N.W.2d at 145 ("The evidence of fifty to sixty-year-old trees at an elevation of 1,426 feet is particularly persuasive").

of trees,[214] but also topographic evidence as shown in aerial photographs[215] and the testimony of persons long familiar with the area.[216] But even government expert testimony is subject to some question when political considerations become involved.[217]

Of course, this does not exhaust the possibilities of proof that will support a vegetation/erosion line test. It does, however, exhaust topics for this book and substantive discussion of title and boundary disputes concerning properties adjacent to underlying navigable water bodies.

---

214. Trees are important because of their relative permanence; if inundated by water for a period of time, they will die. *Determining Natural Ordinary High-water Level, supra,* 384 N.W.2d at 512; *Lindberg, supra,* 381 N.W.2d at 495. The differing requirements for a depth of unsaturated soil can be used to measure the ordinary high-water mark. *Ibid.* The reliability of this method depends on the trees selected for study. *Ibid.*

215. If stereo (overlapping) aerial photographic coverage is available, the photographs can be viewed stereoscopically to obtain a three-dimensional view to discern the shoreline. *Determining Natural Ordinary High-water Level, supra,* 384 N.W.2d at 513; *State of Wisconsin v. Trudeau, supra,* 408 N.W.2d at 344, n.8.

216. In one case an "old-timer recalled throwing peanut shells into the lake from a wagon" while on a lakeshore road. *Determining Natural Ordinary High-water Level, supra,* 384 N.W.2d at 513.

217. *S.D. Wildlife Federation, supra,* 382 N.W.2d at 35 (Wuest, J., dissenting)(contradictory reports prepared by state engineer one of which was prepared after public hearings at which there was "emotional, heated oposition . . .").

GLOSSARY

Accretion: The legal term used to describe the physical process in which the area of littoral or riparian land is increased by the gradual and imperceptible deposition of sand, soil, or other solid material upon the margin of the watercourse. E.g., *County of St. Clair v. Lovingston* (1874) 90 U.S. (23 Wall.) 46, 66–68; *Jefferis v. East Omaha Land Co.* (1890) 134 U.S. 178, 192–194; *City of Los Angeles v. Anderson* (Cal. 1929) 206 Cal. 662, 667; Robillard, W., and Bouman, L., Clark on Surveying and Boundaries (7th Ed., 1997) § 24.02 ; Grimes, *Clark on Surveying and Boundaries* (4th ed., 1976) § 567; 78 Am.Jur.2d Waters § 406, p. 851. See also "Deposition."

Alluvion: Soil that may gradually build up on the bottom, shore, or bank along a watercourse. *County of St. Clair v. Lovingston* (1874) 90 U.S. (23 Wall.) 46, 66; 78 Am.Jur.2d. Waters § 406, p. 851; Robillard, W., and Bouman, L., Clark on Surveying and Boundaries (7th Ed., 1997) § 24.02; Grimes, *Clark on Surveying and Boundaries* (4th ed., 1976), § 567.

Apogean tides: Tides that occur when the moon is in apogee, or farthest from the earth. The tides rise and fall less than usual. H.A. Marmer, *Tidal Datum Planes* (1951) U.S. Coast and Geodetic Survey, Spec. Pub. 135 Rev. Ed., p. 5; Patton, R.S., "Relation of the Tide to Property Boundaries" reprinted in 2 Shalowitz, *Shore and Sea Boundaries* (Dept. Comm. Pub. 10-1 1962), Appendix E, pp. 669–670.

Area elevation curve: A graphic method relating the elevation of the topography as a percentage of the total area.

Avulsion: The legal term that describes the rapid, perceptible and often violent removal or addition to land due to the action of water. E.g., *Jefferis v. East Omaha Land Co.* (1890) 134 U.S. 178, 194; *Nebraska v. Iowa*

The terms defined in this glossary are also defined within the chapters. The authorities for such definition are included as well.

(1892) 143 U.S. 359, 366; *Bauman v. Choctaw-Chickasaw Nations* (10th Cir. 1964) 333 F.2d 785, 789, *cert. den.* (1965) 379 U.S. 965; *State v. Gunther & Shirley Company* (Ariz. 1967) 423 P.2d 352, 356; *Nolte v. Sturgeon* (Okla. 1962) 376 P.2d 616, 620; Robillard, W., and Bouman, L., Clark on Surveying and Boundaries (7th Ed., 1997) § 24.02; Grimes, *Clark on Surveying and Boundaries* (4th ed., 1976), § 567; 78 Am.Jur.2d, Waters § 406, p. 852.

Bank (general): That property which lies between the high- and low-water marks. Porro, "Invisible Boundary—Private and Sovereign Marshland Interests" (1970) 3 Natural Resources Lawyer 512, 516. See also "Shore."

Bank (river): "The water-washed and relatively permanent elevation or acclivity at the outer line of the river bed which separates the bed from the adjacent upland . . . and serves to confine the waters within the bed and to preserve the course of the river." 2 Shalowitz, *Shore and Sea Boundaries* (Dept. Comm. Pub. 10-1 1962), Appendix A, p. 549.

Bed: That property which lies beneath the water of the waterway. Porro, "Invisible Boundary—Private and Sovereign Marshland Interests" (1970) 3 Natural Resources Lawyer 512, 516.

Bench marks: Marks affixed to a permanent object in tidal observations, or along a line of survey to furnish a datum level. *Forgeus v. County of Santa Cruz* (Cal. 1914) 24 Cal.App. 193, 195.

Burden of proof: One must prove that his or her position is correct. E.g., Cal. Evid. Code § 115.

Bureau of Land Management (BLM): Successor to the General Land Office and responsible for the execution of the public land laws and for activities concerning the management and administration of public lands.

Cadastral surveys: Surveys of property boundaries. Such surveys may also include the horizontal positions of physical features.

Call: A visible or natural object or landmark that is named in a grant, deed, or other conveyance as a limit or boundary of the land being conveyed with which points of surveying must correspond. Black's Law Dictionary (4th Ed. 1957) at 256.

Chain of title: A chronological list of the deeds or other instruments of conveyance that originate from the original source of title to the present vested owner. E.g., *Eltman v. Harvey* (N.Y. 1978) 403 N.Y.S.2d 428, 431–432; 1 Ogden's California Real Property Law (1956), p. 709.

Choice-of-law: "In conflicts of law, the question presented in determining what law should govern." Black's Law Dictionary (5th Ed. 1979), p. 219.

Claim of right: Acts of the adverse claimant showing an intent to claim the land against all. Robillard, W., and Bouman, L., Clark on Surveying and Boundaries (7th Ed., 1997) § 22.07; Grimes, *Clark on Surveying and*

Boundaries (4th Ed., 1976) § 538, pp. 699–700; 2 Miller and Starr, California Real Estate Law, § 19.2, pp. 396–397.

Color of title: A written instrument that attempts to convey, but through some defect does not pass, title. Robillard, W., and Bouman, L., Clark on Surveying and Boundaries (7th Ed., 1997) § 22.06; Grimes, *Clark on Surveying and Boundaries* (4th Ed., 1976) § 540; 1 Bowman, Ogden's Revised California Real Property Law (1974) § 4.8, pp. 122–123; *Wright v. Mattison* (1855) 59 U.S. (18 How.) 50.

Commissioner of the General Land Office: Head of the General Land Office.

Conflict of laws: That branch of jurisprudence, arising from the diversity of laws of different nations, states, or jurisdictions, in their application to rights and remedies, which reconciles the inconsistency, or decides which law governs in the particular case, or settles the degree of force to be accorded to the law of another jurisdiction either where it varies from the domestic law, or where the domestic law is silent or not exclusively applicable to the case in point. Black's Law Dictionary (5th Ed. 1979), p. 271.

Current: The horizontal movement of water. H.A. Marmer, *Tidal Datum Planes* (1951) U.S. Coast And Geodetic Survey, Spec. Pub. 135 Rev. Ed., p. 2.

Daily tide: The type of tide in which only one high and one low water happen each day. H.A. Marmer, *Tidal Datum Planes* (1951) U.S. Coast and Geodetic Survey, Spec. Pub. 135 Rev. Ed., p. 9.

Datum: A plane of reference. Shalowitz defines "datum plane" as "[a] surface used as a reference from which heights or depths are reckoned. The plane is called a tidal datum when defined by a phase of the tide. . . ." 1 Shalowitz, Shore and Sea Boundaries (Dept. Comm. Pub. 10-1 1962), p. 286. See also "Tidal datum."

Deliction: The equivalent of erosion, but only when the changes are gradual and imperceptible. *People v. Wm. Kent Estate Co.* (1966) 242 Cal.App.2d 156, 160.

Delta subsidence: The loss of the soil surface caused by the oxidation and decomposition of the underlying peat soils.

Dendrochronology: The study of time span within trees.

Deposition: The act of the deposit of soil or soil particles.

Dereliction: See "Reliction." *Matter of Ownership of Bed of Devils Lake* (N.D. 1988) 423 N.W.2d 141, 143, n.1.

Descriptive reports: Narrative written reports prepared by USCS or USC&GS surveyors that sometimes accompany topographic and hydrographic surveys. 2 Shalowitz, *Shore and Sea Boundaries* (Dept. Comm. Pub. 10-1 1962), Appendix A, p. 559.

Dicta: The part of a judicial decision not necessary for the result of the case. 2 Shalowitz, *Shore and Sea Boundaries* (Dept. Comm. Pub. 10-1 1962), Appendix A, p. 588.

Diurnal inequality: The differences between the morning and afternoon tides. Patton, R.S., "Relation of the Tide to Property Boundaries" reprinted in 2 Shalowitz, *Shore and Sea Boundaries* (Dept. Comm. Pub. 10-1 1962), Appendix E, p. 670.

Equal Footing Doctrine: The constitutional principle that all states admitted into the Union have equal rights of sovereignty. The origin of the Equal Footing Doctrine is explained in Chapter 1.

Equatorial tides: Tides that occur when the moon is near the equator. H.A. Marmer, *Tidal Datum Planes* (1951) U.S. Coast and Geodetic Survey, Spec. Pub. 135 Rev. Ed., p. 5; Patton, R.S., "Relation of the Tide to Property Boundaries" reprinted in 2 Shalowitz, *Shore and Sea Boundaries* (Dept. Comm. Pub. 10-1 1962) Appendix E, p. 670.

Equinoctial tides: Those tides occurring near the times of the equinoxes. *Tide and Current Glossary* (1984 U.S. Dept. of Comm., NOAA, NOS), p. .

Erosion: The opposite of accretion. A legal term describing the gradual and imperceptible wearing away or loss of littoral or riparian land by the action of the water. *Philadelphia Co. v. Stimson* (1911) 225 U.S. 605, 624–627; *Coastal Industrial Water Author. v. York* (Texas 1976) 532 S.W.2d 949, 952; *Jackson v. Burlington Northern, Inc.* (Mont. 1983) 667 P.2d 406, 407; Robillard, W., and Bouman, L., Clark on Surveying and Boundaries (7th Ed., 1997) § 24.02; Grimes, *Clark on Surveying and Boundaries* (4th Ed., 1976) § 567; 78 Am. Jur. 2d Waters § 406, p. 852.

Estuary: A semienclosed coastal body of water that is freely connected to the open sea and within which ocean water is diluted by fresh water from upland river and stream drainages. See *Tide and Current Glossary* (1984 U.S. Dept. of Comm., NOAA, NOS), p. 7.

General Land Office (GLO): An agency within the Department of the Interior responsible for the execution of the public land laws and for activities concerning the management and administration of public lands. It was abolished in 1946 when its functions were combined in the Bureau of Land Management (herein BLM). Robillard, W., and Bouman, L., Clark on Surveying and Boundaries (7th Ed., 1997) § 5.31; Grimes, *Clark on Surveying and Boundaries* (4th Ed., 1976) § 66; Gates, *A History of Public Land Law Development* (Pub. Land Law Rev. Comm. 1968), pp. 127–128. The Commissioner of the General Land Office was the head of that office. *Ibid.*

Geographic location: The particular latitudinal and longitudinal location of a specific, discrete physical feature, that can be definitely marked and measured and later relocated if lost, obliterated, or destroyed. *Kitteridge v.*

Ritter (Iowa 1915) 151 N.W. 1097. Used interchangeably with and the same as the term "Physical location."

Geomorphology: The study of the characteristics, origin, and development of landforms. Hinds, Norman, *Geomorphology* (Prentice-Hall 1943).

Groin: A solid structure that generally lies perpendicular to the shoreline and extends out into the foreshore. Its function is to interrupt the littoral drift and produce the deposition of sand on the updrift side of the groin to widen the beach. E.g., *Muchenberger v. City of Santa Monica* (1929) 206 Cal. 635, 639; *Lummis v. Lilly* (Mass. 1982) 429 N.E.2d 1146, 1148.

Half tide level: See "Mean tide level."

High-water mark: The equivalent of the term "ordinary high-water mark." *Goodtitle v. Kibbe* (1850) 50 U.S. (9 How.) 471, 477–478; *More v. Massini* (1860) 37 Cal. 432, 435; *Borax, Ltd. v. Los Angeles* (1935) 296 U.S. 10, 22–23.

Hydraulic mining: The process by which a bank of gold-bearing earth and rock is excavated by a jet of water, discharged through the converging nozzle of a pipe, under great pressure, the earth and debris being carried away by the same water, through sluices, and discharged on lower levels into the natural streams and water-courses below. *Woodruff v. North Bloomfield Gravel Min. Co.* (9th Cir. 1884) 18 F. 753, 756.

Inland waters: Inland or internal waters encompass all waterways within the land territory. Examples of inland waters are rivers, lakes and bays. Basically, these are lands that are landward of the nation's coastline, the line of ordinary low water in direct contact with the open sea. 1 Shalowitz, Shore and Sea Boundaries (Dept. Comm. Pub. 10-1 1962) § 311, pp. 22–23; *id.* at Appendix A, p. 283.

Internal waters: Same as "Inland waters."

International Great Lakes Datum: The reference system used for expressing elevations in the Great Lakes area. *State of Wisconsin v. Trudeau* (Wisc. 1987) 408 N.W.2d 337, 344 n.7, *cert. den.* (1988) 98 L.Ed.2d 652.

Jetty: A structure of stones, piles, or other materials projecting into the sea so as to protect a harbor or provide a pier or place for the landing of ships or vessels. See *United States v. California* (1980) 447 U.S. 1.

Judicial notice: An accepted legal device that allows courts to accept certain facts as true without any proof. Such facts have been referred to as ". . . facts, particularly with respect to geographical positions, of . . . public notoriety. . . ." *The Planter* (1833) 32 U.S. (7 Pet.) 324, 342; Fed. Rules of Evid. § 201 (b); Calif. Evid. Code § 450-460.

Law of the case: A rule of practice and policy that holds that once a legal issue is resolved at one stage of the litigation, a party is bound to that outcome at later stages and cannot again and again try the same question. *State v. Cain* (1967) 236 A.2d 501.

Legal character: The type or nature of legal title to the land.

Littoral land: Land bordering the ocean, a sea, or a lake. Black's Law Dictionary, (5th Ed., 1979) p. 842.

Meander line: A line described by courses and distances, being a straight line between fixed points or monuments, or a series of connecting straight lines. The line is thus fixed by reason of the difficulty of surveying a course following the sinuosities of the shore, and the impracticability of establishing a fixed boundary along the shifting sands of the ocean. *Den v. Spalding* (1940) 39 Cal.App.2d 623, 627.

Mean higher high water: The average of only the *higher of the high* waters at a location over a 19-year period. H.A. Marmer, *Tidal Datum Planes* (1951) U.S. Coast and Geodetic Survey, Spec. Pub. 135 Rev. Ed., p. 86; 2 Shalowitz, *Shore and Sea Boundaries* (Dept. Comm. Pub. 10-1 1962), p. 581.

Mean high water: The average height of *all the high waters* at a location for a period of 19 years. H.A. Marmer, *Tidal Datum Planes* (1951) U.S. Coast and Geodetic Survey, Spec. Pub. 135 Rev. Ed., p. 86; 2 Shalowitz, *Shore and Sea Boundaries* (Dept. Comm. Pub. 10-1 1962), p. 581. This is true for semidiurnal or mixed tides. There are some special rules for diurnal tides. *Ibid.*

Mean high-water line: The intersection of the tidal datum mean high water with the shore. 2 Shalowitz, *Shore and Sea Boundaries* (Dept. Comm. Pub. 10-1 1962), p. 581; *Swarzwald v. Cooley* (1940) 39 Cal.App.2d 306, 313; *People v. Wm. Kent Estate Co.* (1966) 242 Cal.App.2d 156, 160; *O'Neill v. State Highway Department* (N.J. 1967) 235 A.2d 1, 9.

Mean lower low water: The average height of only the *lower* of the low waters over a 19-year period. H.A. Marmer, *Tidal Datum Planes* (1951) U.S. Coast and Geodetic Survey, Spec. Pub. 135 Rev. Ed., p. 113; 2 Shalowitz, *Shore and Sea Boundaries* (Dept. Comm. Pub. 10-1 1962), p. 581.

Mean low water: The average height of *all the low* waters over a 19-year period. H.A. Marmer, *Tidal Datum Planes* (1951) U.S. Coast and Geodetic Survey, Spec. Pub. 135 Rev. Ed., p. 104; 2 Shalowitz, *Shore and Sea Boundaries* (Dept. Comm. Pub. 10-1 1962), p. 581.

Mean sea level: The mean level of the sea at a particular location determined by averaging the height of the water levels for *all stages* of the tide. This is the primary tidal datum, as all other tidal datums are derived with reference to mean sea level. H.A. Marmer, *Tidal Datum Planes* (1951) U.S. Coast and Geodetic Survey, Spec. Pub. 135 Rev. Ed., p. 45; 2 Shalowitz, *Shore and Sea Boundaries* (Dept. Comm. Pub. 10-1 1962), p. 528.

Mean tide level: This datum is also known as half tide level. It is a tidal datum *midway* between mean high water and mean low water. H.A. Marmer, *Tidal Datum Planes* (1951) U.S. Coast and Geodetic Survey,

Spec. Pub. 135 Rev. Ed., p. 69; 2 Shalowitz, *Shore and Sea Boundaries* (Dept. Comm. Pub. 10-1 1962), p. 568. Mean tide may, and usually does, differ from mean sea level.

Mixed tide: The type of tide in which there are two high and two low waters each day, but there are marked differences in elevation between the two high waters and the two low waters of the day. H.A. Marmer, *Tidal Datum Planes* (1951) U.S. Coast and Geodetic Survey, Spec. Pub. 135 Rev. Ed., p. 9.

Moon's declination: The moon's relation to the earth's equator.

Moon's phase: A regularly recurring aspect of the moon with respect to the amount of illumination produced by the moon; i.e., new moon, full moon.

National Geodetic Vertical Datum (NGVD): Formerly known as Sea-Level Datum 1929. NGVD is a fixed reference datum, or plane, adopted as a standard geodetic datum for elevations determined by leveling. *Tide and Current Glossary* (Dept. of Comm., Nat. Oceanic and Atmos. Admin. 1984), p. 14. In layman's terms, NGVD is the zero (0) point or beginning level of all geodetic elevations. One court defined NGVD as "the fixed vertical reference point from which all land elevations are measured." *State of California, ex rel. Public Works Board v. Southern Pacific Transportation Company* (1983) Sacramento County Superior Court No. 277312, Appendix A, para. B.10.

National Ocean Survey (NOS): Successor to the United States Coast and Geodetic Survey. See "United States Coast Survey."

National Tidal Datum Epoch: The specific 19-year period adopted by the NOS as the official time segment over which tide observations are taken and reduced to obtain mean values for tidal datums. This standardization is necessary because of periodic and secular trends in sea level; i.e., sea level is apparently rising over time. *Tide and Current Glossary* (Dept. of Comm., Nat. Oceanic and Atmos. Admin. 1984), p. 14.

Neap Tides: The tides that rise and fall least, and thus have a lesser range, occur at about the time of the moon's first and third quarters and are known as "neap tides." These tides are produced when the sun and moon are opposite each other. H.A. Marmer, *Tidal Datum Planes* (1951) U.S. Coast And Geodetic Survey, Spec. Pub. 135 Rev. Ed., p. 5; Patton, R.S., "Relation of the Tide to Property Boundaries" reprinted in 2 Shalowitz, *Shore and Sea Boundaries* (Dept. Comm. Pub. 10-1 1962), Appendix E, p. 669.

Ordinary high-water Mark: The waterward property boundary of littoral and riparian lands and the landward property boundary of tidelands and sovereign lands. It is not a scientific term.

Ordinary low-water Mark: The seaward boundary of tidelands and the landward boundary of submerged lands. *E.g., United States v. Pacheco* (1864) 69 U.S. (2 Wall.) 578, 590; *San Francisco Sav. Union v. Irwin* (9th

Cir. 1886) 28 F. 708, 713, *aff'd* (1890) 136 U.S. 578; *People v. California Fish Co.* (1913) 166 Cal. 576, 584.

Perigean tides: Tides that occur when the moon is nearest the earth, or in perigee. The tide range is greater; the tide rises higher and falls lower than usual. H.A. Marmer, *Tidal Datum Planes* (1951) U.S. Coast and Geodetic Survey, Spec. Pub. 135 Rev. Ed., p. 5; Patton, R.S., "Relation of the Tide to Property Boundaries" reprinted in 2 Shalowitz, *Shore and Sea Boundaries* (Dept. Comm. Pub. 10-1 1962), Appendix E, pp. 669–670.

Photogrammetric survey: A survey of a land area utilizing aerial photographs. The survey is reduced to map form by use of stereoscopic or similar equipment. 2 Shalowitz, *Shore and Sea Boundaries* (Dept. Comm. Pub. 10-1 1962), Appendix A, p. 591.

Physical character: The physical features and attributes of land.

Physical location: Used interchangeably with and the same as the term "Geographic location."

Planetable: A surveying and cartographic instrument consisting of a drawing board mounted on a tripod. Attached to the drawing board is an instrument called an "alidade." An alidade is simply a ruler on which a telescope was mounted. 2 Shalowitz, *Shore and Sea Boundaries* (Dept. Comm. Pub. 10-1 1962) § 411, p. 160.

Planimetric map: A map that presents the horizontal position of the features represented. 2 Shalowitz, *Shore and Sea Boundaries* (Dept. Comm. Pub. 10-1 1962), Appendix A, p. 592. In some cases, such maps may show the horizontal position of elevations or contours of equal elevation. However, in strict definitional terms, planimetric surveys or maps omit relief.

Presumption: "An assumption of fact that the law requires to be made from another fact or group of facts found or otherwise established in the [lawsuit]." Calif. Evid. Code § 600(a).

Public domain: " 'Public domain' is equivalent to 'public lands' and these words have acquired a settled meaning in the legislation of this country." *Barker v. Harvey* (1901) 181 U.S. 481, 490. See also "Public lands."

Public lands: The lands of the United States (or of a state) that are subject to sale and disposal under general laws authorizing the disposition and sale of lands. *Barker v. Harvey* (1901) 181 U.S. 481, 490; 73A C. J. S. Public Lands, § 2; *Newhall v. Sanger* (1875) 92 U.S. 761, 763. See also "Public domain."

Public trust: A property interest, an easement, that the public enjoys over tidelands for public purposes and uses. E.g., *Summa Corp. v. California ex rel. State Lands Com'n* (1984) 466 U.S. 198, 205.

Purpresture: An unauthorized encroachment (such as a pier or dock or even hydraulic mining debris) in the bed of or in navigable waters so as to occupy that which should be public open to the enjoyment of the public at large. A purpresture is a particular kind of nuisance. *People v. Gold Run D. & M. Co.* (1884) 66 Cal. 138, 146–147; Black's Law Dictionary (4th Ed. 1957), p. 1401.

Range of tide: The vertical magnitude of the rise and fall of the tide. H.A. Marmer, *Tidal Datum Planes* (1951) U.S. Coast and Geodetic Survey, Spec. Pub. 135 Rev. Ed., p. 4.

Reference datum: The datum to which measurements are referred. For example, on USC&GS hydrographic charts the depths are measured in relation to mean lower low water, the reference datum of such map. See also "National Geodetic Vertical Datum."

Reemergence: The process by which land that once was uncovered by water becomes covered and then uncovered by water. *Arkansas v. Tennessee* (1918) 246 U.S. 158, 174; *Herron v. Choctaw & Chickasaw Nations* (10th Cir. 1956) 228 F.2d 830, 832; *Horry County v. Woodward* (S.C.App. 1984) 318 S.E.2d 584, 587.

Reliction: A legal term that describes the physical process whereby land that was once covered with water becomes exposed or uncovered by the imperceptible recession of the water, usually when the water level is lowered. Robillard, W., and Bouman, L., Clark on Surveying and Boundaries (7th Ed., 1997) § 24.08; Grimes, *Clark on Surveying and Boundaries* (4th Ed., 1976) § 573; *Bear v. United States* (D.C. Neb. 1985) 611 F.Supp.589, 593 n.2, *aff'd* (8th Cir. 1987) 810 F.2d 153; *Flisrand v. Madson* (S.D. 1915) 152 N.W. 796, 798. The terms "reliction" and "accretion" are often used interchangeably. *Bear, supra,* 611 F.Supp. at 593 n.2; *Cal. ex rel. State Lands Com'n v. United States* (9th Cir. 1986) 805 F.2d 857, 860, *cert. den.* (1981) 98 L.Ed.2d 34. An element of permanence of the recession of the water is required by some courts. E.g., *Sapp v. Frazier* (La. 1899) 26 So. 378, 380; *Herschman v. State Department of Natural Resources* (Minn. 1975) 225 N.W.2d 841, 843.

Riparian land: Land along a river or stream. Black's Law Dictionary (5th Ed. 1979), p. 1192.

Rule of property: A settled rule or principle, resting usually on precedents or a course of decisions, regulating the ownership or devolution of property. *Abbott v. City of Los Angeles* (1958) 50 Cal.2d 438, 456–457.

Seabed: The land below the ordinary high-water mark. *United States v. Ray* (5th Cir. 1970) 423 F.2d 16, 20.

Segregation: The process by which public lands are determined no longer subject to disposal under the public land laws. *Utah Div. of State Lands v. United States* (1987) 482 U.S. 193, 107 S.Ct. 2318, 2325. With respect to swamp and overflowed lands, the term means the process by which public lands of the character of swamp and overflowed lands were identified and separated from other public lands.

Semidaily tide: The type of tide in which there are two tidal cycles each day and such tidal cycles resemble each other. H.A. Marmer, *Tidal Datum*

Planes (1951) U.S. Coast and Geodetic Survey, Spec. Pub. 135 Rev. Ed., p. 9.

Semidiurnal tide: See "Semidaily tide."

Shore: The land that is periodically covered and uncovered by the tide; the term is sometimes applied to a river or a pond as synonymous with "bank." *Freeman v. Bellegarde* (108 Cal. 179, 187. See also "Bank."

Sovereign lands: Lands that inure to the states by virtue of their sovereignty pursuant to the Equal Footing Doctrine. This term encompasses both tidelands and lands underlying navigable waters.

Spring tides: During the phases of the new and full moon, when the sun and the moon pull together, the tides rise higher and fall lower and thus have greater range and are known as "spring tides." H.A. Marmer, *Tidal Datum Planes* (1951) U.S. Coast and Geodetic Survey, Spec. Pub. 135 Rev. Ed., p. 5; Patton, R.S., "Relation of the Tide to Property Boundaries" reprinted in 2 Shalowitz, *Shore and Sea Boundaries* (Dept. Comm. Pub. 10-1 1962), Appendix E, p. 669.

State Surveyor General: State official charged with the management and disposition of state lands. *Boone v. Kingsbury* (1928) 206 C. 148, 184-185, *cert. den.* and *app. dism'd. sub. nom Workman v. Boone* (1929) 280 U.S. 517; Cal. Pub. Res. Code § 6102.

Submerged Lands Act: Act of May 22, 1953, 67 Stat. 29, set forth in 43 U.S.C. §§ 1301, et seq. The Submerged Lands Act was enacted as result of the decision in *United States v. California* (1947) 332 U.S. 19, 38. The Submerged Lands Act was enacted to undo the effect of that decision. *United States v. California* (1978) 436 U.S. 32, 37.

Submergence: The reverse of reliction. The legal term denoting the physical process of the gradual and imperceptible disappearance of land under water and the formation of a navigable water body over it. *Coastal Industrial Water Auth. v. York* (Texas 1976) 532 S.W.2d 949, 952–953; *Port Acres Sportsman Club v. Mann* (Texas 1976) 541 S.W.2d 847, 849. See Robillard, W., and Bouman, L., Clark on Surveying and Boundaries (7th Ed., 1997) § 24.14; Grimes, *Clark on Surveying and Boundaries* (4th Ed., 1976) § 579. Some seem to equate erosion and submergence. *Ibid.*; 78 Am.Jur. 2d. Waters, § 421, p. 868.

Subsidence: The lowering of the relative elevation of the entire region.

Thalweg: The channel continuously used for navigation or the middle of the principal channel of navigation. E.g., *Omaha Indian Tribe, Treaty of 1854, etc. v. Wilson* (8th Cir. 1978) 575 F .2d 620, 623, n. 6, *vac.* and *remanded* (1979) 442 U.S. 653.

Tidal datum: "A surface used as a reference from which heights or depths are reckoned when defined by a phase of the tide . . ." 1 Shalowitz, Shore and Sea Boundaries (Dept. Comm. Pub. 10-1 1962), p. 286. Although the term "tidal datum plane" is sometimes also used, that term is redundant as a "datum" is a "plane" of reference. In this work, the more technically correct term "tidal datum" will be used. 2 Shalowitz, *Shore and Sea Boundaries* (Dept. Comm. Pub. 10-1 1962), at p. 611. See also "Datum."

Tidal datum plane: See "Tidal datum."

Tidal prism: The measure of volume of the tide. It can be conceived of as the area or volume of water between the plane of high tide and the plane of low tide within the bay. Gilbert, L. K., *Hydraulic Mining Debris in the Sierra Nevada* USGS Prof. Paper No. 105 (GPO 1917), p. 71.

Tide: The vertical movement of the surface of the water.

Tide curve: A graphic depiction of the rise and fall of the tide. Time is usually represented by the abscissa and height by the ordinate of the graph. *Tide and Current Glossary* (Dept. of Comm., Nat. Oceanic and Atmos. Admin. 1984), p. 22.

Tidelands: Land that is covered and uncovered by the daily rise and fall of the tide. 2 Shalowitz, *Shore and Sea Boundaries* (Dept. Comm. Pub. 10-1 1962), Appendix A, p. 612. Some courts have more precisely defined tidelands as the lands between the lines of mean high water and mean low water. *City of Berkeley v. Superior Court* (1980) 26 Cal.3d 515, 519, n.1, *cert. den. sub.nom Santa Fe Land Improv. Co. v. Berkeley* (1980) 449 U.S. 840.

Tides: The periodic, usually twice daily, rise and fall of the water surface resulting from the gravitational interactions between the earth, moon, and sun. *Tide and Current Glossary* (Dept. of Comm., Nat. Oceanic and Atmos. Admin. 1984), p. 21; H.A. Marmer, *Tidal Datum Planes* (1951) U.S. Coast and Geodetic Survey, Spec. Pub. 135 Rev. Ed., p.1.

Title: The means, method, or evidence of one's ownership of land. 63A Am.Jur.2d Property, § 30; Restatement Property, § 10; *Arraington v. Liscom* (1868) 34 Cal. 365, 385.

Tropic tides: Tides that occur when the earth is farthest from the equator at maximum declination. H.A. Marmer, *Tidal Datum Planes* (1951) U.S. Coast and Geodetic Survey, Spec. Pub. 135 Rev. Ed., p. 5; Patton, R.S., "Relation of the Tide to Property Boundaries" reprinted in 2 Shalowitz, *Shore and Sea Boundaries* (Dept. Comm. Pub. 10-1 1962), Appendix E, p. 670.

Tule: An emergent aquatic plant, a plant rooted in soil with shoots sticking up above the surface of the water. Judicial notice has been taken of where tules will grow. *United States v. Otley* (9th Cir. 1942) 127 F.2d 988, 1000. Note a biblical reference: Moses was found in the tules (aka bulrushes).

Type of tides: The character of the rise and fall of the tide as revealed by an examination of the tide curve. H.A. Marmer, *Tidal Datum Planes* (1951) U.S. Coast and Geodetic Survey, Spec. Pub. 135 Rev. Ed., p. 9.

United States Coast and Geodetic Survey (USC&GS): Successor to the United States Coast Survey. The USC&GS has in turn been succeeded by the National Ocean Survey and the United States Geodetic Survey. See "United States Coast Survey."

United States Coast Survey (USCS): A United States government agency charged with, among other things, the measurement of tides and the mapping of the coastal shoreline of the United States in connection with the preparation of charts for navigation. See generally 2 Shalowitz, *Shore and Sea Boundaries* (Dept. Comm. Pub. 10-1 1962), Chap. 1; *Smith v. State* (Ga. 1981) 282 S.E.2d 76, 79.

United States Geological Survey (USGS): A unit of the Department of the Interior created to assist in the classification of public lands and in the examination of the geological structure, mineral resources, and products of the public domain. 43 U.S.C. § 31(a) and (b); 5 West's Federal Practice Guide (1970) § 5235, p. 21.

United States Surveyor General: Prior to 1925, the United States Surveyor General was the officer in charge of the survey of the public lands in the General Land Office, an agency of the Department of the Interior. Robillard, W., and Bouman, L., Clark on Surveying and Boundaries (7th Ed., 1997) § 5.30; Grimes, *Clark on Surveying and Boundaries* (4th Ed., 1976) § 67. At one time there were 15 United States surveyors general. Gates, *A History of Public Land Law Development* (Pub. Land Law Rev. Comm. 1968), p. 127. For example, the United States surveyor general for California was responsible for public land surveys in California. Do not confuse with state surveyor general, a state official. See "State surveyor general."

Warping: A term that describes the method of reclaiming tidal marshlands by leveeing the marsh, then directing sediment-laden freshwater sources over the land, allowing the sediment to be deposited and the salts to be leached out.

Watercourse: In general, a geomorphic feature that usually consists of water, a bed, and banks and shore. Porro, "Invisible Boundary—Private and Sovereign Marshland Interests" (1970), 3 Natural Resources Lawyer 512, 516.

Weir: A dam or obstruction that backs up or diverts a stream. Ballentine's Law Dictionary (3d Ed. 1949), p. 1364.

APPENDIX A

DELTA MEADOWS DECISION—*STATE V. SOUTHERN PACIFIC*

ENDORSED:

SEP 13 1983

CLERK

Deputy

IN THE SUPERIOR COURT OF THE STATE OF CALIFORNIA

IN AND FOR THE COUNTY OF SACRAMENTO

STATE OF CALIFORNIA, ex rel. STATE PUBLIC WORKS BOARD,

    Plaintiff,

vs.

SOUTHERN PACIFIC TRANSPORTATION COMPANY, etc., et al.,

    Defendants.

No. 277312   DEPT. 17

STATEMENT OF DECISION

This action came regularly on for trial by the Court on June 8, 1982, and continued through and including July 22, 1982. Deputy Attorneys General

Bruce S. Flushman and Richard M. Frank appeared on behalf of plaintiff, State of California, ex rel. State Public Works Board ("plaintiff"). James T. Freeman appeared on behalf of defendant Southern Pacific Transportation Company ("S.P.").

This is an eminent domain action brought by the State of California, acting by and through the Public Works Board, against S.P. and certain other defendants. This action was brought to acquire certain property in the Sacramento-San Joaquin Delta near Walnut Grove and along Snodgrass Slough for a park. A portion of the proposed "Delta Meadows" park property is owned by S.P. ("litigation area"). Plaintiff claims that it has a sovereign ownership interest in the litigation area by virtue of its character as tide and submerged lands held in trust for the benefit of all its citizens.

Except for certain works constructed through a portion of the property for a railroad right-of-way, Delta Meadows is mostly undeveloped.

There are two distinct aspects to this litigation. The first relates to plaintiff's contention, made pursuant to Code of Civil Procedure section 1250.310(c), that the State has a preexisting sovereign interest in the parcel to be condemned and claimed by S.P. The second aspect involves the valuation phase of the condemnation proceeding, at which time the fair market value of the portion of the property found to be owned exclusively by S.P. is determined.

In December 1980, the Court granted plaintiff's motion to sever and try initially the issue relating to the State's sovereign title. Accordingly, the valuation phase of this case has been reserved for a later trial. Since all defendants other than S.P. claim title under leases from S.P., the Court in its pretrial order has excused them from this phase of the case while binding them to the Court's determination in the title phase.

The principal issue in the initial portion of the litigation is whether the State does indeed possess a sovereign interest in all or a portion of the subject property. Stated somewhat differently, does all or a portion of the litigation parcel consist of tidelands to which the State's public trust easement attach? A necessarily-related issue, should the above question be answered affirmatively, is the location of the boundary between those sovereign lands and the swamp and overflowed lands that the parties agree are owned outright by S.P.[1]

S.P. takes the position that none of the property at issue is or ever was tidelands. It further raises the defenses that even if the State is correct that tidelands do exist within the parcel boundaries, plaintiff's claim 1) is barred

---

[1]. The State does not dispute S.P.'s legal title to the swamp and overflowed lands adjoining the alleged tidelands. There is no dispute that S.P. is successor in interest to swamp and overflowed patents issued by the State in the late eighteenth and early nineteenth centuries. Similarly, plaintiff concedes that S.P. has a fee interest in the tidelands at issue as a result of legislation passed over a century ago. However, the State contends that it necessarily reserved and still retains a public trust easement over those tidelands. Relying on *People v. California Fish Co.* (1913) 166 Cal. 579, and related decisions, the State argues that the easement is a fundamental attribute of sovereignty and is incapable of being alienated or extinguished.

by the provisions of Public Resources Code section 6360 and 7552; and 2) has been terminated or abandoned.[2]

It has been agreed between the parties that:

1. Upon California's admission to the Union on September 9, 1850, the State was vested with sovereign title to all lands underlying navigable waters, tide and submerged lands within its borders. These sovereign lands are held by the State in trust for the benefit of its citizens.

2. On September 28, 1850, the United States ceded to California all previously unpatented swamp and overflowed lands within its borders under the Arkansas Swamp Land Act. Such lands were part of the public lands of the United States. Such swamp and overflowed lands were proprietary lands subject to absolute disposition of the State free of any public trust easement.

The Court has heard the testimony of plaintiff's expert witnesses, Dr. John Thompson, Mr. Richard Hansen, Dr. Warren Thompson, Mr. Eugene Begg, Mr. Thomas Whitlow and Mr. Jerry Elliott, as well as the testimony of defendant Southern Pacific's expert witnesses, Albert Hurtado and James Dorsey. The Court has considered that testimony and all the other evidence admitted at trial, together with the pleadings and arguments of counsel including the pretrial and post-trial briefs filed on behalf of plaintiff and defendant Southern Pacific. Having issued its tentative decision on December 27, 1982, and S.P. having requested a Statement of Decision, the Court hereby issues the following Statement of Decision as to the principal controverted issues at trial in accordance with Code of Civil Procedure section 632.

A.

PRIOR FEDERAL AND STATE DETERMINATIONS AS TO THE CHARACTER OF THE LAND AS SWAMP AND OVERFLOWED ARE NOT CONCLUSIVE. SUCH DETERMINATIONS WERE NOT INTENDED TO DISTINGUISH THE BOUNDARY BETWEEN TIDE AND SUBMERGED LANDS AND SWAMP AND OVERFLOWED LANDS. PUBLIC RESOURCES CODE SECTION 7552 DOES NOT REQUIRE THAT THE LITIGATION AREA BE HELD TO BE SWAMP AND OVERFLOWED LAND.

---

2. In its answer to the complaint, S.P. also raised the affirmative defense of estoppel in an effort to bar the State's claim. The Court granted plaintiff's motion to eliminate estoppel as an issue in the case on August 10, 1981, finding that as a matter of law estoppel could not be raised to defeat the State's sovereign interest. (*State of California v. Superior Court (Fogerty)* (1981) 29 Cal.3d 240, 244-247, cert. den. (1981) 454 U.S. 865.)

1. Upon California's admission to the Union on September 9, 1850, California was vested with sovereign title to all lands underlying navigable and tidal waters within its borders. These sovereign lands are held by California in trust for the benefit of its citizens. This is undisputed.

2. On September 28, 1850, by virtue of the Arkansas Swamp Lands Act (Act of September 28, 1850, 9 Stat. 519 (now 43 United States Code section 981, *et seq.*) ("Swamp Lands Act")) the United States granted to California a portion of the public lands of the United States, all previously ungranted swamp and overflowed lands. Such swamp and overflowed lands are held by California as propriety lands subject to California's absolute disposition free of any public trust easement. This is undisputed.

3. Although the Swamp Lands Act was a present grant of all public lands that were swamp and overflowed on September 29, 1850, such public lands must first be identified as falling within the terms of the grant. (*Wright v. Roseberry* (1887) 121 U.S. 488, 509.)

4. Township plats were prepared by the United States for the purpose of depicting the extent and location of the public lands available for disposition by the United States and the identification of swamp and overflowed lands already granted by the United States to California.

5. The identification of swamp and overflowed lands was subject to confusion, as the legal definition of swamp and overflowed lands and tidelands could easily describe the same physical character of land, and it was difficult to physically distinguish between these two legal characters of land. Further, the United States' identification of swamp and overflowed lands purported to locate only the boundary between swamp and overflowed lands and other species of public lands.

6. At the time of the survey of the township in which Delta Meadows was located, federal surveyors did not separate tidelands from swamp and overflowed lands in the Delta Meadows area.

7. By virtue of the Swamp Lands Act, the United States may only convey lands that belong to it. The United States has no power to convey lands which belong to California as tidelands. The jurisdiction of the United States over the public lands therefore does not include tidelands, which are not public lands of the United States. As tidelands came to California on its admission, 19 days before the Swamp Lands Act grant, California already held title to such lands. The Swamp Lands Act could not purport to convey such lands. The United States, in its segregation survey of the township in which the Delta Meadows area was located, had no jurisdiction to determine the extent or the boundary of the State's tidelands.

8. For determinations by United States officials concerning the public lands to be conclusive, such officials must have been acting within their jurisdiction.

9. It remains open to plaintiff to show that the lands in the litigation area were tidelands and not swamp and overflowed lands, and thus not within the jurisdiction of the United States to grant to California. Thus, the United States'

determination of the character of the land in the litigation area as swamp and overflowed is not conclusive.

10. California issued patents to lands incorporating the litigation area purporting to convey such lands to private purchasers as swamp and overflowed.

11. Patents were issued by the State for swamp and overflowed lands based on the applications of purchasers therefor. It was not the State that characterized the land as such but private applicants for purchase. If the papers were regular on their face the State Surveyor General lacked discretion to deny the patent. Nor did he have authority to determine whether the lands were in fact tidelands swamp and overflowed lands, or even submerged lands or other lands not subject to sale at all.

12. Accordingly, the issuance of a State patent does not enhance the United States' authority to convey swamp and overflowed lands to the State. If the subject matter jurisdiction of the United States, its authority to convey the lands as swamp and overflowed, is at issue, questions concerning land character are proper for Court determination. On the other hand, S.P. claims, pursuant to Public Resources Code section 7552 ("section 7552"), that federal and state determinations of the character of the land as swamp and overflowed are conclusive on plaintiff.

13. Public Resources Code section 7552 ("Section 7552") provides, "Lands within this State which are returned by the United States as swamp and overflowed lands, and shown as such on approved township plats, shall, as soon as patents are issued therefor by this State, be held to be of the character so returned. Nothing in this section shall be construed to affect the rights of any homestead or preemption settler claiming under the laws of the United States, nor to prejudice the rights of any settler located upon such lands to perfect title to the lands if permitted under existing laws."

14. The State is bound by the public trust to preserve and protect its public easement and use in tidelands. Statutes purporting to authorize an abandonment of this State sovereign easement must be carefully scanned to ascertain whether or not such was the intent of the Legislature. That intent must be clearly expressed or necessarily implied. It will not be implied if any other inference is reasonably possible. If any interpretation of the statute is reasonably possible which would not involve a destruction of the public use or intention to terminate it in violation of the trust the statute must be given that interpretation.

15. Neither the language nor the legislative history of section 7552 give any consideration whatsoever to the sovereign trust interest of the State in tide or submerged lands, or to the public trust interest under which lands were held. To hold that, as a result of the application of section 7552, the litigation area is swamp and overflowed land would preclude plaintiff from ever proving that the lands identified and returned by the United States as swamp and overflowed and patented as such were, in fact, tide and submerged lands. Such a holding would result in the alienation of the State's sovereign interest

in tide and submerged lands erroneously characterized as swamp and overflowed lands. Alienation of extended amounts of tide and submerged lands through such means is beyond the power of the legislature.

16. No reported decision has cited or relied on section 7552. Cases decided after enactment of section 7552 hold that where there is a conflict between titles derived from the United States and California, the actions of one sovereign cannot bind the other and the question of title should be adjudicated in Court.

17. Also, section 7552 is ambiguous. If lands were returned by the United States as swamp and overflowed and shown as such on United States township plats, and if the State issued a tideland or school land patent for such lands, the statute would preclude the Court from ever considering evidence that such land was in fact tidelands or school lands and not swamp and overflowed.

18. An additional ambiguity in section 7552 is that excepted from its effect are those persons who could perfect title "under existing laws." Such "existing laws" by which a person could perfect title to lands that come within the ambit of section 7552 are nowhere specified. Nor does the statute specify whether such "existing laws" are state or federal, statutory, case or common law.

19. In light of the ambiguity of section 7552 and the respected rule of construction that such statutes must be construed to avoid the abandonment of the State's sovereign trust interest, the Court finds section 7552 to be applicable only when 1) lands are truly swamp and overflowed in character; 2) have been identified and returned as such by the United States; and 3) the State has issued a swamp and overflowed patent therefor. In such a case the such lands are held to be of the character so returned. In this case, S.P.'s reliance on section 7552 is misplaced, however, because the first condition of its application cannot be satisfied. (See part B, *infra*.)

B.

A PORTION OF THE PROPERTY IN DISPUTE IS TIDELANDS AND A PORTION IS SWAMP AND OVERFLOWED LAND.

1. The boundary of the public lands, of which swamp and overflowed lands are a specie, when such lands abut navigable or tidal waters, is the ordinary high water mark; conversely the shoreward boundary of tidelands is the ordinary high water mark. This boundary must be located to determine

which portion of the litigation area was tidelands in which California retains a public trust easement and which portion was swamp and overflowed land owned in fee by S.P.

2. General rules concerning the mapping of geographical features do not provide assistance in locating the property boundary between tidelands and swamp and overflowed lands. Further, United States Coast and Geodetic Survey charts do not locate the high water line in tidal marshes; only the apparent dividing line between the marsh land and water is located.

3. Further, smaller bodies of tidelands included within a larger body of swamp and overflowed lands are not treated as part of the swamplands as there is an inalienable public easement in tidelands.

4. The "ordinary high water mark" is not a physical term nor is there an accepted physical or engineering definition. Case law does not provide a set formula to translate the legal term of "ordinary high water mark" into a physical location in a marsh such as Delta Meadows.

5. In a case where the topography can be established as well as the water level through tidal measurements, the ordinary high water mark is the mean high water line: the intersection of the plane of mean high water and the topography.

6. Plaintiff has established to the Court's satisfaction the two elements of the boundary determination in this case: the water level and the character and elevation of the shore against which that water elevation is to be applied.

7. Plaintiff established that the boundary water course, Snodgrass Slough, is tidal. It further demonstrated that the water levels in Snodgrass Slough are primarily controlled by the water levels in the ocean.

8. Plaintiff established the elevation of mean high water of Snodgrass Slough in the Delta Meadows reach in 1967 using actual tide records obtained from the Department of Water Resources.

9. The elevation of Mean High Water for Snodgrass Slough in the Delta Meadows reach in the 1967 mid tidal epoch year was 2.73 feet above National Geodetic Vertical Datum ("NGVD"), the fixed vertical reference point from which all land elevations are measured.

10. Through the expert testimony of Mr. Elliott and Dr. Warren Thompson, the historic mean high water elevations for the 1850 and 1914 tidal epochs were re-created. These historic water levels were re-created by taking into consideration the factors which influenced the historic water levels in the Sacramento-San Joaquin Delta that would have made such water levels in 1850 and 1914 different from the water levels in 1967. These factors considered were: 1) the Central Valley Project and State Water Project reservoirs and pumping out of the Delta; 2) the deposition of hydraulic mining debris; 3) the long-term rise in mean sea level; 4) Delta reclamation; 5) dredging in the Delta; 6) operation of the Delta Cross-Channel, and 7) any unnatural seepage. Each factor was quantified or otherwise considered. It was established that mean high water for Snodgrass Slough in the Delta Meadows reach in 1914 was 2.39 feet above NGVD and mean high water at that location in 1850 was 2.05 feet above NGVD.

11. Plaintiff also established the location of the perimeter water course, Snodgrass Slough.

12. By expert testimony, the topography of Delta Meadows, including the litigation area, was re-created, as such topography would have existed had the railroad not been constructed in the litigation area.

13. The elevation of mean high water was then located on such topography as a contour line of equal elevation. This established the location of the mean high water line.

14. Corroborating physical evidence is also useful in determining the ordinary high water mark. The soils and vegetation, geomorphological and channel patterns supported and verified the placement of the mean high water line. Also, establishing the ordinary high water mark by using both the statistical technique of determining the elevation of mean high tide and corroborating physical evidence provided by expert witnesses which examined the physical evidence is an acceptable technique.

15. The mean high water line is shown on Exhibit 166 and reflects that a portion of the property was below the ordinary high water mark on September 9, 1850. Such portion of the property is 10.5 percent of the litigation parcel. (Exhibit 170)

16. Defendant initially proved its record title to the litigation area. Thereafter, the burden of proof was on plaintiff to show that a portion of the property was tidelands. Plaintiff made a prima facie case that a portion of the litigation area was tidelands. The burden of producing evidence then shifted to defendant. There is a presumption in favor of sovereign ownership which has not been overcome by any evidence introduced by defendant Southern Pacific. The Court finds that S.P. has failed to introduce any meaningful evidence to contradict plaintiff's evidence regarding the location of the ordinary high water mark in this case.

C.

THE BOUNDARY BETWEEN STATE SOVEREIGN TIDE-LANDS AND S.P.'S SWAMP AND OVERFLOWED LANDS WAS FIXED AS OF SEPTEMBER 9, 1850, AND HAS NOT CHANGED SINCE THAT DATE.

1. As indicated above, the ordinary high water mark separating tidelands from swamp and overflowed lands within the subject property in 1850 is a depicted on Exhibit 166. The question then arises whether that boundary remains fixed as a matter of law or, alternatively, shifts as a result of changes to the water boundaries occurring after 1850.

2. Plaintiff maintains that the boundary between State sovereign lands and private uplands changes as the water boundary is altered by a variety of gradual forces, but does not move in response to sudden or avulsive changes. Plaintiff further contends, and the evidence supports the view, that a multitude of both gradual and avulsive forces served to change the water boundaries of

tidelands within Delta Meadows after 1850. While few precedents directly speak to the issue, plaintiff urges the Court to determine which types of forces—gradual or avulsive—were controlling or dominant at any point in time. Based on this view of the law, plaintiff suggests that the changes to the water boundary were primarily gradual in nature between 1850 and the time of the railroad right-of-way's construction through Delta Meadows at the beginning of the 20th century. According to plaintiff's theory, the legal boundary would therefore continue to move up until that time. Plaintiff further suggests that the railroad construction constitutes an avulsive change that "froze" the boundary between tidelands and swamp and overflowed lands as of 1907.

3. S.P. takes a different view. It agrees that the boundary continues to move after 1850, but disputes plaintiff's distinction between gradual and avulsive changes. Instead, it argues that the Court should adopt a "rule of convenience" and establish the boundary between the State's and S.P.'s lands by reference to present-day conditions.

4. The Court rejects both positions. First, long-standing precedent supports the legal distinction between gradual and avulsive changes to watercourses. Unlike the former, avulsive shifts do not alter the boundary between sovereign and private lands. Instead the boundary remains fixed at the location existing immediately prior to the sudden change. (*City of Long Beach v. Mansell* (1970) 3 Cal.3d 462, 469 (fn. 4); *Arkansas v. Tennessee* (1918) 246 U.S. 158, 173.)

5. The Court is not satisfied that it can determine with the necessary degree of precision the nature of the changes affecting the applicable boundary between 1850 and the advent of the railroad. Considerable testimony was presented regarding the effects of regional flooding, reclamation, hydraulic mining in the Sierra Nevada, the gradual rise in sea level and a host of other factors affecting the boundary over this time span. The testimony indicated that the individual effects of these changes were not specifically discernible. Some of these forces were naturally induced; others by the acts of man. In light of this fact and the absence of any California authority to support plaintiff's "controlling influence" theory, the Court rules that the boundary remains fixed as of the date of California's admission to the Union in 1850.

D.

THE STATE'S PUBLIC TRUST EASEMENT IN TIDELANDS WITHIN THE SUBJECT PROPERTY HAS NOT BEEN EXTINGUISHED OR OTHERWISE TERMINATED.

1. S.P. suggests that even if a portion of the litigation parcel constituted tidelands in 1850 and was therefore subject to the public trust, the State's trust easement has nonetheless been extinguished through intervening circumstances. This argument appears to be based on the fact that these tidelands were originally encompassed within swamp and overflowed patents issued to

S.P.'s predecessors in interest, as well as upon the California Supreme Court's recent decision in *City of Berkeley v. Superior Court* (1980) 26 Cal.3d 515. The Court finds S.P.'s argument unpersuasive.

2. As noted above, public lands sales including the swamp and overflowed land deposition process were subject to many problems. Many acres of sovereign tide and submerged lands in California were erroneously patented into private ownership as a result of these problems. One legislative reaction was to adopt a series of so-called "curative acts" designed to remedy in part the title problems engendered by the defective patents. In a series of statutes commencing in 1860 (Cal.Stats. 1861, ch. 356, p. 363), the Legislature ratified and confirmed prior sales of certain tidelands inadvertently made in connection with patents of swamp and overflowed lands. These acts were generally retroactive in effect and did not encompass submerged lands.

3. These curative acts did not, however, serve to extinguish the State's public trust interest in tidelands encompassed within descriptions of purported swamp and overflowed land patents. This principle was established in the California Supreme Court's decision in *People v. California Fish, supra,* 166 Cal. 577, and reaffirmed in subsequent cases. The Court in *California Fish*, after thoroughly analyzing the process by which the State sold sovereign lands, held that such curative acts 1) made no provision for the separation of tidelands from swamp and overflowed lands; and 2) never gave consideration to the fact that, while various statutes provided for the sale of both types of land, tidelands could only be sold subject to a public easement for commerce, navigation and fishery necessary to protect the paramount interest of the public in such sovereign tidelands. (*Id.* at 589-592.) The Court determined such statutes should not be construed to impair or limit the sovereign power of the State to act in its governmental capacity or to perform its governmental function on behalf of the public unless such an intent specifically appears. (*Id.* at 593.) Finding that because these land sales statutes did not consider such public sovereign interests, *California Fish* stands for the principle that they leave the sovereign public trust easement "unimpaired and unaffected, subject to the future control and regulation of the state."

4. Accordingly, the Court finds that such government actions as issuance of a swamp and overflowed patent or passage of a curative act do not constitute a termination of the public easement for commerce, navigation, fisheries and related trust uses with which sovereign lands are impressed. Such patents or legislation merely convey the title to the soil underlying the waters, subject to the continuing public rights retained under the trust.

5. S.P.'s argument that its filling and construction of improvements on a portion of the property terminates the trust is also incorrect.

6. *City of Berkeley*, relied upon by S.P., in no way compels a contrary result. That decision expressly reaffirms the views first enunciated in *California Fish*. (26 Cal.3d 515, 528.) The Court then applied these principles to hold that the State statutes purporting to sell tidelands in San Francisco Bay *did not* constitute a termination of the trust. (26 Cal.3d 515, 529-532.) The

Court also reaffirmed the long-standing rule that reclamation of tidelands subject to the public trust does not terminate the trust. (*Id.* at 535 (fn. 19).)

7. In *City of Berkeley*, the Supreme Court concluded by noting that it was overruling two prior appellate court decisions, including a California Supreme Court decision, that had ruled that mere sales under a special tideland sales act had terminated the trust. The Court limited application of its ruling to lands along the Berkeley waterfront still subject to tidal action, choosing not to apply its decision retroactively to trust parcels not now subject to the ebb and flow of the tides. (*Id.* at 534.) In balancing the competing public and private interests application of its decision, the Court carefully limited its holding to the facts before it, noting that "implementation of our holding in the manner set forth above makes assumptions which are not valid in every case." (*Id.* at 535-536.) The Court does not interpret this as undermining the precedential effect of the *City of Berkeley*; far from abandoning the principles of *California Fish* as contended by S.P., *City of Berkeley* underscores its continued vitality.

8. The Court further believes that the remedy adopted in *City of Berkeley* cannot be applied here. While S.P. attempts to analogize the provisions of Public Resources Code section 7552 (analyzed above) to the two appellate court decisions overruled in *City of Berkeley*, an important distinction exists. As discussed above, section 7552 does not have the meaning ascribed to it by S.P.; more importantly, no appellate court has ever given it such an interpretation. Unlike *City of Berkeley*, there is therefore no occasion for this Court to concern itself with retroactive application of a judicial decision, contrary to prior appellate decisions, that the public trust continues to exist. For these reasons, the *City of Berkeley* decision fails to support S.P.'s position.

9. Accordingly, the Court finds that the trust with the tidelands identified in this case are impressed has not been extinguished or otherwise terminated.

E.

PUBLIC RESOURCES CODE SECTION 6360 DOES NOT BAR THE STATE'S SOVEREIGN CLAIM; NOR DOES IT REQUIRE THAT THE BOUNDARY BETWEEN THE STATE'S INTEREST AND PRIVATELY-OWNED UPLANDS BE ESTABLISHED AT THE ORDINARY LOW WATER MARK.

1. S.P. argues that, even assuming that tidelands exist within the subject parcel, they are not owned by the State. Instead, S.P. maintains, the boundary between State sovereign lands and S.P.'s private uplands must be fixed at the ordinary *low* water mark under the provisions of Public Resources Code section 6360. Since plaintiff has not attempted to establish the requisite low water mark, the argument concludes, the State's case must fail.

2. This contention is without merit. A review of Public Resources Code section 6360 reveals it to be inapplicable to the lands at issue in this case.

F.

MISCELLANEOUS MATTERS.

Pursuant to California Rules of Court, Rule 232.5, judgment shall be entered in conformance with the terms of this Statement of Decision, but only after the remaining issues in this bifurcated action are resolved.

Plaintiffs shall be awarded its costs of suit applicable to the sovereign title phase of the case.

DATED: September 3, 1983

WILLIAM H. LALLY
JUDGE OF THE SUPERIOR COURT

APPENDIX B

GREAT SALT LAKE SPECIAL MASTER'S REPORT I

Supreme Court of the United States

No. 81, Original

OCTOBER TERM, 1967

State of Utah, Plaintiff

v.

United States of America, Defendant

Report of Special Master

Charles Fahy
Senior Circuit Judge,
Special Master.

		Page
I.	Introduction..	352
II.	Summary of Recommended Conclusions	354
III.	Federal Law Controls......................................	354
IV.	History and General Characteristics of the Great Salt Lake	355
V.	Factual Characteristics of the Reliction Process	357
VI.	The Special Master is unable to Find that the Exposure of the Bed of the Great Salt Lake June 15, 1967, the Date of the Quitclaim Deed, Had Come About by a Gradual and Imperceptible Process.........................	360
VII.	The Rationales of Reliction In Large Measure Those of Accretion, Are Not Supported By the Facts of the Present Case	364
	Findings of Fact..	368
	Conclusions of Law...	370
	Proposed Decree	371
	Appendix A	000

Exhibits P-4, P-5, P-9, P-10, P-11, P-12, P-17, and D-2, D-3 and D-6 are reproduced following Appendix A.
  [not included]

Supreme Court of the United States

No. 81, Original

OCTOBER TERM, 1967

STATE OF UTAH, PLAINTIFF

v.

UNITED STATES OF AMERICA, DEFENDANT

REPORT OF SPECIAL MASTER

To the Chief Justice and Associate Justices of the Supreme Court of the United States:

[1]*
I. INTRODUCTION

This Report of the Special Master is concerned with another stage of the controversy between the State of Utah and the United States over the ownership of certain land, and its content, within the surveyed meander line[1] of the Great Salt Lake.

[2]

Pursuant to the terms of the Act of Congress on June 8, 19[illegible], 80 Stat. 192, as amended, 80 Stat. 349, the Secretary of the Interior on June 15, 1967, by quitclaim deed conveyed to Utah all interests of the United States in these lands, including brines and minerals in solution in the brines, or precipitated or extracted therefrom[2], conditioned upon payment by Utah to the Untied States of the Market value of the interests covered by the deed should it be determined that they actually belonged to the United States. To have that question judicially determined Utah adopted one of the alternative courses authorized by the Act by filing this original suit in the Supreme Court seeking a decree to quiet its title as against the United States to the lands and content in dispute. See section 5(b) of the Act.

Several phases of the over-all controversy have been decided by court. In *Utah v. United States*, 894 U.S. 89 (1969), the Court denied intervention of a private claimant, Morton International, Inc. approving the recommendation of the late Senior Circuit Judge J. Cullen Ganey, Special Master, In *Utah v. United States*, 403 U.S. 9 (1971), again on the basis of the Report of Judge Ganey, the Court held that the Lake was navigable when the state of Utah was admitted to the Union on January 4, 1896, and that, as a consequence, under the equal footing doctrine,[3] title to the bed of the Lake at state-hood

---

* Numbers in brackets reference the page numbers in the original report.
1. The meander line of the Great Salt Lake is the line duly surveyed prior to and in accordance with section 1 of the Act of June 3, 1966, 80 Stat. 192, which provides:

> That the Secretary of the Interior shall within six months of the date of the passage of this Act complete the public land survey around the Great Salt Lake in the State of Utah by closing the meander line of that Lake, following as accurately as possible the mean high water mark of the Great Salt Lake used in fixing the meander line on either side of the unsurveyed area.

2. When "land", "lands", "bed of the Lake" appear in this Report to describe the interests involved the words are used to include the brines and minerals in solution in the brines or precipitated or extracted therefrom.
3. *Pollard's* [illegible], 44 U.S. (8 How.) 212, 221-223, 228-230 (1845); [illegible] *Cattle Co. v. Arizona*, 414 U.S. 313 (1973). Under this doctrine, in the absence of a congressional condition to the contrary, a State, upon its admission to the Union, acquires title to the bed of navigable waters within the State, as had the thirteen other original colonies as successors to the rights of the British Crown.

[3]

vested in Utah. Thereupon the Court, in *Utah v. United States*, 406 U.S. 484, entered into a decree of May 22, 1972, in the form of an injunction against the United States, but in substance quieting in Utah, as against the United States, the title to the bed of the Lake lying below its water's edge on June 15, 1967, the date of the quitclaim deed, with exception not now material, and to the natural resources and living organisms therein or extracted therefrom. The Court also held that Utah is not required to pay the United States for the Interest thus described in the decree.

The decree of May 22, 1971, by its paragraph 8, brings us to the present problem. It provides as follows:

> 8. The basic question yet to be determined in this case is whether prior to June 15, 1967, the claimed doctrine of reliction applied and, if so, whether the doctrine of reliction vests in the United States, an thus divests the State of Utah, of any right, title or interest to any or all of the exposed shorelands situated between the water's edge on June 15, 1967, and the meander line of the Great Salt Lake as duly surveyed prior to or in accordance with [illegible] 1 of the Act of June 8, 1966, 80 Stat. 192. A Special Master will be appointed by the Court to hold such hearings, take such evidence and conduct such proceedings as he deems appropriate and, in due course, to report his recommendations to the Court.

406 U.S., at 484-5. Judge Ganey in the meantime having died The Court appointed the present Special Master, with the authority thus noted. 406 U.S. 940 (1972).

[4]

Following the decree of May 22, 1972, the parties engaged in a cooperative and successful effort to clarify and delineate the issues now considered. This was followed by a hearing before the Special Master at the Untied States Court House in the District of Columbia on February 27, 1978, at which a large number of exhibits were introduced in evidence, and oral testimony adduced. The exhibits incorporate detailed factual data regarding the Lake and its history, the accuracy of which is accepted by the parties[4]. The Briefs followed and the case was taken under submission on October 26, 1973, for preparation of this Report. As it was nearing completion, however, the Court decided *Bonelli Cattle Co. v. Arizona*, 414 U.S. 318 (1978), following which the parties have advised the Special Mater of their views as to the bearing of *Bonelli* upon this case.

It is now important to note that, in accordance with the desire of both the United States and Utah, the basic question now considered is limited to

---

Not at issue in this litigation are the claims of private persons, vendees and patentees, to about 275,000 acres adjacent to the Lake.

4. Reproduced as part of this Report are Plaintiff's Exhibits 4, 5, 9, 10, 11, 12, and 17, and Defendant's Exhibit 2, 3, and 6. The originals of all the Exhibits are on file with the Clerk of the Court.

whether the doctrine of reliction divested Utah of titled to that part of the bed of the Lake at statehood which on June 15, 1967, the date of the quitclaim deed, had become exposed by recession of the waters of the Lake, comprising some 325,000 acres. *Infra*, p. 24. The title to any upland between the bed of the lake at statehood and the meander line is not now considered, the positions of the parties in that regard being reserved pending the answer to the above question.

II. SUMMARY OF RECOMMENDED CONCLUSIONS

The Special Master recommends that the Court hold that the State of Utah has not been divested by reason [5] of the doctrine of reliction of any right, title or interest to that part of the bed of the Great Salt Lake acquired by it at statehood which had become exposed by recession of the waters of the lake to June 15, 1967; that Utah is entitled to a decree quieting its title as against the United States to the area thus described insofar as any claim thereto by the United States rests upon the doctrine of reliction; and that no compensation is due from the State of Utah to the United States under the Act of June 8, 1966, for said area.

III. FEDERAL LAW CONTROLS

Under the recent decision of the Court in *Bonelli, supra*, it seems clear that federal law governs the decision of this case. The land, title to which was in dispute in *Bonelli*, abutted the Colorado River, a navigable stream. Title to the bed of the river had been acquired by Arizona at statehood under the equal-footing doctrine. The riparian owner, Bonelli Cattle Company, held title which stemmed from a federal patent.[5] Prior to the litigation, the river by erosion had inundated part of the Bonelli land, thus extending the river bed eastward. Thereafter, the United States, with the acquiescence of Arizona, rechanneled the river for navigational purposes and caused the eroded area to be reexposed. Bonelli Cattle Company as the riparian landowner claimed the exposed area under the doctrine of accretion. The Court held that this question was to be decided under the federal common law of accretion, not under state law, that the related question of reliction as now presented must also be decided under federal law. As with respect to Arizona in *Bonelli*, so with respect to Utah now, the title of each State accrued under the equal-footing doctrine, and the riparian owner in each case claims title to the upland under the United States.

The Court held in *Bonelli*:

> We continue to adhere to the principle that it is left to the States to determine the rights of riparian owners in beds of navigable streams which,

[6]

under federal law, belong to the State. But this doctrine does not require that state law govern the instant controversy. The issue before us is not

---

[5]. In *Bonelli* the claimant against the State under the doctrine of accretion held title under a federal patent to predecessor in title; but it was not clear whether the land covered by the patent was riparian at the date of the patent. *See* 414 U.S. at 321, n.11.

what rights the State has accorded private owners in lands which the State holds as sovereign; but, rather, how far the State's sovereign right extends under the equal-footing doctrine and the Submerged Lands Act—whether the State retains title to the lands formerly beneath the stream of the Colorado River or whether that title is defeasible by the withdrawal of those waters. As this Court observed in *Borax, Ltd. v. Los Angeles*, 296 U.S. 10, 22 (1935): "The question as to the extent of this federal grant, that is as to the limit of the land conveyed, . . . is necessarily a federal question . . . [I]t involves the ascertainment of the essential basis of a right asserted under federal law."

414 U.S. at 319-20.

While the United States is free to adopt or to acquiesce in the application of state law even though otherwise federal law would govern[6], the Court in *Bonelli* held that it had not done so by the Submerged Lands Act, 48 U.S.C. [illegible] 1801, et. seq., so holding in terms which the Special Master considers applicable as well to the present case:

[7]

The [Submerged Lands] Act did not abrogate the federal law of accretion, but defined lands beneath navigable waters as being those covered by streams as "hereafter modified by accretion, erosion, and reliction." . . . Since the Act does not extend to the States any interest beyond those afforded by this equal-footing doctrine, the State can no more base its claim to lands unnecessary to a navigational purpose on the Submerged Lands Act than on that doctrine. (Footnote omitted.)[7]

414 U.S. at 324-25.

IV. HISTORY AND GENERAL CHARACTERISTICS OF THE GREAT SALT LAKE

The Report of Special Master Ganey, filed October 26, 1970, contains in considerable detail the history and characteristics of the Lake. These are now outlined, beginning with the testimony before the present Special Master on February 27, 1978, of Mr. Theodore Arnow,[8] a joint expert witness. He described the Lake as located entirely within Utah and as a remnant of the ancient Bonneville Lake, one of the last major lakes formed in geological time in the Great Basin.[9] Before it began to dry [8] up over 20,000 years ago

---

6. *United States v. Oregon*, 295 U.S. [illegible], 28 (1935); *Hughs v. Washington*, 389 U.S. 290, 292-93 (1967).
7. By analogous reasoning the reliance on the State upon the Rules of Decision Act, 28 U.S.C. Section 1652, cannot be accepted as a basis for applying state law in light of *Bonelli*, all else aside. It was not applied in either *Hughes v. Washington*, *supra*, or *Bonelli*.
8. Mr. Arnow is the District Chief of the Water Resources Division of the Geological Survey in Utah charged with the responsibility of gathering statistical data regarding the Lake.
9. The Great Basin is located in Western Utah, most of Nevada and smaller parts of other States. Although the climate in the Great Basin is now arid with less precipitation that evaporation, many centuries ago the climate was more humid so that lakes which formed there would in alternate periods dry up and then form once again.

Lake Bonneville covered approximately 35,000 square miles, about the size of Lake Superior and twenty times the area of the Great Salt Lake. The Great Salt Lake is about 70 miles long and 30 miles wide, with a volume of 16,000,000, acre-feet covering about 1,650 square miles. Its maximum depth is 34 feet. The Lake is surrounded by very flat shore-lands except where mountain ranges project upward through the water.[10] The flat shorelands are vast mud flats barren of any vegetation due to the high salinity of the Lake, and are used for nothing except recreation and as evaporation ponds for extracting minerals and salts (Tr. 88.) A visual illustration of the general nature of the mud flats can be obtained from Exhibit P-6 which contains twelve color photographs of the shore as it appeared on February 12, 1972, when covered with snow. The Exhibit also contains twenty-four black and white photographs showing the mud flats as they appeared on October 26, 1972, with no snow. Both sets of photographs contain maps showing the location of each photograph and a description of the view illustrated.

The Lake has no outlet and thus no way of discharging the water which it receives by the inflow of tributary streams and by precipitation. The water level also is influenced by man-made interferences with the rate of inflow.[11] The only "outflow" from the Lake is by evaporation. (Tr. 36.)

[9]

The rate of evaporation and the resulting changes in the elevation of the Lake depend upon wind, temperature, salinity of the water and the surface area of the lake, which is affected by the rate of tributary inflow and the degree of salinity. The periods of maximum inflow, the late fall and the spring, precede the period of maximum evaporation, during the summer. The Lake is continually fluctuating due to these interrelated factors.

Records of the Lake were kept first by the Mormons who arrived in the area in 1847, Mr. Arnow continued. At that time the elevation of the Lake was approximately 4200 feet above sea level.[12] Exhibit P-4. At statehood in 1896 the elevation of the Lake was 4200.2 feet. In February, 1973, at the time of the hearing before the Special Master, the elevation was approximately 4200 feet, within a few inches of the statehood figure. (Tr. 87-88; Exhibit P-4. And see Tr. 18; Post-Trial Brief of Unites States, p. 19, n. 18.) Therefore, in February, 1978, the relation of the waters of the Lake of the adjoining shorelands was about as it was at statehood, quite different from what it was on June 15, 1967, the date of the quitclaim deed.

---

10. Judge Ganey's Report filed [illegible], 1970, states in his finding No. 18, p. 21, in part as follows:

Except for an area at the southeastern shore near the base of the [illegible] Mountains, where the beaches are located, the Lake is surrounded by stretches of salt flats, marshes or bogs, some of which are in pieces several miles in width. (Footnote omitted.)

11. Exhibit P-18 demonstrates that the Lake would be at a higher vertical elevation if man had not interfered with the natural rate of inflow. However, this does not change the nature of the fluctuations of the water level which would still fluctuate as dramatically and frequently but at a higher elevation

12. In Noting at different places in this Report the level or elevation of the Lake the numerals in each instance refer to "above sea level."

Since the Lake has fluctuated as much as thirteen feet from a high of 4205 feet in 1924 to a low of 4192 feet in 1968.[13] The fluctuations have followed no set pattern on either a long range or short-range basis, whether measured at five-year intervals, one-year intervals, six month intervals, monthly **[10]** intervals, or even daily or hourly intervals. See Exhibits P-8–P-17.

Special reference is made by the Special Master to the map marked as Exhibit D-2. The area of the bed of the Lake exposed between statehood and the date of the quitclaim deed is shown in the dark shading, dark blue in the original. The large light area represents the bed at the date of the quitclaim deed, light blue in the original. The area between the bed of the Lake at statehood and the surveyed meander line is the exterior medium shading, green in the original.

V. FACTUAL CHARACTERISTICS OF THE RELICTION PROCESS

The land of a riparian owner may be affected by the action of the water by avulsion erosion, accretion and reliction. Neither avulsion or erosion is involved. The former occurs when there is a sudden or violent change in the course of a stream in which event the title lines remain as before. *County of St. Clair v. Lovingston*, 90 U.S. (23 Wall.) 46, 68 (1874); *Philadelphia Co. v. Stimson*, 223 U.S. 605, 624-25 (19120; *Bonelli, supra*, 414 U.S. at 327. Erosion is the gradual washing away of land by the water, thus extending its bed and simultaneously contracting the riparian land. See *Arkansas v. Tennessee*, 245 U.S. 158, 173 (1918); *United States v. 461.43 Acres of Land*, 222 F.Supp. 55 (N.D. Ohio, 1963).

At common law accretion was the enlargement of the riparian land by the action of the water in gradually and imperceptibly depositing soil. See *Mississippi v. Arkansas*—U.S.—(February 25, 1974). Reliction had this common characteristic of gradual and imperceptible change in the relation of the water to the land, but by uncovering existing land rather than, as in accretion, depositing additional soil. The Court in *Bonelli* affirmed these characteristics of accretion and reliction:

[11]

Federal law recognizes the doctrine of accretion whereby the "grantee of land bounded by a body of navigable water acquires a right to any natural and gradual accretion formed along the shore." *Hughes v. Washington*, 889 U.S. 290, 298 (1967); Accord, *Jones v. Johnson*, 18 How. 150, 156 (1856). When there is a gradual and imperceptible accumulation of land on a navigable riverbank, by way of alluvion or reliction, the riparian owner is the beneficiary of title to the surfaced land:

> "It is the established rule that the riparian proprietor of land bounded by the stream, the banks of which are changed by the gradual and imperceptible process of accretion or erosion, continues to hold the stream as his boundary; if his land is increased he is not accountable for the

---

13. The Lake reached its highest elevation in 1878, prior to statehood, when its level rose to 4212 fee. *See* Exhibit P-4.

gain, and if it is diminished he has no recourse for the loss." *Philadelphia Co. v. Stimson*, 228 U.S. 605, 624 (1912).

414 U.S. at 825-26.

As had been stated in *Hughes v. Washington, supra*, 889 U.S. at 298, land gained by accretion is "by little and little, by small and imperceptible degrees," and belongs to the riparian owner. The Court continued:

> The Court has repeatedly reaffirmed this rule, *County of St. Clair v. Lovingston*, 23 Wall. 46 (1874); *Jeffries v. east Omaha Land Co.*, 134 U.S. 178 (1890), and the soundness of the principle is scarcely open to question. Any other rule would leave riparian owners continually in danger of losing the access to water which is often the most valuable feature of their property, and continually vulnerable to harassing litigation challenging the location of the original water lines. (Footnote omitted.)

The earlier English case of *The King v. Lord Yarborough*, 107 Eng.Rep. 668 (K.B. 1824), states the common law principle. The claim was that the land in dispute "being slowly, gradually, and by imperceptible increase, in long time cast up, deposited and settled by *and from the flux and reflux of the tide and water of the sea*" **[12]** upon the "extremity of the [illegible] lands of the manor, [the projection] hath been formed, and hath been settled, grown, and secured upon . . . the [illegible] lands . . ." *Id*. at 673.

The court there held,

> It is clear upon the evidence, that the land has been formed slowly and gradually in the way mentioned in the plea . . . (a)nd considering the word "imperceptible" in this issue, a connected with the words "show and gradual," we think it must be understood as expressive only of the manner of the accretion, as the words undoubtedly are, and as meaning imperceptible in its progress, not imperceptible after a long lapse of time. . . .

Id. at 674.

It is interesting to compare this statement of 1824 with that of the Supreme Court of Utah in 1971 in *Utah v. Hardy*, 26 Utah 2d 143, 486 P.2d 891. Although Utah law does not govern this case, the United States in its brief before the Special Master recognizes that the opinion of the Supreme Court of Utah defines reliction "in terms perfectly consonant with the federal common law":

> The doctrine of "reliction in the law covers the situation involving the title to the land which emerges from beneath a body of water caused by recession of the waters. The law of reliction is generally the same as that of accretion as it concerns contiguity, imperceptibility and naturalness of the process of the water's recession and the exposure of additional land. The law of reliction, as well as accretion, has evolved over a long period of time and is based upon the rights of a riparian owner to have access to the water adjacent to his property.

26 Utah 2d at 144-45, 486 P.2d at 392-93. The court held, however, after reviewing the "unique and special conditions" affecting Great Salt Lake,

. . . that the recession of the waters from the land [there in question] has not been natural, gradual **[13]** and imperceptible, and that the doctrine of reliction should not be applied.

26 Utah 2d at 145, 486 P.2d at 893.

It is now clear, however, from *Bonelli* that under federal common law though accretion may occur as formerly by the gradual and imperceptible process which evolved under the non-federal common law, it may also occur in a perceptible manner at least where the owner of the bed of the navigable water is a State which acquired title to the bed under the equal-footing doctrine. In this regard the Court in *Bonelli* holds:

> The [earlier] advance of the Colorado's waters divested the title of the upland owners in favor of the State in order to guarantee full public enjoyment of the watercourse. But, when the water receded from the land, there was no longer a public benefit to be protected; consequently, the State, as sovereign, has no need for title. That course of the recession was artificial or that the rate was perceptible, should be of no effect.[14]

414 U.S. at 823-24

[14]

Postponing for consideration under Part VII., *infra*, the view of the Special Master that a reasonably permanent or stable result of the process is essential to a finding of reliction, and recognizing that under *Bonelli* reliction, as accretion, may at times occur perceptively, it seems appropriate first to consider the claim of the Untied States that the recession of the Lake as of June 15, 1967, the date of the quitclaim deed, had occurred by the 'little by little' or gradual and imperceptible process, and for that reason Utah had lost the area in question to the United States under the reliction doctrine.

---

14. We have seen that some decisions refer to accretion or reliction as an addition to land by natural causes:

> A long and unbroken line of decisions of this Court establishes that the grantee of land bounded by a body of navigable water acquires a right to any natural and gradual accretion formed along the shore.

Hughes v. Washington, supra, 889 U.S. at 293, and *see Jones v. Johnston*, 59 U.S. (18 How.) 150, 158 (1856), both referred to in *Bonelli*, 414 U.S. at 325; *Utah v. Hardy, supra*, 26 Utah 2d at 144 –45, 486 P.2d at 392-93. But *see County of St. Clair v. Lovingston, supra*, also cited in *Bonelli* at 327. In the present case, according to the testimony of Mr. Arnow, supported by reference to Exhibit P-18, the following appears:

> . . . the lake would have been 8.7 feet higher at the time of statehood if it weren't for the activities of man in the basin . . . (C)arrying it through 1967 . . . the difference is . . . 5.28 feet higher in 1967 than it actually was, if it weren't for the activities of man.

(Tr. 44.) Except for these activities the bed of the Lake would not have been exposed as of June 15, 1967, to the extent this had occurred. The State of Utah, however, has not pressed before this Special Master reliance upon the effect due to the activities of man.

VI. THE SPECIAL MASTER IS UNABLE TO FIND THAT THE EXPOSURE OF THE BED OF THE GREAT SALT LAKE JUNE 15, 1967, THE DATE OF THE QUITCLAIM DEED, HAD COME ABOUT BY A GRADUAL AND IMPERCEPTIBLE PROCESS

The term "fluctuations" is used herein to mean, unless otherwise indicated, the reactions of the waters of the lake, aside from responses to the wind, to changes in the elevation or level of the Lake. The United States points out that the lake experiences three general types of fluctuations, annual, seasonal and daily. Upon this factual basis the Untied States centers its claim that reliction accounts for the exposure of the shorelands between statehood and June 15, 1967, since, it is said, only the annual fluctuation reflects a change in the ordinary high water mark, which, according to the United States, is that boundary subject to modification by reliction. Thus, we [15] are urged to look only to the annual "fluctuations" of the level of the lake, for daily fluctuation allegedly caused by the wind and is not a change in the level of the Lake, and seasonal fluctuation recurs each year in essentially the same pattern. Exhibit P-4. On the other hand, the annual change in a product of every physical factor affecting the level of the Lake, see pages 8–9, *supra*, and is measured, according to the United States, by differences in the "average yearly stage" of the Lake, which is the average of the levels of the Lake at regular intervals over a 12-month cycle. Exhibit D-6.

Therefore, the United States continues the "average yearly stage" of the lake is the proper standard for measuring the level of the lake. The "average annual change" since statehood has generally been less than 1 foot per year. However, since some of the shorelands are extremely flat, the water moves almost in a horizontal direction and a small change in the level of the lake can inundate or expose hundreds of feet of shoreland. Thus, the annual average change in level of only .69 feet (8.28 inches) would expose or inundate about 50,000 acres of shoreland. Exhibit P-5. The shoreline of the Lake is 850 lineal miles, with the result, it is said, that an average annual movement along the shore is less than 1200 feet, barely 8 feet per day or a little over 1½ inches per hour. The United States accordingly concludes that such a movement is not humanly perceptible as it occurs, adding, "[n]one of the changes in the level of the Great Salt Lake have been of so sudden or violent a character as to be perceptible while the process was going on." Exhibits P-18, D-4.[15]

It should be noted that in distributing the 50,000 acres or shoreland among the 850 lineal miles consideration was [16] not given to the fact that the shorelands are not uniformly affected by change in the level of the lake. The flat shorelands surrounding the Lake are interrupted at different points by small mountain ranges set back at varying distances from the shoreline. For example, on the north side of the lake the Promontory Mountains form a headland of about 30 miles long which juts southward in toward the center of the Lake. See Exhibit P-1. In the southeastern sector of the lake rise the [illegible] Mountains and the Stansbury Mountains. Several mountain clusters

---

15. According to the United States the most extreme change in the boundary occurred between 1906 and 1907 and amounted to less than 15 inches per hour.

also exist within a few miles of the western shore of the Lake. Among these are the Lakeside Mountains and Terrace Mountains. The eastern and northwest shorelands of the Lake, however, are almost entirely dominated by salt and mud flats. See Report of Special Master Ganey, October 26, 1970, at 10-13; Exhibit P-1. The various mountain ranges form an irregular barrier to a uniform spread of the water when the Lake is on the rise. Mr. Arnow refers to differences in consequence of a change of 5 feet in the Lake elevation. For instance, at those points on the west side where the flat shorelands dominate, water could rush inland for 7½ miles, whereas such a change would cause "very little change—no change" where the mountains are.[16] (Tr. at 84)

The movements of the water during any year, and the resulting exposure or inundation of the shorelands, are not mirrored in the annual averages. Thus, an examination [17] of Exhibits P-9 and P-10[17] discloses that the elevations of the Lake on June 1 and November 1, from 1850 to 1972[18] generally ranged between one and two feet within a six-month span, exposing or inundating about 50,000 to 150,000 acres of the shallow mud flats. However, the yearly average of these movements is not their visual movements. Changes in elevation may be gradual and imperceptible but are only a part of a process the whole of which includes the effect upon the shorelands of such changes. If the effect on the shorelands is not imperceptible then the whole of the process is not imperceptible within the meaning of the doctrine of reliction.

The fluctuations at monthly intervals from 1896 to February 1, 1978, are shown on Exhibits P-11 and P-12. Changes in elevation are noted on the former. The latter translate those changes into acreage of shorelands affected. Study of these Exhibits together with Exhibits P-13 and P-14[19] and using as illustrative and period from November 1, 1970 to June 15, 1971, shows that the rise in the level of the Lake flooded over 450,000 acres of the flat shorelands, followed by a recession for three and half months, leaving inundated about 175,000 acres, rising again to inundate some 880,000 acres in the following seven and a half months. The Lake then receded for four and a half months, leaving inundated about some 200,000 acres.

Exhibit P-16 charts the fluctuations for June, 1957. On June 6, for example, the level of the lake rose and fell 10 times. When these actual fluctuations are averaged, however, the net change is zero. For ten days just [18] prior to June 15, 1967, the date of the quitclaim deed, the major fluctuations in elevation ranged between 3 and 6 inches, inundating or exposing between 10,000 and 20,000 acres with each fluctuation.[20]

---

16. When the United States speaks of an average movement along the shore of a little more than 1½ inches per hour there is not reflected the fact that the lake's movement along that part of its shoreline where mountains rise is less than where there are vast stretches of flat land where the water can rush overland almost unimpeded. Utah, however, does not discuss this inaccuracy, as it seems to the Special Master, in the United States' calculations.
17. The data contained in the Exhibits are based on undisputed gauge recordings.
18. Only to June 1, 1972.
19. Exhibits P-14 and P-14, however, are limited to monthly intervals from 1955 to February 1, 1978.
20. The above are a few of many detailed records which have been graphed and charted upon Exhibits placed in the record. Similar graphed chartings of daily and hourly fluctuations in elevation and accompanying effect on acreage are available in the Exhibits, although only some of

If the actual change since statehood in the relation of the waters of the lake to the adjoining land had been as calculated by the United States, only a little over 1½ inches per hour,[21] the position of the United States that the process has been imperceptible would be substantial, but this statement of the rate of change in the relation of the water level to acreage affected does not reflect the actual changes as they occurred. Changes in the level of the Lake are gradual. However, the imperceptibility feature of the reliction is to be judged according to the actual effect on the shorelands of such a change. The unique nature of the area causes a gradual and slight change in the elevation of the Lake to result in a much greater alteration of the relation of the water to the land. This is demonstrated by the statistical data which has been charted from the records, partially analyzed above and more fully disclosed by the Exhibits which chart more fully the constant movements of the waters and

[19]

the effect of these on the land, annually, seasonally, monthly, daily, and hourly.

To place the matter thus disclosed in a light most favorable to the United States is to consider what may be referred to as the net result of the movements of the Lake over a substantial period of time. For example, in the ten years from about 1958 to 1968 (See Exhibits P-12 and D-6), the general movement, due to lowering of the elevation of the Lake, was a recession of its waters from the shorelands, accompanied, however, by rather constant fluctuations up and down as it were, as the general recession continued. At the end of the period there was a very perceptible area which became exposed during the period, but perhaps there had been at no particular moment a separate perceptible component of this total movement of exposure. To accept, this, however, as meeting the imperceptible element of reliction, would in the view of the Special Master be an adaptation of that element of the doctrine to the unique character of the Great Salt Lake rather than to apply the imperceptible element as it has developed in the law. While there are indeed features in the over-all process and its result, thus described, which arguably favor the Court's acceptance of the imperceptible element of the doctrine, nevertheless the continued rise and fall of the elevation of the Lake on an hourly, daily and weekly basis, is unprecedented in the historical development and previous application of this element of the doctrine. This of course is due to the unique character of the area, with the erratic movements of the Lake over the land, almost constantly responding in exaggerated reflexes to slight alterations in the level of its surface.[22] The unique situation [20] resists doc-

---

those specifically referred to are reproduced in this Report. The Special Master does not understand that either the United States or Utah challenges the accuracy of any of the Exhibits, but only their significance on the imperceptibility issue, or, stated otherwise, only their significance on the issue of reliction. There is appended to this Report as Appendix A, calculation of the movements of the lake by periods which have been compiled by the Special Master from the Exhibits. These calculations were approved by the parties with slight modifications.

21. See p. 15, *supra*.

22. The effect of the wind should also be mentioned. Mr. Arnow testified: ". . . we have a pile-up of two feet of water in the first few hours of a storm. And if the wind will persist steadily for a period of days in the same direction, the water level is pushed up ½ a foot and will stay half

trinaire classification. The perceptible net or residual change after a passage of time in the relation of the water to the land, due either to recession or progression of the water, may often be obscured as it occurs by the constant fluctuations of the water. The Special Master, however, is unable to find that the recession of the Lake to its level of June 15, 1967, occurred in a gradual and imperceptible manner.

Moreover, and importantly, even were the recession to June 16, 1967, gradual and imperceptible, the relation of the water to the shorelands at that date was not of the reasonably permanent or stable character essential, in the view of the Special Master, to application of the doctrine of reliction, a matter to be more fully discussed.

VII. THE RATIONALES OF RELICTION, IN LARGE MEASURE THOSE OF ACCRETION, ARE NOT SUPPORTED BY THE FACTS OF THE PRESENT CASE

1. It seems desirable to round all the rationales of the doctrine, with special reference to the Court's treatment of accretion in *Bonelli*. The Court summarized the reasons underlying the common law doctrine as follows:

First, where lands are bounded by water, it may well be regarded as the expectancy of the riparian owners that they should continue to be so bounded. Second, the quality of being riparian, especially to navigable water, may be the land's most valuable feature and is part and parcel of the ownership of the land itself. *Hughes v. Washington, supra*, at 293; *Yates v.* [illegible], 10 Wall. 497, 504 **[21]** (1871). Riparianness also encompasses the vested right to future alluvion, which, is an "essential attribute of the original property." *County of St. Clair v. Lovingston*, 23 Wall. 46, 68 (1874). By requiring that the upland owner suffer the burden of erosion and by giving him the benefit of accretions, riparianness is maintained. Finally, there is a compensation theory at work. Riparian land is at the mercy of the wanderings of the river. Sine a riparian owner is subject to losing land by erosion beyond his control, he should benefit from any addition to his lands by accretions thereto which are equally beyond his control. *Ibid.* the effect of the doctrine of accretion is to give the riparian owner a "fee, determinable upon the occupancy of his soil by the river, and (to afford) the State (a title) to the river bed (which is) likewise a . . . "qualified" fee, "determinable in favor of the riparians upon the abandonment of the bed by the river."

"(Omitted.)

"107 Ariz., at 472, 489 P2d, at 706 (Lockwood, J., dissenting), quoting, *State v. R.E. Janes Gravel Co.*, 175 B.W.2d 739, 741, (Tax.Civ.App. 1943), rev'd on other grounds sub nom, *Morales v. State*, 142 Tex. 559, 180 S.W. 2d 144 (1944).

414 U.S. at 326.

Reverting to the factual situation of *Bonelli*, the Court pointed out tin addition that riparian owner, "because of the navigational servitude" is at the

---

a foot higher for several days." (Tr. 47) Also, "the higher the wind the higher the evaporation," and "the higher the temperature the higher the evaporation."

mercy of "governmental forces which may similarly affect the riparian quality of his estate":

> Accordingly, where land cast up in the Federal Governments exercise of the servitude is not related to furthering the navigational or related public interests, the accretion doctrine should provide a disposition of the land as between the riparian owner and the State. See *Michaelson v. Silver Beach Assn.*, 842 Mass. 251, 178 N.E.2d 273 (1961).

414 U.S. at 829.

The navigational and related public [22] interests of a State have required that a State retain title to and control of navigable waterways as an element of its sovereignty. This has proved historically necessary because,

> "Such waters . . . are incapable of ordinary and private occupation, cultivation and improvement; and their natural and primary uses are public in their nature, for highways of navigation and commerce, domestic and foreign, and for the purpose of fishing . . ." *Shively v. Bowlby, supra*, at 11.

414 U.S. at 322.

However, once the rechanneliztion project was accomplished the land exposed as a consequence, formerly part of the riverbed, was "no longer . . . incapable of ordinary and private occupation. . ." and there was, as well, "no longer a public benefit to be protected; consequently, the State, as sovereign, had no need for title." 414 U.S. at 323-24. Thus, *Bonelli* supplements the traditional doctrine of accretion by requiring an appraisal of the state interests affected by the exposure, compared with the interests of the riparian owner.

2. The foregoing considerations underlying the doctrine of accretion, and, insofar as the facts of the present case permit, the doctrine of reliction, subsume, in the view of the Special Master, a reasonable permanence or stability in the change which has occurred in the reliction of the water to the land. The United States in its brief before the Special Master urges that a more precise statement would be that the land formation must be "not clearly temporary." The situation in *Bonelli* created by the channeling project was "not clearly temporary." Indeed it was of a permanent character.

The generally uniform characterization of the common law doctrine of reliction and accretion has always seemed to contemplate a result substantially permanent; thus, the land "hath formed, and hath been settled, grown and accrued,' the language of *The King Yarborough, supra*. The omission of the case law to refer [23] uniformly to permanence is understandably due to this quality being implied as a result of the gradual and imperceptible process. The situation should be no different where the claim rests upon a process which may have been perceptible. Both accretion and reliction involve an addition to the riparian land, ordinarily by a deposit of soil in the case of accretion, by the exposure of land previously under water in the case of reliction. The factor of "addition" implies reasonable stability or permanence. The criterion of "little by little, by small and imperceptible degrees," *Jones v. Jones, supra*, 59 U.S. at 155, had reference to "land gained from the sea." Which implicitly assumes something more than a temporary condition of the "soil that had gradually been deposited." In *Sapp v. Frasier*, 61 LaAnn. 1718,

26 So. 878 (1899), it was said that "... 'reliction' is land added to a front tract by the permanent uncovering of the waters," and as used by the English law, (it) meant when the sea shrank back below the usual watermark, and remained there." 26 So. at 880. *State v. Longyear Holding Co.*, 224 Minn. 451, 29 N.W. 2d 667 (1947) is to the same effect:

> It is also clear that before a riparian owner can claim title to lands as a result of relictions, such reliction must be of a permanent nature, without the possibility of the water again filling in or covering the relicted area. (Emphasis in original.)

29 N.W.2d at 667.

Derelicted or relicted land is land added by the recession of the water leaving a portion of the bed dry.

Fontenelle v. Omaha Tribe of Nebraska, 298 F.Supp. 855 (D.Neb. 1969), aff'd. 430 F.2d 143 (8th Cir. 1970).

Landquist, *Artificial Additions to Riparian Land: Extending the Doctrine of Accretion*, 14 ARIZ.L.REV. 816, 321 (1972). And see, *Utah v. Hardy, supra.*

Though the use of "permanent" is a necessary quality of the change sometimes in opinions where the [24] doctrine is not applied because the change has been so clearly temporary as in *Ferrand v. Madson*, 36 S.D. 457, 162 N.W. 796, 798 (1916) and *Hillebrand v. Knapp*, 65 S.D. 414, 274 N.W. 821, 829 (1937), none of the decisions which omit such reference is inconsistent with the idea of permanence or stability. There is nothing in the Court's application of federal common law to the situation in *Bonelli* which indicates any departure from the non-federal common law in this respect. All references to the change in relation of the Colorado River to the shoreland indicate the result was a permanent addition to the riparian land.

3. The special facts of the present case bearing upon the question of permanence are now considered.

At the date of the quitclaim deed the elevation of the Lake was 4194.9 feet (Tr.8), a drip of 5.8 feet in its level from January 4, 1896, the date of statehood. At that time its elevation was 4200.2 feet. See Report of Special Master Ganey, October 26, 1970, at 29. This lowering of the level of the Lake had exposed by June 15, 1967 an area of about 325,000 acres, shown on Exhibit D-2 in the dark shading. Since June 15, 1967, the level of the Lake has risen to within a few inches of its statehood level, reinundating this area which had become exposed after statehood. Thus as of February 27, 1973, the date of hearing before the Special Master, nearly all the land now in dispute had been resubmerged. See P.9, *supra.*

This development is part of the physical situation respecting the lake and adjoining land, demonstrated by data available to the Court and bearing upon the question to be decided. Whether interests the United States owned June 15, 1967, and described in the deed, passed to Utah; but whether any of this land which had become exposed after statehood was then owned by the United States depends upon whether the doctrine of reliction [25] had divested Utah of its prior ownership. That question, in the opinion of the Special Master, cannot be answered so as to divest Utah of title to the area in question unless the condition in June 15, 1967, was a reasonably permanent or stable one. This element of the doctrine is illuminated by the continuity of the Lakes history since June 15, 1967.

The importance of the post June 15, 1967 date is that it demonstrates that the situation had not prior to or on June 15, 1967, reached a state of stability or reasonable permanence. While the court's decree on May 22, 1972, states the basic question to be whether "prior to June 15, 1967, the claimed doctrine of reliction applies," the facts bearing upon whether the doctrine applies prior to that date include data with respect to the [illegible] of the waters of the lake to the land on and subsequent to June 15, 1967. This evidence discloses a continuing movement which has reinundated the area, thus reconstituting the bed of the Lake just as it was at statehood.

It does not seem that the issue of possibly divesting Utah of title to the area should be determined by freezing the situation as it was at some moment on or prior to June 15, 1967, when it has developed that the Lake on that date was in a rising movement which, with interim up and down fluctuations, has continued to the time of the making of the record in this case. See Exhibit P-4.

The doctrines of accretion and reliction contemplate ambulation in title boundaries; but the valuable features riparian ownership, particularly those incident to maintaining access to the water, and the compensation theory referred to in *Bonelli, supra*, 414 U.S. at 326, seem to the Special Master to envisage a situation different from the special relation of the waters of the Great Salt Lake to the riparian land. Such relation seems inconsistent with the stability of which should pertain to a change **[26]** in title by operation of law. In providing for payment by Utah to the United States of such interests as the United States might be found to have conveyed to the State by the quitclaim deed of June 15, 1967, the statute of June 8, 1966 is indicative of a congressional assumption that such payment would be required only if the situation at the date was reasonably permanent in nature rather than temporary, as the history of the Lake has demonstrated it to have been.[23]

4. Under a dominant principle of *Bonelli* the reinundation during the period of this litigation of almost the entire statehood bed of the Great Salt Lake confirms the Special Master in his recommendation adverse to application of the reliction doctrine, because the public benefit of Utah entitled to protection has moved along with the reinundating waters of the Lake, whereas in the *Bonelli* case Arizona's public benefit to be protected faded away as the waters of the Colorado River receded from the land.

Note is taken of the *Statement of the State of Utah With Respect to Bonelli v. Arizona*, filed January 28, 1974, which refers to various interests of Utah in the disputed area as substantial. Among these is that the exercise of its navigational servitude over the Lake would be threatened by a holding that

---

23. Should the Court hold that the doctrine of reliction had directed Utah of title to the land claimed by the United States under that doctrine, a question, not factually addressed by the parties on the present record, may arise as to the exact boundary lakeward of the relicted lands on June 15, 1967.

the disputed land is not part of the bed of the Lake. Reference also is made to the brines, salts and minerals in solution. These minerals are extracted by pumping water from the Lake into "settling ponds" on the salt or mud flats, where solar evaporation causes the minerals to precipitate on the bottom of the ponds. *See* Tr. At 88, 81.

[27]

The doctrine of reliction does not seem to furnish a sound basis for resolution as of June 15, 1967, of the respective interests of the two sovereigns in and about this unusual body of water.

FINDINGS OF FACT[24]

1. The Great Salt Lake is a large body of water surrounded in greater part by shorelands which are so flat that a slight change in the elevation of the Lake causes the water, except where the shore is mountainous, to spread over a large area quite out of proportion to the change in elevation of the surface of the Lake if the change is a rise, and vice versa if a fall.

2. Only a relatively small part of the immediate shoreland is mountainous. See Exhibit P-1; Report of Special Master Ganey, October 26, 1970, at 10-18.

3. Streams of clear water are the chief tributary sources of the Lake, supplemented by precipitation. The Lake has no outlet. Its elevation accordingly is governed by the inflow of the streams, by precipitation, and by evaporation, the latter depending upon climate, the area [28] of the surface of the Lake at different times, and the degree of salinity of the waters of the Lake.

4. As the name of the Lake implies the water is salty. The degree of salinity, which in turn affects the rate of evaporation, depends in part upon the relation of fresh water inflow and precipitation to the degree of evaporation. The latter depends largely upon the relative aridity of the climate at any particular time.

5. Man-made interferences with the rate of tributary inflow over the years has affected the elevation of the Lake. The elevation would have been 8.7 feet higher at the time of statehood and 5.28 feet higher in 1967 if it were not for the activities of man.

6. The various factors above described and their interaction one upon another cause a continuous change in the elevation of the Lake. This causes a

---

24. The parties are in agreement as to the physical characteristics and history of the movements of the Lake in relation to the shorelands, represented by the statistical data and charts contained in the Exhibits. This data as set forth in the Report and, also, the testimony of Mr. Arrow and Mr. Hewitt, may therefore be considered as undisputed findings. The present Findings of Fact are limited to those basic or ultimate findings which the Special Master believes are determinative of the conclusion to be reached. They reflect in final form the interim findings and conclusions set forth in the course of his Report.

The fact that the Report in important respects is discursive of the evidence in its legal consequences, and is not confined to factual findings, the Special Master considered justified by the nature of the case.

constant fluctuation of its waters, in both a receding and inundating character with respect to the flat shorelands.

7. At some periods in the history of the Lake, of which accurate records have been kept, made available to the Court in the record complied before the Special Master, the general trend in the relation of the waters of the Lake to the shorelands has been a recession from the line which defined the bed of the Lake at statehood, when title to the bed vested in Utah. These periods of general recession, exposing land which was part of the bed at statehood, have always been accompanied by "up and down" fluctuations which did not interrupt the general recession.

8. At other periods in the history of the Lake the opposite of the above has occurred; that is, the general direction of the movement of the waters has been to inundate or reinundate large areas of shoreland previously exposed, accompanied by similar smaller fluctuations "up and down" within the general direction of inundation or reinundation.

[29]

9. The situation described in findings 7 and 8 is illustrated by Exhibit P-4 made a part of this Report, and grows out of the variations which occur among the several factors which contribute to the movements of the waters in relation to the shorelands at different times.

10. On January 4, 1896, the date of statehood of Utah, the elevation of the Lake was 4200.2 feet. On June 15, 1967, the date of the quitclaim deed, it was 4194.9 feet, a recession of 5.3 feet. This had resulted as of June 15, 1967, in the exposure of an estimated 325,000 acres of land which was part of the bed of the Lake at statehood.

11. On June 15, 1967, as part of a process of movement of the waters of the Lake in relation to the shorelands which began in 1968, the elevation of the Lake was rising and has continued in a general rising movement to the time of the hearing before the Special Master in February, 1973, at which time nearly all the land exposed as described in Finding 10 had been reinundated so that the bed of the Lake was about as at statehood; that is, the land which at statehood had constituted the bed of the Lake was again almost entirely covered by the waters of the Lake.

12. The Lake experiences three general types of fluctuations, annual, seasonal and daily, although there are also measurable monthly and hourly changes. The annual change is the product of every physical factor affecting the level of the Lake on a yearly basis. The average of changes in the level of the Lake at regular intervals over a 12-month cycle since statehood has been about .69 feet; that is, the average annual change of the level has been about 8.28 inches, a change which, considered alone, would inundate about 60,000 acres of shoreland due to the fact that much of the shoreland is so flat that the water moves almost in a horizontal direction.

[30]

13. The shoreline of the Lake is 350 lineal miles. If the shoreline were of about equal flatness and the 50,000 acres referred to in Finding 12 were distributed equally among the lineal miles, the average movement along the shore would be about 1200 feet annually, about 3 feet per day, or a little over 1-1/2 inches per hour. These calculations do not reflect the actual movements

of the water as they occur but are based on averages calculated subsequent to the actual movements.

14. The shorelands are not of equal flatness or width of flatness, due to the mountains which are adjacent to some parts of the Lake, shown on Exhibit P-1, and other mountains which border the Lake. The movement of the waters along the lineal miles of shoreland is affected by mountains some of which, although not at the edge of the water or obtruding upward in it, border the shorelines at a distance near enough to affect the spread of the waters in a manner not reflected in the calculation described in Finding 19.

15. By reason of the conditions described in Finding 14 above the calculations reflected in Finding 13 that the average movement along the shore would be about 1200 feet annually, about three feet per day, or little more than 1-1/2 inches per hour, are to an unascertained degree an under estimate of the rate of movement over extensive areas of shore, a rate of movement which the Special Master cannot find on this record to be imperceptible.

16. The continuous rise and fall of the elevation of the Lake, though gradual, is reflected in a continuous movement of the waters to and fro across the shorelands, apart from the effects of the wind. These movements are often perceptible.

17. The net amount of land exposed or inundated over a substantial period of time is perceptible in [31] comparison with the situation at the beginning of such period, although one may not be able at any particular moment to perceive a separate component of the resulting exposure or inundation.

18. The constant fluctuations[25] of the Lake may often obscure the progress, as it occurs, of a recession or inundation which is readily perceived after the passage of time, but the Special Master is unable to find that the progress of such recession or inundation is at a rate which would be imperceptible as it occurs.

19. The land referred to as exposed at the time of the quitclaim deed, which at statehood was part of the bed of the Lake, was not an addition of a reasonably permanent or stable character to the uplands, title to which is in the United States. The land in question has been almost entirely reinundated by a spreading of the waters of the Lake by a movement which began in 1969, was in progress at the date of the quitclaim deed, and continued to the time of the hearing in this case before the Special Master in February, 1973.

CONCLUSIONS OF LAW

1. The question considered is to be decided under the federal common law doctrine of reliction.

2. Whether under that doctrine any interests were owned by the United States in the lands lying between the outer edge of the bed of the Lake at statehood and its bed underwater on June 15, 1967, which passed to the State of Utah by the quitclaim deed of the United States of June 15, 1967, depends upon whether the exposure of the land referred to, either by a perceptible or a gradual and imperceptible process, constituted a reasonably permanent or stable addition to the riparian land which was [32] upland from the bed of the Lake at statehood, title to which was in the United States.

---

25. The use of "fluctuations" in this finding does not exclude the affect of the wind.

3. The exposure of the lands referred to occurred in the course of such unique changes in the relation of the waters of the Lake to the shorelands as not to come within the doctrine of reliction. These changes were not at the date of the quitclaim deed of such a reasonably permanent or stable character as to warrant application of the doctrine.

4. The public benefit of Utah which is entitled to protection has accompanied the reinundation of the bed of the Lake to approximately its extent at statehood.

5. The law of reliction has not divested the State of Utah of title to the lands described.

6. The State of Utah is entitled to a decree quieting its title as against the United States to the bed of Great Salt Lake at the date of statehood.

7. The State of Utah is not required to pay the United States for the land covered by Great Salt Lake and below the boundary line of the Lake's bed as of January 4, 1896.

8. The United States is the riparian owner which would be entitled to the benefit of the doctrine of reliction were it applicable to the land the title to which is the subject matter of this Report. The Lake, its bed, and the adjoining land were ceded by Mexico to the United States in 1848 under the Treaty of Guadeloupe Hidalgo, 9 Stat. 922. See Report of Judge Ganey, October 26, 1970, p. 9.

PROPOSED DECREE

It is ordered, adjudged and decreed that:

1. The United States of America, its departments and agencies, are enjoined, subject to any regulations which the Congress may impose in the interest of navigation [33] or pollution control, from asserting against the State of Utah any claim of right, title and interest:

a) to any of the exposed shorelands situated between the edge of the waters of the Great Salt Lake on June 15, 1967, and the bed of the Lake on January 4, 1896, when Utah became a State, with the exception of any lands within the Bear River Migratory Bird Refuge and the Weber Basin federal reclamation project;

(b) to the natural resources and living organisms in or beneath any of the exposed shorelands of the Great Salt Lake delineated in (a) above; and

(c) to the natural resources and living organisms either within the waters of the Great Salt Lake, or extracted therefrom, as delineated in (a) above.

2. The State of Utah is not required to pay the United States, through the Secretary of the Interior, for the exposed shorelands, including any minerals, delineated in paragraph 1 above of this decree.

3. There remains the question whether any lands within the meander line of the Great Salt Lake (as duly surveyed prior to or in accordance with section I of the Act of June 3, 1966, 80 Stat. 192), and conveyed by quitclaim deed to the State of Utah, included any federally owned uplands above the bed of the Lake on the date of statehood (January 4, 1896) which the United States still owned prior to the conveyance to Utah.[26] In the absence of agreement

---

26. As appears from p. 4 of the Special Master's Report the parties have reserved their position with respect to this question.

between the parties disposing of the above question or of the necessity for further proceedings with respect thereto, the Special Master is directed to hold such hearings, take such evidence, and **[34]** conduct such proceedings with respect to that question as he deems appropriate and, in due course, to report his recommendation to the Court.

4. The prayer of the United States of America in its answer to the State of Utah's Complaint that this Court "confirm, declare and establish that the United States is the owner of all right, title and interest in all of the lands described in Section 2 of the Act of June 3, 1966, 80 Stat. 192, as amended by the Act of August 23, 1988, 80 Stat. 349, and that the State of Utah is without any right, title or interest in such lands, save for the right to have these lands conveyed to it by the United States, and to pay for them, in accordance with the provisions of the Act of June 3, 1966, as amended," is denied.

Respectfully submitted,

CHARLES FAHY,
Senior Circuit Judge,
Special Master.

[35]
APPENDIX A

Statistical Data as to Rise and Fall of Elevation
of Lake by Periods, with Effect on Bordering Lands

1. 1873–1906

At the time of Utah's statehood, the Lake was in a period of general decline which had begun in 1873. In 1873 the Lake's average level was 4211.20 feet and in 1896 it had fallen to an average level of 4201.10 feet. It continued to fall until in 1906 it reached an elevation of an annual average of 4196.83 feet. Thus from 1873 to 1906, the Lake's elevation declined approximately 14.3 feet, and from 1896 to 1906, 4.27 feet. [Exhibits D-3 and D-6.]

The surface area of the Lake receded from an approximate average annual area of 1,554,000 acres in 1873 to approximately 1,120,000 acres in 1896 and 835,700 acres in 1906, with a recession between 1873 and 1906 of 284,300 acres of land. [Exhibits D-3 and P-5.]

2. 1906–1910

From 1906 to 1910 the Lake rose 6.10 feet, from its average annual elevation of 4196.83 feet to 4202.93 feet and increased its surface area by approximately 371,800 acres, from 835,700 acres to 1,207,500 acres.

3. 1906–1920

From 1910 to 1920 the Lake declined 1.71 feet in elevation from an average elevation of 4202.93 feet to 4201.22 feet, and receding 81,700 acres, from a surface area of 1,207,500 acres to 1,125,800 acres. [Exhibits D-3 and P-5.]

[36]
4. 1920–1924

From 1920 to 1924 the Lake rose 3.09 feet in elevation from an average of 4201.22 feet to 4204.31 feet, and its surface area increased by 143,700 acres from 1,125,800 acres to 1,269,500 acres. [Exhibits D-3 and P-5.]

5. 1924–1936

From 1924 to 1936 the Lake declined 9.54 feet in elevation from an average of 4204.31 feet to 4194.77 feet, and its surface area decreased 556,700 acres, from 1,269,500 acres to 702,800 acres. [Exhibits D-3 and P-5.]

6. 1936–1953

From 1936 to 1953 the Lake rose 5.17 feet in elevation from an average of 4194.77 feet to 4199.94 feet, and its surface area increased 335,600 acres from 702,800 acres to 1,058,400 acres. [Exhibits D-3 and P-5.]

7. 1953–1963

From 1953 to 1963 the Lake declined 7.2 feet in elevation from an average elevation of 4199.94 feet to 4192.22 feet, and its surface area decreased by 438,800 acres from 1,058,400 acres to 619,600 acres. [Exhibits D-3 and P-5.]

8. 1963–1973 Hearing

From 1963 to the 1973 Hearing the Lake has been generally rising. In 1963 the elevation of lake was 4192.22 feet, the average level. In 1967, the average elevation of the Lake was 4194.01 feet, an increase of 1.79 feet. By 1967 the Lake's surface area had increased 52,800 acres from an average 619,600 acres to 672,400 acres. If the Lake has reached 4200 feet, it would have risen 7.78 feet since 1963 and its surface area would have increase 442,400 acres to 1,062,000 acres. [Exhibits D-3 and P-5.]

INDEX

Accretion:
    artificial, property boundary effect, 96, 268
    choice-of-law, 269
    contiguity required, 255
    definition, 92, 129
    distinguished from alluvion, 92
    distinguished from erosion, 92
    element, imperceptible, 93
    exceptions to rule, 96
    legal term, 93
    origin of rule, 256
    presumption, 99, 263–264
    property boundary effect, 95–96
    purpose of rule, 253
    relation to avulsion, 97–99
    relation to re-emergence, 133
    relation to reliction, 93, 97
    theory, 95, 253
Accretion—estuary:
    artificial, property boundary effect, 173–175
    Civil Code section 1014, 174
Accretion—lakes:
    applicability, 294–295
    artificial, property boundary effect, 311–314
    element, imperceptibility, 316–317
    element, permanence, 318
    presumption, 316
    property boundary effect, 313, 319
    relation to reliction, 313, 317–318
Accretion—open coast:
    artificial, policy basis for rule, 131
    artificial, proof, 136
    artificial, property boundary effect, 130–131
    burden of proof, 137
    definition, 129
    filling, property boundary effect, 132–133
    gradual and imperceptible, described, 129
    presumption, 134
    proof, 135–136
    property boundary effect, 129
    relation to deliction, 129
    relation to re-emergence, 133
    relation to reliction, 133
Accretion—rivers:
    application of doctrine, 256–257, 261–263
    artificial, 268–270
    basis for rule, 269
    burden of proof, 263–264
    burden of proof, effect, 267
    described, 255, 270
    element, identifiable land not required, 256, 261–262
    element, imperceptibility, 256, 265–267
    island rule, 271
    landowner caused, property boundary effect, 268
    origin of rule, 256
    presumption in favor of, 263–264
    proof, experts, 247
    property boundary effect, 251–252, 255–257, 261–262
    rapidity, 264–266
    relation of accretion to avulsion, 252, 254–267, 270
Act of 1861, 25, 155
    described, 155
    purpose, 204
Administrative construction, 16, 289
    estoppel, compared, 289
Admiralty:
    courts, explained, 237
    jurisdiction, extent, 205

374 INDEX

Adverse possession:
  applicability to sovereign lands, 39–41, 158–159
  applicability to United States, 41, 159
  applicable to private lands, 42
  burden of proof, 264
  character of occupancy, 37
  claim of right, defined, 36
  color of title, defined, 36
  elements, 35
  evidence of hostile possession, 36
  governed by state law, 66, 159, 312
  payment of taxes, 37
  proof, 40
  relation to location of ordinary high-water mark, 321
  Tidelands cases, 40
  voluntary tax assessments, 37
Aerial photographs:
  evidence of changes, 200
  interpretation, 187, 219–222, 226, 228–229
  sources, 135–136
  stereo, use, 325
  use, 217–219, 221, 223, 225–226, 228–229, 294, 324–325
  use in photogrammetric surveys, 225
Agreed boundaries:
  compare to prescriptive easement, 28
  elements, 38, 322
  inapplicability, 38, 322
  proof, 279
Alluvion:
  artificial, 268
  defined, 92–93, 254, 326
  relation to accretion, 92, 95, 256, 261
  relation to artificial accretion, 130
  relation to avulsion, 256
  relation to reliction, 93, 313
Articles of Confederation, 23
Avulsion:
  artificial, 316
  defined, 93, 259–260, 326
  legal term, 93
  no presumption, 100, 263
  property boundary effect, 97, 131
  purpose of rule, 253–254, 260
  relation to accretion, 97–99, 257
  relation to artificial accretion, 131, 257
  relation to reliction, 316
  relation to revulsion, 93
  theory, 253
  time, element, 98
Avulsion—estuary:
  artificial, 173, 209
  defined, 173
  proof, 173
  property boundary effect, 208
  relation to accretion, 209
  relation to artificial accretion, 175
  relation to artificial reliction, 175
Avulsion—lakes, 316
  artificial, 316
  relation to reliction, 313, 316–317
  time, element, 316
Avulsion—open coast:
  applicability of doctrine, 134–135, 138
  burden of proof, 137
  hurricanes, 134–135
  no presumption, 134
  proof, 135, 138
Avulsion—rivers, 275
  application of doctrine, 257–258, 260, 262
  artificial, 270
  burden of proof, effect, 235, 263–264, 266–267
  compared to erosion, 266
  described, 257, 265, 270
  evidence, 267, 279
  identifiable land, element, 256, 260–262, 279
  legal term, 257
  no presumption, 263
  proof, 266, 279–280
  property boundary effect, 251–252, 254, 257–260, 262, 270
  public ownership of bed, 254
  relation of avulsion to accretion, 252, 254, 256–257, 260–262, 266–267, 270
  relation of avulsion to erosion, 260, 266–267
  right to reclaim, 271
  time, element, 260, 267, 270
  witnesses, 266–267
Bay Conservation and Development Commission:
  boundary location, 72
  jurisdiction, 72
Beaches:
  coastal development permits, nexus, 102
  coastal processes, 126–127
  groins, effect, 130
  implied dedication, 102
  private, 101–102
  public trust, 102
  seasonal fluctuations, 79, 127
  value, 101
  winter profile, 128

Bench marks:
  defined, 109
Board of Land Commissioners:
  appeal, 7
  authority, 7
  function, 6
  survey after confirmation, 7
Board of Tideland Commissioners:
  grants, effect of filling, 34
  grants, validated, 34, 103
Boundary line agreements:
  effect of constitutional prohibition of sale of tidelands, 28–29
  elements, 28–29
  purpose, 28
  validity, 28
Boundary location:
  non-navigable waters, 236
  purpose, 71–72
  purposes, Bay Conservation and Development Commission, 72
  purposes, political jurisdiction, 72
  purposes, Rivers and Harbors Act, 71
Burden of proof:
  accretion, when, 264
  avulsion, when, 137, 267
  challenge to existing conditions, 99
  challenge to patents, 88, 157
  change in physical location, process, 263
  choice-of-law, effect, 254
  coastal processes, 139
  defined, 99, 327
  effect, 138, 231, 235, 263–264
  fraudulent survey, 298
  importance, 88, 99, 139, 263–264, 279
  material change, 267, 279
  navigability, 244
  navigability, effect of meandering, 241, 244
  party initiating suit, 139, 231
  party not in possession, 99
  quiet title, 99, 139
  reasons for rules, 98
  shifting, 231
Bureau of Land Management, 15
  effect of decisions, 155
  function, 327
  instructions, 82
  source of information, 135
  successor to General Land Office, 8, 11
  survey of public lands, 15
Cadastral surveys:
  explained, 216
  defined, 327

Chain of title:
  defined, 3, 327
  purpose, 44, 108, 151, 186, 259
  use in case, 216
Choice-of-law:
  accretion, 269
  adverse possession, 66, 159, 312
  adverse possession, state law, 159
  basis for application of state law, 67
  Bonelli rule, 53, 55
  branch of conflict of laws, 46
  Corvallis rule, 53, 55, 61
  defined, 47
  erosion, 269
  federal common law, 57–58, 66
  federal common law, reliction, 312–314, 317–319
  federal courts, application of state law, 47
  Hughes case, 55, 60
  Humboldt Spit case, 54, 59–62
  Humboldt Spit case, criticism, 61–62, 65
  importance, 49, 130, 254, 312
  international relations, basis for federal law, 60–62
  Interstate Boundary Compacts, effect, 51–53
  interstate boundary, absence of compact, 51
  interstate political boundaries, 49–50, 58
  Mexican land grants, 166
  navigable waters, 161
  need for nationwide federal rule, 58–59, 67
  ordinary high water mark, 273
  property disputes, federal law, 54–55, 58, 60, 62–64
  property disputes, state law, 47, 53, 55–56, 63–64, 66
  reliction, 269
  state courts, not apply federal law, 47
  Submerged Lands Act, basis for federal law, 61–64
  tidelands, extent, 161
  Wilson case, 54, 56–59, 63
Clean Water Act:
  jurisdiction, extent, 205, 238
  navigability, 238
  permits, effect of public trust, 25
Coastal mapping:
  hydrographic survey, 118
  interpretation, 182–183
  property boundary determination, use, 118, 182–183
  purpose, 118
  reference datum, purpose, 118
  topographic survey, 118

376 INDEX

Commissioner of the General Land Office:
  described, 8
  identification of swamp and overflowed lands, 19, 22
  instructions, 77
Common law public trust:
  authority to alienate, 30–31
  constitutional prohibition, compared, 31–34
  described, 30–31
  effect, 29, 31
Conflict of laws:
  defined, 47
Constitutional prohibition on sale of tidelands, 27–28
  applicability to sale to United States, 27
  applicable to filled tidelands, 27
  boundary line agreements, effect, 28–29
  common law public trust, compared, 31–34
  described, 27
  Mansell case, exception, 43
Corps of Engineers:
  404 permits, 25
  field notes, 216–217
  jurisdiction, Clean Water Act, 205, 238
  jurisdiction, Rivers and Harbors Act, 71, 205
  mapping, 178, 216–217, 219–220, 224–225, 227, 229
  reports, 60
  source of information, 136
  surveys, 280
Curative acts:
  defects, alleged, 160
  described, 156, 160
  effect, 156, 160–161
  example, 160
  origin, 160
  swamp and overflowed lands, 21
Datums, 110
  calculation, 211
  coastal mapping, 118–119, 125
  conflicts, 220, 225
  conflicts, resolution, 225
  construction of mean high-water line, 119
  conversion factor, 225
  geodetic, 212
  International Great Lakes Datum, 324
  length of record, 117
  National Geodetic Vertical Datum, 212
  National Tidal Datum Epoch, 212
  reference, 110, 212
  reference, importance, 220
  Sea-Level Datum, 212
  tidal, 115, 117

  tidal bench marks, 109, 212, 327
Deed interpretation:
  bank equivalent to river, 275
  low water mark, 259
  priority of calls, 86
Definitions:
  accretion, 92, 129
  alidade, 179
  alluvion, 92–93, 254, 326
  area elevation curve, 229
  avulsion, 93, 173, 326
  bank, 79, 84, 274–275, 297
  bank-shore equivalent, 84
  bed, 79, 274–275, 300
  bench mark, 109
  burden of proof, 99, 327
  cadastral survey, 216
  coast line, 124
  current, 110
  declination, 112
  deliction, 129, 328
  dendrochronology, 279
  dereliction, 328
  descriptive reports, 181
  dicta, 122
  erosion, 92, 129, 328–329
  estuary, 141
  geographic location, 48
  geomorphology, 142
  hydraulic mining, 199
  inland waters, 63, 330
  judicial notice, 206
  laches, 39
  lake, 293, 300
  law of the case, 291
  legal character, 147
  littoral land, 69
  mean high-water line, 119
  National Geodetic Vertical Datum, 212, 332
  National Tidal Datum Epoch, 212
  peg books, 185
  photogrammetric survey, 225
  physical location, 48
  planimetric survey, 215
  potomologist, 279
  presumption, 99, 333
  range of tides, 111
  re-emergence, 92, 334
  reliction, 92–93, 264, 313, 334
  revulsion, 93
  riparian land, 69, 334
  rule of property, 289
  seabed, 79
  shore, 79, 84, 105, 297

shorezone, 285
submergence, 335
territorial sea, 103
thalweg, 248
tidal prism, 171
tide curve, 113
tidelands, 26, 161
tides, 110
United States Geological Survey, 210
warping, 173
watercourse, 79
Delta:
   Delta Cross-Channel, 200
   flood control, 200
   geology, 196–198
   marshes, described, 197–198
   modifications to regime, 198
Dereliction—lakes:
   described, 313
   relation to accretion, 313
Descriptive reports:
   described, 181
Digitization, 136, 216
Equal Footing Doctrine:
   basis of sovereign title, 55–56, 63, 72, 149, 235
   effect of Submerged Lands Act, 62
   federal question, 161
   navigability federal question, 237
   Northwest Ordinance, 23
   origin, 23
Erosion:
   choice-of-law, 269
   distinguished from accretion, 92
   exceptions to rule, 96
   gradual and imperceptible, element, 93
   legal term, 93
   presumption, 99
   property boundary effect, 95–96
   relation to avulsion, 97, 99
   relation to re-emergence, 133
   relation to submergence, 93
Erosion—estuary:
   artificial erosion, 174
   Civil Code section 1014, 174
   property boundary effect, 172
Erosion—open coast:
   artificial, property boundary effect, 132
   definition, 129
   presumption, 134
   property boundary effect, 129, 133, 135, 328
   relation to accretion, 129
   relation to avulsion, 135
   relation to re-emergence, 133
Erosion—rivers:
   compared to avulsion, 266
   legal term, 257
   rapidity, 264–266
   relation to avulsion, 260, 264, 266–267, 270
   relation to re-emergence, 261
Estoppel:
   administrative construction, compared, 289
   applicability to land title, 39
   applicability to sovereign lands, 39–41, 158–159, 289
   applicable to private lands, 42
   constructive fraud, 39, 42
   elements, 39
   governmental, 42–44
   governmental, proof, 42
   judicial, 41
   Tidelands cases, 40
Expert witnesses:
   aerial photographs, interpretation, 187, 219
   basis for opinion, 172, 184, 213–214, 216, 224, 227–228
   botanical taxonomists, 187
   botanists, 184, 187
   cartographer, 181, 293
   cartographic symbols interpretation, 219
   civil engineer, 211, 213, 216, 219, 224–228
   coastal oceanographer, 207, 211–213, 218–220, 224, 229–230
   conflicts, 247
   credibility, 137, 247, 278–279
   effect of reclamation, 214–215
   fees, 247
   forester, 279
   form of testimony, 217, 230–231
   geologist, 226–229
   geomorphologist, 279
   historic conditions, 213
   historical geographer, 207, 213, 222, 227
   independent verification, 212
   instructions, 231
   interrelationship, 213, 215, 217, 227–228, 230
   lake boundary case, 324
   location of features, 216
   location of ordinary high-water mark, use, 78
   map intepretation, 220
   mistakes, 139–140, 231
   navigability, 245
   plant ecologist, 184, 187, 220–222, 230
   plant taxonomist, 184

Expert witnesses (*continued*):
  potamologist, 279
  purposes, 137, 246–247, 252
  qualifications, 140, 247
  river movement, 247
  site investigations, 216
  soils scientist, 187, 222–223, 226–227, 229–230
  standard for opinion, 140
  surveyor, 201, 203, 207, 217, 219–220, 226, 228
  testimony, use, 137
  tidality, 207
  types, 138, 176, 184, 196, 201, 207, 211, 213, 217, 219
Federal quiet title:
  statute of limitation, 159
General Land Office:
  approval of surveys, 8
  described, 329
  effect of decisions, 155
  function, 7–8, 11, 21
  identification of swamp and overflowed lands, 155
  instructions, 82, 296
  records, 135
  records of prior sovereign land grants, 136
Geneva Convention:
  concerns international boundaries, 61, 125
Geographic Information System (GIS), 137, 186, 195, 216
Global Positioning System (GPS), 195
Hoffman, Ogden, 7
Homestead Act, 17
Hurricanes, property boundary effect, 134–135
Hydraulic mining:
  described, 199
  effect, 171, 199–200, 227
Hydrographic survey:
  described, 118
  purpose, 182
Imperceptibility:
  described, 129, 265–266
  discussed, 256, 316
  lack of witness, 266–267
Implied dedication:
  elements, 102
Indian title:
  burden of proof, 63
  impact on choice-of-law, 58
  United States' trust obligation, 57
International boundary:
  Geneva Convention use, 61

International Great Lakes Datum:
  described, 324
Interstate boundaries:
  absence of compact, 51
  choice-of-law, 49–50, 58
  effect of private title litigation, 52–53
  effect of property boundary principles, 72
Interstate boundary compacts:
  effect on titles, 51–53, 261
  rivers, 261
Island rule, 258, 271
Judicial notice:
  described, 206
  drought, 206
  navigability, 293
  rainfall varies, 300
  river stages, 233
  seasonal nature of rivers, 233
  tule growth, 221
Jus privatum:
  alienable, 33, 147
  described, 33–34, 147
  extinguished, 272
  jus publicum, compared, 33, 147
Jus publicum:
  curative acts, effect on, 156
  described, 33, 147
  incapable of alienation, 147
  jus privatum, compared, 33, 147
Laches:
  defined, 39
  applicability to sovereign lands, 39, 41, 158
Lakes:
  attributes, 294, 300
  bank, defined, 297
  bed ownership, 285–290
  defined, 293
  geographic extent, 292
  lakebed, defined, 300
  littoral disputes, nature, 283
  public trust, 289
  regime, 300–301
  regime, importance, 303–305
  shore, defined, 297
  surveying instructions, 303
  terminal, described, 290, 295
Land grants, 6
  confirmation, 6, 7
  Board of Land Commissioners, 6–7
  boundary descriptions, physical features, 164
  challenge, 167
  claim, effect of failure to file, 9
  claim filing, 6–7

decree, effect on boundaries, 164
effect of confirmatory patent, 8, 166
former sovereigns, 4–5
international law, 6, 164
Land Claims Act, 6
location, 164
purpose of survey, 8, 165
recognition, 6
records of proceedings, 7, 136
survey, 7, 164
Land title:
  basic principles, 2
  chain of title, 3, 108
  doctrine of discovery, 3
  marshes, 148, 151
  original source of title, 2, 81
  United States successor, 3
Law of the case:
  described, 291
Levees:
  formation, 197
  foundation, 218
Louisiana Purchase, 4
Luna Bar litigation, 264
Maps:
  digitization, 136
  GIS, 136
  sources, 135
Marshes:
  applicability of property boundary principles, 172
  comparative features, 219
  Delta, described, 197–198
  dynamic physical character, 148
  extent, San Francisco Bay, 142
  geomorphology, 219
  historic character, proof, 178–179, 181–187, 219–220
  land title, 151
  meadowlands equivalent, 142
  proof of character, difficulties, 177
  reclamation, effect, 171
  reclamation, methods, 173
  regime, 170, 187
  regime, elements, 170–171
  salt marsh not legal term, 147
  uses, 142
  vegetation, 184
Meander line:
  accuracy, 216, 297, 306
  cartographic depiction, 86
  defined, 85, 331
  erroneous, 89, 299
  erroneous, burden of proof, 298
  evidence of navigability, 240–241, 244
  evidence of physical feature, 86, 89, 241, 295–296, 299
  example, 86
  inaccuracy, 88, 186, 216, 306
  instructions, 82, 84–85, 185, 240, 296–297
  lake, 297
  ordinary high-water mark location, 81, 275, 296–299, 306–307
  practice, 88
  property boundary effect, 84–87, 89–90, 185, 275, 296–299, 306
  purpose, 13, 82, 85–87, 89, 296
  result, 85
  township plats, 87, 298
Mexican land grants:
  "Bay" boundary, 167
  "Bay" boundary, ordinary high water mark, 166
  boundary descriptions, physical features, 164
  boundary location, 166
  challenge, 166
  confirmation, 6–7
  extent, 163
  juridical possession, 164
  location, 164
  Mexican law, applicability, 166
  patent, effect, 8–9, 165
  purpose, 164
  survey, 7–8
  survey, purpose, 7–9, 165
  surveying instructions, 167
Missouri River:
  turbulent nature, 50
National Geodetic Vertical Datum:
  explained, 212
National Ocean Survey:
  coastal mapping, purpose, 118
  described, 109
  tidal bench marks, 212
  tidal datum calculation, 212
National Tidal Datum Epoch:
  defined, 212
Nautical charts:
  coastal mapping, purpose, 118
  construction, 118
  datum, mean lower low water line, 125
  reference datum, 118
Navigability:
  basis for State sovereign title, 72, 234–235, 240–241
  burden of proof, 244
  Commerce Clause test, 238

380 INDEX

Navigability (*continued*):
  Commerce Clause vs. title test, 238
  condition of watercourse, 241–242
  date to establish, 241
  defined, 72, 161, 236
  difference between regimes, 95
  effect of isolation of watercourse, 241
  effect of later changes, 242
  effect of meandering, 240–241, 244
  effect of tidality, 204, 287, 289, 291
  effect on property boundary, 236
  effect on public trust, 163
  elements, 240–243
  federal law determines, 237
  federal test, 237–238, 291–292
  judicial notice, 293
  legislative declaration, 240
  methods of, 241
  navigability in fact test, 161
  origin, 162, 235, 237–238
  proof, 204, 241, 243–245, 293, 321
  proof, admissibility, 245
  purpose of test, 235, 292
  question of fact, 240
  recreational boating test, 239
  state and federal tests compared, 239, 292
  state tests, 238–239, 244, 291–292
  state tests, purpose, 238
  state tests, relevance, 238
  susceptibility, 242–243
  tests compared, 162
  tidal test, 161
Navigable waters:
  acts of admission, 236
  basis for State sovereign title, 102, 149, 161, 234–236, 288
  Clean Water Act, jurisdiction, 205, 238
  defined, 72, 161
  difference between regimes, 95
  effect of tidality, 161–163, 169, 204–205, 236, 287, 289, 291
  extent, 287
  federal law determines, 161, 237
  inclusion of non-navigable area, 240
  navigability, state and federal tests compared, 292
  navigable, defined, 236–237
  Northwest Ordinance, 236
  principle of navigability, purpose, 235
  proof of navigability, 204
  Rivers & Harbors Act, 205
  scope, 292
  subject to public trust, 286, 290
  Submerged Lands Act definition, 124

  surveying instructions, 82–84, 185, 240
  surveying instructions, meander line, 85
  surveying methods, 84
  watercourse, described, 79
Non-navigable waters:
  property boundary location, 236
  public rights, 238–239
Northwest Ordinance of 1787:
  origin of equal footing, 23
  public rights in navigable waters, 236
Ordinary high-water mark—estuary:
  "Bay" boundary, 165
  concept, 145
  headland-to-headland rule, 169
  legal character vs. physical character, 147
  location, cultural information, 186
  location, mean high water line, 169
  location, proof, 177–179, 181–187, 209–215, 217–230
  location, USCS/USC&GS maps, 182–183
  mean high-water line, adoption, 146, 168–169, 301
  mean high-water line, *Borax* case, 119–120
  proof, 170
Ordinary high-water mark—general, 77
  ambulatory boundary, 90, 94
  ambulatory boundary, purpose, 94
  Borax case, 74
  boundary of state sovereign lands, 73, 272
  boundary, swamp and overflowed land, 148
  boundary, tidelands, 149
  components, 78–81, 209
  definition, 74, 78
  difference between regimes, 73, 78
  dynamic nature, 79–80
  equivalent terms, 73–74, 103
  headland-to-headland rule, 74
  importance of location, 72, 75, 76
  location, evidence, 78, 89
  location, grant surveys, 168–169
  location, meander line, 81
  location, methods, 78, 80–81
  mean high water line, adoption, 207
  neap tide rule, 76, 77
  no definite physical location, 74–75, 77–78
  not scientific term, 210
  ocean beach, 79
  origin, 77
  origin, English common law, 76
  origin, French law, 76
  origin, Mexican law, 76
  origin, Roman law, 76
  origin, Spanish law, 76
  purpose, 75

seasonal fluctuations, 79
surveying instructions, 77
uncertainty, 73–74, 77, 80–81, 108
vegetation line, 74, 81
Ordinary high-water mark—rivers:
bank, river, equivalent terms, 275
equivalent term, bank, 272
evidence, meander line, 275
evidence, soil, 275
evidence, stages or flows, 275–276
Howard test, 273–276
Howard test, shortcomings, 274
legal term, not scientific term, 272
location, 272–274, 276
location, federal law determines, 273
location, flexibility, 276
location, highest level, 274
location, proof, 278–280
Ordinary high-water mark—lakes:
ambulatory boundary, 311
bank, equivalent, 297
boundary of state sovereign lands, 288, 294–295
location, agreed boundaries inapplicable, 322
location, flat terrain, 88
location, importance of regime, 303–305
location, indefinite, 296, 299–301
location, meander line, 296–299, 306–307
location, new natural doctrine, 321
location, physical indicia, 298
location, proof, 302, 310–311
location, use of water level measurements, 306–307
location, vegetation/erosion line, 298, 301–305, 307
location, visual, 296
location, water level, prescription, 321
location. meander line, 298–299
mean high-water, undefined, 297
proof, 307, 321–325
vegetation line adopted, 324
vegetation line, rejected, 323
Ordinary high-water mark—open coast:
accretion, effect, 129
ambulatory boundary, 128
boundary of state sovereign lands, 102
equivalent terms, shore, 105
erosion, effect, 129
location, development, 104–105
location, uncertainty in use of tidal measurements, 110
location, use of tidal measurements, 106–107, 109

location, use of USC&GS mapping, 109
location, use of USCS mapping, 109
location, use of vegetation line, 107
location, vegetation line uncertain, 107
location, vegetation line, inapplicable, 107, 120–121
location, winter beach profile, 128
mean high-water line, adoption, 119–121, 301
mean high-water line, *Borax* case, 119–120
mean high-water line, debate, 119
mean high-water line, defined, 119
mean high-water line, legislative use, 119
neap tide rule, 104, 106–109, 114–115, 119–120, 122
neap tide rule, rejected, 120–122
neap tide rule, unlawful conveyance, 122
purpose, 121
shore, described, 105
uncertainty, 105, 108
Ordinary low-water mark:
definition, 332
location, 122
origin, 77
state land grants, 77, 205
Submerged Lands Act, 77
Ordinary low-water mark—open coast:
1947 California Decision, 123
1963 California Decision, 124–125
applicability of property boundary principles, 126
base point of territorial sea, 103
boundary of submerged lands, 102, 122
definition, 123, 308
importance of location, 102, 103
location, 123, 125–126
location, confusion, 125
location, descriptive approach, 125
location, landform fluctuation, 126–128
location, mean low-water line, 125–126
seaward boundary of tidelands, 122
Ordinary low-water mark—rivers:
boundary of state sovereign lands, 272
evidence, stages or flows, 277
location, 277–278, 308
public trust boundary, 272
Ordinary low-water mark—lakes:
boundary of state sovereign lands, 288–289, 294
definition, 308–309
importance of location, 294, 311
location, 307–311
location, difficulties, 308
location, proof, 310–311

Ordinary low-water mark—lakes (*continued*):
  location, rationale, 309
  location, use of water level measurements, 309–310
  state land grants, 288–289
Paramount rights doctrine:
  basis for United States' claim, 123
  basis of *1947 California Decision*, 62, 123–124
  Submerged Lands Act, effect on, 62
Patents:
  burden of proof, overturning, 88
  effect, 157
  incorporation of township plat, 82, 85
Peg books, 185
Planetable:
  alidade, described, 179
  described, 179
  method, described, 179
Planimetric survey:
  explained, 215
Prescriptive easement, 38
  compare to agreed boundary, 38
Presumption:
  accretion/erosion, basis, 263
  accretion/erosion, in favor of, 98–100, 134, 263–264
  application, 264
  avulsion, none, 98–99, 134, 263
  basis, 263
  defined, 99, 333
  effect of, 134
  inapplicability, 264
  overcoming, 135, 263–264, 316
  property boundary unchanged, 100
  reasons for, 99, 263
  sovereign ownership, in favor of, 231, 284
  when operates, 263
Private Land Claims Act, 6
  authority of Board of Land Commissioners, 7
  claims presentation, 6
  effect of confirmatory patent, 8, 165
  effect of failure to file claim, 9, 167–168
  location of records, 7
  right of appeal, 7
  statute of limitations, 7, 167–168
  survey of confirmed grants, 7, 164–165
Proof:
  accuracy of surveys, 216
  aerial photographs, 135, 220, 325
  agreed boundaries, 279, 321
  artificial accretion, 136
  avulsion, 138, 279
  borings, 187
  botanical surveys, 187
  change in location, 246–247
  change in regime, 172, 176
  coastal processes, 139
  computer model, 214
  effect of reclamation, 214–215
  exhibits, 186
  expert witnesses, 137
  field notes, 221
  geographic location, 280
  GIS, 136, 186, 195, 216
  historic accounts, 178, 221
  historic character, marshes, 178–179, 181–187, 210
  historic lake levels, difficulty, 320
  historic maps, 135
  historic mean high-water line, 228–230
  historic physical character, 217
  historic topography, 218–220, 224–228
  historic vegetation, 220–222
  historic water level, 213–215
  identifiable land, 279
  map digitization, 136
  navigability, 293
  ordinary high-water mark, historic, 209–215, 217–230
  ordinary high-water mark, lakes, 302, 307, 310–311, 324
  ordinary high-water mark, lake stage levels, 321–322
  ordinary high-water mark, lakes, vegetation line adopted, 324
  ordinary high-water mark, lakes, vegetation line rejected, 323
  ordinary high-water mark, meander line, 299
  ordinary high-water mark, rivers, 278–280
  ordinary low-water mark, lakes, 310–311
  photogrammetric survey, 225
  physical character, photographs, 219
  physical investigations, 187
  practical approach, 280
  quantum required, 139–140
  soil types, 222–223
  soils map, 223
  soils maps, government, 223
  sources of information, 135–136, 186, 293
  standard, 280
  surveys, 187
  tidality, 206–207, 212
  township surveys, 217
  trees, age, 325
  use of field books, 219

use of historic maps, 138, 178–179, 181–187
vegetation map, 221–222
watercourse location, 215–217
Property boundary principles:
accretion, applicability to lakes, 294–295
accretion, applicability to rivers, 256–258, 260–263
accretion, burden of proof, 267
accretion, civil law rule, 95
accretion, distinguished from erosion, 92
accretion, effect, 95–96, 129, 172, 174, 251–252, 255–257, 261–262, 313, 319
accretion, English rule, 94–95
accretion, exceptions to rule, 96
accretion, presumption in favor of, 99, 134, 263–264, 316
accretion, purpose of rule, 253
ambulatory boundary, 90, 94, 97, 128, 208, 311
ambulatory boundary, basis, 69, 94
ambulatory boundary, disadvantage, 69
applicability to marshes, 172
applicability to open coast, 129
applicability to ordinary low-water mark, 126
artificial accretion, 173–174
artificial accretion, effect, 130–131, 174–175, 209, 268–270
artificial accretion, landowner caused, 268
artificial avulsion, effect, 173, 209, 270
artificial erosion, effect, 132, 174
artificial reliction, effect, 174, 312, 314–315
artificial submergence, effect, 174, 320–321
avulsion, applicability to open coast, 134–135, 138
avulsion, effect, 97, 131, 208, 251–252, 254, 257–260, 262, 270
basis, 90
case study, 90–91
change, controlling influence theory, 175, 208
change, controlling influence theory rejected, 209
change, nature, 208
date to establish location of boundary, 215
differences between regimes, 233
disputes, nature, 70–71
effect of burden of proof, 263–264
effect of filling, 132
effect of fluctuation of the landform, 127
effect of nature of physical regime, 98
erosion, distinguished from accretion, 92
erosion, effect, 95–96, 129, 133, 172, 328
erosion, exceptions to rule, 96
erosion, presumption in favor of, 99, 134
gradual and imperceptible, elements, 93, 129, 256, 265–267, 315–319
importance of boundary determination, 72
importance of facts, 255
importance of understanding, 91, 248–249, 251–252
island rule, 258, 271
legal terms, 93
nature of physical regime, 256
new natural, 176
ordinary high-water mark, legal term, 70
ordinary high-water mark, mean high-water line, 119–121, 207
ordinary high-water mark, physical location, 104–105
ordinary low-water mark, grant to, 73
re-emergence, application, 261
re-emergence, effect, 97, 133, 261, 271
relation of accretion to avulsion, 97–99, 209, 252, 254–267, 270
relation of accretion to deliction, 129
relation of accretion to erosion, 92
relation of accretion to re-emergence, 133
relation of accretion to reliction, 97, 133, 313, 317–318
relation of alluvium to accretion, 92, 95
relation of alluvion to artificial accretion, 130
relation of alluvion to avulsion, 256
relation of alluvion to reliction, 93, 313
relation of avulsion to reliction, 313
relation of avulsion to revulsion, 93
relation of erosion to avulsion, 97, 99, 135, 260, 264, 266–267, 270
relation of erosion to re-emergence, 261
relation of re-emergence to accretion, 96
relation of re-emergence to erosion, 96
relation of re-emergence to reliction, 92, 261
relation of reliction to accretion, 93, 313
relation of reliction to alluvion, 93
relation of reliction to avulsion, 313
relation of reliction to re-emergence, 92
relation of reliction to submergence, 93
relation of submergence to reliction, 93
reliction, artificial, 269
reliction, effect, 97, 133, 172, 174, 295, 312–319
right to reclaim, 190
rivers, 246

384 INDEX

Property boundary principles (*continued*):
　submergence, effect, 97, 172, 320
　use of natural features, 68
Public domain:
　equivalent to public lands, 4
Public lands:
　500,000 acre grant, 17
　Alaska acquisition, 4
　authority of Congress, 16
　cessions from the original states, 4
　cessions from the Original States, 273
　defined, 4
　differences from private, 76
　disposition, 9, 16–17
　doctrine of discovery, 3
　effect of survey, 14
　equivalent to public domain, 4
　Florida cession, 4
　Gadsden purchase, 4
　grants to States, 17
　internal improvements, 17
　islands, 258
　land grant colleges, 17
　later aquisitions, 4
　Louisiana Purchase, 4
　original source of title, 10
　preemption, 17
　prior sovereign land grants, effect of patent for, 8, 165
　prior sovereign land grants, survey of, 8
　prior sovereigns land grants not public lands, 6, 165
　railroad grants, 17
　sales, 16
　school land grants, 17
　seminaries of learning grant, 17
　state sovereign lands not part, 9
　surveying, 10–13
　swamp and overflowed lands grant, 17–22, 24
　Texas annexation, 4
　Treaty of Guadalupe-Hidalgo, 4
　Treaty with England, 4
　United States acquisition, 4
　unsurveyed lands, 14
　withdrawal, 9
Public trust:
　adverse possession, effect, 41, 159
　authority to alienate, 30–31, 72
　boundary, 72, 102, 145, 169
　boundary, effect, 145
　common law, 29, 31–32, 122
　common law, defined, 31
　common law, purpose, 31
　curative acts, effect, 156, 161
　defined, 24, 72, 102, 333
　dry sand area, 102
　effect, 24–25, 33–35, 73, 146, 169, 194, 230, 289–291
　effect of filling, 34
　effect on improvements, 35
　effect on United States, 27
　equitable estoppel, effect, 42–43, 158
　exercise, no compensation, 25, 87
　geographic extent, 24, 102, 163, 236, 272, 287, 289–290
　grants to ordinary low water mark, 272
　lakes, 289
　Mexican law, 166
　nonnavigable waters, 24
　Northwest Ordinance, 236
　not applicable to artificial reservoir, 289
　not limited solely to navigable waterways, 163
　obligations of state, 33, 286
　origin, Roman law, 76
　Private Land Claims Act, 168
　recognition by Supreme Court, 163
　re-examination of grants, 169
　reserved easement, 24, 33–35, 72–73, 102, 191, 194, 272, 289
　restraint on alienation, 25–26, 29, 72, 102, 122, 147–148, 286
　rivers, 236
　rule of statutory interpretation, 33, 286, 289
　state sovereign lands, 24, 26–27, 29, 32, 40, 158, 162, 285
　statutes of limitation, effect, 158
Reclamation:
　effect on tides, 171
　right to reclaim, 190
Rectangular system of public land surveying:
　advantages, 14
　base line, 12
　fractional sections, 13
　fraud, 15
　inaccuracy, 15
　initial point, 12
　location of township by numbering system, 13
　lots, 13
　meandering, 13
　numbering of ranges, 12
　numbering of sections, 13
　numbering of townships, 12
　principal meridian, 12
　range line, 12
　sections, 13

shortcomings, 14
subdivision, 12
subdivisions, 13
township lines, 12
township survey, 201
township, incorporated in patent, 82, 85
townships, 13
Re-emergence:
 defined, 92, 334
 property boundary effect, 97
 relation to accretion, 96
 relation to erosion, 96
 relation to reliction, 92
Re-emergence—open coast:
 property boundary effect, 133
Re-emergence—rivers:
 application, 261
 property boundary effect, 261, 271
 relation to reliction, 261
Reference datum:
 described, 118
Reliction:
 artificial, property boundary effect, 269
 choice-of-law, 269
 equivalent to dereliction, 92
 glacio-isostatic uplift, 174
 gradual and imperceptible, elements, 93, 315–316
 legal term, 93
 property boundary effect, 95, 97, 133
 relation to accretion, 93
 relation to alluvion, 93
 relation to re-emergence, 92
 relation to submergence, 93
Reliction—estuary:
 artificial reliction, 174
 Civil Code section 1014, 174
 property boundary effect, 172, 174
Reliction—lakes:
 applicability, 295, 312
 artificial reliction, 314–315
 artificial reliction, basis, 315
 artificial reliction, drainage, 314–315
 artificial reliction, filling, 314–315
 defined, 313
 discussion in Great Salt Lake case, 317
 elements, 312–313
 elements, imperceptibility, 315–319
 elements, imperceptibility, averaging, 319
 elements, natural, 319
 elements, permanence, 315, 317–319
 intentional drainage effect, 314
 property boundary effect, 312, 314, 317, 319

 property boundary effect, federal rule, 312–313, 319
 relation to accretion, 313
 relation to avulsion, 313
 relation to dereliction, 313
Reliction—open coast:
 property boundary effect, 133
Reliction—rivers:
 defined, 264
Reservations:
 pre-statehood, 248
Revulsion:
 defined, 93
 relation to avulsion, 93
Riparian land ownership:
 boundary disputes, 69
 boundary location principles, 92–98, 133–134, 174, 246–249, 251–262, 266–267, 269, 271, 309, 313, 319
 boundary location, importance, 71–72, 75, 232, 234
 center of channel boundary, 257
 contiguity, 255
 deed interpretation, 259
 islands, 258
 meander line, effect, 85, 87, 299
 meander line, purpose, 86, 89
 Mexican law, rights, 167
 non-navigable waters, 236
 ordinary high-water mark boundary, 73–74, 77, 166–167, 272
 ordinary low-water mark boundary, 73, 77, 257, 277, 288
 presumption of accretion, 100
 public trust, effect, 25, 73, 87
 right to reclaim, 271
 rights, 86, 167, 253, 260, 309, 315
Rivers:
 property boundary principles, 246
 public trust, 236
 seasonal nature, 233
 unpredictability, 233
Rivers and Harbors Act:
 jurisdiction, 71, 205
Rule of property:
 claimed, 34, 289
 defined, 34, 289, 334
 failure to assert public trust, 290
School land grants, 17
Seminaries of learning grant, 17
Shipwrecks, 283
Shorezone:
 described, 285
 value, 285

State land grants:
  administration, lax, 25–26
  challenges, 26
  consideration of public trust purposes, 32, 285–286, 288–289
  construction, 33
  effect, 155, 204
  geographic extent, 32
  limitations on disposal, 26–35
  ordinary low-water mark, 73, 126, 205
  power of State, 35
  pre-admission, U.S. holds trust, 72
  purpose, 25
  role of State Surveyor General, 32
  surveys, inaccuracy, 216
  swamp and overflowed lands, 25, 154, 156, 201, 203
  swamp and overflowed surveys, 201, 203
  swamp and overflowed surveys, fraud, 203
  thalweg grants, 73
State land sale statutes, 25
  Act of 1861 (California), 155
  consideration of public trust, 32–33, 285–286, 288–289
  construction, 33
  double right, 33
  geographic extent, 32
  purpose, 33, 147, 157
  statutory construction, 32
  swamp and overflowed, salt marsh and tidelands, 32
State sovereign lands:
  adverse possession not applicable, 39–41
  alienable, 102–103
  authority to alienate, 72, 288
  avulsion, consequences, 254
  Board of Tideland Commissioner's grants, 103
  boundary, ordinary high-water mark, 72, 76, 149, 205, 272, 294–295
  boundary, ordinary low-water mark, 126, 205, 272, 288–289, 294
  boundary, thalweg, 257
  confirmed by Submerged Lands Act, 23, 285
  constitutionally based title, 23
  described, 23, 72
  differs from other State lands, 30
  differs from private ownership, 41
  double right, 33, 147, 157
  Equal Footing Doctrine is basis, 23, 72, 285
  estoppel not applicable, 39–41, 158
  geographic extent, 287, 291
  islands, 258
  lake bed ownership, 285–290
  naked fee, 35
  ordinary high-water mark, 288
  pre-admission, U.S. holds trust, 72
  pre-statehood reservations, 9–10, 248
  subject to public trust, 23–24, 29, 33, 72–73, 102–103, 145, 191, 194, 230, 236, 285–290
  submerged lands, alienable, 103
State Surveyor General:
  State land grants, role, 32
Statutes of limitation:
  applicability to sovereign lands, 158
  federal, 41, 159
Statutory construction:
  avoid impairment of public trust, 33, 161, 286, 289
  Civil Code section 830, 288
  legal system history, 287
  state land sales statutes, 33
Statutory prohibition on sale of tidelands, 27
  two mile, 28
Submerged Lands Act, 23, 335
  basis for application of federal law, 61–64
  coastline, defined, 125
  confirmation of state sovereign title, 55, 61–62, 64, 77, 124, 149, 285
  effect on choice-of-law, 54, 61
  effect on lands underlying non-tidal navigable waters, 62
  effect, undo *1947 California Decision*, 124
  exceptions to confirmation, accretions, 61
  extent of confirmation, 124
  Geneva Convention, 125
  impact on paramount rights doctrine, 62
  inland waters, defined, 63
  intent, 54, 62, 124
  lands affected, 124
  lands beneath navigable waters, defined, 124
  line of ordinary low water, 125
  ordinary low-water mark, 77
  origin, 54, 62
  seaward boundary, 124
  use of mean lower low water line, 125
Submergence:
  definition, 93
  gradual and imperceptible, elements, 93
  legal term, 93
  property boundary effect, 97
  relation to erosion, 93
  relation to reliction, 93

Submergence—estuary:
  artificial erosion, 174
  new natural, 176
  property boundary effect, 172
Submergence—lakes:
  artificial, 320–321
  property boundary effect, 320–321
Subsidence:
  calculation, 226–227
  explained, 226
Supreme Court:
  original jurisdiction, 40
Survey of public lands:
  accuracy, 83, 185, 216
  background, 10–11
  boundaries along navigable watercourses, 82–85
  burden of proof, overturn, 88
  difficulties, physical, 201
  effect, 13, 82, 203
  field notes, 185, 201, 217
  field notes, purpose, 201
  information, physical character, 186
  instructions, 185
  instructions, use, 185
  Land Ordinance, 11
  meandering, 13, 82, 84–86
  Mexican land grants, 164
  patents, incorporation of township plat, 82, 85
  peg books, 185
  physical character, information, 185
  purpose, 185
  rectangular system, 13
  surveyor, authority, 203
  township survey, 201
Surveying instructions:
  General Land Office, 82, 296
  lakes, 303
  mean high-water elevation, lakes, 297
  meandering, 82, 84–85, 185, 240, 296–297
  Mexican land grants, 167
  navigable waters, 82–85, 185, 240
  ordinary high-water mark, 77
  swamp and overflowed lands, 157
Swamp and overflowed land grant:
  1861 Act (California), 22, 157, 203
  1866 Act, 21
  Arkansas Swamp Lands Act, 18
  character of lands, evidence, 20, 170, 223
  characteristics, 19, 104, 150–151
  conclusiveness of government actions, 155–156, 160, 203

  curative acts, 21
  date of grant, 154
  date to determine character, 170, 204
  date, importance, 177
  Delta lands characterized, 198
  effect of patenting process, 156
  extent of grant, 22
  grant to States, 152
  identification, 20, 152, 154, 201
  identification by States, 21–22, 155, 157
  identification, difficulties, 19–20, 104, 149–151, 154, 157, 201, 223
  identification, effect, 157
  identification, methods, 22
  lands, characteristics, 104
  marshes, 148
  Mexican land grant confirmation, effect, 167
  present grant, 18, 152
  purpose, 18
  salt marsh, equivalent, 147
  segregation, 20, 152
  state patenting, 156
  surveying instructions, 157
  swamp and overflowed lands, described, 19
  waterward boundary, ordinary high-water mark, 148
Territorial sea:
  basepoint, 103
  defined, 103
Thalweg, 248
  boundary, 257
  defined, 90, 335
  island rule, 271
  state lands grants to, 73
Tidal datums:
  basis, 117
  calculation, 211–212
  explained, 115
  mean high water, 118
  mean high water, fixed plane, 126
  mean higher high water, 118
  mean low water, 117
  mean lower low water, 117
  mean lower low water, nautical charts, 118
  mean sea level, 117
  mean tide level, 118
  purpose, 118
  record length, 117, 211
  records, historic, 211
  reference datum, defined, 118
  tidal bench marks, 212
  use in coastal mapping, 118

Tidality:
  descriptions, 205
  expert witnesses, 207
  importance, 204
  necessity, 161, 204–205
  proof, 206–207, 212
Tidelands:
  boundary, ordinary high-water mark, 149
  characteristics, 105, 150
  condemnation, 27
  constitutional prohibition, 27–29
  defined, 26, 161
  double right, 147, 157
  extent, 161–163
  federal law, defines, 161
  identification, difficulties, 149–151
  navigable waters, necessity, 161–163, 204
  restraint on alienation, 27–35
  salt marsh, equivalent, 147
  seaward boundary, 122, 125
  statutory prohibition on alienation, 27–28
Tidelands cases:
  adverse possession not applicable, 40
  discussed, 123–125
  estoppel not applicable, 40
Tides:
  18.6 year cycle, 117
  apogean, 112
  averaging, need for long term, 116, 211
  bench marks, 109
  defined, 110
  distinguised from current, 110
  diurnal inequality, explained, 113
  equatorial, 113
  equinoctial, defined, 106
  explained, 110–111, 113–115
  kinds, recognized by law, 106
  methods of tidal observation, 116
  National Tidal Datum Epoch, 212
  neap, 107–108
  neap, confusion, 114–115, 119–120
  neap, defined, 106
  neap, explained, 111–112, 115
  observations, long term, 116–117
  perigean, 112
  range, 111–113
  spring, 106, 111
  spring, defined, 106
  spring, explained, 112
  tidal cycle, explained, 113, 117
  tidal datums, explained, 115, 117–118
  tidal prism, 171
  tide curve, described, 113
  tide gauge, 116
  tide staff, 116
  tide stations, 116, 211
  tide stations, primary, 116
  tide tables, 109–110
  time, 113
  tropic, 113
  type, 113–114
  United States Coast & Geodetic Survey, 109
  United States Coast & Geodetic Survey, role, 116
Topographic survey:
  described, 118, 179, 184
  descriptive reports, 181
  interpretation, 181, 184
  maps, cartographic symbols, 181, 184
  maps, high-water line, 181
  maps, physical character, 184
  maps, pin pricks, 184
  planetable, 179
  purpose, 182
  scale, importance, 226
  use, 224
Treaty of Guadalupe-Hidalgo, 4
United States Coast & Geodetic Survey:
  bench marks, 109
  coastal mapping, 109
  coastal mapping, high-water line, 181
  coastal mapping, interpretation, 181, 184
  coastal mapping, physical character, 184
  coastal mapping, property boundary use, 178–179, 181–184
  coastal mapping, purpose, 118, 182
  datum plane, topography, 118
  described, 109
  descriptive reports, 181
  instructions, high-water line, 181
  planetable, use, 179
United States Coast Survey:
  base points, surveys, 12
  bench marks, 109
  cartographic symbols, 181
  coastal mapping use for property boundary, 179
  coastal mapping, high-water line, 181
  coastal mapping, interpretation, 181, 184
  coastal mapping, physical character, 184
  coastal mapping, pin pricks, 184
  coastal mapping, property boundary use, 75, 210
  coastal mapping, purpose, 118, 182
  coastal mapping, symbolization, 181, 184
  coastal mapping, use for property boundary, 178–179, 181–184

described, 109
descriptive reports, 181
headland-to-headland rule, 74, 169, 210
instruction, shoreline mapping, 181
instructions, high water line, 181
planetable, use, 179
topographic survey map, described, 179, 184
United States Geodetic Survey:
  described, 109
United States Geological Survey:
  described, 210
  mapping, property boundary use, 210
United States Surveyor General:
  contrast with State Surveyor General, 7
  described, 7, 337
  segregation of swamp and overflowed lands, 22
  survey approval, 88
  survey of confirmed land grants, 7
Von Geldern, Otto, 105
Water projects, 198–199
Witnesses:
  instructions, 278

THIS BOOK MAY NOT
LEAVE THE LIBRARY!

Wentworth - Alumni Library